T0180056

LONDON MATHEMATICAL SOCIETY LECTURE NOTE SERIES

Managing Editor: Professor M. Reid, Mathematics Institute, University of Warwick, Coventry CV4 7AL, United Kingdom

The titles below are available from booksellers, or from Cambridge University Press at www.cambridge.org/mathematics

London Mathematical Society Lecture Notes series: 378

Probability and Mathematical Genetics

Edited by

N. H. BINGHAM
Imperial College London

C. M. GOLDIE
University of Sussex

CAMBRIDGE
UNIVERSITY PRESS

CAMBRIDGE
UNIVERSITY PRESS

University Printing House, Cambridge CB2 8BS, United Kingdom

One Liberty Plaza, 20th Floor, New York, NY 10006, USA

477 Williamstown Road, Port Melbourne, VIC 3207, Australia

314-321, 3rd Floor, Plot 3, Splendor Forum, Jasola District Centre, New Delhi - 110025, India

103 Penang Road, #05-06/07, Visioncrest Commercial, Singapore 238467

Cambridge University Press is part of the University of Cambridge.

It furthers the University's mission by disseminating knowledge in the pursuit of education, learning and research at the highest international levels of excellence.

www.cambridge.org
Information on this title: www.cambridge.org/9780521145770

© Cambridge University Press 2010

First published 2010

A catalogue record for this publication is available from the British Library

ISBN 978-0-521-14577-0 Paperback

Additional resources for this publication at www.prob.org.uk

Contents

Contributors

David J. Aldous   *University of California at Berkeley*
Graeme K. Ambler   *University of Cambridge*
Andrew D. Barbour   *University of Zürich*
John D. Biggins   *University of Sheffield*
Nicholas H. Bingham   *Imperial College London*
Rosie Cornish   *University of Bristol*
Daryl J. Daley   *Australian National University and University of Melbourne*
Aurore Delaigle   *University of Melbourne and University of Bristol*
Peter Donnelly   *University of Oxford*
Warren J. Ewens   *University of Pennsylvania*
Alexander V. Gnedin   *University of Utrecht*
Charles M. Goldie   *University of Sussex*
Peter J. Green   *University of Bristol*
David R. Grey   *University of Sheffield*
Robert C. Griffiths   *University of Oxford*
Geoffrey Grimmett   *University of Cambridge*
Peter G. Hall   *University of Melbourne and University of California at Davis*
Chris Haulk   *University of California at Berkeley*
Peter H. Haynes   *University of Cambridge*
Viet Ha Hoang   *Nanyang Techological University, Singapore*
Frank P. Kelly   *University of Cambridge*
Wilfrid S. Kendall   *University of Warwick*
Sir John [J. F. C.] Kingman   *University of Bristol*
Stephen Leslie   *University of Oxford*
James R. Norris   *University of Cambridge*
Adam J. Ostaszewski   *London School of Economics*
Jim Pitman   *University of California at Berkeley*
Carol L. Robinson   *Loughborough University*

Vadim Shcherbakov   *Moscow State University*
Bernard W. Silverman   *University of Oxford*
Dario Spanó   *University of Warwick*
Simon Tavaré   *University of Cambridge*
Stanislav Volkov   *University of Bristol*
Geoffrey A. Watterson   *Monash University*
Peter Whittle   *University of Cambridge*
David Williams   *Swansea University*
Ruth J. Williams   *University of California at San Diego*
Konstantinos C. Zygalakis   *University of Oxford*

Preface

John Frank Charles Kingman was born on 28th August 1939, a few days before the outbreak of World War II. This Festschrift is in honour of his seventieth birthday.

John Kingman was born in Beckenham, Kent, the son of the scientist Dr F. E. T. Kingman FRSC and the grandson of a coalminer. He was brought up in north London, where he attended Christ's College, Finchley. He was an undergraduate at Cambridge, where at age 19 at the end of his second year he took a First in Part II of the Mathematical Tripos, following it with a Distinction in the graduate-level Part III a year later, for his degree. He began postgraduate work as a research student under Peter Whittle, but transferred to David Kendall in Oxford when Peter left for Manchester in 1961, returning to Cambridge when Kendall became the first Professor of Mathematical Statistics there in 1962.

John's early work was on *queueing theory*, a subject he had worked on with Whittle, but was also an interest of Kendall's. His lifelong interest in *mathematical genetics* also dates back to this time (1961). His next major interest was in Markov chains, and in a related matter—what happens to Feller's theory of recurrent events in continuous time. His first work here dates from 1962, and led to his landmark 1964 paper on regenerative phenomena, where we meet (Kingman) p-functions. This line of work led on to his celebrated characterisation of those p-functions that are diagonal Markov transition probabilities (1971), and to his book, *Regenerative Phenomena* (1972). Meanwhile, he had produced his work on queues in *heavy traffic* (1965). His work on subadditivity began in 1968, and led to the *Kingman subadditive ergodic theorem* of 1973. His genetic interests led to his book *Mathematics of Genetic Diversity* of 1980, and his famous paper on the (Kingman) coalescent of 1982. Later work includes his book *Poisson Processes* of 1993. Other interests include

Spitzer's identity and its connection with queues, the subject of his *The Algebra of Queues* of 1966.

John began his academic career in Cambridge, as Assistant Lecturer (1962-64) and Lecturer (1964–65), with a fellowship at his undergraduate college, Pembroke (1961–65). He left for a Readership at the University of Sussex, where he was promoted to Professor at the very early age of 26 in 1966, the year in which he published his first book, *Introduction to Measure and Probability*, with S. James Taylor. He left Sussex to be Professor at Oxford from 1969–85. He was elected a Fellow of the Royal Society in 1971 at age 31. He was made a Foreign Associate of the US National Academy of Sciences in 2007.

We all know very good mathematicians who could not run a corner sweetshop, let alone a mathematics department, still less a university. On the other hand, mathematicians who are not very bad at administration are often very good at it. John Kingman is a shining example of the latter category. This led to his secondment, while at Oxford, to chair the Science Board of the Science Research Council (1979–81), and later to serve as Chairman of the Science and Engineering Research Council (1981–85), for which he was knighted in 1985. It led also to John's career change in 1985, when he became Vice-Chancellor of the University of Bristol, serving a remarkable sixteen years until 2001. He then served for 5 years as Director of the Isaac Newton Institute for Mathematical Sciences in Cambridge. In 2000 he became the first chairman of the Statistics Commission, overseeing the Office of National Statistics.

John Kingman is the only person who has been President of both the Royal Statistical Society (1987–89) and the London Mathematical Society (1990–92). He has also served as President of the European Mathematical Society (2003–06). He received the LMS Berwick Prize in 1967, the RSS Guy Medal in silver in 1981, and the RS Royal Medal in 1983 (for his work on queueing theory, regenerative phenomena and mathematical genetics). He holds a number of honorary doctorates. He does not hold a PhD, being Mr Kingman until he was made Professor Kingman at Sussex, later taking a Cambridge ScD.

John Kingman's mathematical work is remarkable for both its breadth and its depth. But what shines out from everything he does, whether his written papers and books or his lectures and seminars, is *lucidity*. Kingman is always clear, and lucid. This even extends to his handwriting— small, neat and beautifully legible. The Wiley typesetters who set his 1972 book worked from his handwritten manuscript, which they said was easier to work from than most authors' typescripts. During his Oxford years, the secretaries there revered him: they were not used to Chairmen of the Mathematical Institute whose desks were tidy, who handled paperwork promptly, and who would give a decision in real time, rather than procrastinate.

John Kingman has been blessed since his marriage in 1964 in the love and support of his distinguished wife Valerie Cromwell; they have a son and a daughter, who are now acquiring distinction themselves. They now live in retirement in Bristol and London.

While probabilists may regret the loss to probability theory of John's years in administration rather than mathematics, this is offset by the continuing impact of his most important work, whether in queueing theory and heavy traffic, Markov chains and regenerative phenomena (the subject of some of his most recent papers, where he has successfully solved some problems that had remained open since his own work of thirty years ago), subadditive ergodic theory or mathematical genetics and the coalescent. Indeed, the intense concentration of effort on the genetics side associated with the Human Genome Project has thrown in to ever higher relief the fundamental importance of Kingman's work in this area. The editors and contributors to this volume take pleasure in dedicating this book to him, on the occasion of his seventieth birthday.

N. H. Bingham and C. M. Goldie, December 2009.

Acknowledgements All contributions to this collection have been refereed. The editors are most grateful to the referees for their efforts, particularly in those cases where a fast response had to be requested.

The *Bibliography of J. F. C. Kingman* (pp. 1–16) was compiled and arranged by Charles Goldie and Jim Pitman. The editors are grateful to Jim for his collaboration, and also thank John Kingman for providing details of his publications that made the task much easier.

The photograph for the Frontispiece is reproduced by courtesy of the Isaac Newton Institute for Mathematical Sciences, Cambridge.

Bibliography of J. F. C. Kingman

Compiled by Charles M. Goldie and Jim Pitman[a]

1 Books authored

[B1] Kingman, J. F. C., and Taylor, S. J. 1966. *Introduction to Measure and Probability.* Cambridge: Cambridge University Press.

[B2] Kingman, J. F. C. 1966. *On the Algebra of Queues.* Methuen's Supplementary Review Series in Applied Probability, vol. 6. London: Methuen & Co. Ltd. Reprint of [M36].

[B3] Kingman, J. F. C. 1972. *Regenerative Phenomena.* Wiley Series in Probability and Mathematical Statistics. London: John Wiley & Sons.

[B4] Kingman, J. F. C. 1980. *Mathematics of Genetic Diversity.* CBMS-NSF Regional Conference Series in Applied Mathematics, vol. 34. Philadelphia, PA: Society for Industrial and Applied Mathematics (SIAM).

[B5] Kingman, J. F. C. 1993. *Poisson Processes.* Oxford Studies in Probability, vol. 3. Oxford: Oxford University Press.

[a] This bibliography was prepared using the BibServer system developed by Jim Pitman with the assistance of NSF Award 0835851, Bibliographic Knowledge Network.

[B6] Kingman, J. F. C. 2002. *Procesy Poissona*. Wydawnictwo Naukowe PWN. Polish translation of [B5] by Bobrowski, A.

2 Books edited

[E1] Ibragimov, I. A., and Linnik, Yu. V. 1971. *Independent and Stationary Sequences of Random Variables*. Groningen: Wolters-Noordhoff Publishing. With a supplementary chapter by I. A. Ibragimov and V. V. Petrov. Translation from the Russian edited by J. F. C. Kingman.

[E2] Bodmer, W. F., and Kingman, J. F. C. (eds). 1983. *Mathematical Genetics: Proceedings of a Royal Society Meeting held in London, April 20, 1983*. London: Royal Society.

[E3] Kingman, J. F. C., and Reuter, G. E. H. (eds). 1983. *Probability, Statistics and Analysis*. London Math. Soc. Lecture Note Ser., vol. 79. Cambridge: Cambridge University Press. Papers dedicated to David G. Kendall on the occasion of his sixty-fifth birthday.

[E4] Kendall, D. G., with the assistance of Kingman, J. F. C. and Williams, D. (ed). 1986. *Analytic and Geometric Stochastics: Papers in Honour of G. E. H. Reuter*. Adv. in Appl. Probab., vol. 18 (Supplement). Sheffield: Applied Probability Trust.

3 Mathematical articles

[M1] Kingman, J. F. C. 1960. A note on the axial symmetry of the disturbance. *Proc. R. Soc. Lond. Ser. A Math. Phys. Sci.*, **258**, 87–89.

[M2] Yih, Chia-Shun (with an appendix by Kingman, J. F. C.). 1960. Instability of a rotating liquid film with a free surface. *Proc. R. Soc. Lond. Ser. A Math. Phys. Sci.*, **258**, 63–89. Expanded version of [M1].

[M3] Kingman, J. F. C. 1961. A convexity property of positive matrices. *Quart. J. Math. Oxford Ser. (2)*, **12**, 283–284.

[M4] Kingman, J. F. C. 1961. A mathematical problem in population genetics. *Proc. Cambridge Philos. Soc.*, **57**, 574–582.

[M5] Kingman, J. F. C. 1961. On an inequality in partial averages. *Quart. J. Math. Oxford Ser. (2)*, **12**, 78–80.

[M6] Kingman, J. F. C. 1961. The ergodic behaviour of random walks. *Biometrika*, **48**, 391–396.

[M7] Kingman, J. F. C. 1961. The single server queue in heavy traffic. *Proc. Cambridge Philos. Soc.*, **57**, 902–904.

[M8] Kingman, J. F. C. 1961. Two similar queues in parallel. *Ann. Math. Statist.*, **32**, 1314–1323.

[M9] Kingman, J. F. C. 1962. On queues in heavy traffic. *J. Roy. Statist. Soc. Ser. B*, **24**, 383–392.

[M10] Kingman, J. F. C. 1962. On queues in which customers are served in random order. *Proc. Cambridge Philos. Soc.*, **58**, 79–91.

[M11] Kingman, J. F. C. 1962. Some inequalities for the queue $GI/G/1$. *Biometrika*, **49**, 315–324.

[M12] Kingman, J. F. C. 1962. Spitzer's identity and its use in probability theory. *J. Lond. Math. Soc. (2)*, **37**, 309–316.

[M13] Kingman, J. F. C. 1962. The effect of queue discipline on waiting time variance. *Proc. Cambridge Philos. Soc.*, **58**, 163–164.

[M14] Kingman, J. F. C. 1962. The imbedding problem for finite Markov chains. *Z. Wahrscheinlichkeitstheorie verw. Gebiete*, **1**, 14–24.

[M15] Kingman, J. F. C. 1962. The use of Spitzer's identity in the investigation of the busy period and other quantities in the queue $GI/G/1$. *J. Austral. Math. Soc. A*, **2**, 345–356.

[M16] Kingman, J. F. C. 1963. A continuous time analogue of the theory of recurrent events. *Bull. Amer. Math. Soc.*, **69**, 268–272.

[M17] Kingman, J. F. C. 1963. Ergodic properties of continuous-time Markov processes and their discrete skeletons. *Proc. Lond. Math. Soc. (3)*, **13**, 593–604.

[M18] Kingman, J. F. C. 1963. On continuous time models in the theory of dams. *J. Austral. Math. Soc. A*, **3**, 480–487.

[M19] Kingman, J. F. C. 1963. On inequalities of the Tchebychev type. *Proc. Cambridge Philos. Soc.*, **59**, 135–146.

[M20] Kingman, J. F. C. 1963. Poisson counts for random sequences of events. *Ann. Math. Statist.*, **34**, 1217–1232.

[M21] Kingman, J. F. C. 1963. Random walks with spherical symmetry. *Acta Math.*, **109**, 11–53.

[M22] Kingman, J. F. C. 1963. The exponential decay of Markov transition probabilities. *Proc. Lond. Math. Soc. (3)*, **13**, 337–358.

[M23] Kingman, J. F. C. 1964. A martingale inequality in the theory of queues. *Proc. Cambridge Philos. Soc.*, **60**, 359–361.

[M24] Kingman, J. F. C. 1964. A note on limits of continuous functions. *Quart. J. Math. Oxford Ser. (2)*, **15**, 279–282.

[M25] Kingman, J. F. C. 1964. Metrics for Wald spaces. *J. Lond. Math. Soc. (2)*, **39**, 129–130.

[M26] Kingman, J. F. C. 1964. On doubly stochastic Poisson processes. *Proc. Cambridge Philos. Soc.*, **60**, 923–930.

[M27] Kingman, J. F. C., and Orey, Steven. 1964. Ratio limit theorems for Markov chains. *Proc. Amer. Math. Soc.*, **15**, 907–910.

[M28] Kingman, J. F. C. 1964. Recurrence properties of processes with stationary independent increments. *J. Austral. Math. Soc. A*, **4**, 223–228.

[M29] Kingman, J. F. C. 1964. The stochastic theory of regenerative events. *Z. Wahrscheinlichkeitstheorie verw. Gebiete*, **2**, 180–224 (1964).

[M30] Kingman, J. F. C. 1965. Linked systems of regenerative events. *Proc. Lond. Math. Soc. (3)*, **15**, 125–150.

[M31] Kingman, J. F. C. 1965. Mean free paths in a convex reflecting region. *J. Appl. Probab.*, **2**, 162–168.

[M32] Kingman, J. F. C. 1965. Some further analytical results in the theory of regenerative events. *J. Math. Anal. Appl.*, **11**, 422–433.

[M33] Kingman, J. F. C. 1965. Stationary measures for branching processes. *Proc. Amer. Math. Soc.*, **16**, 245–247.

[M34] Kingman, J. F. C. 1965. The heavy traffic approximation in the theory of queues. Pages 137–169 of: Smith, Walter L., and Wilkinson, William E. (eds), *Proceedings of the Symposium on Congestion Theory, University of North Carolina 1964*. Chapel Hill, NC: Univ. North Carolina Press. With discussion and response.

[M35] Kingman, J. F. C. 1966. An approach to the study of Markov processes. *J. Roy. Statist. Soc. Ser. B*, **28**, 417–447. With discussion and response.

[M36] Kingman, J. F. C. 1966. On the algebra of queues. *J. Appl. Probability*, **3**, 285–326.

[M37] Kingman, J. F. C. 1967. Additive set functions and the theory of probability. *Proc. Cambridge Philos. Soc.*, **63**, 767–775.

[M38] Kingman, J. F. C. 1967. An inequality involving Radon–Nikodym derivatives. *Proc. Cambridge Philos. Soc.*, **63**, 195–198.

[M39] Kingman, J. F. C. 1967. Completely random measures. *Pacific J. Math.*, **21**, 59–78.

[M40] Kingman, J. F. C. 1967. Markov transition probabilities I. *Z. Wahrscheinlichkeitstheorie verw. Gebiete*, **7**, 248–270.

[M41] Kingman, J. F. C. 1967. Markov transition probabilities II, Completely monotonic functions. *Z. Wahrscheinlichkeitstheorie verw. Gebiete*, **9**, 1–9.

[M42] Kingman, J. F. C. 1968. Markov transition probabilities III, General state spaces. *Z. Wahrscheinlichkeitstheorie verw. Gebiete*, **10**, 87–101.

[M43] Kingman, J. F. C. 1968. Markov transition probabilities IV, Recurrence time distributions. *Z. Wahrscheinlichkeitstheorie verw. Gebiete*, **11**, 9–17.

[M44] Kingman, J. F. C., and Robertson, A. P. 1968. On a theorem of Lyapunov. *J. Lond. Math. Soc. (2)*, **43**, 347–351.

[M45] Kingman, J. F. C. 1968. On measurable p-functions. *Z. Wahrscheinlichkeitstheorie verw. Gebiete*, **11**, 1–8.

[M46] Kingman, J. F. C. 1968. Some recent developments in the theory of Markov chains. Pages 71–79 of: *Selected Statistical Papers, I*. Amsterdam: Mathematisch Centrum. From the European Meeting of Statisticians, 1968.

[M47] Kingman, J. F. C. 1968. The ergodic theory of subadditive stochastic processes. *J. Roy. Statist. Soc. Ser. B*, **30**, 499–510.

[M48] Kingman, J. F. C. 1969. An ergodic theorem. *Bull. Lond. Math. Soc.*, **1**, 339–340. Addendum in [M52].

[M49] Kingman, J. F. C. 1969. Markov population processes. *J. Appl. Probab.*, **6**, 1–18.

[M50] Kingman, J. F. C. 1969. Random secants of a convex body. *J. Appl. Probab.*, **6**, 660–672.

[M51] Kingman, J. F. C. 1970. A class of positive-definite functions. Pages 93–109 of: Gunning, R. C. (ed), *Problems in Analysis (Lectures at the Sympos. in Honor of Salomon Bochner, Princeton Univ., Princeton, NJ, 1969)*. Princeton, NJ: Princeton Univ. Press.

[M52] Kingman, J. F. C. 1970. Addendum: "An ergodic theorem". *Bull. Lond. Math. Soc.*, **2**, 204. Addendum to [M48].

[M53] Kingman, J. F. C. 1970. An application of the theory of regenerative phenomena. *Proc. Cambridge Philos. Soc.*, **68**, 697–701.

[M54] Kingman, J. F. C. 1970. Inequalities in the theory of queues. *J. Roy. Statist. Soc. Ser. B*, **32**, 102–110.

[M55] Kingman, J. F. C. 1970. Stationary regenerative phenomena. *Z. Wahrscheinlichkeitstheorie verw. Gebiete*, **15**, 1–18.

[M56] Kingman, J. F. C. 1971. Markov transition probabilities V. *Z. Wahrscheinlichkeitstheorie verw. Gebiete*, **17**, 89–103.

[M57] Kingman, J. F. C. 1972. On random sequences with spherical symmetry. *Biometrika*, **59**, 492–494.

[M58] Kingman, J. F. C. 1972. Regenerative phenomena and the characterization of Markov transition probabilities. Pages 241–262 of: Le Cam, L., Neyman, J., and Scott, E. L. (eds), *Proceedings of the Sixth Berkeley Symposium on Mathematical Statistics and Probability (Univ. California, Berkeley, Calif., 1970/1971), Vol. III: Probability Theory*. Berkeley, CA: Univ. California Press.

[M59] Kingman, J. F. C. 1972. Semi-*p*-functions. *Trans. Amer. Math. Soc.*, **174**, 257–273.

[M60] Kingman, J. F. C. 1973. An intrinsic description of local time. *J. Lond. Math. Soc. (2)*, **6**, 725–731.

[M61] Kingman, J. F. C. 1973. Homecomings of Markov processes. *Adv. in Appl. Probab.*, **5**, 66–102.

[M62] Burville, P. J., and Kingman, J. F. C. 1973. On a model for storage and search. *J. Appl. Probab.*, **10**, 697–701.

[M63] Kingman, J. F. C. 1973. On the oscillation of *p*-functions. *J. Lond. Math. Soc. (2)*, **6**, 747–752.

[M64] Kingman, J. F. C. 1973. Some algebraic results and problems in the theory of stochastic processes with a discrete time parameter. Pages 315–330 of: Kendall, D. G., and Harding, E. F. (eds), *Stochastic Analysis (a Tribute to the Memory of Rollo Davidson)*. London: John Wiley & Sons.

[M65] Kingman, J. F. C. 1973. Subadditive ergodic theory. *Ann. Probab.*, **1**, 883–909. With discussion and response.

[M66] Kingman, J. F. C., and Williams, David. 1973. The combinatorial structure of non-homogeneous Markov chains. *Z. Wahrscheinlichkeitstheorie verw. Gebiete*, **26**, 77–86.

[M67] Kingman, J. F. C. 1974. On the Chapman-Kolmogorov equation. *Philos. Trans. Roy. Soc. London Ser. A*, **276**, 341–369.

[M68] Kingman, J. F. C. 1974. Regeneration. Pages 389–406 of: Gani, J.,

Sarkadi, K., and Vincze, I. (eds), *Progress in Statistics (European Meeting of Statisticians, Budapest, 1972)*. Colloq. Math. Soc. János Bolyai, vol. 9. Amsterdam: North-Holland.

[M69] Kingman, J. F. C. 1975. A property of the derivatives of Markov transition probabilities. *Quart. J. Math. Oxford Ser. (2)*, **26**, 121–128.

[M70] Kingman, J. F. C. 1975. Anticipation processes. Pages 201–215 of: Gani, J. (ed), *Perspectives in Probability and Statistics (Papers in Honour of M. S. Bartlett on the Occasion of his 65th Birthday)*. Sheffield: Applied Probability Trust.

[M71] Kingman, J. F. C. 1975. Geometrical aspects of the theory of non-homogeneous Markov chains. *Math. Proc. Cambridge Philos. Soc.*, **77**, 171–183.

[M72] Kingman, J. F. C. 1975. Markov models for spatial variation. *The Statistician*, **24**, 167–174.

[M73] Kingman, J. F. C. 1975. Random discrete distributions. *J. Roy. Statist. Soc. Ser. B*, **37**, 1–22. With discussion and response.

[M74] Kingman, J. F. C. 1975. The first birth problem for an age-dependent branching process. *Ann. Probab.*, **3**(5), 790–801.

[M75] Kingman, J. F. C. 1976. Coherent random walks arising in some genetical models. *Proc. R. Soc. Lond. Ser. A Math. Phys. Eng. Sci.*, **351**, 19–31.

[M76] Kingman, J. F. C. 1976. Subadditive processes. Pages 167–223 of: Hennequin, P.-L. (ed), *École d'Été de Probabilités de Saint-Flour, V– 1975*. Lecture Notes in Math., vol. 539. Berlin: Springer-Verlag.

[M77] Kingman, J. F. C. 1977. A note on multi-dimensional models of neutral mutation. *Theor. Population Biology*, 285–290.

[M78] Kingman, J. F. C. 1977. On the properties of bilinear models for the balance between genetic mutation and selection. *Math. Proc. Cambridge Philos. Soc.*, **81**(3), 443–453.

[M79] Kingman, J. F. C. 1977. Remarks on the spatial distribution of a reproducing population. *J. Appl. Probab.*, **14**(3), 577–583.

[M80] Kingman, J. F. C. 1977. The asymptotic covariance of two counters. *Math. Proc. Cambridge Philos. Soc.*, **82**(3), 447–452.

[M81] Kingman, J. F. C. 1977. The population structure associated with the Ewens sampling formula. *Theor. Population Biology*, **11**(2), 274–283.

[M82] Kingman, J. F. C. 1978. A simple model for the balance between selection and mutation. *J. Appl. Probab.*, **15**(1), 1–12.

[M83] Kingman, J. F. C. 1978. Random partitions in population genetics. *Proc. R. Soc. Lond. Ser. A Math. Phys. Eng. Sci.*, **361**, 1–20.

[M84] Kingman, J. F. C. 1978. The dynamics of neutral mutation. *Proc. Roy. Soc. London Ser. A*, **363**, 135–146.

[M85] Kingman, J. F. C. 1978. The representation of partition structures. *J. Lond. Math. Soc. (2)*, **18**(2), 374–380.

[M86] Kingman, J. F. C. 1978. Uses of exchangeability. *Ann. Probab.*, **6**(2), 183–197.

[M87] Kingman, J. F. C. 1982. Exchangeability and the evolution of large populations. Pages 97–112 of: Koch, G., and Spizzichino, F. (eds), *Exchangeability in Probability and Statistics*. Amsterdam: North-Holland. Proceedings of the International Conference on Exchangeability in Probability and Statistics, Rome, 6th-9th April, 1981, in honour of Professor Bruno de Finetti.

[M88] Kingman, J. F. C. 1982. On the genealogy of large populations. Pages 27–43 of: Gani, J., and Hannan, E. J. (eds), *Essays in Statistical Science: Papers in Honour of P. A. P. Moran*. J. Appl. Probab., Special Volume 19A. Sheffield: Applied Probability Trust.

[M89] Kingman, J. F. C. 1982. Queue disciplines in heavy traffic. *Math. Oper. Res.*, **7**(2), 262–271.

[M90] Kingman, J. F. C. 1982. The coalescent. *Stochastic Process. Appl.*, **13**(3), 235–248.

[M91] Kingman, J. F. C. 1982. The thrown string. *J. Roy. Statist. Soc. Ser. B*, **44**(2), 109–138. With discussion and response.

[M92] Kingman, J. F. C. 1983. Three unsolved problems in discrete Markov theory. Pages 180–191 of: Kingman, J. F. C., and Reuter, G. E. H. (eds), *Probability, Statistics and Analysis*. London Math. Soc. Lecture Note Ser., vol. 79. Cambridge: Cambridge Univ. Press. D.G. Kendall 65th birthday volume [E3].

[M93] Kingman, J. F. C. 1984. Present position and potential developments: some personal views. Probability and random processes. *J. Roy. Statist. Soc. Ser. A (General)*, **147**, 233–244.

[M94] Kingman, J. F. C. 1985. Random variables with unsymmetrical linear regressions. *Math. Proc. Cambridge Philos. Soc.*, **98**(2), 355–365.

[M95] Kingman, J. F. C. 1986. The construction of infinite collections of random variables with linear regressions. Pages 73–85 of: *Analytic and Geometric Stochastics: Papers in Honour of G. E. H. Reuter*. Adv. in Appl. Probab., vol. 18 (Supplement). Sheffield: Applied Probability Trust.

[M96] Kingman, J. F. C. 1988. Random dissections and branching processes. *Math. Proc. Cambridge Philos. Soc.*, **104**(1), 147–151.

[M97] Kingman, J. F. C. 1988. Typical polymorphisms maintained by selection at a single locus. *J. Appl. Probab.*, **25A**, 113–125. Special volume: A Celebration of Applied Probability.

[M98] Kingman, J. F. C. 1989. Maxima of random quadratic forms on a simplex. Pages 123–140 of: Anderson, T. W., Athreya, K. B., and Iglehart, D. L. (eds), *Probability, Statistics, and Mathematics: Papers in Honor of Samuel Karlin*. Boston, MA: Academic Press.

[M99] Kingman, J. F. C. 1990. Some random collections of finite subsets. Pages 241–247 of: Grimmett, G. R., and Welsh, D. J. A. (eds), *Disorder in Physical Systems*. Oxford Sci. Publ. New York: Oxford Univ. Press. Volume in honour of John M. Hammersley on the occasion of his 70th birthday.

[M100] Kingman, J. F. C. 1996. Powers of renewal sequences. *Bull. Lond. Math. Soc.*, **28**(5), 527–532.

[M101] Kingman, J. F. C. 1999. Martingales in the OK Corral. *Bull. Lond. Math. Soc.*, **31**(5), 601–606.

[M102] Kingman, J. F. C. 2000. Origins of the coalescent: 1974–1982. *Genetics*, **156**, 1461–1463.

[M103] Kingman, J. F. C. 2002. Stochastic aspects of Lanchester's theory of warfare. *J. Appl. Probab.*, **39**(3), 455–465.

[M104] Kingman, J. F. C., and Volkov, S. E. 2003. Solution to the OK Corral model via decoupling of Friedman's urn. *J. Theoret. Probab.*, **16**(1), 267–276.

[M105] Kingman, J. F. C. 2004. Extremal problems for regenerative phenomena. *J. Appl. Probab.*, **41A**, 333–346. Stochastic Methods and Their Applications, a Festschrift for Chris Heyde.

[M106] Kingman, J. F. C. 2004. Powers and products of regenerative phenomena. *Aust. N. Z. J. Stat.*, **46**(1), 79–86. Festschrift in honour of Daryl Daley.

[M107] Kingman, J. F. C. 2004. *The Poisson–Dirichlet Distribution and the Frequency of Large Prime Divisors*. Preprint NI04019. Isaac Newton Institute for Mathematical Sciences, Cambridge.

[M108] Kingman, J. F. C. 2006. Poisson processes revisited. *Probab. Math. Statist.*, **26**(1), 77–95.

[M109] Kingman, J. F. C. 2006. Progress and problems in the theory of regenerative phenomena. *Bull. Lond. Math. Soc.*, **38**(6), 881–896.

[M110] Kingman, J. F. C. 2006. *Spectra of Positive Matrices and the Markov Group Conjecture*. Preprint NI06031. Isaac Newton Institute for Mathematical Sciences, Cambridge.

[M111] Kingman, J. F. C. 2009. The first Erlang century—and the next. *Queueing Syst.*, **63**, 3–12.

4 Abstracts

[A1] Kingman, J. F. C. 1962. The imbedding problem for finite Markov chains. *Notices Amer. Math. Soc.*, **9**(February), 35. Abstract 62T–2, presented by title.

[A2] Kingman, J. F. C. 1962. Fourier integral representations for Markov transition probabilities (preliminary report). *Ann. Math. Statist.*, **33**(2), 832. Abstract 6, presented by title.

[A3] Kingman, J. F. C. 1964. Recurrent events and completely monotonic sequences. *Ann. Math. Statist.*, **35**(1), 460. Abstract 3, presented by title.

[A4] Kingman, J. F. C. 1977. The thrown string: an unsolved problem. *Adv. in Appl. Probab.*, **9**(3), 431. Abstract for the Buffon Bicentenary Symposium on Stochastic Geometry and Directional Statistics, Lake Sevan, Erevan, Armenia, 13-18 September 1976.

[A5] Kingman, J. F. C. 1979. Deterministic and stochastic models in population genetics. *Adv. in Appl. Probab.*, **11**(2), 264. Abstract for 8th Conference on Stochastic Processes and Applications, Canberra, 6-10 July 1978.

5 Book reviews authored

[R1] Kingman, J. F. C. 1963. Review of *Mathematical Theories of Traffic Flow*, by Haight, Frank A., 1963. *Zentralblatt Math.* 0127.37202

[R2] Kingman, J. F. C. 1964. Review of *Generalized Markovian Decision Processes. Part I: Model and Method*, by Leve, G. de, 1964. *Math. Rev.* **30** #4295(MR0174088).

[R3] Kingman, J. F. C. 1964. Review of *Generalized Markovian Decision Processes. Part II: Probabilistic Background*, by Leve, G. de, 1964. *Math. Rev.* **31** #798(MR0176526).

[R4] Kingman, J. F. C. 1964. Review of *Probability Theory*, by Loève, Michel. *J. Roy. Statist. Soc. Ser. A (General)*, **127**(1), 127–128.

[R5] Kingman, J. F. C. 1964. Review of *Studies in Mathematical Analysis and Related Topics; Essays in Honour of George Pólya*, by Szegö, G. (ed). *Rev. Int. Statist. Inst.*, **32**(3), 325.

[R6] Kingman, J. F. C. 1964. Review of *Statistical Analysis; Ideas and Methods*, by Lewis, E. Vernon. *Math. Gazette*, **48**(365), 328.

[R7] Kingman, J. F. C. 1964. Review of *Integration, Measure and Probability*, by Pitt, H. R. *Proc. Edinb. Math. Soc. (2)*, **14**(01), 81.

[R8] Kingman, J. F. C. 1964. Review of *General Stochastic Processes in the Theory of Queues*, by Beneš, Václav E. *J. Lond. Math. Soc.*, **39**(1), 381.

[R9] Kingman, J. F. C. 1964. Review of *Ergodic Theory, Parts I and II*, by Jacobs, K. *J. Lond. Math. Soc.*, **39**(1), 380.

[R10] Kingman, J. F. C. 1965. Review of *Introduction to Probability and Statistics from a Bayesian Viewpoint. Part I: Probability*, by Lindley, D. V., 1965. *Math. Rev.* **29** #5348(MR0168083).

[R11] Kingman, J. F. C. 1965. Review of *Introduction to Probability and Statistics from a Bayesian Viewpoint. Part II: Inference*, by Lindley, D. V., 1965. *Math. Rev.* **29** #5349(MR0168084).

[R12] Kingman, J. F. C. 1965. Review of *Packing and Covering*, by Rogers, C. A. *Nature*, **205**, 738.

[R13] Kingman, J. F. C. 1965. Review of *Inequalities on Distribution Functions*, by Godwin, H. J. *The Statistician*, **15**(1), 88–89.

[R14] Kingman, J. F. C. 1965. Review of *Convex Transformations of Random Variables*, by Zwet, W. R. van. *J. Roy. Statist. Soc. A (General)*, **128**(4), 597–598.

[R15] Kingman, J. F. C. 1966. Review of *Ergodic Theory and Information*, by Billingsley, Patrick. *J. Roy. Statist. Soc. A (General)*, **129**(3), 472.

[R16] Kingman, J. F. C. 1966. Review of *Green's Function Methods in Probability Theory*, by Keilson, Julian. *J. Roy. Statist. Soc. A (General)*, **129**(1), 119–120.

[R17] Kingman, J. F. C. 1966. Review of *Mathematical Foundations of the Calculus of Probability*, by Neveu, Jacques. *J. Roy. Statist. Soc. A (General)*, **129**(3), 475–476.

[R18] Kingman, J. F. C. 1966. Review of *Bibliography on Time Series and Stochastic Processes: An International Team Project*, by Wold, Herman O. A. *J. Roy. Statist. Soc. A (General)*, **129**(2), 295.

[R19] Kingman, J. F. C. 1966. Review of *Mathematical Theory of Connecting Networks and Telephone Traffic*, by Beneš, Václav E. *J. Roy. Statist. Soc. A (General)*, **129**(2), 295–296.

[R20] Kingman, J. F. C. 1967. Review of *An Introduction to Probability Theory and its Applications, vol. II*, by Feller, William. *J. Roy. Statist. Soc. A (General)*, **130**(1), 109.

[R21] Kingman, J. F. C. 1967. Review of *A First Course in Stochastic Processes*, by Karlin, Samuel. *J. Lond. Math. Soc.*, **42**(1), 367.

[R22] Kingman, J. F. C. 1967. Review of *Processus stochastiques et mouvement brownien*, by Lévy, P. *J. Lond. Math. Soc.*, **42**(1), 190.

[R23] Kingman, J. F. C. 1968. Review of *Priority Queues*, by Jaiswal, N. K., 1968. *Math. Rev.* **38** #5307(MR0237014).

[R24] Kingman, J. F. C. 1968. Review of *The Theory of Gambling and Statistical Logic*, by Epstein, R. A. *Math. Gazette*, **52**(382), 426.

[R25] Kingman, J. F. C. 1968. Review of *The Theory of Random Clumping*, by Roach, S. A., 1968. *Math. Rev.* **37** #5908(MR0230346).

[R26] Kingman, J. F. C. 1968. Review of *Stationary Random Processes*, by Rozanov, Yu. A. *J. Lond. Math. Soc.*, **43**(1), 574.

[R27] Kingman, J. F. C. 1969. Random Processes. Review of *Stochastic Integrals*, by McKean, H. P. *Nature*, 219.

[R28] Kingman, J. F. C. 1969. Review of *Convergence of Probability Measures*, by Billingsley, P. *J. Roy. Statist. Soc. C (Applied Statistics)*, **18**(3), 282.

[R29] Kingman, J. F. C. 1969. Review of *Convergence of Probability Measures*, by Billingsley, P. *Rev. Int. Statist. Inst.*, **37**(3), 322.

[R30] Kingman, J. F. C. 1969. Review of *An Introduction to Probability Theory*, by Moran, P. A. P. *J. Roy. Statist. Soc. A (General)*, **132**(1), 106.

[R31] Kingman, J. F. C. 1969. Review of *Denumerable Markov Chains*, by Kemeny, John G., Snell, J. Laurie and Knapp, Anthony W. *Math. Gazette*, **53**(384), 206.

[R32] Kingman, J. F. C. 1969. Review of *The Single Server Queue*, by Cohen, J. W., 1969. *Math. Rev.* **40** #3628(MR0250389).

[R33] Kingman, J. F. C. 1970. Review of *Generalized Markovian Decision Processes. Applications*, by Leve, G. de, Tijms, H. C. and Weeda, P. J., 1970. *Math. Rev.* **42** #4187(MR0269291).

[R34] Kingman, J. F. C. 1970. Russian Probability. Review of *Probability Theory: Basic Concepts, Limit Theorems, Random Processes*, by Prohorov, Yu. V. and Rozanov, Yu. A. *Nature*, **225**, 1169.

[R35] Kingman, J. F. C. 1970. Review of *Introduction to Mathematical Probability Theory*, by Eisen, Martin. *J. Roy. Statist. Soc. A (General)*, **133**(1), 98.

[R36] Kingman, J. F. C. 1970. Review of *A Brief Introduction to Probability Theory*, by Hoyt, John P. *Math. Gazette*, **54**(387), 90.

[R37] Kingman, J. F. C. 1970. Review of *Information Theory and Statistics*, by Kullback, Solomon. *Math. Gazette*, **54**(387), 90.

[R38] Kingman, J. F. C. 1970. Review of *The Single Server Queue*, by Cohen, J. W. *Bull. Lond. Math. Soc.*, **2**(3), 356.

[R39] Kingman, J. F. C. 1970. Review of *Stochastic Convergence*, by Lukacs, Eugene. *Bull. Lond. Math. Soc.*, **2**(2), 246.

[R40] Kingman, J. F. C. 1971. Review of *Lecture Notes on Limit Theorems for Markov Chain Transition Probabilities*, by Orey, Steven, 1971. *Math. Rev.* **48** #3123(MR0324774).

[R41] Kingman, J. F. C. 1971. Review of *Probability on Discrete Sample Spaces with Applications*, by Scheerer, Anne E. *J. Roy. Statist. Soc. A (General)*, **134**(1), 91.

[R42] Kingman, J. F. C. 1971. Review of *The Ergodic Theory of Markov Processes*, by Foguel, S. R., and *Stationary Stochastic Processes*, by Hida, T. *Math. Gazette*, **55**(394), 479–480.

[R43] Kingman, J. F. C. 1972. Review of *Numerical Methods in Markov Chains and Bulk Queues*, by Bagchi, T. P. and Templeton, J. G. C., 1972. *Math. Rev.* **49** #9976(MR0345237).

[R44] Kingman, J. F. C. 1972. Review of *An Introduction to Probability Theory and its Applications, vol. II, 2nd ed.*, by Feller, W. *J. Roy. Statist. Soc. A (General)*, **135**(3), 430.

[R45] Kingman, J. F. C. 1972. Review of *The Analysis of Binary Data*, by Cox, D. R. *Math. Gazette*, **56**(395), 67–68.

[R46] Kingman, J. F. C. 1972. Review of *Theory of Probability*, by Levine, Arnold. *Math. Gazette*, **56**(396), 177.

[R47] Kingman, J. F. C. 1972. Review of *A Course in Probability Theory*, by Chung, K. L., *Introduction to Probability and Statistics*, by Goldman, M., and *Probability and Statistical Inference*, by Krutchkoff, R. G. *Math. Gazette*, **56**(395), 67.

[R48] Kingman, J. F. C. 1972. Review of *Dynamic Probabilistic Systems. Volume 1: Markov Models and Volume 2: Semi-Markov and Decision Processes*, by Howard, Ronald A. *J. Roy. Statist. Soc. A (General)*, **135**(1), 152–153.

[R49] Kingman, J. F. C. 1973. Fourier Probabilists. Review of *Fourier Analysis in Probability Theory*, by Kawata, Tatsuo. *Nature*, **243**, 245–246.

[R50] Kingman, J. F. C. 1973. Review of *Martingales à temps discret*, by Neveu, Jacques. *J. Roy. Statist. Soc. A (General)*, **136**(4), 624–625.

[R51] Kingman, J. F. C. 1973. Review of *Proceedings of the Sixth Berkeley Symposium on Mathematical Statistics and Probability. Volume II: Probability Theory*, by Le Cam, Lucien M. and Neyman, Jerzy and Scott, Elizabeth L. (eds). *J. Roy. Statist. Soc. A (General)*, **136**(3), 450–451.

[R52] Kingman, J. F. C. 1974. Strong Probability. Review of *Stochastic Analysis: A Tribute to the Memory of Rollo Davidson*, by Kendall, D. G. and Harding E. F. (eds). *Nature*, **248**, 87.

[R53] Kingman, J. F. C. 1974. Review of *Stochastic Processes and the Wiener Integral*, by Yeh, J. *Bull. Lond. Math. Soc.*, **6**(2), 251.

[R54] Kingman, J. F. C. 1974. Review of *Introduction to Queueing Theory*, by Cooper, R. B. *Bull. Lond. Math. Soc.*, **6**(1), 105–106.

[R55] Kingman, J. F. C. 1975. Review of *Random Sets and Integral Geometry*, by Matheron, G. *Bull. Amer. Math. Soc.*, **81**, 844–847.

[R56] Kingman, J. F. C. 1975. Review of *Theory of Probability, A Critical Introductory Treatment, vol. 1*, by Finetti, Bruno de (English translation by Machi, A. and Smith, A. F. M.). *J. Roy. Statist. Soc. A (General)*, **138**(1), 98–99.

[R57] Kingman, J. F. C. 1975. Review of *Asymptotic Expansions and the Deficiency Concept in Statistics*, by Albers, W. *J. Roy. Statist. Soc. A (General)*, **138**(4), 577–578.

[R58] Kingman, J. F. C. 1975. Review of *Conditional Probability Distributions*, by Tjur, Tue. *J. Roy. Statist. Soc. A (General)*, **138**(4), 578.

[R59] Kingman, J. F. C. 1975. Review of *Random Processes*, by Rosenblatt, Murray. *J. Roy. Statist. Soc. A (General)*, **138**(3), 435.

[R60] Kingman, J. F. C. 1975. Review of *Approximate Stochastic Behaviour of n-Server Service Systems with Large n*, by Newell, G. F. *Rev. Int. Statist. Inst.*, **43**(2), 248–249.

[R61] Kingman, J. F. C. 1975. Review of *Probability Theory: A Historical Sketch*, by Maistrov, L. E. and Kotz, Samuel. *J. Roy. Statist. Soc. A (General)*, **138**(2), 267–268.

[R62] Kingman, J. F. C. 1975. Review of *Letters on Probability*, by Rényi, Alfréd. *Rev. Int. Statist. Inst.*, **43**(2), 249.

[R63] Kingman, J. F. C. 1975. Review of *Mathematics and Statistics: Essays in Honour of Harald Bergström*, by Jagers, P. and Rôde, L. (eds). *Rev. Int. Statist. Inst.*, **43**(3), 370.

[R64] Kingman, J. F. C. 1976. Review of *Discrete-Parameter Martingales*, by Neveu, Jacques. *J. Roy. Statist. Soc. A (General)*, **139**(4), 547–548.

[R65] Kingman, J. F. C. 1976. Review of *Theory of Probability, A Critical Introductory Treatment, vol. 2*, by Finetti, Bruno de (English translation by Machi, A. and Smith, A. F. M.). *J. Roy. Statist. Soc. A (General)*, **139**(3), 403.

[R66] Kingman, J. F. C. 1976. Review of *The Theory of Stochastic Processes II*, by Gihman, I. I. and Skorohod, A. V. *Bull. Lond. Math. Soc.*, **8**(3), 330.

[R67] Kingman, J. F. C. 1976. Review of *The Theory of Stochastic Processes I*, by Gihman, I. I. and Skorohod, A. V. *Bull. Lond. Math. Soc.*, **8**(3), 326.

[R68] Kingman, J. F. C. 1977. Review of *Stochastic Processes in Queueing Theory*, by Borovkov, A. A. *Bull. Amer. Math. Soc.*, **83**, 317–318.

[R69] Kingman, J. F. C. 1977. Review of *Problems of Analytical Statistics*, by Linnik, Y. V. *J. Roy. Statist. Soc. A (General)*, **140**(4), 545.

[R70] Kingman, J. F. C. 1977. Review of *Stochastic Population Theories*, by Ludwig, Donald. *Rev. Int. Statist. Inst.*, **45**(1), 99–100.

[R71] Kingman, J. F. C. 1977. Review of *The Mathematical Theory of Infectious Diseases*, by Bailey, Norman T. J. *Rev. Int. Statist. Inst.*, **45**(1), 95–96.

[R72] Kingman, J. F. C. 1977. Review of *Elements of Probability Theory*, by Fortet, Robert. *Math. Gazette*, **61**(417), 237.

[R73] Kingman, J. F. C. 1977. Review of *Integration in Hilbert Space*, by Skorohod, A. V. *Bull. Lond. Math. Soc.*, **9**(2), 233.

[R74] Kingman, J. F. C. 1979. Review of *Information and Exponential Families in Statistical Theory*, by Barndorff-Nielsen, O. *J. Roy. Statist. Soc. A (General)*, **142**(1), 67.

[R75] Kingman, J. F. C. 1979. Review of *Infinitely Divisible Point Processes*, by Matthes, K., Kerstan, J. and Mecke, J. *J. Roy. Statist. Soc. A (General)*, **142**(2), 263.

[R76] Kingman, J. F. C. 1979. Review of *Introduction to Probability and Measure*, by Parthasarathy, K. R. *J. Roy. Statist. Soc. A (General)*, **142**(3), 385–386.

[R77] Kingman, J. F. C. 1979. Review of *Probability Theory: Independence, Interchangeability, Martingales*, by Chow, Y. S. and Teicher, H., and *Diffusions, Markov Processes and Martingales, Vol. 1: Foundations*, by Williams, D. *J. Roy. Statist. Soc. A (General)*, **142**(4), 509.

[R78] Kingman, J. F. C. 1980. Review of *Random Walks with Stationary Increments and Renewal Theory*, by Berbee, H. C. P. *J. Roy. Statist. Soc. A (General)*, **143**(3), 373.

[R79] Kingman, J. F. C. 1980. Review of *Reversibility and Stochastic Networks*, by Kelly, F. P. *Eur. J. Oper. Res.*, **4**(5), 358–359.

[R80] Kingman, J. F. C. 1980. Review of *Multi-dimensional Diffusion Processes*, by Stroock, D. W. and Varadhan, S. R. S. *Bull. Lond. Math. Soc.*, **12**(2), 139–140.

[R81] Kingman, J. F. C. 1981. Review of *Finite Markov Processes and their Applications*, by Iosifescu, Marius. *Bull. Lond. Math. Soc.*, **13**(3), 250–251.

[R82] Kingman, J. F. C. 1985. Review of *Science and Politics*, by Bogdanor, V. *Government and Opposition*, **20**(3), 422–424.

[R83] Kingman, J. F. C. 1987. Review of *Brownian Motion and Stochastic Flow Systems*, by Harrison, J. Michael. *The Statistician*, **36**(1), 66–67.

6 Discussion contributions

[D1] Kingman, J. F. C. 1964. Discussion of *Some statistical problems connected with crystal lattices*, by Domb, C. *J. Roy. Statist. Soc. Ser. B*, **26**, 392-93.

[D2] Kingman, J. F. C. 1965. Discussion of *A Bayesian significance test for multinomial distributions*, by Good, I. J. *J. Roy. Statist. Soc. Ser. B*, **29**, 425.

[D3] Kingman, J. F. C. 1965. Discussion of *Spacings*, by Pyke, R. *J. Roy. Statist. Soc. Ser. B*, **27**, 438-39.

[D4] Kingman, J. F. C. 1968. Discussion of *A generalisation of Bayesian inference*, by Dempster, A. P. *J. Roy. Statist. Soc. Ser. B*, **30**, 241-42.

[D5] Kingman, J. F. C. 1969. Discussion of *Stochastic models of capital investment*, by Plackett, R.L. *J. Roy. Statist. Soc. Ser. B*, **31**, 20-21.

[D6] Kingman, J. F. C. 1971. Discussion of *Spline transformations: three new diagnostic aids for the statistical data-analyst*, by Boneva, L. I., Kendall, D. G. and Stefanov, I. *J. Roy. Statist. Soc. Ser. B*, **33**, 55.

[D7] Kingman, J. F. C. 1973. Discussion of *Central limit analogues for Markov population processes*, by McNeil, D. R. and Schach, S. *J. Roy. Statist. Soc. Ser. B*, **35**, 15-17.

[D8] Kingman, J. F. C. 1975. Discussion of *Statistics of directional data*, by Mardia, K. V. *J. Roy. Statist. Soc. Ser. B*, **37**, 376-77.

[D9] Kingman, J. F. C. 1977. Discussion of *Modelling spatial patterns*, by Ripley, B. D. *J. Roy. Statist. Soc. Ser. B*, **39**, 195-96.

[D10] Kingman, J. F. C. 1977. Discussion of *Spatial contact models for ecological and epidemic spread*, by Mollison, D. *J. Roy. Statist. Soc. Ser. B*, **39**, 319.

[D11] Kingman, J. F. C. 1978. Discussion of *Operational research in the health and social services*, by Duncan, I. B. and Curnow, R. N. *J. Roy. Statist. Soc. Ser. A (General)*, **141**, 183-84.

[D12] Kingman, J. F. C. 1978. Discussion of *Some problems in epidemic theory*, by Gani, J. *J. Roy. Statist. Soc. Ser. A (General)*, **141**, 342.

[D13] Kingman, J. F. C. 1978. Discussion of *The inverse Gaussian distribution and its statistical application—a review*, by Folks, J. L. and Chhikara, R. S. *J. Roy. Statist. Soc. Ser. B*, **40**, 281.

[D14] Kingman, J. F. C. 1979. Discussion of *On the reconciliation of probability assessments*, by Lindley, D. V., Tversey, A. and Brown, R. V. *J. Roy. Statist. Soc. Ser. A (General)*, **142**, 171.

[D15] Kingman, J. F. C. 1981. Discussion of *Revising previsions: a geometric interpretation*, by Goldstein, M. *J. Roy. Statist. Soc. Ser. B*, **43**, 120-21.

[D16] Kingman, J. F. C. 1983. Discussion of *The analysis of library data*, by Burrell, Q. L and Cane, V.R. *J. Roy. Statist. Soc. Ser. A (General)*, **145**, 463-64.

[D17] Kingman, J. F. C. 1984. Discussion of *Statistical inference of phylogenies*, by Felsenstein, J. *J. Roy. Statist. Soc. Ser. A (General)*, **146**, 264-5.

[D18] Kingman, J. F. C. 1985. Discussion of *A symposium on stochastic networks*, by Kelly, F. P., Mitrani, I. and Whittle, P. *J. Roy. Statist. Soc. Ser. B*, **47**, 415-17.

[D19] Kingman, J. F. C. 1990. Discussion of *The skills challenge of the Nineties*, by Moore, P. G. *J. Roy. Statist. Soc. Ser. A (Statistics in Society)*, **153**, 284.

[D20] Kingman, J. F. C. 1999. Discussion of *Recent common ancestors of all present-day individuals*, by Chang, J. T. *Adv. in Appl. Probab.*, **31**, 1027–1035.

[D21] Kingman, J. F. C. 2000. Discussion of *The philosophy of statistics*, by Lindley, D. V. *J. Roy. Statist. Soc. Ser. D (The Statistician)*, **49**, 326-27.

7 Other contributions

[C1] Davidson, R. 1973. Smith's phenomenon, and "jump" *p*-functions. Pages 234–247 of: *Stochastic Analysis (a Tribute to the Memory of Rollo Davidson)*. London: John Wiley & Sons. From an incomplete manuscript, edited by J. F. C. Kingman, G. E. H. Reuter and D. S. Griffeath.

[C2] Kingman, J. F. C. 1983. Introductory Remarks (to [C3]). *Proc. Roy. Soc. Lond. Ser. B (Biological Sciences)*, **219**, 221–222.

[C3] Bodmer, W. F., and Kingman, J. F. C. (eds). 1983. Abstracts of Papers from the Discussion Meeting on Mathematical Genetics. *Proc. Roy. Soc. Lond. Ser. A*, **390**, 217–220.

[C4] Kingman, J. F. C. 1986. Obituary: Harald Cramér: 1893–1985. *J. Roy. Statist. Soc. Ser. A (General)*, **149**(2), 186.

[C5] Kingman, J. F. C. 1986. Science and the Public Purse, the 1985 *Government and Opposition*/Leonard Schapiro Public Lecture at the London School of Economics, 7 November 1985. *Government and Opposition*, **21**(1), 3–16.

[C6] Department of Education and Science Committee of Inquiry into the Teaching of the English Language (chaired by Sir J. F. C. Kingman). 1988. *Report of the Committee of Inquiry into the Teaching of the English Language*.

[C7] Kingman, J. F. C., Durbin, J., Cox, D. R., and Healy, M. J. R. 1988. Statistical Requirements of the AIDS Epidemic. *J. Roy. Statist. Soc. A (Statistics in Society)*, **151**(1), 127–130. A Statement by the RSS.

[C8] Kingman, J. F. C. 1989. Statistical responsibility (RSS Presidential Address). *J. Roy. Statist. Soc. Ser. A (Statistics in Society)*, **152**(3), 277–285.

[C9] Kingman, J. F. C. 1993. The Pursuit of Truth. *The Times Higher Education Supplement*. June 18, 1993.

[C10] Kingman, J. F. C. 1993. *Truth in the University*. The E. H. Young Lecture, 27 April 1993. Univ. Bristol.

[C11] Kingman, J. F. C. 1995. Double Take. *The Times Higher Education Supplement.* 22 Sept. 1995.

[C12] Kingman, J. F. C. 1998. Statistics, Science and Public Policy. Pages 7–12 of: Herzberg, A., and Krupka, I. (eds), *Statistics, Science and Public Policy.* Kingston, Ontario: Queen's University.

[C13] Kingman, J. F. C. 1999. Scientific Shortfall. *The Times Higher Education Supplement.* 12 Nov. 1999.

[C14] Kingman, J. F. C. 2009. David George Kendall, 15 January 1918–23 October 2007. *Biographical Memoirs of the Fellows of the Royal Society,* **55**, 121–138.

8 Interviews and Biographies

[I1] Drazin, Philip. 2001. Interview with Sir John Kingman. *Mathematics Today,* **37**(3), 70–71. A publication of the Institute of Mathematics and its Applications, Southend-on-Sea, UK.

[I2] Körner, Tom. 2002. Interview with Sir John Kingman. *European Mathematical Society Newsletter,* **37**, 14–15.
http://www.emis.de/newsletter/newsletter43.pdf

[I3] 2006. *Sir John Kingman, FRS.* Isaac Newton Institute for Mathematical Sciences, Cambridge. Biographical Sketch.
http://www.newton.cam.ac.uk/history/kingman.html

[I4] O'Connor, J. J., and Robertson, E. F. 2009. *John Frank Charles Kingman.* MacTutor History of Mathematics Biographies. School of Mathematics and Statistics, University of St Andrews, Scotland.
http://www-groups.dcs.st-and.ac.uk/~history/Mathematicians/Kingman.html

[I5] Kingman, J. F. C. 2010. A fragment of autobiography, 1957–1967. Chap. 1 of: Bingham, N. H., and Goldie, C. M. (eds), *Probability and Mathematical Genetics: Papers in Honour of Sir John Kingman.* London Math. Soc. Lecture Note Ser. Cambridge: Cambridge Univ. Press.

1

A fragment of autobiography, 1957–1967

J. F. C. Kingman

It is a great honour to have been given this fine collection, and I am most grateful to the Editors and to my friends who have contributed to it. It is poor thanks to inflict on them this self-centred account, and I have tried to mitigate the offence by limiting it to a particular decade, which begins as a new student arrives at Pembroke College in Cambridge to start a degree course in mathematics, and ends with him as a professor in the new University of Sussex. It was obviously a crucial time in my own mathematical life, but it happens also to have been a very interesting period in the development of probability theory in Britain, in which I was fortunate to play a junior part.

With little reluctance I have restricted myself to the mathematical aspects of my life. The reader will seek in vain for any details of my transition from a schoolboy to a happy husband soon to be a proud father; the latter conditions have proved lasting.

Of course the story goes back long before 1957. It might be said to have started in about 1920, when Charles Kingman, a miner on the small but prosperous North Somerset coalfield, summoned his two sons William and Frank for a serious talk. Charles came of a family of Mendip villagers, his grandfather having been the carter of Ston Easton. Kingman is a commoner surname in the USA than in England, because most of the family emigrated to the American colonies in the 17th century, but obviously a few overslept and missed the boat.

The message that William and Frank heard from their father was that whatever they did in life they must not go down the mine. William became an excellent baker, and ironically his lungs suffered as badly from flour as they would have done from coal dust. My father Frank (born in 1907) won an award that enabled him to enter the young University of Bristol (chartered 1909), and for six years he commuted by train between the family home in Midsomer Norton and the University, ending up with a first class BSc in chemistry and a PhD in catalytic adsorption.

Armed with this qualification, Frank was able to move to the distinguished Colloid Science Laboratory in Cambridge, to undertake what we now call post-doctoral work under Sir Eric Rideal. It was an exciting environment, with colleagues from varying disciplines like H. W. Melville, who later founded the Science Research Council, and J. S. Mitchell, a future Regius Professor of Physic. Nowadays such an experience might well have led to an academic career, but university posts were scarce, and Frank entered the Scientific Civil Service, working at the Fuel Research Station in east London.

At first he lived with an uncle, but he soon met, and in 1938 married, a girl called Maud Harley who came of a London family, her father a successful gentlemen's hairdresser. They settled in the southern suburb of West Wickham, and produced two sons, myself and my younger brother Robert (who in due course followed his father in reading chemistry at Bristol). I was born six days before the outbreak of war in 1939, and therefore grew up in the grim time of war-torn London. The area around West Wickham was heavily bombed, but fortunately we were spared, and I started school towards the end of the war.

Because of my father's scientific occupation, he was not called up to fight, so that our family remained together. One of the values that my parents shared was a firm belief in education, and a concern that their sons should work hard to develop any talents they might possess. Thus I was taught to read and write at home before starting school, and thereafter was given every encouragement to take advantage of the teaching provided. My most vivid memory was of a lovely middle-aged teacher Mrs Underhay, who allowed her pupils to walk home with her after school to practise as we walked the multiplication table. I got to the 16-times table before graduating to the wonderfully complex weights and measures of that pre-metric age.

Shortly after the war, my father was transferred to the Fire Research Station, which was in Elstree north of London, so we moved in 1949 to the northern suburb of Mill Hill, and the next year I entered Christ's College Finchley, which had been founded in 1857 as a church school, but had later become a secondary grammar school funded by the local authority. Tragically my mother died of breast cancer in 1951, and my father was left to bring up two teenage boys on his own.

Christ's College was an excellent school of a type that has now almost disappeared. It was ruled by an old-fashioned headmaster H. B. Pegrum, an Oxford man who believed in academic excellence, team games, clean living and the Church of England. I found that I could share in the first of these, aided by a competitive instinct that found no expression on the sports field, where I was notably incompetent. The fourth was somewhat incongruous, since the school drew largely from the Jewish community of Finchley and Edgware, but this cultural mix was beneficial for me because I was pitted against boys with a strong family belief in learning.

I found that a good memory and quick thinking made me good at subjects that required no coordination of hand and eye, but mathematics was the one I enjoyed. The senior mathematics master was H. W. Turl, a man of broad interests who taught economics and law as well. He encouraged me to think of myself as a possible future mathematician, and he gained a powerful ally in the head of chemistry after I had dropped the Kipps apparatus and filled the laboratories with the smell of rotten eggs.

Turl had a sardonic style that commended him to cynical schoolboys. Faced one day with some laboured calculation I had produced, he advised me: "Kingman, let me tell you, mathematicians are lazy". This injunction, to find the right way to solve a problem, if possible without grinding through avoidable algebra, was probably the first piece of real mathematical insight I encountered, and half a century of experience has proved its worth.

It was assumed, by both the school and my father, that I would go on to a university to read mathematics. The most popular universities for science students from Christ's College were Imperial College London and Bristol, but Pegrum persuaded me and my father that I should drop all subjects except mathematics and physics in the sixth form, take the final examinations a year early, and then try for a Cambridge scholarship. My father, who had nostalgic memories of his time at Cambridge, found this advice congenial. I had little idea of what would be involved, but when I looked at past papers for the scholarship examination, I found them much more interesting than those I had seen in school examinations. However, both he and I were more than surprised when a telegram arrived from Pembroke College telling me that I had been awarded a Major Scholarship.

Thus it was that, in October 1957, I presented myself at the porter's lodge of Pembroke to study mathematics. Cambridge is full of splendid ancient structures, some like King's College Chapel built of stone, some of flesh and blood, and others less material but no less important. Of the last group, the Mathematical Tripos is pre-eminent. Founded in 1748 as probably the first written degree examination in Christendom, it has graded some of the finest British brains into its three classes of Wrangler, Senior Optime and Junior Optime. Until 1909 the three classes were each listed in order of merit, and to be Senior Wrangler was to be hailed by the whole university. Even after the order ceased to be published, it was known informally, and there was fierce competition between and within colleges.

This competitive aspect has attracted criticism, but I am not convinced that its effects were altogether bad. It encouraged examiners to set questions that demanded more than the regurgitation of standard proofs and fashionable jargon, and put a premium on solving hard problems. If one wanted to do well in the Tripos examinations, one attended the lectures and tested one's understanding by trying to answer questions in recent past papers. If this failed, one took difficulties to the weekly supervision, at which the problem might or might not be resolved. It was a demanding regime, but one that made allowance for different levels of ability, and that challenged the better students to develop their abilities to the full.

At that time the Mathematical Tripos was divided into three parts. Most students took Part I at the end of their first year, and Part II at the end of their third. Success in Part II allowed them to graduate BA, after which they either left Cambridge or spent a fourth year taking Part III. It was however considered that someone who had gained a college scholarship had probably covered most of the Part I syllabus, and could start at once on Part II, taking that examination at the end of the second year. It was not allowed to graduate in less than three years, so the third year could be spent on Part III or indeed on some other subject.

The lectures for Part II were very variable both in style and in competence. Some of the lecturers were world class mathematicians—Hoyle, Atiyah, Cartwright for instance. Others were unknowns who had published nothing since their PhD theses. Some were brilliant lecturers, while others would never have survived the modern world of teaching quality assessment. The correlation between these two classifications was not statistically significant. The syllabuses were old-fashioned, because any change had to be agreed by the whole faculty. Lecturers had to stick to the syllabus, because they did not in general set the examination questions. Thus it was very difficult to introduce new approaches, still less to introduce 'new' branches such as functional analysis.

Much therefore depended on the supervisions, which were organised on a college basis. The normal pattern was for the students to be grouped in pairs of similar ability. Each pair would have two supervisions each week, one in pure and one in applied mathematics. These might be given by a fellow of the college, or a research student, or by one of a diffuse cloud of mathematicians whose only source of subsistence was to offer such teaching to colleges in need.

I had chosen Pembroke for personal reasons unconnected with its teachers. In fact the only teaching fellow was a delightful old gentleman Robert Stoneley, a geophysicist who was an applied mathematician of a very old school. He regarded vector calculus as a very dubious invention, and tensors as unfit for civilised discourse. For pure mathematics, we were sent up the Newmarket Road to a charming man who was probably a better teacher than we realised, but who certainly did not inspire.

At the end of my first year, my luck changed. The Master of Pembroke, Sir Sydney Roberts, retired and was succeeded by Sir William Hodge, the Lowndean Professor of Astronomy and Geometry. Hodge had been a professorial fellow for many years, and as such had not taught in the College, but he had been instrumental, with the Tutor Tony Camps, in raising the academic standards and introducing such things as research fellowships. He now made a decisive move by attracting his former student Michael Atiyah from Trinity. It is impossible to understand how Trinity, of which Atiyah later became Master, allowed this to happen, but it was a stroke of good fortune for Pembroke mathematics and for me in particular.

In the first year I had been paired with Raymond Lickorish, whose forte was (and is) geometry, but Atiyah taught him alone, and I teamed up with John Bather, later to be my professorial successor at Sussex. Supervisions with Michael Atiyah were a quite remarkable experience, and the opportunity for close contact with one of the greatest mathematicians of our time (though we did not then know it) was an immense privilege. He had a habit of saying, when we took him a tricky question from a past Tripos paper: "Of course, the examiner hasn't really asked the right question". Only gradually did we realise that, as well as being a way of avoiding an awkward issue, Michael was educating us in mathematical priorities and developing our mathematical taste.

Where was probability in all this? Nowhere in the lecture list for the first year of Part II, but there was a Part I course in Statistics given by a Mr Lindley. The splendidly introverted convention of the Cambridge lecture list was that this prefix meant that the lecturer had no Cambridge doctorate. Had he been a Cambridge PhD he would have been listed as Dr D. V. Lindley. This would distinguish him from a Dr Lindley with no initials, denoting the holder of the higher ScD degree. Anyway, I had heard that Lindley was a good lecturer, so I went to this course although it would not contribute to my Part II.

The lectures, on basic probability and the standard statistical tests and estimators (Lindley had not yet met Bayes on the road to Damascus) were first rate, clear, well organised and full of interest. Lindley worked in the Statistical Laboratory, which meant that unlike most of the lecturers he had access to a duplicator and was able to issue duplicated notes, an unheard-of innovation.

Since I was not particularly well off, I thought I should try to get a job in the summer vacation after my first year. Through a friend of my father in Mill Hill, I heard that the Post Office Engineering Research Station at Dollis Hill, not far away, took on university students in the summer in a very junior temporary capacity, and I successfully applied for a job lasting some eight weeks. I was allocated to RE1, a small research group giving advice on mathematics and computing to the engineers working on a range of problems in the telecommunications field. After proving that I was not good at making the tea, I was given a handbook of common Fourier series which was known to be riddled with errors, and which I was told to revise. This had the useful effect of giving me one Tripos question which I could answer at sight, since there was always a function whose Fourier series we had to compute, and I had all the easy examples by heart.

After a couple of weeks, I graduated to some simple modelling the group was doing of the demand for telephone cable pairs to homes. In those days not every family had a telephone; if you applied for one you were lucky if there were enough cables in your road. Our models were very simple, and involved probability calculations little more advanced than those in Lindley's lectures. But it was enough to rouse my interest, and when I went back to Cambridge I looked at the available lectures to see if I could discover more about this sort of mathematics.

The word 'probability' did not figure in the list. Later I discovered that this was at the insistence of Sir Harold Jeffreys, who had just retired from the Plumian chair and who had laid down that the only Theory of Probability that could be taught was that contained in his book with that title. For Jeffreys, probability was a sort of objective subjective concept, Bayesian inference where the prior distribution was determined by considerations of invariance; no one but Jeffreys wanted to teach that. However, there was a course entitled 'Random Variables', which seemed promising not least because it was given by Mr Lindley.

It proved even better than Lindley's Part I course. From an axiomatic foundation which was a much watered down version of Kolmogorov, it proceeded to Markov chains and some of their important applications. In particular, Lindley used examples from the theory of queues. What his audience did not know was that Lindley had been the first referee for the paper that D. G. Kendall read to the Royal Statistical Society (RSS) in 1951, and that in proposing the vote of thanks to Kendall, Lindley had announced his famous identity which became the main basis of subsequent work on single server queues.

John Bather shared my enthusiasm for this course, and we asked Stoneley and Atiyah whether it counted as pure or applied for the purposes of supervision. Neither, was the unhelpful response, and when we protested they arranged for Dennis Lindley himself to give us a few special supervisions. These were in their own way as fascinating and challenging as Atiyah's, and they gave both of us a lifelong interest in probability.

At the end of my second year I returned to Dollis Hill, and told the head of RE1 (H. J. Josephs, whose main interest was the life and work of Oliver Heaviside) of my interest in queueing theory. He saw this as an opportunity to keep me out of mischief, and told me to go to the library and read the run of the Bell System Technical Journal, where I would find much queueing theory in its historic context of teletraffic theory. Thus I encountered the work of authors such as Pollaczek, Khinchin, Jensen, Palm, and others in the teletraffic world, as well as mathematicians such as Kendall, Lindley, Takács and the like.

Reporting back to Josephs, I was told that much of the mathematical work was irrelevant to the real world. For instance, the assumption that customers are served in order of arrival was not applicable to calls arriving at a telephone exchange, because the equipment could not store waiting calls in order. So I tried to solve simple queues with service in random order, and eventually found a formal solution for $M/G/1$ for this discipline. Although the solution was pretty uncomputable, it did have a useful approximation in heavy traffic, when the traffic intensity is just less than its critical value. This made me wonder whether there was a general phenomenon of robust approximation in heavy traffic, and so it turned out.

Back to Cambridge and Part III of the Mathematical Tripos. Part III is a complete contrast to the staid Part II. Every lecturer gets the opportunity to lecture on a topic of his or her own choosing, and to set the questions for the summer examination, and the result is a wonderful array of courses across the whole sweep of current mathematics. Dennis Lindley, Morris Walker and Violet Cane lectured on statistics, and Peter Whittle, fresh from New Zealand, on stochastic processes. But these made up only a fraction of the number of courses one was expected to take, so I found myself listening to Hodge on differential geometry, Fred Hoyle on relativity, George Batchelor on fluid dynamics and, most remarkable of all, Paul Dirac on quantum mechanics. To hear Dirac explaining how he had predicted the spin of the electron was an experience not to be missed. Even that, however, did not convert me to mathematical physics, and I remained firm in my preference for probability.

One other course, which I added only as an afterthought, had consequences much later. Sir Ronald Fisher, the father of modern statistics, had for many years been the Arthur Balfour Professor of Genetics, with his own department unconnected with the Mathematics Faculty. He had however given a Part III course on mathematical genetics, and when he retired he gave his voluminous lecture notes to his colleague A. R. G. Owen. George Owen was a nervous man whose main interest was poltergeists, and he gave the impression that he was always frightened of something (possibly Fisher). He lectured very badly, by turning over Fisher's notes and picking out topics at random for our delight.

One such topic was the Mandel–Scheuer inequality, which asserts in the spirit of Fisher's Fundamental Theorem of Natural Selection that, in a large randomly mating population evolving under selection at a single locus, the mean fitness increases from one generation to the next. It is not biologically important because it does not generalise to multilocus selection, but it is a nice bit of mathematics, and Owen dropped a hint that it would appear in the examination. The proof was however too complicated to memorise, and I managed to find a much simpler one, which I could remember. After the examination it was suggested that I might publish my proof, and this drew me to the attention of population geneticists in Australia and California, with whom I was later to collaborate.

My assumption however was still that I would embark on a PhD in stochastic processes, and my Part III performance was good enough to win me a grant, so I asked Lindley to be my supervisor. He declined because he had just been appointed a professor in the University College of Wales in Aberystwyth, which he thought would be a less good environment than Cambridge. I therefore joined the Statistical Laboratory in

1960, as a PhD student of Peter Whittle, and set to work to build on the research I had started at Dollis Hill.

Peter was a supervisor with a light touch, who led by example. He rarely gave direct advice, but to see the way he attacked the most difficult problems, with a combination of intuition and rigorous analysis, was to learn how to do applied mathematics at the highest level. I liked too the way he visibly enjoyed mathematics, not a chore but a delight.

I was fortunate too in my fellow students. The Laboratory was housed, for reasons it would take too long to relate, in a basement of the new chemical laboratories on Lensfield Road. Two rooms were allocated to research students, and we decided that one would be for smokers and one for non-smokers. The latter was occupied by Roger Miles (just finishing his thesis on Poisson flats), Bob Loynes (who had proved beautiful theorems about general queueing systems) and me. The other room was so full of pipe smoke that its occupants were not visible, but Hilton Miller and John Bather sometimes came up for air. It was a small enough group for us to learn a lot from each other, and I benefited particularly from Bob's very perceptive approach to queues.

Like most probabilists of my generation, I had been greatly influenced by William Feller's *Introduction to Probability Theory and its Applications*, of which Volume 1 had appeared in 1950. I was particularly taken by his use of recurrent events to analyse discrete time Markov chains, an approach which D. G. Kendall had used in his influential 1951 and 1953 papers on queues. But my own work on queues had been much more concerned with continuous time, and I wondered whether Feller's technique could be made to work with continuous time chains, such as those I was reading about in Kai Lai Chung's 1960 book *Markov Chains with Stationary Transition Probabilities*. In fact, a formula analogous to Feller's fundamental equation could be found in Maurice Bartlett's *Stochastic Processes*, and I had used it for the random service problem.

I was pursuing such thoughts alongside more concrete problems when Peter Whittle took me by surprise by announcing that he was to move to Manchester to succeed Bartlett in the statistics chair there. His departure, following that of Lindley, would leave a very small statistics group in Cambridge, and clearly I should have a change of scene. I would have been happy to follow Peter to Manchester, but he suggested that I should take the opportunity to ask David Kendall if I could work with him in Oxford.

With some daring, for DGK was an acronym to conjure with, I wrote to him, and received a charming and welcoming reply. I could pursue my Cambridge PhD for at least a year in 'the other place', or transfer to an Oxford DPhil. But fate had another surprise in store. One summer evening after dinner I was playing croquet on the Pembroke lawn with college friends, when I became aware that Sir William Hodge was standing in gown and mortarboard watching our game. The standard of play was not such as to justify such an exalted spectator, but after a few minutes he called me over to tell me that I had just been elected to a research fellowship of the College. I missed the next shot and lost the game.

I had not realised that I was being considered for such a position, and was more than surprised by the news. I explained that I was committed to a move to Oxford, but the Master was not put off. Not only would the College allow me to hold my fellowship at a distance, but Pembroke had a sister college in Oxford which would treat me as one of its own research fellows.

Thus it was that in the autumn of 1961 I arrived in Oxford, with a desk in the Mathematical Institute, high table rights at The Queen's College, and David Kendall as my PhD supervisor.

The Institute was something of which there was no equivalent in Cambridge, offering space to mathematicians of all varieties, including the professors and other teaching staff, but also to research students and visitors from other universities. It was ruled in theory by E. C. Titchmarsh, the Savilian Professor of Geometry who carried the wonderful title of Curator of the Institute, but in practice by a formidable administrator Rosemary Schwerdt. Rosemary told me I had to visit the Curator to be given a key; the interview took place in his room which resembled a rather fusty summerhouse, and it would be hard to say which of us was the more shy and tongue tied.

My desk was in an annexe which seemed to be used as a quarantine station for new arrivals from other places like Cambridge. In particular, there was my old teacher Atiyah, who had moved to a readership in preparation for his inevitable succession to the Savilian chair. I also found to my delight two probabilists visiting from the USA, Steven Orey and Don Iglehart, with whom I enjoyed fruitful collaborations.

David Kendall however, like other mathematicians whose teaching posts were primarily college based, had no place in the Institute, and worked in his rooms in Magdalen College, with beautiful views over the deer park. He was very welcoming, but also frightening, and I looked on my weekly visits to Magdalen with both delight and trepidation. His standards of both substance and presentation were high, and he decided in particular that my mathematical writing was sloppy and imprecise.

He also made it clear that his enthusiasm for queueing theory had waned. He thought that the decade since his RSS paper had seen an explosion of papers giving useless formal solutions of routine problems. I hope he found my efforts a little above the general level, but he certainly did not encourage me to continue to concentrate on queues. He was much more impressed by my thoughts on extending Feller's recurrent events to continuous time, and he made a number of excellent suggestions which helped me enormously in formulating what became the theory of regenerative phenomena.

I knew of course of his famous collaboration with Harry Reuter on the general theory of continuous time Markov processes with a countable infinity of states, and of the deep work that their student David Williams had done in developing their methods. David had left Oxford to work with Harry in Durham, but he returned for his DPhil viva by David Edwards and Steven Orey. I was allowed to attend this as a fly on the wall (but in proper academic dress), and I realised how much I had to learn before I could apply my ideas to general Markov theory. DGK told me that Kolmogorov had posed the problem of characterising the diagonal transition probabilities of Markov chain; I knew that they were standard p-functions (which I could characterise), but the question of which p-functions could arise from Markov chains was much more delicate. I was sure that my methods were the right ones to answer the question, but it took me nearly another decade to achieve the result.

Once a week, at David Kendall's suggestion, he and I would join Steven and Don for an austere supper in Halifax House, the Oxford attempt at a social centre for postgraduates. One of the issues we discussed was how slow British mathematicians had been to embrace the rigorous probability theory that had grown out of the work of Kolmogorov and spread to the USA, France and elsewhere. There were active groups of applied probabilists in London under Bartlett and Cox, in Manchester with Whittle, Birmingham with Daniels, and elsewhere, but nothing comparable on the pure side. David was determined to spread the Kolmogorov gospel, and he and Harry invented the Stochastic Analysis Group (StAG) to give some coherence and visibility to the scattered disciples.

StAG was to be a very informal group of like-minded enthusiasts, sharing ideas and meeting as opportunities arose, for instance as splinter groups at the annual British Mathematical Colloquium. The London Mathematical Society was persuaded to hold one of its Durham instructional conferences on probability, at which the lecturers included Kendall's student David Edwards, by now specialising in functional analysis but still interested in probability, and the London analyst James Taylor. James and I were encouraged to write a book suitable for final year undergraduates, basing probability firmly on measure theory.

The existence of StAG gave great encouragement to young mathematicians taking their first steps in the field, and gave a more visible presence to the rigorous approach. For myself, I was excited by these new developments, but I still felt an affinity with the more pragmatic approach in which I had been brought up. Kai Lai Chung once said that mathematicians are more interested in building fire stations than in fighting fires, but I enjoyed using the hose as well as designing the headquarters.

Among the interesting mathematical figures I came to know in Oxford, one of the most memorable was George Temple, the Sedleian Professor of Natural Philosophy. He was a Fellow of Queen's, and the egalitarian world of the high table made it easy for a young research student to talk with a great authority on applied mathematics. George had a very broad range of mathematical interests, and by that time he was lecturing mainly on generalised functions, which to the pure mathematicians were distributions in the sense of Laurent Schwartz. But he was worth listening to on almost any subject, and I learned much from his genial wisdom. Many years later, I was privileged to attend the service in Quarr Abbey on the Isle of Wight, at which George took his final vows as a Benedictine monk.

All this time, I was of course in touch with my Cambridge college, and heard news of stirring developments. The applied mathematicians were at last tired of being scattered around the University, and were forming themselves into a Department of Applied Mathematics and Theoretical Physics (DAMTP) under an energetic leader, George Batchelor, who was determined to find a building in which his colleagues could have proper offices and facilities. The pure mathematicians could see the likely outcome as being that they would lose out, and were inching reluctantly towards a departmental structure in self-defence. Meanwhile, the University was under pressure from the RSS, which pointed out that Cambridge was the only major UK university that had never had a professor of statistics, and which helped to raise some funds to establish a chair.

What I did not know was that the two RSS representatives on the board of electors to the new chair, Maurice Kendall (no relation) and Egon Pearson, had seen the importance of the first professor being someone of real mathematical weight, and had set their sights on the appointment of DGK. Thus it came as a double surprise when David told me that he had agreed to move to Cambridge to head the Statistical Laboratory, and that he wanted me to go with him as an Assistant Lecturer. (Things were done like that in those days, with no nonsense of applications or interviews; the post to which I was appointed was the lectureship which had remained vacant when Whittle left, downgraded because of my junior status.)

So October 1962 found me back in Lensfield Road, with a room of my own and with all the jobs which neither David, nor the much more senior lecturers Morris Walker and Violet Cane, wanted to do. One of these was to give the Part I Statistics lectures. Most courses were given in parallel by two lecturers, but this was a luxury the statisticians could not afford, and I had all the first year mathematicians, well over 200, in Lecture Room A of the Arts Schools. Mercifully, I had kept my copy of Lindley's duplicated notes

Another job was rather different. The University Press was moving to new buildings out by the railway station, and Batchelor had persuaded the University to allocate part of the vacated site, on Silver Street, to DAMTP. Meanwhile the pure mathematicians had persuaded Sir William Hodge to head a new Department of Pure Mathematics and Mathematical Statistics (DPMMS), and Hodge applied his considerable political skill to finding a similar home for his department. He found an unexpected ally in his fellow Scot, the chemist Sir Alexander Todd. If the statisticians could be extruded from Lensfield Road, the space would be filled with chemists.

Thus it was that the paper warehouse of the Press was allocated to DPMMS, and Hodge set up a committee to plan the new home, with himself and two colleagues Frank Smithies and Ian Cassels, offering Kendall a place to represent the statisticians, which David filled with his new Assistant Lecturer. Here it was that I first learned the lessons of academic politics, especially that 'les absents ont toujours tort'. Of course it helped that I was a fellow of the college of which Hodge was Master, and he was less ruthless with me than if I had been from another college.

The conversion of the warehouse took a long time, and I did not stay long enough to benefit from my labours. The quarters in chemistry soon filled up with the visitors attracted by the new Professor Kendall, and his enthusiasm and vigour breathed life into what had been a demoralised group. He had had few research students in Oxford, but now very able mathematicians came to study for their PhD in pure and applied probability, Daryl Daley from Melbourne, John Haigh, Peter Lee, David Mannion, Jane Speakman. I shared with David in their supervision.

One casualty was my own PhD. Not only were the tasks of the junior member of the staff somewhat heavy, but my research studentship was superseded by the university post, and I should have had to meet the supervision and examination fees myself. I was reconciled to remaining Mr Kingman, partly because the doctoral title had always in my family been attached to my father. And so it remained until much later my daughter qualified in medicine.

Among the overseas visitors at a more senior level the one I remember best was Alfred Rényi from Budapest. He came with a great reputation in both probability and number theory, and proved to be a great source of ideas, problems and anecdotes, not least from his collaboration with Paul Erdős.

One day in early 1963 a knock on my door introduced someone I knew well by reputation but had never met, Joe Gani over from Canberra. His opening words were "How would you like to go to Western Australia?". This turned out not to be a threat of transportation; he had been commissioned to find a young statistician who would spend the (English) summer in the Perth winter teaching in the mathematics department of the University of Western Australia. I could give a graduate course on whatever I liked, so long as I also gave a final year undergraduate course on either sampling or experimental design. Joe dismissed my protestations that I knew nothing about either subject, so I found myself enjoying the Australian winter and instructing bemused students in the finer points of balanced incomplete block designs. At a party a real statistician delivered the killer question in broad Australian: "Have you ever designed an experiment?".

The most valuable aspect of my Australian trip was the opportunity to visit other universities, in Adelaide, Melbourne, Canberra and Sydney, on my way home. It was particularly good to meet Pat Moran at ANU, whose work on genetics as well as other areas of probability and statistics I much admired. He knew of my proof of the Mandel–Scheuer inequality, and suggested some other genetical problems. It was through him that I later came to collaborate with Warren Ewens and Geoff Watterson at Monash on genetic diversity in large populations, work that led to the useful concept of the coalescent.

Continuing to the east, I landed in San Francisco and spent a few days in Stanford. There I called on Sam Karlin, who like Moran talked about genetics as well as his work on Markov chains. I was to return to Stanford for a longer stay in 1968, to enjoy Sam's hospitality and that of Kai Lai Chung.

The early 1960s were of course the great period of university expansion in Britain, when newly established universities were scrabbling for academic staff. One such was the University of Sussex, sited between Brighton and Lewes in a beautiful fold of the South Downs. Its first Vice-Chancellor, John (later Lord) Fulton, had taken full advantage of the opportunity to recruit staff of the calibre of David Daiches and Asa Briggs, and the science side was led by the Oxford physicist Roger Blin-Stoyle. They saw the need for a strong mathematical presence, and enlisted the help of George Temple in finding possible leaders. He recommended Bernard Scott and Gilford Ward as professors of pure and applied mathematics respectively, but statistics proved more difficult.

Perhaps George remembered his conversations with the young Cambridge probabilist, for in 1964 Bernard Scott approached me to ask if I would want to join his group. What Bernard did not know was that my future wife Valerie Cromwell, a Fellow of Newnham College, was being offered a lectureship in history by Asa Briggs, so that his proposition had unusual attractions. Both Kendall and Hodge advised me against the risks of joining a new university, but their advice seemed to me not entirely disinterested, and was ignored. I became Reader in Mathematics and Statistics on April Fool's Day 1965, and Professor a year later.

Sussex was naturally a great contrast to both Cambridge and Oxford. In an ancient university you do what you did last year unless there are overwhelming reasons for change. (The Oxford don asked to support a measure of reform: "Reform, aren't things bad enough already?") In Sussex there were no precedents, and one always had a blank sheet of paper. It was easier to strike out in a different direction than to copy the practices of more traditional universities. As a result many good ideas were put into effect, but so were many bad ones, and the University was not always good at telling the difference.

One of the new ideas was that all students in the School of Social Sciences must take and pass a course on elementary statistics, which it was my job to deliver. In the year before I arrived the course had been given by Walter Ledermann, an algebraist who is I think the best teacher of mathematics I have encountered. Walter assured me that it was a very difficult assignment, and told me of the female student who had come to him in tears, sure that she would fail and have to leave the University. He had tried to discover her problem, and eventually she sobbed "Oh, Dr Ledermann, it is so difficult, I was away from school the day they did decimals". I did my best to temper the mathematical wind to the shorn lambs, but I doubt if I matched the Ledermann skill, or his gentle courtesy.

My main task at Sussex was to try to build up a group in statistics to complement the already lively groups of pure and applied mathematicians. We were very fortunate to attract John Scott from Oxford to lead the more applied statistics, and he proved a delightful colleague. He was to remain at Sussex, rising to be Pro-Vice-Chancellor, until he emigrated to Australia in 1977 as Vice-Chancellor of La Trobe University. We were joined by my Pembroke contemporary John Bather, and on the probability side by John Haigh and Charles Goldie.

John Fulton retired as Vice-Chancellor in 1967, and as a member of Senate I was involved in the choice of his successor. This turned out to be a foregone conclusion, since it had always been intended that Fulton would be succeeded by Asa Briggs. I was allowed to record my dissent, because I believed that the University would have benefited at that stage from a fresh look by an outsider not so intimately involved in the founding decisions. The subsequent history of Sussex does I think bear out that view, and the University has perhaps not altogether lived up to its early promise.

As it transpired, however, I was not to be directly involved in that history. The University of Oxford had at last registered the loss that it had sustained when David Kendall moved to Cambridge. Encouraged by Maurice Bartlett, who had become Professor of Biomathematics in 1967, it had agreed to create a new chair dedicated to the mathematics of probability. I did not know anything of this in 1967, or indeed until I was told that I had been elected as the first professor. Still less did I know that I only had some twelve years left as a full time mathematician, and that this would be followed by twenty years in other posts, when mathematics would become a hobby and an aid to sanity.

Meanwhile probability theory was establishing its place in the mainstream of British mathematics. I had left Cambridge too soon to be involved in the flowering of the subject there under David Kendall's enthusiastic leadership. In 1965 Rollo Davidson started his PhD, and the remarkable five years of research in stochastic analysis and geometry cut short by his tragic death. David Williams was in Cambridge from 1966 to 1972, when he left for a chair in Swansea before returning in 1985 to succeed David Kendall. And Peter Whittle's return in 1967, to the newly established Churchill Professorship of Mathematics for Operational Research, balanced the pure and applied aspects, particularly when he was joined by Frank Kelly.

A similar alliance between pure and applied probability took place in London in 1965, when Harry Reuter moved from Durham to join David Cox's already strong group at Imperial College. That same year Joe Gani came to Sheffield from Canberra, and founded both the joint school with Manchester, and the Applied Probability Trust, whose journals have established themselves as major publications on the world stage. In fact the distinction between the pure and applied approaches, between the spirit of Kolmogorov and the spirit of Bartlett, has largely disappeared. Queueing theorists talk happily of martingales and Lévy processes, and the Itô calculus forms the basis for applications from biology to mathematical finance.

Moreover, sophisticated probabilistic ideas have proved their usefulness in the development of statistical methodology, and excellent probabilists like Peter Green and Bernard Silverman have made important advances in modern statistics. In this they stand in the DGK tradition, for David's research in Cambridge was largely in problems of inference, in particular in his development of Shape Theory.

This phenomenon of confluence is a characteristic of modern mathematics, where boundaries between different areas have come to be regarded as increasingly irrelevant, and even as obstacles to the advance of the subject and its many applications. That is why I was shocked to find, when I returned to Cambridge in 2001, that the divide between DAMTP and DPMMS was as real as when I had left 36 years earlier. Most British universities had either, like Oxford, never had such an institutional apartheid, or had abandoned it in favour of a recognition of the essential unity of mathematics.

My own attitude has always been to enjoy mathematical problems both for their own sake, and because they can often be applied to a wide range of other disciplines. William Morris famously told us to "have nothing in your houses that you do not know to be useful, or believe to be beautiful". In mathematics, usefulness may take decades or even centuries to manifest itself. On the other hand, beauty or elegance can often be appreciated at once, and shared with like-minded colleagues. We should therefore invert the Morris dictum, and argue that we should have nothing in our mathematics that we do not know to be beautiful, or believe to be useful.

A note on sources

Detailed references would be out of place in an informal history such as this, and the reader will easily link what I have said about my own work with the bibliography on page 1. I have written at greater length about David Kendall in the *Biographical Memoirs of the Royal Society* (Volume 55 (2009), 121–138). Something of the story of my interest in genetics can be found in [M102]. Finally, I strongly recommend Peter Whittle's lively history of the Cambridge Statistical Laboratory, which can be found on the Laboratory's website together with a fascinating series of group photographs of the members through the years.

2

More uses of exchangeability: representations of complex random structures

David J. Aldous[a]

Abstract

We review old and new uses of exchangeability, emphasizing the general theme of exchangeable representations of complex random structures. Illustrations of this theme include processes of stochastic coalescence and fragmentation; continuum random trees; second-order limits of distances in random graphs; isometry classes of metric spaces with probability measures; limits of dense random graphs; and more sophisticated uses in finitary combinatorics.

AMS subject classification (MSC2010) 60G09, 60C05, 05C80

1 Introduction

Kingman's write-up [44] of his 1977 Wald Lectures drew attention to the subject of exchangeability, and further indication of the topics of interest around that time can be seen in the write-up [3] of my 1983 Saint-Flour lectures. As with any mathematical subject, one might expect some topics subsequently to wither, some to blossom and new topics to emerge unanticipated. This Festschrift paper aims, in informal lecture style,

(a) to recall the state of affairs 25 years ago (sections 2.1–2.3, 3.1);
(b) to briefly describe three directions of subsequent development that

a Department of Statistics, Evans Hall, University of California at Berkeley, Berkeley, CA 94720-3860; aldous@stat.berkeley.edu; http://www.stat.berkeley.edu/users/aldous. Research supported by N.S.F. Grant DMS-0704159.

have recently featured prominently in monographs [16, 43, 52] (sections 2.4, 3.1–3.2);

(c) to describe 3 recent rediscoveries, motivated by new theoretical topics outside mainstream mathematical probability, of the theory of representations of partially exchangeable arrays (sections 2.5, 5.1–5.2);

(d) to emphasize a general program that has interested me for 20 years. It doesn't have a standard name, but let me here call it *exchangeable representations of complex random structures* (section 4).

The survey focusses on mathematical probability; although the word *Bayesian* appears several times, I do not attempt to cover the vast territory of explicit or implicit uses of exchangeability in Bayesian statistics, except to mention here its use in *hierarchical models* [20, 36].

This article is very much a bird's-eye view. Of the monographs mentioned above, let me point out Pitman's *Combinatorial Stochastic Processes* [52], which packs an extraordinary number of detailed results into 200 pages of text and exercises. Exchangeability is a recurring theme in [52], which covers about half of the topics we shall mention (and much more, not related to exchangeability), and so [52] is a natural starting place for the reader wishing to get to grips with details.

2 Exchangeability

2.1 de Finetti's theorem

I use *exchangeability* to mean, roughly, 'applications of extensions of de Finetti's theorem'. Let me assume only that the reader is familiar with the definition of an exchangeable sequence of random variables

$$(Z_i, i \geq 1) \stackrel{d}{=} (Z_{\pi(i)}, i \geq 1) \text{ for each finite permutation } \pi$$

and with the common verbal statement of de Finetti's theorem:

An infinite exchangeable sequence is distributed as a mixture of i.i.d. sequences.

Scanning the graduate-level textbooks on measure-theoretic probability on my bookshelves, the theorem makes no appearance in about half, a brief appearance in others [18, 31, 46] and only three [22, 35, 42] devote much more than a page to the topic.

A remarkable feature of de Finetti's theorem is that there are many ways to state essentially the same result, depending on the desired emphasis. This feature is best seen when you state results more in words rather than symbols, so that's what I shall do. Take a nice space S, and either don't worry what 'nice' means, or assume $S = \mathbb{R}$. Write $\mathcal{P}(S)$ for the space of probability measures on S. Write μ for a typical element of $\mathcal{P}(S)$ and write α for a typical *random* element of $\mathcal{P}(S)$, that is a typical *random measure*. When we define an infinite exchangeable sequence of S-valued random variables we are really defining an *exchangeable measure* (Θ, say) on $\mathcal{P}(S^\infty)$, where Θ is the distribution of the sequence.

Functional analysis viewpoint. The subset $\{\mu^\infty = \mu \times \mu \times \mu \times \ldots :$ $\mu \in \mathcal{P}(S)\} \subset \mathcal{P}(S^\infty)$ is the set of extreme points of the convex set of exchangeable elements of $\mathcal{P}(S^\infty)$, and the identification

$$\Theta(\cdot) = \int_{\mathcal{P}(S)} \mu^\infty(\cdot) \, \Lambda(d\mu)$$

gives a bijection between probability measures Λ <u>on</u> $\mathcal{P}(S)$ (that is, $\Lambda \in \mathcal{P}(\mathcal{P}(S))$) and exchangeable measures Θ in $\mathcal{P}(S^\infty)$.

Probability viewpoint. Here are successively more explicit versions of the same idea. Let $(Z_i, \; 1 \le i < \infty)$ be exchangeable S-valued.

(a) Conditional on the tail (or invariant or exchangeable) σ-field of the sequence (Z_i), the random variables Z_i are i.i.d.
(b) There exists a random measure α such that, conditional on $\alpha = \mu$, the random variables Z_i are i.i.d.(μ).
(c) Giving $\mathcal{P}(S)$ the topology of weak convergence, the empirical measure $F_n = F_n(\omega, \cdot) = n^{-1} \sum_{i=1}^{n} 1_{(Z_i(\omega) \in \cdot)}$ converges a.s. to a limit random measure $\alpha(\omega, \cdot)$ satisfying (b).

Theoretical statistics viewpoint. In contexts where a frequentist would assume data are i.i.d.(μ) from an unknown distribution μ, a Bayesian would put a prior distribution Λ on possible μ; so de Finetti's theorem is saying that the Bayesian assumption is logically equivalent to the assumption that the data $(Z_i, i \ge 1)$ are exchangeable. Note a mathematical consequence. There is a posterior distribution $\Lambda_n(\omega, \cdot) \in \mathcal{P}(\mathcal{P}(S))$ for Λ given $(Z_1, \ldots, Z_n)$, and an extension of (c) above is

(d) $\Lambda_n(\omega, \cdot) \to \delta_{\alpha(\omega, \cdot)}$ a.s. in $\mathcal{P}(\mathcal{P}(S))$.

Such results are historically often used as a starting point for philosophical and mathematical discussion of consistency/inconsistency of frequentist and Bayesian methods, inspired of course by Bruno de Finetti himself.

But there's more! Later we encounter at least two further, somewhat different, viewpoints: explicit constructions (section 2.7), and our central theme of using exchangeability to describe complex structures (sections 3 and 4). This theme is related to the general features that

- exchangeable-like properties are preserved under weak convergence;
- parallel to representation theorems for infinite exchangeable-like structures, are convergence theorems giving necessary and sufficient condition for finite exchangeable-like structures to converge in distribution to an infinite such structure.

In the setting of de Finetti's theorem, the condition for finite exchangeable sequences $\mathbf{X}^{(n)} = (X_1^{(n)}, \ldots, X_n^{(n)})$ to converge in distribution to an infinite exchangeable sequence $\mathbf{X}$ is

$$\alpha_n \overset{d}{\to} \alpha \text{ on } \mathcal{P}(S)$$

where α is the 'directing' random measure for $\mathbf{X}$ in (b) above, and α_n is the empirical distribution of $(X_1^{(n)}, \ldots, X_n^{(n)})$. Note that when we talk of convergence in distribution to infinite sequences or arrays, we mean w.r.t. product topology, i.e. convergence of finite restrictions.

2.2 Exchangeability, 25 years ago

Here I list topics from the two old surveys [44, 3], for the purpose of saying a few words about those topics I will *not* mention further, while pointing to sections where other topics will be discussed further.

Classical topics not using de Finetti's theorem.

(a) Combinatorial aspects for classical stochastic processes, e.g. ballot theorems: [56].

(b) Weak convergence for 'sampling without replacement' processes (e.g. [17] Thm 24.1).

Variants of de Finetti's theorem. Several variants were already classical, for instance:

(c) Schoenberg's[1] theorem ([3](3.6)) for the special case of spherically symmetric sequences;

(d) the analogous representation ([3](3.9)) in the setting of two sequences $(X_i, 1 \leq i < \infty;\ Y_j, 1 \leq j < \infty)$ whose joint distribution is invariant under finite permutations of either;

(e) the *selection property* ([44] p.188), that the exchangeability hypothesis in de Finetti's theorem can be weakened to the assumption

$$(X_1, X_2, \ldots X_n) \overset{d}{=} (X_{k_1}, X_{k_2}, \ldots, X_{k_n})$$

$$\text{for all } 1 \leq k_1 < k_2 < \cdots < k_n.$$

Other variants had been developed in the 1970s, for instance:

(f) the analog for continuous-time processes with exchangeable *increments* [40];

(g) Kingman's paintbox theorem for exchangeable random partitions; see section 3.1.

Finite versions. The general forms of de Finetti's theorem and some classical variants can be proved by comparing sampling with and without replacement. This method [24] also yields finite-n variants.

Mathematical population genetics, the coalescent and the Poisson–Dirichlet distribution. Exchangeability is involved in this large circle of ideas, developed in part by Kingman in the 1970s, which continues to prove fruitful in many ways. For the population genetics aspects of Kingman's work see the article by Ewens and Watterson [33] in this volume; also the *Kingman coalescent* fits into the more general *stochastic coalescent* material in section 3.2.

The subsequence principle. The idea emerged in the 1970s that, from any tight sequence of random variables, one can extract a subsequence which is close to exchangeable, close enough that one can prove analogs of classical limit theorems (CLT and LIL, for instance) for the subsequence. General versions of this principle were established in [1, 15], which pretty much killed the topic.

---

[1] Persi Diaconis observes that the result is hard to deduce from Schoenberg [54] and should really be attributed to Freedman [34].

Sufficient statistics and mixtures of Markov chains. One can often make a Bayesian interpretation of 'sufficient statistic' in terms of some context-dependent invariance property [25]. Somewhat similarly, one can characterize mixtures of Markov chains via the property that transition counts are sufficient statistics [23].

2.3 Partially exchangeable arrays

The topic, emerging around 1980, of *partially exchangeable arrays*, plays a role in what follows and so requires more attention. Take a measurable function $f : [0,1]^2 \to S$ which is symmetric, in the sense $f(x,y) \equiv f(y,x)$. Take $(U_i, i \geq 1)$ i.i.d. Uniform$(0,1)$ and consider the array

$$X_{\{i,j\}} := f(U_i, U_j) \tag{2.1}$$

indexed by the set $\mathbb{N}_{(2)}$ of unordered pairs $\{i,j\}$. The exchangeability property of (U_i) implies what we shall call the *partially exchangeable* property for the array:

$$(X_{\{i,j\}}, \{i,j\} \in \mathbb{N}_{(2)}) \overset{d}{=} (X_{\{\pi(i),\pi(j)\}}, \{i,j\} \in \mathbb{N}_{(2)})$$
$$\text{for each finite permutation } \pi. \tag{2.2}$$

Note this is a weaker property than the 'fully exchangeable' property for the countable collection $(X_{\{i,j\}}, \{i,j\} \in \mathbb{N}_{(2)})$, because the permutations of $\mathbb{N}_{(2)}$ which are of the particular form $\{i,j\} \to \{\pi(i), \pi(j)\}$ for a finite permutation π of $\mathbb{N}$ are only a subset of all permutations of $\mathbb{N}_{(2)}$.

Aside from construction (2.1), how else can one produce an array with this partially exchangeable property? Well, an array with i.i.d. entries has the property, and so does the trivial case where all entries are the same r.v. After a moment's thought we realize we can combine these ideas as follows.

Take a function $f : [0,1]^4 \to S$ such that $f(u, u_1, u_2, u_{12})$ is symmetric in (u_1, u_2), and then define

$$X_{\{i,j\}} := f(U, U_i, U_j, U_{\{i,j\}}) \tag{2.3}$$

where all the r.v.s in the families $U, (U_i, i \in \mathbb{N}), (U_{\{i,j\}}, \{i,j\} \in \mathbb{N}_{(2)})$ are i.i.d. Uniform$(0,1)$. Then the resulting array $\mathbf{X} = (X_{\{i,j\}})$ is partially exchangeable.

Finding oneself unable to devise any other constructions, it becomes natural to conjecture that every partially exchangeable array has a representation (in distribution) of form (2.3). This was proved by Hoover

[39] and (in the parallel non-symmetric setting) by Aldous [2], the latter proof having been substantially simplified due to a personal communication from Kingman.

Constructions of partially exchangeable arrays appear in Bayesian statistical modeling; see e.g. the family of copulae introduced in [19] in the context of a semi-parametric model for Value at Risk.

2.4 Fast forward

Such *partially exchangeable representation theorems* were the state of the art in the 1984 survey [3]. They were subsequently extended systematically by Kallenberg, both for arrays and analogs such as exchangeable-increments continuous-parameter processes, and for the *rotatable matrices* to be mentioned in section 2.6, during the late 1980s and early 1990s. The whole topic of representation theorems is the subject of Chapters 7–9 of Kallenberg's 2005 monograph [43]. Not only does this monograph provide a canonical reference to the theorems, but also its introduction provides an excellent summary of the topic.

In the particular setting above we have

Theorem 2.1 (Partially Exchangeable Representation Theorem) *An array* $\mathbf{X}$ *which is partially exchangeable, in the sense* (2.2), *has a representation in the form* (2.3).

This is one of the family of results described carefully in Chapter 7 of [43]. There are analogous results for higher-parameter arrays (X_{ijk}), and for arrays in which the 'joint exchangeability' assumption (2.2) is replaced by a 'separate exchangeability' assumption for non-symmetric arrays $(X_{i,j}, 1 \leq i, j < \infty)$:

$$(X_{i,j}, 1 \leq i, j < \infty) \stackrel{d}{=} (X_{\pi_1(i),\pi_2(j)}, 1 \leq i, j < \infty)$$

for finite permutations π_1, π_2.

One aspect of this theory is surprisingly subtle, and that is the issue of *uniqueness* of representing functions f. In representation (2.3), if we take Lebesgue-measure-preserving maps ϕ_0, ϕ_1, ϕ_2 from $[0,1]$ to $[0,1]$, then the arrays $\mathbf{X}$ and $\mathbf{X}^*$ obtained from f and from $f^*(u, u_1, u_2, u_{12}) := f(\phi_0(u), \phi_1(u_1), \phi_1(u_2), \phi_2(u_{12}))$ must have the same distribution. But this is not the only way to make arrays have the same distribution: there are other ways to construct measure-preserving transformations of $[0,1]^4$, and (because measure-preserving transformations are not invertible in general) one needs to insert randomization variables. (I thank the referee for correcting a blunder in my first draft, and for the comment "this may be a major reason why non-standard analysis is effective here".) For an explicit statement of the uniqueness result in two dimensions see [41] and for higher dimensions see [43].

Relative to the Big Picture of Mathematics, this theory of partial exchangeability was perhaps regarded during 1980–2005 as a rather small niche inside mathematical probability—and ignored outside mathematical probability. So it is ironic that around 2004-8 it was rediscovered in at least three different contexts outside mainstream mathematical probability. Let me say one such context right now and the others later (sections 5.1 and 5.2).

2.5 Isometry classes of metric spaces with probability measures

The definition of *isometry* between two metric spaces (S_1, d_1) and (S_2, d_2) contains an 'if there exists ...' expression. Asking for a *characterization* of metric spaces up to isometry is asking for a scheme that associates some notion of 'label' to each metric space in such a way that two metric spaces are isometric if and only if they have the same label. I am not an expert on this topic, but I believe there is no known such characterization.

But suppose instead we consider 'metric spaces with probability measure', (S_1, d_1, μ_1) and (S_2, d_2, μ_2), and require the isometry to map μ_1 to μ_2. It turns out there is now a conceptually simple characterization. Given (S, d, μ), take i.i.d.(μ) random elements $(\xi_i, 1 \leq i < \infty)$ of S, form the array (of form (2.1))

$$X_{\{i,j\}} = d(\xi_i, \xi_j); \ \{i, j\} \in \mathbb{N}_{(2)} \tag{2.4}$$

and let Ψ be the distribution of the infinite random array. It is obvious that, for two isometric 'metric spaces with probability measure', we get the same Ψ, and the converse is a simple albeit technical consequence of the uniqueness part of Theorem 2.1, implying:

'metric spaces with probability measure' are characterized
up to isometry by the distribution Ψ. (2.5)

This result was given by Vershik [57], as one rediscovery of part of the general theory of partial exchangeability.

2.6 Rotatable arrays and random matrices with symmetry properties

In Theorem 2.1 we described $\mathbf{X} = (X_{ij})$ as an *array* instead of a *matrix*, partly because of the extension to higher-dimensional parametrizations and partly because we never engage matrix multiplication. Now regarding $\mathbf{X}$ as a matrix, one can impose stronger 'matrix-theoretic' assumptions and ask for characterizations of the random matrices satisfying such assumptions. One basic case, *rotatable matrices*, is where the $n \times n$ restrictions are invariant in distribution under the orthogonal group, and the characterization goes back to [2]. Two other cases (I thank the referee for suggesting (ii) and the subsequent remark) are

(i) non-negative definite jointly exchangeable arrays: [30, 51];
(ii) rotatable *Hermitian* matrices [50], motivated indirectly by problems in quantum mechanics and thereby related to the huge literature on semicircular laws.

Returning to the basic case of rotatable matrices, for the higher-dimensional analogs the basic representations are naturally stated in terms of multiple Wiener–Itô integrals, which form the fundamental examples of rotatable random functionals. Such multiple Wiener–Itô integrals are also a basic tool in [49], a subject with important applications to analysis.

2.7 Revisiting de Finetti's theorem

Returning to a previous comment, the theory of partially exchangeable representation theorems reminds us that one can take a similar view of de Finetti's theorem itself, to add to the list in section 2.1.

Construction viewpoint. Given a measurable function $f : [0,1]^2 \to S$ and i.i.d. Uniform$(0,1)$ random variables $(U; U_i, i \geq 1)$, the process $(Z_i, i \geq 1)$ defined by $Z_i = f(U, U_i)$ is exchangeable, and every exchangeable process arises (in distribution) in this way from some f.

3 Using exchangeability to describe complex structures

Here is my attempt at articulating the first part of the central theme of this paper.

One way of examining a complex mathematical structure is to sample i.i.d. random points and look at some form of induced substructure relating the random points.

The idea being that the i.i.d. sampling induces some kind of 'exchangeability' on the distribution of the substructure, when the substructure is regarded as an object in its own right.

The 'isometry' result (2.5) nicely fits this theme—the substructure is simply the induced metric on the sampled points. The rest of the present paper seeks to illustrate that this, admittedly very vague, way of looking at structures can indeed be useful, conceptually and/or technically. Let us mention here two prototypical examples (which will reappear later) of what a 'substructure' might be. Given k vertices $v(1), \ldots, v(k)$ in a graph, one can immediately see two different ways to define an induced substructure.

(i) The induced subgraph on vertices $1, \ldots, k$: there is an edge (i, j) iff the original graph has an edge $(v(i), v(j))$.

(ii) The distance matrix: $d(i, j)$ is the number of edges in the shortest path from $v(i)$ to $v(j)$.

But before considering graph theoretic examples, let us explain with hindsight how Kingman's work on exchangeable random partitions fits this theme.

3.1 Exchangeable random partitions and Kingman's paintbox theorem

The material here is covered in detail in Pitman [52] Chapters 2–4.

Given a discrete sub-probability distribution, one can write the probabilities in decreasing order as $p_1 \geq p_2 \geq \ldots > 0$ and then write $p_{(\infty)} := 1 - \sum_j p_j \geq 0$ to define a probability distribution $\mathbf{p}$. Imagine objects $1, 2, 3, \ldots$ each independently being colored, assigned color j with probability p_j or assigned with probability $p_{(\infty)}$ a unique color (different from that assigned to any other object). Then consider the resulting 'same color' equivalence classes as a random partition of $\mathbb{N}$. So a realization of this process might be

$$\{1, 5, 6, 9, 13, \ldots\}, \quad \{2, 3, 8, 11, 15, \ldots\}, \quad \{4\}, \quad \{7, 23, \ldots\}, \quad \ldots \ldots; \quad (3.1)$$

sets in the partition are either infinite or singletons. This *paintbox*($\mathbf{p}$)

distribution on partitions is exchangeable in the natural sense. *Kingman's paintbox theorem*, an analog of de Finetti's theorem, states that every exchangeable random partition of $\mathbb{N}$ is distributed as a mixture over **p** of paintbox(**p**) distributions.

We mentioned in section 2.1 as a general feature that, associated with a representation theorem like this, there will be a convergence theorem. Here are two slightly different ways of looking at the convergence theorem in the present setting. Suppose that for each n we are given an arbitrary random (or non-random) partition $\mathcal{G}(n)$ of $\{1, 2, \ldots, n\}$. For each $k < n$ sample without replacement k times from $\{1, 2, \ldots, n\}$ to get $U_n(1)$, $\ldots$, $U_n(k)$, consider the induced partition on the sampled elements $U_n(1)$, $\ldots$, $U_n(k)$, and relabel these elements as 1, $\ldots$, k to get a random partition $\mathcal{S}(n, k)$ of $\{1, 2, \ldots, k\}$. This random partition $\mathcal{S}(n, k)$ is clearly exchangeable. If there is a limit

$$\mathcal{S}(n, k) \xrightarrow{d} \mathcal{S}_k \text{ as } n \to \infty \qquad (3.2)$$

(the set of all possible partitions of $\{1, 2, \ldots, k\}$ is finite, so there is nothing technically sophisticated here) then the limit $\mathcal{S}_k$ is exchangeable; and if a limit (3.2) exists for all k then the family $(\mathcal{S}_k, 1 \leq k < \infty)$ specifies the distribution of an exchangeable random partition of $\mathbb{N}$, to which Kingman's paintbox theorem can be applied.

The specific phrasing above was chosen to fit a general framework in section 4.1 later, but here is a more natural phrasing. For any random partition of $\{1, \ldots, n\}$ write $\mathbf{F}^{(n)} = (F_1^{(n)}, F_2^{(n)}, \ldots)$ for the *ranked empirical frequencies*, the numbers $n^{-1} \times$ (sizes of sets in partition) in decreasing order. For a paintbox(**p**) distribution the SLLN implies $\mathbf{F}^{(n)} \to$ **p** a.s., and so Kingman's paintbox theorem implies that for any infinite exchangeable random partition Π, the limit $\mathbf{F}^{(n)} \to \mathbf{F}$ exists a.s. and is the 'directing random measure' (conditional on $\mathbf{F} = \mathbf{p}$ the distribution of Π is paintbox(**p**)). Now suppose for each n we have an *exchangeable* random partition $\Pi^{(n)}$ of $\{1, 2, \ldots, n\}$ and write $\mathbf{F}^{(n)}$ for its ranked empirical frequencies. The *convergence theorem* states that the sequence $\Pi^{(n)}$ converges in distribution (meaning its restriction to $\{1, \ldots, k\}$ converges, for each k) to some limit Π, which is necessarily some infinite exchangeable random partition with some directing random measure $\mathbf{F}$, if and only if $\mathbf{F}^{(n)} = (F_j^{(n)}, 1 \leq j < \infty) \xrightarrow{d} \mathbf{F} = (F_j, 1 \leq j < \infty)$.

A final important idea is *size-biased order*. In the context of exchangeable random partitions this just means writing the components in a realization, as at (3.1), starting with the component containing element 1, then the component containing the least element not in the first component, and so on. In the infinite case, the frequencies $\mathbf{F}^* = (F_1^*, F_2^*, \ldots)$ of the components in size-biased order are just a random permutation of the frequencies $\mathbf{F}$ given by Kingman's paintbox theorem. In the paintbox$(\mathbf{p})$ case, replacing non-random $\mathbf{F} = \mathbf{p}$ by random $\mathbf{F}^*$ is perhaps merely complicating matters, but in the general case of random $\mathbf{F}$ it is often more natural to work with the size-biased order than with the ranked order. For instance, the size-biased order codes information such as

$$\mathbb{E}(F_1^*)^m = \mathbb{E} \sum_{i \geq 1} F_i^{m+1}.$$

I am highlighting these 'structural' results as part of my overall theme, but in many ways the concrete examples are more interesting. The one-parameter Poisson–Dirichlet(θ) family was already recognized 25 years ago as a mathematically canonical family of measures arising in several different contexts: the Ewens sampling formula in neutral population genetics, the 'Chinese restaurant process' construction, a construction via subordinators, the size-biased order of asymptotic frequencies is the GEM distribution; and special cases arise as limits of component sizes in random permutations and in random mappings. Subsequently the two-parameter Poisson–Dirichlet(α, θ) distribution introduced by Pitman–Yor [53] was shown to possess many analogous properties. The paper [37] in this volume gives the flavor of current work in this direction.

Now partitions are rather simple structures, and the paintbox theorem (which can be derived from de Finetti's theorem) isn't so convincing as an illustration of the theme 'using exchangeability to describe complex structures'. The theme becomes more visible when we consider the more complex setting of partitions evolving over time, and this setting arises naturally in the following context.

3.2 Stochastic coalescence and fragmentation

The topic of this section is treated in detail in Bertoin [16], the third monograph in which exchangeability has recently featured prominently. The topic concerns models in which, at each time, unit mass is split into clusters of masses $\{x_j\}$. One studies models of dynamics under which clusters split (*stochastic fragmentation*) or merge (*stochastic coalescence* or *coagulation*[2]) according to some random rules.

---

[2] The word *coagulation*, introduced in German in [55], sounds strange to the native English speaker to whom it suggests blood clotting; *coalescence* seems a more apposite English word.

Conceptually, states are unordered collections $\{x_j\}$ of positive numbers with $\sum_j x_j = 1$. What is a good mathematical representation of such states? The first representation one might devise is as vectors in decreasing order, say $\mathbf{x}^\downarrow = (x_1, x_2, \ldots)$. But this representation has two related unsatisfactory features; fragmenting one cluster forces one to relabel the others; and given the realizations at two times, one can't identify a particular cluster at the later time as a fragment of a particular cluster at the earlier time. These difficulties go away if we think instead in terms of sampling 'atoms' and tracking which cluster they are in. A uniform random atom will be in a mass-x cluster of a configuration $\mathbf{x}^\downarrow$ with probability x; sampling atoms $i = 1, 2, 3, \ldots$ and taking the partition of $\{1, 2, 3, \ldots\}$ into 'atoms of the same cluster' components gives an exchangeable random partition Π with paintbox($\mathbf{x}^\downarrow$) distribution.

Thus instead of representing a process as $(\mathbf{X}^\downarrow(t), 0 \leq t < \infty)$ we can represent it as a partition-valued process $(\Pi(t), 0 \leq t < \infty)$ which tracks the positions of (i.e. the clusters containing) particular atoms. For fixed t, both $\Pi(t)$ and $\mathbf{X}^\downarrow(t)$ give the same information about the cluster masses—and note that clusters in $\Pi(t)$ automatically appear in size-biased order. But as processes in t, $(\Pi(t), 0 \leq t < \infty)$ gives more information than $(\mathbf{X}^\downarrow(t), 0 \leq t < \infty)$, and in particular avoids the unsatisfactory features mentioned above.

Now in one sense this is merely a technical device, but I find it does give some helpful insights.

The basic general stochastic models. In the basic model of *stochastic fragmentation*, different clusters evolve independently, a mass-x cluster splitting at some stochastic rate λ_x into clusters whose relative masses $(x_j/x, \ j \geq 1)$ follow some probability distribution $\mu_x(\cdot)$. (So the model neglects detailed 3-dimensional geometry; the shape of a cluster is assumed not to affect its propensity to split, and different clusters do not interact). Especially tractable is the *self-similar* case where $\mu_x = \mu_1$ and $\lambda_x = x^\alpha$ for some *scaling exponent* α. Such processes are closely related to classical topics in theoretical and applied probability—the log-masses form a continuous time branching random walk, and the mass of the cluster containing a sample atom forms a continuous-time Markov process on state space $(0, 1]$.

The basic model for *stochastic coalescence* is to have n particles, initially in single particle clusters of masses $1/n$, and let clusters merge according to a kernel $\kappa(x, x')$ indicating the rate (probability per unit time) at which a typical pair of clusters of masses x and x' may merge.

For fixed n this is just a finite-state continuous-time Markov chain, but it is natural to study $n \to \infty$ limits, and there are two different regimes. On the time-scale where typical clusters contain $O(1)$ particles, i.e. have mass $O(1/n)$, there is an intuitively natural *hydrodyamical limit* (law of large numbers), that is differential equations for the relative proportions $y_i(t)$ of i-particle clusters in the $n \to \infty$ limit. This *Smoluchowski coagulation equation* has a long history in several areas of science such as physical chemistry, as indicated in the survey [6]. Recent theoretical work has made rigorous the connection between the stochastic and deterministic models, and part of this is described in [16] Chapter 5. A different limit regime concerns the time-scale when the largest clusters contain order n particles, i.e. have mass of order 1. In this limit we have real-valued cluster sizes evolving over time $(-\infty, \infty)$ and 'starting with dust' at time $-\infty$, that is with the largest cluster mass $\to 0$ as $t \to -\infty$ (just as, in the basic fragmentation model, the largest cluster mass $\to 0$ as $t \to +\infty$) and $\to 1$ as $t \to +\infty$. (So these models incidentally provide novel examples within the classical topic of *entrance boundaries* for Markov processes).

Finally recall *Kingman's coalescent*, as a model of genealogical lines of descent within neutral population genetics, which (with its many subsequent variations) has become a recognized topic within mathematical population genetics—see e.g. Wakeley [58]. Although the background story is different, it can mathematically be identified with the constant rate $(\kappa(x, x') = 1)$ stochastic coalescent in the present context.

Discussion and special cases. There are three settings above (fragmentation; discrete-particle coalescence; continuous-mass coalescence) which one might formalize differently, but the advantage of the 'exchangeable random partition' set-up is that each can be represented as a partition-valued process $(\Pi(t))$. Intuitively, coalescence and fragmentation are time-reversals of each other, and it is noteworthy that

(i) there are several fascinating examples of special models where a precise duality relation exists and is useful (see e.g. section 4.4 (iv));
(ii) but there seems to be no *general* precise duality relationship within the usual stochastic models.

In the general models the processes $(\Pi(t))$ are all Markov, as processes on partitions of $\mathbb{N}$. One can consider their restrictions $(\Pi_k(t))$ to partitions of $\{1, \dots, k\}$, i.e. consider masses of the components containing k

sampled atoms. In general $(\Pi_k(t))$ will not be Markov, but it is Markov in special cases, which are therefore particularly tractable. One such case ([16] section 3.1) is *homogeneous fragmentation*, where each cluster has the same splitting rate ($\lambda_x \equiv \lambda_1$). Another such case ([16] Chapter 4) is the elegant general theory of *exchangeable coalescents*, which eliminates the 'only binary merging' aspect of Kingman's coalescent, and is interpretable as $n \to \infty$ limit genealogies of more general models in population genetics.

4 Construction of, and convergence to, infinite random combinatorial objects

4.1 A general program

de Finetti's theorem refers specifically to *infinite* sequences. Of course we can always try to view an infinite object as a limit of finite objects, and in the '25 years ago' surveys such convergence ideas were explicit in the context of weak convergence for 'sampling without replacement' processes, in finite versions such as [24], and in some other contexts, such as Kingman's theory of exchangeable random partitions. I previously stated the first part of our central theme as

> One way of examining a complex mathematical structure is to sample i.i.d. random points and look at some form of induced substructure relating the random points

which assumes we are *given* the complex structure. But now the second and more substantial part of the theme is that we can often use exchangeability in the *construction* of complex random structures as the $n \to \infty$ limits of random finite n-element structures $\mathcal{G}(n)$.

> Within the n-element structure $\mathcal{G}(n)$ pick k random elements, look at the induced substructure on these k elements—call this $\mathcal{S}(n, k)$. Take a limit (in distribution) as $n \to \infty$ for fixed k, any necessary rescaling having been already done in the definition of $\mathcal{S}(n, k)$—call this limit $\mathcal{S}_k$. Within the limit random structures ($\mathcal{S}_k, 2 \le k < \infty$), the k elements are exchangeable, and the distributions are consistent as k increases and therefore can be used to *define* an infinite structure $\mathcal{S}_\infty$.

Where one can implement this program, the random structure $\mathcal{S}_\infty$ will for many purposes serve as a $n \to \infty$ limit of the original n-element structures. Note that $\mathcal{S}_\infty$ makes sense as a rather abstract object, via the Kolmogorov extension theorem, but in concrete cases one tries to identify it with some more concrete construction.

4.2 First examples

To invent a name for the program above, let's call it *exchangeable representations of complex random structures*. Let me first mention three examples.

1. Our discussion (section 3.1) of exchangeable random partitions fits the program but is atypically simple, in that the limit $\mathcal{S}_\infty$ is visibly the same kind of object (an exchangeable random partition) as is the finite object $\mathcal{G}(n)$. But when we moved on to coalescence and fragmentation *processes* in section 3.2, our 'exchangeability' viewpoint prompts consideration of the limit process as the partition-valued process $(\Pi(t))$, which is rather different from the finite-state Markov processes arising in coalescence for finite n.

2. A conceptually similar example arises in the technically more sophisticated setting of measure-valued diffusions (μ_t). In such processes the states are probability measures μ on some type-space, representing a 'continuous' infinite population. But one can alternately represent μ via an infinite i.i.d. sequence of samples from μ, and thereby represent the state more directly as a discrete countable infinite population $(Z(i), i \geq 1)$ and the process as a particle process $(Z_t(i), i \geq 1)$. This viewpoint was emphasized in the *look-down construction* of Kurtz–Donnelly [28, 29].

3. As a complement to the characterization (2.5) of metric spaces with a probability measure (p.m.), we can define a notion of *convergence* of such objects, say of finite metric spaces with p.m.s (S_n, d_n, μ_n) to a continuous limit $(S_\infty, d_\infty, \mu_\infty)$. The definition is simply that we have weak convergence

$$\mathbf{X}^n \overset{d}{\to} \mathbf{X}^\infty$$

of the induced random arrays defined as at (2.4):

$$X^n_{\{i,j\}} = d_n(\xi^n_i, \xi^n_j); \ \{i,j\} \in \mathbb{N}_{(2)}$$

for i.i.d. $(\xi^n_i, i \geq 1)$ with distribution μ_n. This definition provides an intriguing complement to the more familiar notion of Gromov–Hausdorff distance between compact metric spaces.

Let us move on to the fundamental setting where the program gives a substantial payoff, the formalization of *continuum random trees* as rescaled limits of finite random trees.

4.3 Continuum random trees

The material here is from the old survey [4].

Probabilists are familiar with the notion that rescaled random walk converges in distribution to Brownian motion. Now in the most basic case—simple symmetric RW of length n—we are studying the uniform distribution on a combinatorial set, the set of all 2^n simple walks of length n. So what happens if we study instead the uniform distribution on some other combinatorial set? Let us consider the set of n-vertex trees. More precisely, consider either the set of rooted labeled trees (Cayley's formula says there are n^{n-1} such trees), or the set of rooted ordered trees (counted by the Catalan number $\frac{1}{n}\binom{2n-2}{n-1}$), and write $\mathcal{T}_n$ for the uniform random tree.

Trees fit nicely into the 'substructure' framework. Vertices $v(1)$, ..., $v(k)$ of a tree define a spanning (sub)tree. Take each maximal path $(w_0, w_1, \ldots, w_\ell)$ in the spanning tree whose intermediate vertices have degree 2, and contract to a single edge of length ℓ. Applying this to k independent uniform random vertices from a n-vertex model $\mathcal{T}_n$, then rescaling edge-lengths by the factor $n^{-1/2}$, gives a tree we'll call $\mathcal{S}(n, k)$. We visualize such trees as in Figure 4.1, vertex $v(i)$ having been relabeled as i.

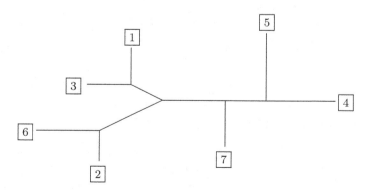

Figure 4.1 A leaf-labeled tree with edge-lengths.

In the two models for $\mathcal{T}_n$ mentioned above, one can do explicit calculations of the distribution of $\mathcal{S}(n,k)$, and use these to show that in distribution there is an $n \to \infty$ limit $\mathcal{S}_k$ which (up to a model-dependent scaling constant we'll ignore in this informal exposition) is the following distribution.

(i) The state space is the space of trees with k leaves labeled 1, 2, ..., k and with unlabeled degree-3 internal vertices, and where the $2k - 3$ edge-lengths are positive real numbers.

(ii) For each possible topological shape, the chance that the tree has that particular shape and that the vector of edge-lengths $(L_1, \ldots, L_{2k-3})$ is in $([l_i, l_i + dl_i], 1 \le i \le 2k - 3)$ equals $s \exp(-s^2/2)dl_1 \ldots dl_{2k-3}$, where $s = \sum_i l_i$.

One can check from the explicit formula what must be true from the general program, that for fixed k the distribution is exchangeable (in labels 1, ..., k), and the distributions are consistent as k increases (that is, the subtree of $\mathcal{S}_{k+1}$ spanned by leaves 1, ..., k is distributed as $\mathcal{S}_k$).

So some object $\mathcal{S}_\infty$ exists, abstractly—but what is it, more concretely? A *real tree* is a metric space with the 'tree' property that between any two points v and w there is a unique path. This implicitly specifies a length measure λ such that the metric distance $d(v,w)$ equals the length measure of the set of points on the path from v to w. When a real tree is equipped with a mass measure μ of total mass 1, representing a method for picking a vertex at random, I call it a *continuum tree*. We will consider random continuum trees—which I call continuum random trees or CRTs because it sounds better!—and the Portmanteau Theorem below envisages realizations of $\mathcal{S}_\infty$ as being equipped with a mass measure.

Returning to the n-vertex random tree models $\mathcal{T}_n$, by assigning 'mass' $1/n$ to each vertex we obtain the analogous 'mass measure' on the vertices, used for randomly sampling vertices. The next result combines existence, construction and convergence theorems. The careful reader will notice that some details in the statements have been omitted.

The Portmanteau Theorem [4, 5]

1. **Law of spanning subtrees.** There exists a particular *Brownian CRT* which agrees with $\mathcal{S}_\infty$ in the following sense. Take a realization of the Brownian CRT, then pick k i.i.d. vertices from the mass measure, and consider the spanning subtree on these k vertices. The unconditional law of this subtree is the law in (ii) above.

2. **Construction from Brownian excursion.** Consider an excursion-type function $f : [0,1] \to [0,\infty)$ with $f(0) = f(1) = 0$ and $f(x) > 0$ elsewhere. Use f to define a continuum tree as follows. Define a pseudo-metric on $[0,1]$ by: $d(x,y) = f(x) + f(y) - 2\min(f(u) : x \leq u \leq y)$, $x \leq y$. The continuum tree is the associated metric space, and the mass measure is the image of Lebesgue measure on $[0,1]$. Using this construction with standard Brownian excursion (scaled by a factor 2) gives the Brownian CRT.

3. **Line-breaking construction.** Cut the line $[0,\infty)$ into finite segments at the points of a non-homogeneous Poisson process of intensity $\lambda(x) = x$. Build a tree by inductively attaching a segment $[x_i, x_{i+1}]$ to a uniform random point of the tree built from the earlier segments. The tree built from the first $k - 1$ segments has the law (ii) above. The metric space closure of the tree built from the whole half-line is the Brownian CRT, where the mass measure is the a.s. weak limit of the empirical law of the first k cut-points.

4. **Weak limit of conditioned critical Galton–Watson branching processes and of uniform random trees.** Take a critical Galton–Watson branching process where the offspring law has finite non-zero variance, and condition on total population until extinction being n. This gives a random tree. Rescale edge-lengths to have length $n^{-1/2}$. Put mass $1/n$ on each vertex. In a certain sense that can be formalized, the $n \to \infty$ weak limit of these random trees is the Brownian CRT (up to a scaling factor). This result includes as special cases the two combinatorial models $\mathcal{T}_n$ described above.

4.4 Complements to the continuum random tree

More recent surveys by Le Gall [45] and by Evans [32] show different directions of development of the preceding material over the last 15 years. For instance

(i) the *Brownian snake* [45], which combines the genealogical structure of random real trees with independent spatial motions.

(ii) Diffusions on real trees: [32] Chapter 7.

(iii) *Continuum-tree valued diffusions.* There are several natural ways to define Markov chains on the space of n-vertex trees such that the stationary distribution is uniform. Since the $n \to \infty$ rescaled limit of the stationary distribution is the Brownian CRT, it is natural to conjecture that the entire rescaled process can be made to converge to some continuum-tree valued diffusion whose stationary distribution is the Brownian CRT. But this forces us to engage a question that was deliberately avoided in the previous section: what exactly is the space of all continuum trees, and when should we consider two such trees to be the same? This issue is discussed carefully in [32], based on the notion of the Gromov–Hausdorff space of all compact spaces. Two specific continuum-tree valued diffusions are then studied in Chapters 5 and 9 of [32].

(iv) Perhaps closer to our 'exchangeability' focus, a surprising aspect of CRTs is their application to stochastic coalescence. For $0 < \lambda < \infty$ split the Brownian CRT into components at the points of a Poisson process of rate λ along the skeleton of the tree. This gives a vector $Y(\lambda) = (Y_1(\lambda), Y_2(\lambda), \ldots)$ of masses of the components, which as λ increases specifies a fragmentation process. Reversing the direction of time by setting $\lambda = e^{-t}$ provides a construction of the (standard) additive coalescent [8], that is the stochastic coalescent (section 3.2) with kernel $\kappa(x, y) = x + y$ 'started from dust'. This result is non-intuitive, and notable as one of a handful of precise instances of the conceptual duality between stochastic coalescence and fragmentation. Also surprisingly, there are different ways that the additive coalescent can be 'started from dust', and these can also be constructed via fragmentation of certain inhomogeneous CRTs [10]. This new family of CRTs satisfies analogs of the Portmanteau Theorem, and in particular there is an explicit analog of the formula (ii) in section 4.3 for the distribution of the subtree $\mathcal{S}_k$ spanned by k random vertices [9]. This older work is complemented by much current work, the flavor of which can be seen in [38].

(v) A function $F : \{1, \ldots, n\} \to \{1, \ldots, n\}$ defines a directed graph with edges $(i, F(i))$, and the topic *random mappings* studies the graph derived from a random function F. One can repeat the section 4.1 general program in this context. Any k sampled vertices define an induced substructure, the subgraph of edges $i \to F(i) \to F(F(i)) \to \ldots$ reachable from some one of the sampled vertices. Analogously to Figure 4.1, contract paths between sampled vertices/junctions to single edges, to obtain (in the $n \to \infty$ limit) a graph $\mathcal{S}_k$ with edge-lengths, illustrated in Figure 4.2. The theory of $n \to \infty$ limits of random mappings turns out to be closely related to that of random trees; the approach based on studying the consistent family $(\mathcal{S}_k, k \geq 1)$ was developed in Aldous–Pitman [11].

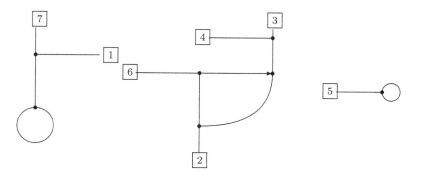

Figure 4.2 The substructure $\mathcal{S}_k$ of a random mapping.

4.5 Second-order distance structure in random networks

In the context (section 4.3) of continuum random trees the substructure was distances between sampled points. At first sight one might hope that in many models of size-n random networks one could repeat that analysis and find an interesting limit structure. But the particular feature of the models in section 4.3 is 'first order randomness' of the distance D_n between two random vertices; D_n/ED_n has a non-constant limit distribution, leading to the randomness in the limit structure. Other models tend to fall into one of two categories. For geometric networks (vertices having positions in $\mathbb{R}^2$; route-lengths as Euclidean length) the route-length tends to grow as constant c times Euclidean distance, so any limit structure reflects only the randomness of sampled vertex positions and the constant c, not any more interesting properties of the network. In non-geometric (e.g. Erdős–Rényi random graph) models, D_n tends to be first-order constant. So counter-intuitively, we don't know any other first-order random limit structures outside the 'somewhat tree-like' context.

Understanding *second*-order behavior in spatial models is very challenging—for instance, the second order behavior of first passage percolation times remains a longstanding open problem. But one can get second order results in simple 'random graph' type models, and here is the basic example (mentioned in Aldous and Bhamidi [7] as provable by the methods of that paper). The probability model used for a random n-vertex network $\mathcal{G}_n$ starts with the complete graph and assigns independent Exponential(rate $1/n$) random lengths $L_{ij} = L_{ji} = L_e$ to the $\binom{n}{2}$ edges $e = (i, j)$. In this model $ED_n = \log n + O(1)$ and $\text{var}(D_n) = O(1)$, and there is second-order behavior—a non-constant limit distribution for $D_n - \log n$.

Now fix $k \geq 3$ and write $D_n(i, j) \overset{d}{=} D_n$ for the distance between vertices i and j. We expect a joint limit

$$(D_n(1, 2) - \log n, \dots, D_n(1, k) - \log n) \overset{d}{\to} (D(1, 2), \dots, D(1, k)) \quad (4.1)$$

and it turns out the limit distribution is

$$(D(1, 2), \dots, D(1, k)) \overset{d}{=} (\xi_1 + \eta_{12}, \dots, \xi_1 + \eta_{1k})$$

where ξ_1 has the double exponential distribution

$$\mathbb{P}(\xi \leq x) = \exp(-e^{-x}), \quad -\infty < x < \infty,$$

the η_{1j} have logistic distribution

$$\mathbb{P}(\eta \leq x) = \frac{e^x}{1+e^x}, \quad -\infty < x < \infty$$

and (here and below) the r.v.s in the limits are independent. Now we can go one step further: we expect a joint limit for the array

$$(D_n(i, j) - \log n, 1 \leq i < j \leq k) \overset{d}{\to} (D(i, j), 1 \leq i < j \leq k)$$

and it turns out that the joint distribution of the limit is

$$(D(i, j), 1 \leq i < j \leq k) \overset{d}{=} (\xi_i + \xi_j - \xi_{ij}, 1 \leq i < j \leq k)$$

where the limit r.v.s all have the double exponential distribution. Of course the limit here must fit into the format of the partially exchangeable representation theorem (Theorem 2.1), and it is pleasant to see an explicit function f.

5 Limits of finite deterministic structures

Though we typically envisage limiting random structures arising as limits of finite *random* structures, it also makes sense to consider limits of finite *deterministic* structures. Let me start with a trivial example. Suppose that for each n we have a sequence $b_{n,1}, \dots, b_{n,n}$ of n bits (binary digits), and write p_n for the proportion of 1s. For each k and n, sample

k random bits from $b_{n,1}, \ldots, b_{n,n}$ and call the samples $X_{n,1}, \ldots, X_{n,k}$. Then, rather obviously, the property

$$(X_{n,1}, \ldots, X_{n,k}) \overset{d}{\to} (k \text{ independent Bernoulli}(p))$$

$$\text{as } n \to \infty; \text{ for each } k$$

is equivalent to the property $p_n \to p$.

Now in one sense this illustrates a big limitation to the whole program —sampling a substructure might lose most of the interesting information in the original structure! But a parallel procedure in the deterministic graph setting (next section) does get more interesting results, and more sophisticated uses are mentioned in section 5.2.

5.1 Limits of dense graphs

Suppose that for each n we have a graph G_n on n vertices. Write p_n for the proportion of edges, relative to the total number $\binom{n}{2}$ of possible edges. We envisage the case $p_n \to p \in (0,1)$.

For each n let $(U_{n,i}, i \geq 1)$ be i.i.d. uniform on $1, \ldots, n$. Consider the infinite $\{0,1\}$-valued matrix $\mathbf{X}^n$:

$$X_{i,j}^n = 1((U_{n,i}, U_{n,j}) \text{ is an edge of } G_n).$$

When $n \gg k^2$ the k sampled vertices $(U_{n,1}, \ldots, U_{n,k})$ of G_n will be distinct and the $k \times k$ restriction of $\mathbf{X}^n$ is the incidence matrix of the induced subgraph $\mathcal{S}(n,k)$ on these k vertices. Suppose there is a limit random matrix $\mathbf{X}$:

$$\mathbf{X}^n \overset{d}{\to} \mathbf{X} \text{ as } n \to \infty \tag{5.1}$$

in the usual product topology, that is

$$(X_{i,j}^n, \ 1 \leq i, j \leq k) \overset{d}{\to} (X_{i,j}, \ 1 \leq i, j \leq k) \text{ for each } k.$$

(Note that by compactness there is always a *subsequence* in which such convergence holds.) Now each $\mathbf{X}^n$ has the partially exchangeable property (2.2), and the limit $\mathbf{X}$ inherits this property, so we can apply the representation theorem (Theorem 2.1) to describe the possible limits. In the $\{0,1\}$-valued case we can simplify the representation. First consider a function of form (2.3) but not depending on the first coordinate—that is, a function $f(u_i, u_j, u_{\{i,j\}})$. Write

$$q(u_i, u_j) = \mathbb{P}(f(u_i, u_j, u_{\{i,j\}}) = 1).$$

The distribution of a $\{0,1\}$-valued partially exchangeable array of the special form $f(U_i, U_j, U_{\{i,j\}})$ is determined by the symmetric function $q(\cdot, \cdot)$, and so for the general form (2.3) the distribution is specified by a probability distribution over such symmetric functions.

This all fits our section 4.1 general program. From an arbitrary sequence of finite deterministic graphs we can (via passing to a subsequence if necessary) extract a 'limit infinite random graph' $\mathcal{S}_\infty$ on vertices 1, 2, ..., defined by its incidence matrix $\mathbf{X}$ in the limit (5.1), and we can characterize the possible limits. But what is a more concrete interpretation of the relation between $\mathcal{S}_\infty$ and the finite graphs (G_n)? To a probabilist the verbal expression of (5.1)

> the restriction $\mathcal{S}_k$ of $\mathcal{S}_\infty$ to vertices 1, ..., k is distributed as the $n \to \infty$ limit of the induced subgraph of G_n on k random vertices

is clear enough, but here is a translation into more graph-theoretic language, following [26]. For finite graphs F, G write $t(F, G)$ for the proportion of all mappings from vertices of F to vertices of G that are graph homomorphisms, i.e. map adjacent vertices to adjacent vertices. Suppose F has k vertices, and we label them arbitrarily as 1, ..., k. Take the subgraph $G[k]$ of G on k randomly sampled vertices, labeled 1, ..., k, and note that whether we sample with or without replacement makes no difference to $n \to \infty$ limits. Then $t(F, G)$ is the probability that F is a subgraph of $G[k]$. Now write $t_=(F, G)$ for the probability that $F = G[k]$. For fixed k, a standard inclusion-exclusion argument shows that, for a sequence (G_n), the existence of either family of limits

$$\lim_n t(F, G_n) \text{ exists, for each graph } F \text{ on vertices } \{1, \ldots, k\}, \quad (5.2)$$

$$\lim_n t_=(F, G_n) \text{ exists, for each graph } F \text{ on vertices } \{1, \ldots, k\}, \quad (5.3)$$

implies existence of the other family of limits.

In our program, the notion of $\mathcal{S}_\infty$ being the limit of G_n was defined by (5.1), which is equivalent to requiring existence of limits (5.3) for each k, in which case the limits are just $\mathbb{E}t_=(F, \mathcal{S}_k)$. And as indicated above, the partially exchangeable representation theorem (Theorem 2.1) characterizes the possible limit structures $\mathcal{S}_\infty$. A recent line of work in graph theory, initiated by Lovász and Szegedy [48], started by defining convergence in the equivalent way via (5.2) and obtained the same characterization. This is the second recent rediscovery of special cases of partially exchangeable representation theory. Diaconis and Janson [26] give a very clear and detailed account of the relation between the two settings, and Diaconis–Holmes–Janson [27] work through to an explicit description of the possible limits for a particular subclass of graphs called *threshold graphs*. Of course the line of work started in [48] has been developed further to produce new and interesting results in graph theory—see e.g. [21].

5.2 Further uses in finitary combinatorics

The remarkable recent survey by Austin [12] gives a more sophisticated treatment of the theory of representations of jointly exchangeable arrays, with the goal ([12] section 4) of clarifying connections between that theory and topics involving limits in finitary combinatorics, such as those in our previous section. I don't understand this material well enough to do more than copy a few phrases, as follows. Section 4.1 of [12] gives a general discussion of 'extraction of limit objects', somewhat parallel to our section 4.1, but with more detailed discussion of different possible precise mathematical structures. The paper continues, describing connections with the 'hypergraph regularity lemmas' featuring in combinatorial proofs of Szemerédi's Theorem, and with the structure theory within ergodic theory that Furstenberg developed for his proof of Szemerédi's Theorem. A subsequent technical paper Austin–Tao [13] applies such methods to the topic of hereditary properties of graphs or hypergraphs being testable with one-sided error; informally, this means that if a graph or hypergraph satisfies that property 'locally' with sufficiently high probability, then it can be modified into a graph or hypergraph which satisfies that property 'globally'.

6 Miscellaneous comments

1. To get an idea of the breadth of the topic, *Mathematical Reviews* created an 'exchangeability' classification 60G09 in 1984, which has attracted around 300 items; *Google Scholar* finds around 350 citations of the survey [3]; and the overlap is only around 50%. The topics in this paper, centered around structure theory—theory and applications of extensions of de Finetti's theorem—are in fact only a rather small part of this whole. In particular the 'exchangeable pairs' idea central to Stein's method [14] is really a completely distinct field.

2. Our central theme involved exchangeability, but one can perhaps view it as part of a broader theme:

a mathematical object equipped with a probability measure is sometimes a richer and more natural structure than the object by itself.

For instance, elementary discussion of fractals like the Sierpiński gasket view the object as a set in $\mathbb{R}^2$, but it comes equipped with its natural 'uniform probability distribution' which enables richer questions—the measure of small balls around a typical point, for example. Weierstrass's construction of a continuous nowhere differentiable function seems at first sight artificial—where would such things arise naturally?—but then the fact that the Brownian motion process puts a probability measure on such functions indicates one place where they do arise naturally. Analogously the notion of *real tree* (section 4.3) may seem at first sight artificial—how might such objects arise naturally?—but then realizing they arise as limits of random finite trees indicates one place where they do arise naturally. Of course the underlying structure 'a space with a metric and a measure' arises in many contexts, for example (under the name *metric measure space*) in the context of differential geometry questions [47].

Acknowledgments I thank Persi Diaconis, Jim Pitman and an anonymous referee for very helpful comments.

References

[1] Aldous, D. J. 1977. Limit theorems for subsequences of arbitrarily-dependent sequences of random variables. *Z. Wahrscheinlichkeitstheorie verw. Gebiete*, **40**(1), 59–82.

[2] Aldous, D. J. 1981. Representations for partially exchangeable arrays of random variables. *J. Multivariate Anal.*, **11**, 581–598.

[3] Aldous, D. J. 1985. Exchangeability and related topics. Pages 1–198 of: Hennequin, P. L. (ed), *École d'Été de Saint-Flour XIII—1983*. Lecture Notes in Math., vol. 1117. Berlin: Springer-Verlag.

[4] Aldous, D. J. 1991. The continuum random tree II: an overview. Pages 23–70 of: Barlow, M. T., and Bingham, N. H. (eds), *Stochastic Analysis*. Cambridge: Cambridge Univ. Press.

[5] Aldous, D. J. 1993. The continuum random tree III. *Ann. Probab.*, **21**, 248–289.

[6] Aldous, D. J. 1999. Deterministic and stochastic models for coalescence (aggregation and coagulation): a review of the mean-field theory for probabilists. *Bernoulli*, **5**, 3–48.

[7] Aldous, D. J., and Bhamidi, S. 2007. *Edge Flows in the Complete Random-Lengths Network*. http://arxiv.org/abs/0708.0555.

[8] Aldous, D. J., and Pitman, J. 1998. The standard additive coalescent. *Ann. Probab.*, **26**, 1703–1726.

[9] Aldous, D. J., and Pitman, J. 1999. A family of random trees with random edge lengths. *Random Structures & Algorithms*, **15**, 176–195.

[10] Aldous, D. J., and Pitman, J. 2000. Inhomogeneous continuum random trees and the entrance boundary of the additive coalescent. *Probab. Theory Related Fields*, **118**, 455–482.

[11] Aldous, D. J., and Pitman, J. 2002. Invariance principles for non-uniform random mappings and trees. Pages 113–147 of: Malyshev, V. A., and Vershik, A. M. (eds), *Asymptotic Combinatorics with Applications to Mathematical Physics*. Dordrecht: Kluwer. Available at http://www.stat.berkeley.edu/tech-reports/594.ps.Z.

[12] Austin, T. 2008. On exchangeable random variables and the statistics of large graphs and hypergraphs. *Probab. Surv.*, **5**, 80–145.

[13] Austin, T., and Tao, T. 2010. On the testability and repair of hereditary hypergraph properties. *Random Structures & Algorithms*. To appear; published online January 28.

[14] Barbour, A. D., and Chen, L. H. Y. (eds). 2005. *Stein's Method and Applications*. Lecture Notes Series, vol. 5. Singapore: Singapore Univ. Press, for Institute for Mathematical Sciences, National University of Singapore.

[15] Berkes, I., and Péter, E. 1986. Exchangeable random variables and the subsequence principle. *Probab. Theory Related Fields*, **73**(3), 395–413.

[16] Bertoin, J. 2006. *Random Fragmentation and Coagulation Processes*. Cambridge Stud. Adv. Math., vol. 102. Cambridge: Cambridge Univ. Press.

[17] Billingsley, P. 1968. *Convergence of Probability Measures*. New York: John Wiley & Sons.

[18] Billingsley, P. 1995. *Probability and Measure*. Third edn. Wiley Ser. Probab. Math. Stat. New York: John Wiley & Sons.

[19] Bingham, N. H., Kiesel, R., and Schmidt, R. 2003. A semi-parametric approach to risk management. *Quant. Finance*, **3**(6), 426–441.

[20] Blei, D. M., Ng, A. Y., and Jordan, M. I. 2003. Latent Dirichlet allocation. *J. Machine Learning Res.*, **3**, 993–1022.

[21] Borgs, C., Chayes, J. T., Lovász, L., Sós, V. T., and Vesztergombi, K. 2008. Convergent sequences of dense graphs. I. Subgraph frequencies, metric properties and testing. *Adv. Math.*, **219**(6), 1801–1851.

[22] Chow, Y. S., and Teicher, H. 1997. *Probability Theory: Independence, Interchangeability, Martingales*. Third edn. Springer Texts Statist. New York: Springer-Verlag.

[23] Diaconis, P., and Freedman, D. A. 1980a. de Finetti's theorem for Markov chains. *Ann. Probab.*, **8**(1), 115–130.

[24] Diaconis, P., and Freedman, D. A. 1980b. Finite exchangeable sequences. *Ann. Probab.*, **8**(4), 745–764.

[25] Diaconis, P., and Freedman, D. A. 1984. Partial exchangeability and sufficiency. Pages 205–236 of: *Statistics: Applications and New Directions (Calcutta, 1981)*. Calcutta: Indian Statist. Inst.

[26] Diaconis, P., and Janson, S. 2008. Graph limits and exchangeable random graphs. *Rend. Mat. Appl. (7)*, **28**(1), 33–61.

[27] Diaconis, P., Holmes, S., and Janson, S. 2009. *Threshold Graph Limits and Random Threshold Graphs.* http://arxiv.org/abs/0908.2448.

[28] Donnelly, P., and Kurtz, T. G. 1996. A countable representation of the Fleming-Viot measure-valued diffusion. *Ann. Probab.*, **24**(2), 698–742.

[29] Donnelly, P., and Kurtz, T. G. 1999. Particle representations for measure-valued population models. *Ann. Probab.*, **27**(1), 166–205.

[30] Dovbysh, L. N., and Sudakov, V. N. 1982. Gram-de Finetti matrices. *Zap. Nauchn. Sem. Leningrad. Otdel. Mat. Inst. Steklov. (LOMI)*, **119**, 77–86, 238, 244–245. Problems of the Theory of Probability Distribution, VII.

[31] Durrett, R. 2005. *Probability: Theory and Examples.* Third edn. Statistics/Probability Series. Pacific Grove, CA: Brooks/Cole.

[32] Evans, S. N. 2008. *Probability and Real Trees.* Lecture Notes in Math., vol. 1920. Berlin: Springer-Verlag.

[33] Ewens, W. J., and Watterson, G. A. 2010. Kingman and mathematical population genetics. In: Bingham, N. H., and Goldie, C. M. (eds), *Probability and Mathematical Genetics: Papers in Honour of Sir John Kingman.* London Math. Soc. Lecture Note Ser. Cambridge: Cambridge Univ. Press.

[34] Freedman, D. A. 1962. Invariants under mixing which generalize de Finetti's theorem. *Ann. Math. Statist*, **33**, 916–923.

[35] Fristedt, B., and Gray, L. 1997. *A Modern Approach to Probability Theory.* Probab. Appl. Ser. Boston, MA: Birkhäuser.

[36] Gelman, A., Carlin, J. B., Stern, H. S., and Rubin, D. B. 2004. *Bayesian Data Analysis.* Second edn. Texts Statist. Sci. Ser. Boca Raton, FL: Chapman & Hall/CRC.

[37] Gnedin, A. V., Haulk, C., and Pitman, J. 2010. Characterizations of exchangeable partitions and random discrete distributions by deletion properties. In: Bingham, N. H., and Goldie, C. M. (eds), *Probability and Mathematical Genetics: Papers in Honour of Sir John Kingman.* London Math. Soc. Lecture Note Ser. Cambridge: Cambridge Univ. Press.

[38] Haas, B., Pitman, J., and Winkel, M. 2009. Spinal partitions and invariance under re-rooting of continuum random trees. *Ann. Probab.*, **37**(4), 1381–1411.

[39] Hoover, D. N. 1979. *Relations on Probability Spaces and Arrays of Random Variables.* Preprint, Institute of Advanced Studies, Princeton.

[40] Kallenberg, O. 1973. Canonical representations and convergence criteria for processes with interchangeable increments. *Z. Wahrscheinlichkeitstheorie verw. Gebiete*, **27**, 23–36.

[41] Kallenberg, O. 1989. On the representation theorem for exchangeable arrays. *J. Multivariate Anal.*, **30**(1), 137–154.

[42] Kallenberg, O. 1997. *Foundations of Modern Probability.* Probab. Appl. (N.Y.). New York: Springer-Verlag.

[43] Kallenberg, O. 2005. *Probabilistic Symmetries and Invariance Principles.* Probab. Appl. (N.Y.). New York: Springer-Verlag.

[44] Kingman, J. F. C. 1978. Uses of exchangeability. *Ann. Probab.*, **6**(2), 183–197.

[45] Le Gall, J.-F. 2005. Random trees and applications. *Probab. Surv.*, **2**, 245–311 (electronic).

[46] Loève, M. 1978. *Probability Theory II*. Fourth edn. Grad. Texts in Math., vol. 46. New York: Springer-Verlag.

[47] Lott, J. 2007. Optimal transport and Ricci curvature for metric-measure spaces. Pages 229–257 of: *Surveys in Differential Geometry*. Surv. Differ. Geom., vol. 11. Somerville, MA: Int. Press.

[48] Lovász, L., and Szegedy, B. 2006. Limits of dense graph sequences. *J. Combin. Theory Ser. B*, **96**(6), 933–957.

[49] Nualart, D. 2006. *The Malliavin Calculus and Related Topics*. Second edn. Probab. Appl. Ser. Berlin: Springer-Verlag.

[50] Olshanski, G., and Vershik, A. 1996. Ergodic unitarily invariant measures on the space of infinite Hermitian matrices. Pages 137–175 of: *Contemporary Mathematical Physics*. Amer. Math. Soc. Transl. Ser. 2, vol. 175. Providence, RI: Amer. Math. Soc.

[51] Panchenko, D. 2009. *On the Dovbysh-Sudakov Representation Result*. http://front.math.ucdavis.edu/0905.1524.

[52] Pitman, J. 2006. *Combinatorial Stochastic Processes*. Lecture Notes in Math., vol. 1875. Berlin: Springer-Verlag.

[53] Pitman, J., and Yor, M. 1997. The two-parameter Poisson-Dirichlet distribution derived from a stable subordinator. *Ann. Probab.*, **25**(2), 855–900.

[54] Schoenberg, I. J. 1938. Metric spaces and positive definite functions. *Trans. Amer. Math. Soc.*, **44**(3), 522–536.

[55] Smoluchowski, M. von. 1916. Drei Vorträge über Diffusion, Brownsche Bewegung und Koagulation von Kolloidteilchen. *Physik. Z.*, **17**, 557–585.

[56] Takács, L. 1967. *Combinatorial Methods in the Theory of Stochastic Processes*. New York: John Wiley & Sons.

[57] Vershik, A. M. 2004. Random metric spaces and universality. *Uspekhi Mat. Nauk*, **59**(2(356)), 65–104.

[58] Wakeley, J. 2008. *Coalescent Theory: An Introduction*. Greenwood Village, CO: Roberts & Co.

3

Perfect simulation using dominated coupling from the past with application to area-interaction point processes and wavelet thresholding

Graeme K. Ambler[a] and Bernard W. Silverman[b]

Abstract

We consider perfect simulation algorithms for locally stable point processes based on dominated coupling from the past, and apply these methods in two different contexts. A new version of the algorithm is developed which is feasible for processes which are neither purely attractive nor purely repulsive. Such processes include multiscale area-interaction processes, which are capable of modelling point patterns whose clustering structure varies across scales. The other topic considered is nonparametric regression using wavelets, where we use a suitable area-interaction process on the discrete space of indices of wavelet coefficients to model the notion that if one wavelet coefficient is non-zero then it is more likely that neighbouring coefficients will be also. A method based on perfect simulation within this model shows promising results compared to the standard methods which threshold coefficients independently.

Keywords coupling from the past (CFTP), dominated CFTP, exact simulation, local stability, Markov chain Monte Carlo, perfect simulation, Papangelou conditional intensity, spatial birth-and-death process

AMS subject classification (MSC2010) 62M30, 60G55, 60K35

[a] Department of Medicine, Addenbrooke's Hospital, Hills Road, Cambridge CB2 0QQ; graeme@ambler.me.uk

[b] Smith School of Enterprise and Environment, Hayes House, 75 George Street, Oxford OX1 2BQ; bernard.silverman@stats.ox.ac.uk

1 Introduction

Markov chain Monte Carlo (MCMC) is now one of the standard approaches of computational Bayesian inference. A standard issue when using MCMC is the need to ensure that the Markov chain we are using for simulation has reached equilibrium. For certain classes of problem, this problem was solved by the introduction of coupling from the past (CFTP) (Propp and Wilson, 1996, 1998). More recently, methods based on CFTP have been developed for perfect simulation of spatial point process models (see for example Kendall (1997, 1998); Häggström et al. (1999); Kendall and Møller (2000)).

Exact CFTP methods are therefore attractive, as one does not need to check convergence rigorously or worry about burn-in, or use complicated methods to find appropriate standard errors for Monte Carlo estimates based on correlated samples. Independent and identically distributed samples are now available, so estimation reduces to the simplest case. Unfortunately, this simplicity comes at a price. These methods are notorious for taking a long time to return just one exact sample and are often difficult to code, leading many to give up and return to nonexact methods. In response to these issues, in the first part of this paper we present a dominated CFTP algorithm for the simulation of locally stable point processes which potentially requires far fewer evaluations per iteration than the existing method in the literature (Kendall and Møller, 2000).

The paper then goes on to discuss applications of this CFTP algorithm, in two different contexts, the modelling of point patterns and nonparametric regression by wavelet thresholding. In particular it will be seen that these two problem areas are much more closely related than might be imagined, because of the way that the non-zero coefficients in a wavelet expansion may be modelled as an appropriate point process.

The structure of the paper is as follows. In Section 2 we discuss perfect simulation, beginning with ordinary coupling from the past (CFTP) and moving on to dominated CFTP for spatial point processes. We then introduce and justify our perfect simulation algorithm. In Section 3 we first review the standard area-interaction process. We then introduce our multiscale process, describe how to use our new perfect simulation algorithm to simulate from it, and discuss a method for inferring the parameter values from data, and present an application to the Redwood seedlings data. In Section 4 we turn attention to the wavelet regression problem. Bayesian approaches are reviewed, and a model introduced which incorporates an area-interaction process on the discrete space of indices of wavelet coefficients. In Section 5 the application of our perfect simulation algorithm in this context is developed. The need appropriately to modify the approach to increase its computational feasibility is addressed, and a simulation study investigating its performance on standard test examples is carried out. Sections 3 and 5 both conclude with some suggestions for future work.

2 Perfect simulation

2.1 Coupling from the past

In this section, we offer a brief intuitive introduction to the principle behind CFTP. For more formal descriptions and details, see, for example, Propp and Wilson (1996), MacKay (2003, Chapter 32) and Connor (2007).

Suppose we wanted to sample from the stationary distribution of an irreducible aperiodic Markov chain $\{Z_t\}$ on some (finite) state space X with states $1, \ldots, n$. Intuitively, if it were possible to go back an infinite amount in time and start the chain running, the chain would be in its stationary distribution when one returned to the present (i.e. $Z_0 \sim \pi$, where π is the stationary distribution of the chain).

Now, suppose we were to set not one, but n chains $\{Z_t^{(1)}\}, \ldots, \{Z_t^{(n)}\}$ running at a fixed time $-M$ in the past, where $Z_{-M}^{(i)} = i$ for each chain $\{Z_t^{(i)}\}$. Now let all the chains be coupled so that if $Z_s^{(i)} = Z_s^{(j)}$ at any time s then $Z_t^{(i)} = Z_t^{(j)} \ \forall t \geq s$. Then if all the chains ended up in the same state j at time zero (i.e. $Z_0^{(i)} = j \ \forall i \in X$), we would know that whichever state the chain passing from time minus infinity to zero was in at time $-M$, the chain would end up in state j at time zero. Thus the state at time zero is a sample from the stationary distribution provided M is large enough for coalescence to have been achieved for the realisations being considered.

When performing CFTP, a useful property of the coupling chosen is that it be *stochastically monotone* as in the following definition.

Definition 2.1 Let $\{Z_t^{(i)}\}$ and $\{Z_t^{(j)}\}$ be two Markov chains obeying the same transition kernel. Then a coupling of these Markov chains is stochastically monotone with respect to a partial ordering $\preceq$ if whenever $Z_t^{(i)} \preceq Z_t^{(j)}$, then $Z_{t+k}^{(i)} \preceq Z_{t+k}^{(j)}$ for all positive k.

Whenever the coupling used is stochastically monotone and there are maximal and minimal elements with respect to $\preceq$ then we need only simulate chains which start in the top and bottom states, since chains starting in all other states are sandwiched by these two. This is an important ingredient of the dominated coupling from the past algorithm introduced in the next section.

Although attempts have been made to generalise CFTP to continuous state spaces (notably Murdoch and Green (1998) and Green and Murdoch (1998), as well as Kendall and Møller (2000), discussed in Section 2.2), there is still much work to be done before exact sampling becomes universally, or even generally applicable. For example, there are no truly general methods for processes in high, or even moderate, dimensions.

2.2 Dominated coupling from the past

Dominated coupling from the past was introduced as an extension of coupling from the past which allowed the simulation of the area-interaction process (Kendall, 1998), though it was soon extended to other types of point processes and more general spaces (Kendall and Møller, 2000). We give the formulation for locally stable point processes.

Let x be a spatial point pattern in some bounded subset $S \subset \mathbb{R}^n$, and u a single point $u \in S$. Suppose that x is a realisation of a spatial point process X with density f with respect to the unit rate Poisson process. The *Papangelou conditional intensity* λ_f is defined by

$$\lambda_f(u; x) = \frac{f(x \cup \{u\})}{f(x)};$$

see, for example, Papangelou (1974) and Baddeley et al. (2005). If the process X is locally stable, then there exists a constant λ such that $\lambda_f(u; x) \leq \lambda$ for all finite point configurations $x \subset S$ and all points $u \in S \setminus x$.

The algorithm given in Kendall and Møller (2000) is then as follows.

1 Obtain a sample of the Poisson process with rate λ.
2 Evolve a Markov process $D(T)$ *backwards* until some fixed time $-T$, using a birth-and-death process with death rate equal to 1 and birth rate equal to λ. The configuration generated in step 1 is used as the initial state.
3 Mark all of the points in the process with U[0,1] marks. We refer to the mark of point x as $P(x)$.

4 Recursively define upper and lower processes, U and L as follows. The initial configurations at time $-T$ for the processes are

$$U_{-T}(-T) = \{x : x \in D(-T)\};$$
$$L_{-T}(-T) = \{\mathbf{0}\}.$$

5 Evolve the processes *forwards* in time to $t = 0$ in the following way.

Suppose that the processes have been generated up until a given time, u, and suppose that the next birth or death to occur after that time happens at time t_i. If a **birth** happens next then we accept the birth of the point x in U_{-T} or L_{-T} if the point's mark, $P(x)$, is less than

$$
\begin{aligned}
&\min\left\{\frac{\lambda_f(x; X)}{\lambda} : L_{-T}(u) \subseteq X \subseteq U_{-T}(u)\right\} \text{ or} \\
&\max\left\{\frac{\lambda_f(x; X)}{\lambda} : L_{-T}(u) \subseteq X \subseteq U_{-T}(u)\right\}
\end{aligned}
\tag{2.1}
$$

respectively, where x is the point to be born.

If, however, a **death** happens next then if the event is present in either of our processes we remove the dying event, setting $U_{-T}(t_i) = U_{-T}(u) \setminus \{x\}$ and $L_{-T}(t_i) = L_{-T}(u) \setminus \{x\}$.

6 Define $U_{-T}(u+\varepsilon) = U_{-T}(u)$ and $L_{-T}(u+\varepsilon) = L_{-T}(u)$ for $u < u+\varepsilon < t_i$.

7 If U_{-T} and L_{-T} are identical at time zero (i.e. if $U_{-T}(0) = L_{-T}(0)$), then we have the required sample from X, the locally stable point process of interest. If not, go to step 2 and repeat, extending the underlying Poisson process back to $-(T+S)$ and generating additional $U[0,1]$ marks (keeping the ones already generated).

This algorithm involves calculation of $\lambda(u; X)$ for each configuration that is both a subset of $U(T)$ and a superset of $L(T)$. Since calculation of $\lambda(u; X)$ is typically expensive, this calculation may be very costly. The method proposed in Section 2.3 uses an alternative version of step 5 which requires us only to calculate $\lambda(u; X)$ for upper and lower processes.

The more general form given in Kendall and Møller (2000) may be obtained from the above algorithm by replacing the evolving Poisson process $D(T)$ with a general dominating process on a partially ordered space $(\Omega, \preceq)$ with a unique minimal element $\mathbf{0}$. The partial ordering in the above algorithm is that induced by the subset relation $\subseteq$. Step 5 is replaced by any step which preserves the crucial *funnelling property*

$$L_{-T}(u) \preceq L_{-(T+S)}(u) \preceq U_{-(T+S)}(u) \preceq U_{-T}(u) \tag{2.2}$$

for all $u < 0$ and T, $S > 0$ and the *sandwiching relations*

$$L_{-T}(u) \preceq X_{-T}(u) \preceq U_{-T}(u) \preceq D(u) \quad \text{and} \tag{2.3}$$
$$L_{-T}(t) = U_{-T}(t) \quad \text{if} \ \ L_{-T}(s) = U_{-T}(s) \tag{2.4}$$

for $s \leq t \leq 0$. In equation (2.3), $X_{-T}(u)$ is the Markov chain or process from whose stationary distribution we wish to sample.

2.3 A perfect simulation algorithm

Suppose that we wish to sample from a locally stable point process with density

$$p(X) = \alpha \prod_{i=1}^{m} f_i(X), \tag{2.5}$$

where $\alpha \in (0, \infty)$ and $f_i : \mathfrak{R}^f \to \mathbb{R}$ are positive-valued functions which are monotonic with respect to the partial ordering $\preceq$ induced by the subset relation[1] and have uniformly bounded Papangelou conditional intensity:

$$\lambda_{f_i}(u; \mathbf{x}) = \frac{f_i(\mathbf{x} \cup \{u\})}{f_i(\mathbf{x})} \leq K. \tag{2.6}$$

When the conditional intensity (2.6) can be expressed in this way, as the product of monotonic interactions, then we shall demonstrate that the crucial step of the Kendall–Møller algorithm may be re-written in a form which is computationally much more efficient, essentially by dealing with each factor separately.

Clearly

$$\lambda_p(u; \mathbf{x}) \leq \lambda = \prod_{i=1}^{m} \max_{X, \{x\}} \lambda_{f_i}(x; X) \tag{2.7}$$

for all u and $\mathbf{x}$, and λ is finite. Thus we may use the algorithm in Section 2.2 to simulate from this process using a Poisson process with rate λ as the dominating process.

However, as previously mentioned, calculation of $\lambda_p(u; \mathbf{x})$ is typically expensive, increasing at least linearly in $n(\mathbf{x})$. Thus to calculate the expressions in (2.1), we must in general perform $2^{n(U_{-T}(t_i)) - n(L_{-T}(t_i))}$

---

[1] That is, configurations x and y satisfy $x \preceq y$ if $x \subseteq y$.

of these calculations, making the algorithm non-polynomial. In practice it is clearly not feasible to use this algorithm in all but the most trivial of cases, so we must look for some way to reduce the computational burden in step 5 of the algorithm.

This can be done by replacing step 5 with the following alternative.

5' Evolve the processes *forwards* in time to $t = 0$ in the following way.

Suppose that the processes have been generated up a given time, u, and suppose that the next birth or death to occur after that time happens at time t_i. If a **birth** happens next then we accept the birth of the point x in U_{-T} or L_{-T} if the point's mark, $P(x)$, is less than

$$\prod_{i=1}^{m} \left[\max \left\{ \lambda_{f_i}(u; U(T)), \lambda_{f_i}(u; L(T)) \right\} / \lambda \right] \quad \text{or} \quad (2.8)$$

$$\prod_{i=1}^{m} \left[\min \left\{ \lambda_{f_i}(u; U(T)), \lambda_{f_i}(u; L(T)) \right\} / \lambda \right] \quad (2.9)$$

respectively, where x is the point to be born.

If, however, a **death** happens next then if the event is present in either of our processes we remove the dying event, setting $U_{-T}(t_i) = U_{-T}(u) \setminus \{x\}$ and $L_{-T}(t_i) = L_{-T}(u) \setminus \{x\}$.

Lemma 2.2 *Step 5' obeys properties* (2.2), (2.3) *and* (2.4), *and is thus a valid dominated coupling-from-the-past algorithm.*

Proof Property (2.3) follows by noting that

$$(2.9) \leq \lambda_p(u; X) \leq (2.8) \leq 1.$$

Property (2.4) is trivial. Property (2.2) follows from the monotonicity of the f_i. □

Theorem 2.3 *Suppose that we wish to simulate from a locally stable point process whose density $p(X)$ with respect to the unit-rate Poisson process is representable in form* (2.5). *Then by replacing Step 5 by Step 5' it is possible to bound the necessary number of calculations of $\lambda_p(u; X)$ per iteration in the dominated coupling-from-the-past algorithm independently of $n(X)$.*

Proof Step 5' clearly involves only a constant number of calculations, so by Lemma 2.2 above and Theorem 2.1 of Kendall and Møller (2000), the result holds. □

In the case where it is possible to write $p(X)$ in form (2.5) with $m = 1$, Step 5' is identical to Step 5. This is the case for models which are either purely attractive or purely repulsive, such as the standard area-interaction process discussed in Section 3.1. It is not the case for the multiscale process discussed in Section 3.2, or the model for wavelet coefficients discussed in Section 4.2.

The proof of Theorem 2.1 in Kendall and Møller (2000) does not require that the initial configuration of L_{-T} be the minimal element $\mathbf{0}$, only that it be constructed in such a way that properties (2.2), (2.3) and (2.4) are satisfied. Thus we may refine our method further by modifying step 4 so that the initial configuration of L_{-T} is given by

$$L_{-T}(-T) = \left\{ x \in D(-T) : P(x) \leq \prod_{i=1}^{m} \left[\min_{X,\{x\}} \lambda_{f_i}(x; X) / \lambda \right] \right\}, \quad (2.10)$$

which clearly satisfies the necessary requirements.

3 Area-interaction processes

3.1 Standard area-interaction process

There are several classes of model for stochastic point processes, for example simple Poisson processes, cluster processes such as Cox processes, and processes defined as the stationary distribution of Markov point processes, such as Strauss processes (Strauss, 1975) and area-interaction processes (Baddeley and Lieshout, 1995). The area-interaction point process is capable of producing both moderately clustered and moderately ordered patterns depending on the value of its clustering parameter. It was introduced primarily to fill a gap left by the Strauss point process (Strauss, 1975), which can produce only ordered point patterns (Kelly and Ripley, 1976).

The general definition of the area-interaction process depends on a specification of the *neighbourhood* of any point in the space χ on which the process is defined. Given any $x \in \chi$ we denote by $B(x)$ the neighbourhood of the point x. Given a set $X \subseteq \chi$, the neighbourhood $U(X)$ of X is defined as $\bigcup_{x \in X} B(x)$. The general area-interaction process is then defined by Baddeley and van Lieshout (1995) as follows.

Let χ be some locally compact complete metric space and $\mathfrak{R}^f$ be the space of all possible configurations of points in χ. Suppose that m is a finite Borel regular measure on χ and $B : \chi \to \mathcal{K}$ be a myopically continuous function (Matheron, 1975), where $\mathcal{K}$ is the class of all compact subsets of χ. Then the probability density of the general area-interaction process is given by

$$p(X) = \alpha \lambda^{N(X)} \gamma^{-m\{U(X)\}} \quad (3.1)$$

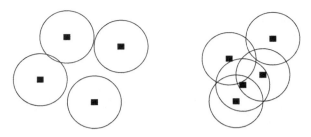

Figure 3.1 An example of some events together with circular 'grains' G. The events in the above diagram would be the actual members of the process. The circles around them are to show what the set $X \oplus G$ would look like. If γ were large, the point configuration on the right would be favoured, whereas if γ were small, the configuration on the the left would be favoured.

with respect to the unit rate Poisson process, where $N(X)$ is the number of points in configuration $X = \{x_1, \ldots, x_{N(X)}\} \in \mathfrak{R}^f$, α is a normalising constant and $U(X) = \bigcup_{i=1}^{N(X)} B(x_i)$ as above.

In the spatial point-process case, for some fixed compact set G in $\mathbb{R}^d$, the neighbourhood $B(x)$ of each point x is defined to be $x \oplus G$. Here $\oplus$ is the Minkowski addition operator, defined by $A \oplus B = \{a + b : a \in A, b \in B\}$ for sets A and B. So the resulting area-interaction process has density

$$p(X) = \alpha \lambda^{N(X)} \gamma^{-m(X \oplus G)} \tag{3.2}$$

with respect to the unit-rate Poisson process, where α is a normalising constant, $\lambda > 0$ is the *rate* parameter, $N(X)$ is the number of points in the configuration X, $\gamma > 0$ is the *clustering* parameter. Here $0 < \gamma < 1$ is the *repulsive* case, while $\gamma > 1$ is the *attractive* case. The case $\gamma = 1$ reduces to the homogeneous Poisson process with rate λ. Figure 3.1 gives an example of the construction when G is a disc.

3.2 A multiscale area-interaction process

The area-interaction process is a flexible model yielding a good range of models, from regular through total spatial randomness to clustered. Unfortunately it does not allow for models whose behaviour changes at different resolutions, for example repulsion at small distances and attraction at large distances. Some examples which display this sort of behaviour are the distribution of trees on a hillside, or the distribution of zebra in a patch of savannah. A physical example of large scale attraction and small scale repulsion is the interaction between the strong nuclear force and the electro-magnetic force between two oppositely charged particles. The physical laws governing this behaviour are different from those governing the behaviour of the area-interaction class of models, though they may be sufficiently similar so as to provide a useful approximation.

We propose the following model to capture these types of behaviour.

Definition 3.1 The *multiscale area-interaction process* has density

$$p(X) = \alpha \lambda^{N(X)} \gamma_1^{-m(X \oplus G_1)} \gamma_2^{-m(X \oplus G_2)}, \tag{3.3}$$

where α, λ and $N(X)$, are as in equation (3.2); $\gamma_1 \in [1, \infty)$ and $\gamma_2 \in (0, 1]$; and G_1 and G_2 are balls of radius r_1 and r_2 respectively.

The process is clearly Markov of range $\max\{r_1, r_2\}$. If $G_1 \supset G_2$, we will have small-scale repulsion and large-scale attraction. If $G_1 \subset G_2$, we will have small-scale attraction and large-scale repulsion.

Theorem 3.2 *The density* (3.3) *is both measurable and integrable.*

This is a straightforward extension of the proof of Baddeley and van Lieshout (1995) for the standard area-interaction process; for details, see the Appendix of Ambler and Silverman (2004b).

3.3 Perfect simulation of the multiscale process

Perfect simulation of the multiscale process (3.3) is possible using the method introduced in Section 2.3. Since (3.3) is already written as a product of three monotonic functions with uniformly bounded Papangelou conditional intensities, we need only substitute into equations (2.7–2.10) as follows.

Substituting into equation (2.7), we find that the rate of a suitable dominating process is

$$\lambda \gamma_2^{-m(G_2)}.$$

The initial configurations of the upper and lower processes U and L are then found by simulating this process, thinning with a probability of

$$\gamma_1^{-m(G_1)} \gamma_2^{m(G_2)}$$

for L.

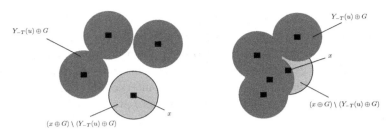

Figure 3.2 Another look at Figure 3.1 with some shading added to show the process of simulation. Dark shading shows $Y_{-T}(u) \oplus G$ where $Y_{-T}(u)$ is the state of either U or L immediately before we add the new event and G could be either G_1 or G_2. Light shading shows the amount added if we accept the new event. In the configuration on the left, $x \oplus G = (x \oplus G) \setminus (Y_{-T}(u) \oplus G)$, so that the attractive term in (3.4) or (3.5) will be very small, whereas the repulsive term will be large. In the configuration on the right we are adding very little area to $(Y_{-T}(u) \oplus G)$ by adding the event, so the attractive term will be larger and the repulsive term will be smaller.

As U and L evolve towards time 0, we accept points x in U with probability

$$\gamma_1^{-m((x \oplus G_1) \setminus U_{-T}(u) \oplus G_1)} \gamma_2^{m(G_2) - m((x \oplus G_2) \setminus L_{-T}(u) \oplus G_2)} \tag{3.4}$$

and accept events in L whenever

$$P(x) \leq \gamma_1^{-m((x \oplus G_1) \setminus L_{-T}(u) \oplus G_1)} \gamma_2^{m(G_2) - m((x \oplus G_2) \setminus U_{-T}(u) \oplus G_2)}. \tag{3.5}$$

Figure 3.2 gives examples of the construction $(x \oplus G) \setminus Y_{-T}(u) \oplus G$.

3.4 Redwood seedlings data

We take a brief look at a data set which has been much analysed in the literature, the Redwood seedlings data first considered by Strauss (1975). We examine a subset of the original data chosen by Ripley (1977) and later analysed by Diggle (1978) among others. The data are plotted in Figure 3.3. We wish to model these data using the multiscale model we have introduced. The right pane of Figure 3.3 gives the estimated point process L-function[2] of the data, defined by $L(t) = \sqrt{\pi^{-1} K(t)}$ where K is the K-function as defined by Ripley (1976, 1977).

---

[2] There is no connection between the point process L-function and the use of the notation L elsewhere in this paper for the lower process in the CFTP algorithm; the clash of notation is an unfortunate result of the standard use of L in both contexts. Nor does either use of L refer to a likelihood.

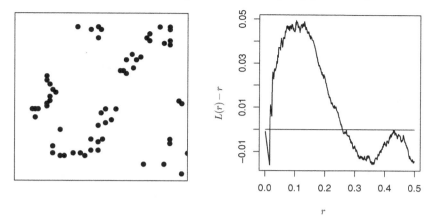

Figure 3.3 Redwood seedlings data. Left: The data, selected by Ripley (1977) from a larger data set analysed by Strauss (1975). Right: Plot of the point-process L-function for the redwood seedlings. There seems to be interaction at 3 different scales: (very) small scale repulsion followed by attraction at a moderate scale and then repulsion at larger scales.

From this plot we estimate values of R_1 and R_2 as 0.07 and 0.013 respectively, giving repulsion at small scales and attraction at moderate scales. It also seems that there is some repulsion at slightly larger scales, so it may be possible to use $R_2 = 0.2$ and to model the large-scale interaction rather than the small-scale interaction as we have chosen.

Experimenting with various values for the remaining parameters, we chose values $\gamma_1 = 2000$ and $\gamma_2 = 10^{-200}$. The value $\lambda = 0.118$ was chosen to give about 62 points in each realisation, the number in the observed data set. The remarkably small value of γ_2 was necessary because the value of R_2 was also very small. It is clear from these numbers that it would be more natural to define γ_1 and γ_2 on a logarithmic scale. Figure 3.4 shows point process L- and T-function plots for 19 simulations from this model, providing approximate 95% Monte-Carlo confidence envelopes for the values of the functions. It can be seen that on the basis of these functions, the model appears to fit the data reasonably well. The T-function, defined by Schladitz and Baddeley (2000), is a third order analogue of the K-function, and for a Poisson process $T(r)$ is proportional to r^4; in Figure 3.4 the function is transformed by taking the fourth root of a suitable multiple and then subtracting r, in order

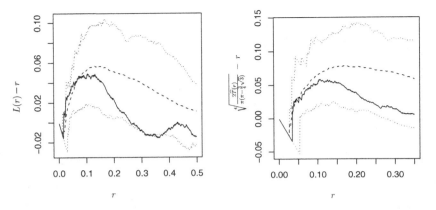

Figure 3.4 Point process L- and transformed T-function plots of the redwood seedlings data. Left: L-function plots of the data together with simulations of the multiscale model with parameters $R_1 = 0.07$, $R_2 = 0.013$, $\lambda = 0.118$, $\gamma_1 = 2000$ and $\gamma_2 = 10^{-200}$. Dotted lines give an envelope of 19 simulations of the model, the solid line is the redwood seedlings data and the dashed line is the average of the 19 simulations. Right: the corresponding plots for the transformed T-function.

to yield a function whose theoretical value for a Poisson process would be zero.

The plots show several things: firstly that the model fits reasonably well, but that it is possible that we chose a value of R_1 which was slightly too large. Perhaps $R_1 = 0.06$ would have been better. Secondly, it seems that the large-scale repulsion may be an important factor which should not be ignored. Thirdly, in this case we have gained little new information by plotting the T-function—the third-order behaviour of the data seems to be similar in nature to the second-order structure.

3.5 Further comments

The main advantage of our method for the perfect simulation of locally stable point processes is that it allows acceptance probabilities to be computed in $O(n)$ instead of $O(2^n)$ steps for models which are neither purely attractive nor purely repulsive. Because of the exponential dependence on n, the algorithm of Kendall and Møller (2000) is not feasible in these situations.

It is clear that in practice it is possible to extend the work to more general multiscale models. For example, the sample L-function of the redwood seedlings might, if the sample size were larger, indicate the appropriateness of a three-scale model

$$p(X) = \alpha \lambda^{N(X)} \gamma_1^{-m(X \oplus G_1)} \gamma_2^{-m(X \oplus G_2)} \gamma_3^{-m(X \oplus G_3)}. \qquad (3.6)$$

The proof given in the Appendix of Ambler and Silverman (2004b) can easily be extended to show the existence of this process, and (3.6) is also amenable to perfect simulation using the method of Section 2.3. Because of the small size of the redwood seedlings data set a model of this complexity is not warranted, but the fitting of such models, and even higher order multiscale models in appropriate circumstances, would be an interesting topic for future research.

Another topic is the possibility of fitting parameters by a more systematic approach than the subjective adjustment approach we have used. Ambler and Silverman (2004b) set out the possibility of using pseudo-likelihood (Besag, 1974, 1975, 1977; Jensen and Møller, 1991) to estimate the parameters λ, γ_1 and γ_2 for given R_1 and R_2. However, this method has yet to be implemented and investigated in practice.

4 Nonparametric regression by wavelet thresholding

4.1 Introductory remarks

We now turn to our next theme, nonparametric regression. Suppose we observe

$$y_i = g(t_i) + \varepsilon_i. \qquad (4.1)$$

where g is an unknown function sampled with error at regularly spaced intervals t_i. The noise, ε_i is assumed to be independent and Normally distributed with zero mean and variance σ^2.

The standard wavelet-based approach to this problem is based on two properties of the wavelet transform:

1. a large class of 'well-behaved' functions can be sparsely represented in wavelet space;
2. the wavelet transform maps independent identically distributed noise to independent identically distributed wavelet coefficients.

These two properties combine to suggest that a good way to remove noise from a signal is to transform the signal into wavelet space, discard all of the small coefficients (i.e. threshold), and perform the inverse transform. Since the true (noiseless) signal had a sparse representation in wavelet space, the signal will essentially be concentrated in a small number of large coefficients. The noise, on the other hand, will still be spread evenly among the coefficients, so by discarding the small coefficients we must have discarded mostly noise and will thus have found a better estimate of the true signal.

The problem then arises of how to choose the threshold value. General methods that have been applied in the wavelet context are SureShrink (Donoho and Johnstone, 1995), cross-validation (Nason, 1996) and false discovery rates (Abramovich and Benjamini, 1996). In the BayesThresh approach, Abramovich et al. (1998) propose a Bayesian hierarchical model for the wavelet coefficients, using a mixture of a point mass at 0 and a $N(0, \tau^2)$ density as their prior. The marginal posterior median of the population wavelet coefficient is then used as the estimate. This gives a thresholding rule, since the point mass at 0 in the prior gives non-zero probability that the population wavelet coefficient will be zero.

Most Bayesian approaches to wavelet thresholding model the coefficients independently. In order to capture the notion that nonzero wavelet coefficients may be in some way clustered, we allow prior dependency between the coefficients by modelling them using an extension of the area-interaction process as defined in Section 3.1 above. The basic idea is that if a coefficient is nonzero then it is more likely that its neighbours (in a suitable sense) are also non-zero. We then use an appropriate CFTP approach to sample from the posterior distribution of our model.

4.2 A Bayesian model for wavelet thresholding

Abramovich et al. (1998) consider the problem where the true wavelet coefficients are observed subject to Gaussian noise with zero mean and some variance σ^2,

$$\widehat{d}_{jk} | d_{jk} \sim N(d_{jk}, \sigma^2),$$

where $\widehat{d}_{jk}$ is the value of the noisy wavelet coefficient (the data) and d_{jk} is the value of the true (noiseless) coefficient.

Their prior distribution on the true wavelet coefficients is a mixture of a Normal distribution with zero mean and variance dependent on the level of the coefficient, and a point mass at zero as follows:

$$d_{jk} \sim \pi_j N(0, \tau_j^2) + (1 - \pi_j)\delta(0), \tag{4.2}$$

where d_{jk} is the value of the kth coefficient at level j of the discrete wavelet transform, and the mixture weights $\{\pi_j\}$ are constant within each level. An alternative formulation of this can be obtained by introducing auxiliary variables $Z = \{\zeta_{jk}\}$ with $\zeta_{jk} \in \{0,1\}$ and independent hyperpriors

$$\zeta_{jk} \sim \text{Bernoulli}(\pi_j). \tag{4.3}$$

The prior given in equation (4.2) is then expressed as

$$d_{jk}|Z \sim N(0, \zeta_{jk}\tau_j^2). \tag{4.4}$$

The starting point for our extension of this approach is to note that Z can be considered to be a point process on the discrete space, or lattice, χ of indices (j,k) of the wavelet coefficients. The points of Z give the locations at which the prior variance of the wavelet coefficient, conditional on Z, is nonzero. From this point of view, the hyperprior structure given in equation (4.3) is equivalent to specifying Z to be a Binomial process with rate function $p(j,k) = \pi_j$.

Our general approach will be to replace Z by a more general lattice process ξ on χ. We allow ξ to have multiple points at particular locations (j,k), so that the number ξ_{jk} of points at (j,k) will be a nonnegative integer, not necessarily confined to $\{0,1\}$. We will assume that the prior variance is proportional to the number of points of ξ falling at the corresponding lattice location. So if there are no points, the prior will be concentrated at zero and the corresponding observed wavelet will be treated as pure noise; on the other hand, the larger the number of points, the larger the prior variance and the less shrinkage applied to the observed coefficient. To allow for this generalisation, we extend (4.4) in the natural way to

$$d_{jk}|\xi \sim N(0, \tau^2 \xi_{jk}), \tag{4.5}$$

where τ^2 is a constant.

We now consider the specification of the process ξ. While it is reasonable that the wavelet transform will produce a sparse representation, the time-frequency localisation properties of the transform also make it natural to expect that the representation will be clustered in some sense. The existence of this clustered structure can be seen clearly in Figure 4.1, which shows the discrete wavelet transform of several common test functions represented in the natural binary tree configuration. With this clustering in mind, we model ξ as an area-interaction process on the space χ. The choice of the neighbourhoods $B(x)$ for x in χ will be discussed below. Given the choice of neighbourhoods, the process will

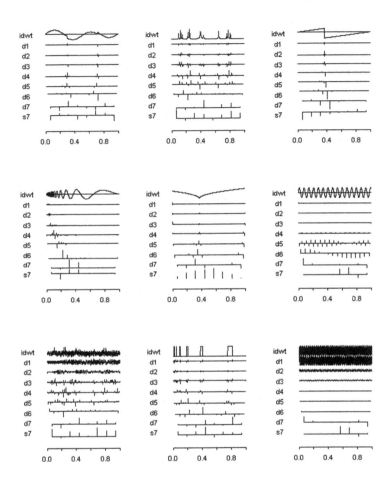

Figure 4.1 Examples of the discrete wavelet transform of some test functions. There is clear evidence of clustering in most of the graphs. The original functions are shown above their discrete wavelet transform each time.

be defined by

$$p(\xi) = \alpha \lambda^{N(\xi)} \gamma^{-m\{U(\xi)\}} \qquad (4.6)$$

where $p(\xi)$ is the intensity relative to the unit rate independent auto-

Poisson process (Cressie, 1993). If we take $\gamma > 1$ this gives a clustered configuration. Thus we would expect to see clusters of large values of d_{jk} if this were a reasonable model—which is exactly what we do see in Figure 4.1.

A simple application of Bayes's theorem tells us that the posterior for our model is

$$p(\xi, \mathbf{d}|\widehat{\mathbf{d}}) = p(\xi) \prod_{j,k} p(d_{jk}|\xi_{jk}) \prod_{j,k} p(\widehat{d}_{jk}|d_{jk}, \xi_{jk})$$

$$= \alpha \lambda^{N(\xi)} \gamma^{-m\{U(\xi)\}} \prod_{j,k} \frac{\exp(-d_{jk}^2/2\tau^2\xi_{jk})}{\sqrt{2\pi\tau^2\xi_{jk}}} \prod_{j,k} \frac{\exp\{-(\widehat{d}_{jk} - d_{jk})^2/2\sigma^2\}}{\sqrt{2\pi\sigma^2}}$$

$$= \alpha \lambda^{N(\xi)} \gamma^{-m\{U(\xi)\}} \prod_{j,k} \frac{\exp\{-d_{jk}^2/2\tau^2\xi_{jk} - (\widehat{d}_{jk} - d_{jk})^2/2\sigma^2\}}{\sqrt{2\pi\tau^2\xi_{jk}}\sqrt{2\pi\sigma^2}}. \qquad (4.7)$$

Clearly (4.7) is not a standard density. In Section 5.1 we show how the extension of the coupling-from-the-past algorithm described in Section 2.3 enables us to sample from it.

4.3 Completing the specification

We first note that in this context χ is a discrete space, so the technical conditions required in Section 3.1 of $m(\cdot)$ and $B(\cdot)$ are trivially satisfied.

In order to complete the specification of our area-interaction prior for ξ, we need a suitable interpretation of the neighbourhood of a location $x = (j, k)$ on the lattice χ of indices (j, k) of wavelet coefficients. This lattice is a binary tree, and there are many possibilities. We decided to use the parent, the coefficient on the parent's level of the transform which is next-nearest to x, the two adjacent coefficients on the level of x, the two children and the coefficients adjacent to them, making a total of nine coefficients (including x itself). Figure 4.2 illustrates this scheme, which captures the localisation of both time and frequency effects. Figure 4.2 also shows how we dealt with boundaries: we assume that the signal we are examining is periodic, making it natural to have periodic boundary conditions in time. If $B(x)$ overlaps with a frequency boundary we simply discard those parts which have no locations associated with them. The simple counting measure used has $m\{B(x)\} = 9$ unless x is in the bottom row or one of the top two rows.

Other possible neighbourhood functions include using only the parent, children and immediate sibling and cousin of a coefficient as $B(x)$, or a variation on this taking into account the length of support of the wavelet

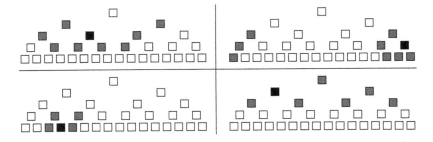

Figure 4.2 The four plots give examples of what we used as $B(\cdot)$ for four different example locations showing how we dealt with boundaries. Grey boxes are $B(x) \setminus \{x\}$ for each example location x, while x itself is shown as black.

used. Though we have chosen to use periodic boundary conditions, our method is equally applicable without this assumption, with appropriate modification of $B(x)$.

5 Perfect simulation for wavelet curve estimation

5.1 Exact posterior sampling for lattice processes

In this section, we develop a practical approach to simulation from a close approximation to the posterior density (4.7), making use of coupling from the past. One of the advantages of the Normal model we propose in Section 4.2 is that it is possible to integrate out d_{jk} and work only with the lattice process ξ. Performing this calculation, we see that equation (4.7) can be rewritten as

$$p(\xi|\widehat{\mathbf{d}}) = p(\xi) \prod_{j,k} \frac{\exp\left\{-\widehat{d}_{jk}^2/2(\sigma^2 + \tau^2\xi_{jk})\right\}}{\sqrt{2\pi(\sigma^2 + \tau^2\xi_{jk})}},$$

by the standard convolution properties of normal densities. We now see that it is possible to sample from the posterior by simulating only the process ξ and ignoring the marks $\mathbf{d}$. This lattice process is amenable to perfect simulation using the method of Section 3.3 above. Let

$$f_1(\xi) = \lambda^{N(\xi)},$$
$$f_2(\xi) = \gamma^{-m\{U(\xi)\}},$$

$$f_3(\xi) = \prod_{j,k} \exp\{-\widehat{d}_{jk}^2 \, / \, 2(\sigma^2 + \tau^2 \xi_{jk})\} \text{ and}$$

$$f_4(\xi) = \prod_{j,k} \left\{2\pi(\sigma^2 + \tau^2 \xi_{jk})\right\}^{-1/2}.$$

Then

$$\lambda_{f_1}(u; \xi) = \lambda,$$

$$\lambda_{f_2}(u; \xi) = \gamma^{-m\{B(u) \backslash U(\xi)\}} \leq 1,$$

$$\lambda_{f_3}(u; \xi) = \exp\left\{ \frac{\widehat{d}_u^2 \tau^2}{2(\sigma^2 + \tau^2 \xi_u)\{\sigma^2 + \tau^2(\xi_u + 1)\}} \right\}$$

$$\leq \exp\left\{ \frac{\widehat{d}_u^2 \tau^2}{2\sigma^2(\tau^2 + \sigma^2)} \right\} \text{ and}$$

$$\lambda_{f_4}(u; \xi) = \left\{ \frac{\tau^2 \xi_u + \sigma^2}{\tau^2(\xi_u + 1) + \sigma^2} \right\}^{1/2} \leq 1.$$

By a slight abuse of notation, in the second and third equations above we use u to refer both to the point $\{u\}$ and the location (j, k) at which it is found. The functions $f_1, \ldots, f_4$ are also monotone with respect to the subset relation, so all of the conditions for exact simulation using the method of Section 2.3 are satisfied.

In the spatial processes considered in detail in Section 3.3, the dominating process had constant intensity across the space χ. In the present context, however, it is necessary in practice to use a dominating process which has a different rate at each lattice location, and then use location-specific maxima and minima rather than global maxima and minima. Because we can now use location-specific, rather than global, maxima and minima, we can initialise upper and lower processes that are much closer together than would have been possible with a constant-rate dominating process. This has the consequence of reducing coalescence times to feasible levels. A constant-rate dominating process would not have been feasible due to the size of the global maxima, so this modification to the method of Section 3.3 is essential; see Section 5.3 for details. Chapter 5 of Ambler (2002) gives some other examples of dominating processes with location-specific intensities.

The location-specific rate of the dominating process D is

$$\lambda_{jk}^{dom} = \lambda e^{\widehat{d}_{jk}^2 \tau^2 / 2\sigma^2(\tau^2 + \sigma^2)} \tag{5.1}$$

for each location (j, k) on the lattice. The lower process is then started as a thinned version of D. Points are accepted with probability

$$P(x) = \gamma^{-M(\chi)} \left(\frac{\sigma^2}{\tau^2 + \sigma^2} \right)^{1/2} \times \exp \left\{ -\frac{\widehat{d}_x^2 \tau^2}{2\sigma^2(\tau^2 + \sigma^2)} \right\},$$

where $M(\chi) = \max_\chi[m\{B(x)\}]$. The upper and lower processes are then evolved through time, accepting points as described in Section 2.3 with probability

$$\frac{1}{\lambda_{jk}^{dom}} \lambda_{f_1}(u; \xi^{\mathrm{up}}) \lambda_{f_2}(u; \xi^{\mathrm{up}}) \lambda_{f_3}(u; \xi^{\mathrm{low}}) \lambda_{f_4}(u; \xi^{\mathrm{up}})$$

for the upper process and

$$\frac{1}{\lambda_{jk}^{dom}} \lambda_{f_1}(u; \xi^{\mathrm{low}}) \lambda_{f_2}(u; \xi^{\mathrm{low}}) \lambda_{f_3}(u; \xi^{\mathrm{up}}) \lambda_{f_4}(u; \xi^{\mathrm{low}})$$

for the lower process. The remainder of the algorithm carries over in the obvious way. There are still some issues to be addressed due to very high birth rates in the dominating process, and this will be done in Section 5.3.

5.2 Using the generated samples

Although $\mathbf{d}$ was integrated out for simulation reasons in Section 4.2 it is, naturally, the quantity of interest. Having simulated realisations of $\xi|\widehat{\mathbf{d}}$ we then generate $\mathbf{d}|\xi, \widehat{\mathbf{d}}$ for each realisation ξ generated in the first step. The sample median of $\mathbf{d}|\xi, \widehat{\mathbf{d}}$ gives an estimate for $\mathbf{d}$. The median is used instead of the mean as this gives a thresholding rule, defined by Abramovich et al. (1998) as a rule giving $p(d_{jk} = 0|\widehat{\mathbf{d}}) > 0$.

We calculate $p(\mathbf{d}|\xi, \widehat{\mathbf{d}})$ using logarithms for ease of notation. Assuming that $\xi_{jk} \neq 0$ we find

$$\log p(d_{jk}|\widehat{d}_{jk}, \xi_{jk} \neq 0) = \log p(d_{jk}|\xi_{jk} \neq 0) + \log p(\widehat{d}_{jk}|d_{jk}, \xi_{jk} \neq 0) + C$$

$$= \frac{-d_{jk}^2}{2\tau^2 \xi_{jk}} + \frac{-(\widehat{d}_{jk} - d_{jk})^2}{2\sigma^2} + C_1$$

$$= -\frac{(\sigma^2 + \tau^2 \xi_{jk}) \left(d_{jk} - \frac{\tau^2 \xi_{jk} \widehat{d}_{jk}}{\sigma^2 + \tau^2 \xi_{jk}} \right)^2}{2\sigma^2 \tau^2 \xi_{jk}} + C_2$$

where C, C_1 and C_2 are constants. Thus

$$d_{jk}|\widehat{d}_{jk}, \xi_{jk} \neq 0 \sim N \left(\frac{\tau^2 \xi_{jk} \widehat{d}_{jk}}{\sigma^2 + \tau^2 \xi_{jk}}, \frac{\sigma^2 \tau^2 \xi_{jk}}{\sigma^2 + \tau^2 \xi_{jk}} \right).$$

When $\xi_{jk} = 0$ we clearly have $p(d_{jk}|\xi_{jk}, \widehat{d}_{jk}) = \delta(0)$.

5.3 Dealing with large and small rates

We now deal with some approximations which are necessary to allow our algorithm to be feasible computationally. Recall from equation (5.1) that if the maximum data value d_{jk} is twenty times larger in magnitude than the standard deviation of the noise (a not uncommon event for reasonable noise levels) then we have

$$\lambda_{dom} = \lambda e^{400\sigma^2\tau^2/2\sigma^2(\tau^2+\sigma^2)}$$
$$= \lambda e^{200\tau^2/(\tau^2+\sigma^2)}.$$

Now unless τ is significantly smaller than σ, this will result in enormous birth rates, which make it necessary to modify the algorithm appropriately. To address this issue, we noted that the chances of there being no live points at a location whose data value is large (resulting in a value of λ_{dom} larger than e^4) is sufficiently small that for the purposes of calculating $\lambda_{f_2}(u;\xi)$ for nearby locations it can be assumed that the number of points alive was strictly positive.

This means that we do not know the true value of ξ_{jk} for the locations with the largest values of d_{jk}. This leads to problems since we need to generate d_{jk} from the distribution

$$d_{jk}|\xi_{jk}, \widehat{d}_{jk} \sim N\left(\frac{\tau^2\xi_{jk}\widehat{d}_{jk}}{\sigma^2 + \tau^2\xi_{jk}}, \frac{\sigma^2\tau^2\xi_{jk}}{\sigma^2 + \tau^2\xi_{jk}}\right),$$

which requires values of ξ_{jk} for each location (j,k) in the configuration. To deal with this issue, we first note that, as $\xi_{jk} \to \infty$,

$$\frac{\tau^2\xi_{jk}\widehat{d}_{jk}}{\sigma^2 + \tau^2\xi_{jk}} \longrightarrow \widehat{d}_{jk}$$

monotonically from below, and

$$\frac{\tau^2\xi_{jk}\sigma^2}{\sigma^2 + \tau^2\xi_{jk}} \longrightarrow \sigma^2,$$

also monotonically from below. Since σ is typically small, convergence is very fast indeed. Taking $\tau = \sigma$ as an example we see that even when $\xi_{jk} = 5$ we have

$$\frac{\tau^2\xi_{jk}\widehat{d}_{jk}}{\sigma^2 + \tau^2\xi_{jk}} = \frac{5}{6}\widehat{d}_{jk}$$

and

$$\frac{\tau^2 \xi_{jk} \sigma^2}{\sigma^2 + \tau^2 \xi_{jk}} = \frac{5}{6} \sigma^2.$$

We see that we are already within $\frac{1}{6}$ of the limit. Convergence is even faster for larger values of τ.

We also recall that the dominating process gives an upper bound for the value of ξ_{jk} at every location. Thus a good estimate for d_{jk} would be gained by taking the value of ξ_{jk} in the dominating process for those points where we do not know the exact value. This is a good solution but is unnecessary in some cases, as sometimes the value of λ_{dom} is so large that there is little advantage in using this value. Thus for exceptionally large values of λ_{dom} we simply use $N(\widehat{d}_{jk}, \sigma^2)$ numbers as our estimate of d_{jk}.

5.4 Simulation study

We now present a simulation study of the performance of our estimator relative to several established wavelet-based estimators. Similar to the study of Abramovich et al. (1998), we investigate the performance of our method on the four standard test functions of Donoho and Johnstone (1994, 1995), namely 'Blocks', 'Bumps', 'Doppler' and 'Heavisine'. These test functions are used because they exhibit different kinds of behaviour typical of signals arising in a variety of applications.

The test functions were simulated at 256 points equally spaced on the unit interval. The test signals were centred and scaled so as to have mean value 0 and standard deviation 1. We then added independent $N(0, \sigma^2)$ noise to each of the functions, where σ was taken as $1/10$, $1/7$ and $1/3$. The noise levels then correspond to root signal-to-noise ratios (RSNR) of 10, 7 and 3 respectively. We performed 25 replications. For our method, we simulated 25 independent draws from the posterior distribution of the d_{jk} and used the sample median as our estimate, as this gives a thresholding rule. For each of the runs, σ was set to the standard deviation of the noise we added, τ was set to 1.0, λ was set to 0.05 and γ was set to 3.0.

The values of parameters σ and τ were set to the true values of the standard deviation of the noise and the signal, respectively. In practice it will be necessary to develop some method for estimating these values. The value of λ was chosen to be 0.05 because it was felt that not many

Table 5.1 *Average mean-square errors ($\times 10^4$) for the area-interaction BayesThresh (AIBT), SureShrink (SS), cross-validation (CV), ordinary BayesThresh (BT) and false discovery rate (FDR) estimators for four test functions for three values of the root signal-to-noise ratio. Averages are based on 25 replicates. Standard errors are given in parentheses.*

RSNR	Method	Test functions			
		Blocks	Bumps	Doppler	Heavisine
	AIBT	25 (1)	84 (2)	49 (1)	32 (1)
	SS	49 (2)	131 (6)	54 (2)	66 (2)
10	CV	55 (2)	392 (21)	112 (5)	31 (1)
	BT	344 (10)	1651 (17)	167 (5)	35 (2)
	FDR	159 (14)	449 (17)	145 (5)	64 (3)
	AIBT	56 (3)	185 (5)	87 (3)	52 (2)
	SS	98 (3)	253 (10)	99 (4)	94 (4)
7	CV	96 (3)	441 (25)	135 (6)	54 (3)
	BT	414 (11)	1716 (21)	225 (6)	57 (2)
	FDR	294 (18)	758 (27)	253 (9)	93 (4)
	AIBT	535 (21)	1023 (15)	448 (18)	153 (6)
	SS	482 (13)	973 (45)	399 (14)	147 (3)
3	CV	452 (11)	914 (34)	375 (13)	148 (6)
	BT	860 (24)	2015 (37)	448 (12)	140 (4)
	FDR	1230 (52)	2324 (88)	862 (31)	148 (3)

of the coefficients would be significant. The value of γ was chosen based on small trials for the heavisine and jumpsine datasets.

We compare our method with several established wavelet-based estimators for reconstructing noisy signals: SureShrink (Donoho and Johnstone, 1994), two-fold cross-validation as applied by Nason (1996), ordinary BayesThresh (Abramovich et al., 1998), and the false discovery rate as applied by Abramovich and Benjamini (1996).

For test signals 'Bumps', 'Doppler' and 'Heavisine' we used Daubechies' least asymmetric wavelet of order 10 (Daubechies, 1992). For the 'Blocks' signal we used the Haar wavelet, as the original signal was piecewise constant. The analysis was carried out using the freely available R statistical package. The WaveThresh package (Nason, 1993) was used to perform the discrete wavelet transform and also to compute the SureShrink, cross-validation, BayesThresh and false discovery rate estimators.

The goodness of fit of each estimator was measured by its average mean-square error (AMSE) over the 25 replications. Table 5.1 presents the results. It is clear that our estimator performs extremely well with respect to the other estimators when the signal-to-noise ratio is moderate or large, but less well, though still competitively, when there is a small signal-to-noise ratio.

5.5 Remarks and directions for future work

Our procedure for Bayesian wavelet thresholding has used the naturally clustered nature of the wavelet transform when deciding how much weight to give coefficient values. In comparisons with other methods, our approach performed very well for moderate and low noise levels, and reasonably competitively for higher noise levels.

One possible area for future work would be to replace equation (4.5) with

$$d_{jk}|\xi \sim N(0, \tau^2(\xi_{jk})^z),$$

where z would be a further parameter. This would modify the number of points which are likely to be alive at any given location and thus also modify the tail behaviour of the prior. The idea behind this suggestion is that when we know that the behaviour of the data is either heavy or light tailed, we could adjust z to compensate. This could possibly also help speed up convergence by reducing the number of points at locations with large values of d_{jk}.

A second possible area for future work would be to develop some automatic methods for choosing the parameter values, perhaps using the method of maximum pseudo-likelihood (Besag, 1974, 1975, 1977).

Finally, it would be of obvious interest to find an approach which made the approximations of Section 5.3 unnecessary and allowed for true CFTP to be preserved.

6 Conclusion

This paper, based on Ambler and Silverman (2004a,b), has drawn together a number of themes which demonstrate the way that modern computational statistics has made use of work in applied probability and stochastic processes in ways which would have been inconceivable not many decades ago. It is therefore a particular pleasure to dedicate it to John Kingman on his birthday!

References

Abramovich, F., and Benjamini, Y. 1996. Adaptive thresholding of wavelet coefficients. *Comput. Statist. Data Anal.*, **22**, 351–361.

Abramovich, F., Sapatinas, T., and Silverman, B. W. 1998. Wavelet thresholding via a Bayesian approach. *J. Roy. Statist. Soc. Ser. B*, **60**, 725–749.

Ambler, G. K. 2002. *Dominated Coupling from the Past and Some Extensions of the Area-Interaction Process.* Ph.D. thesis, Department of Mathematics, University of Bristol.

Ambler, G. K., and Silverman, B. W. 2004a. *Perfect Simulation for Bayesian Wavelet Thresholding with Correlated Coefficients.* Tech. rept. Department of Mathematics, University of Bristol. http://arXiv.org/abs/0903.2654v1 [stat.ME].

Ambler, G. K., and Silverman, B. W. 2004b. *Perfect Simulation of Spatial Point Processes using Dominated Coupling from the Past with Application to a Multiscale Area-Interaction Point Process.* Tech. rept. Department of Mathematics, University of Bristol. http://arXiv.org/abs/0903.2651v1 [stat.ME].

Baddeley, A. J., and Lieshout, M. N. M. van. 1995. Area-interaction point processes. *Ann. Inst. Statist. Math.*, **47**, 601–619.

Baddeley, A. J., Turner, R., Møller, J., and Hazelton, M. 2005. Residual analysis for spatial point processes (with Discussion). *J. Roy. Statist. Soc. Ser. B*, **67**, 617–666.

Besag, J. E. 1974. Spatial interaction and the statistical analysis of lattice systems. *J. Roy. Statist. Soc. Ser. B*, **36**, 192–236.

Besag, J. E. 1975. Statistical analysis of non-lattice data. *The Statistician*, **24**, 179–195.

Besag, J. E. 1977. Some methods of statistical analysis for spatial data. *Bull. Int. Statist. Inst.*, **47**, 77–92.

Connor, S. 2007. Perfect sampling. In: Ruggeri, F., Kenett, R., and Faltin, F. (eds), *Encyclopedia of Statistics in Quality and Reliability.* New York: John Wiley & Sons.

Cressie, N. A. C. 1993. *Statistics for Spatial Data.* New York: John Wiley & Sons.

Daubechies, I. 1992. *Ten Lectures on Wavelets.* Philadelphia, PA: SIAM.

Diggle, P. J. 1978. On parameter estimation for spatial point processes. *J. Roy. Statist. Soc. Ser. B*, **40**, 178–181.

Donoho, D. L., and Johnstone, I. M. 1994. Ideal spatial adaption by wavelet shrinkage. *Biometrika*, **81**, 425–455.

Donoho, D. L., and Johnstone, I. M. 1995. Adapting to unknown smoothness via wavelet shrinkage. *J. Amer. Statist. Assoc.*, **90**, 1200–1224.

Green, P. J., and Murdoch, D. J. 1998. Exact sampling for Bayesian inference: towards general purpose algorithms (with discussion). Pages 301–321 of: Bernardo, J. M., Berger, J. O., Dawid, A. P., and Smith, A. F. M. (eds), *Bayesian Statistics 6*. Oxford: Oxford Univ. Press.

Häggström, O., Lieshout, M. N. M. van, and Møller, J. 1999. Characterisation results and Markov chain Monte Carlo algorithms including exact simulation for some spatial point processes. *Bernoulli*, **5**, 641–658.

Jensen, J. L., and Møller, J. 1991. Pseudolikelihood for exponential family models of spatial point processes. *Ann. Appl. Probab.*, **1**, 445–461.

Kelly, F. P., and Ripley, B. D. 1976. A note on Strauss's model for clustering. *Biometrika*, **63**, 357–360.

Kendall, W. S. 1997. On some weighted Boolean models. Pages 105–120 of: Jeulin, D. (ed), *Advances in Theory and Applications of Random Sets*. Singapore: World Scientific.

Kendall, W. S. 1998. Perfect simulation for the area-interaction point process. Pages 218–234 of: Accardi, L., and Heyde, C. C. (eds), *Probability Towards 2000*. New York: Springer-Verlag.

Kendall, W. S., and Møller, J. 2000. Perfect simulation using dominated processes on ordered spaces, with applications to locally stable point processes. *Adv. in Appl. Probab.*, **32**, 844–865.

MacKay, D. J. C. 2003. *Information Theory, Inference, and Learning Algorithms*. Cambridge: Cambridge Univ. Press.

Matheron, G. 1975. *Random Sets and Integral Geometry*. New York: John Wiley & Sons.

Murdoch, D. J., and Green, P. J. 1998. Exact sampling from a continuous state space. *Scand. J. Statist.*, **25**, 483–502.

Nason, G. P. 1993. *The WaveThresh Package: Wavelet Transform and Thresholding Software for S-Plus and R*. Available from Statlib.

Nason, G. P. 1996. Wavelet shrinkage using cross-validation. *J. Roy. Statist. Soc. Ser. B*, **58**, 463–479.

Papangelou, F. 1974. The conditional intensity of general point processes and an application to line processes. *Z. Wahrscheinlichkeitstheorie verw. Geb.*, **28**, 207–226.

Propp, J. G., and Wilson, D. B. 1996. Exact sampling with coupled Markov chains and applications to statistical mechanics. *Random Structures & Algorithms*, **9**, 223–252.

Propp, J. G., and Wilson, D. B. 1998. How to get a perfectly random sample from a generic Markov chain and generate a random spanning tree of a directed graph. *J. Algorithms*, **27**, 170–217.

Ripley, B. D. 1976. The second-order analysis of stationary point processes. *J. Appl. Probab.*, **13**, 255–266.

Ripley, B. D. 1977. Modelling spatial patterns (with Discussion). *J. Roy. Statist. Soc. Ser. B*, **39**, 172–212.

Schladitz, K., and Baddeley, A. J. 2000. A third order point process characteristic. *Scand. J. Statist.*, **27**, 657–671.

Strauss, D. J. 1975. A model for clustering. *Biometrika*, **62**, 467–475.

4

Assessing molecular variability in cancer genomes

Andrew D. Barbour[a] and Simon Tavaré[b]

Abstract

The dynamics of tumour evolution are not well understood. In this paper we provide a statistical framework for evaluating the molecular variation observed in different parts of a colorectal tumour. A multi-sample version of the Ewens Sampling Formula forms the basis for our modelling of the data, and we provide a simulation procedure for use in obtaining reference distributions for the statistics of interest. We also describe the large-sample asymptotics of the joint distributions of the variation observed in different parts of the tumour. While actual data should be evaluated with reference to the simulation procedure, the asymptotics serve to provide theoretical guidelines, for instance with reference to the choice of possible statistics.

AMS subject classification (MSC2010) 92D20; 92D15, 92C50, 60C05, 62E17

1 Introduction

Cancers are thought to develop as clonal expansions from a single transformed, ancestral cell. Large-scale sequencing studies have shown that cancer genomes contain somatic mutations occurring in many genes; cf. Greenman et al. [9], Sjöblom et al. [20], Shah et al. [16]. Many of these

[a] Institut für Mathematik, Universität Zürich, Winterthurerstrasse 190, CH–8057 Zürich, Switzerland; A.D.Barbour@math.uzh.ch
[b] DAMTP and Department of Oncology, Cancer Research UK Cambridge Research Institute, Li Ka Shing Centre, Robinson Way, Cambridge CB2 0RE; st321@cam.ac.uk

mutations are thought to be passenger mutations (those that are not driving the behaviour of the tumour), and some are pathogenic driver mutations that influence the growth of the tumour. The dynamics of tumour evolution are not well understood, in part because serial observation of tumour growth in humans is not possible.

In an attempt to better understand tumour growth and structure, a number of evolutionary approaches have been described. Merlo et al. [15] give an excellent overview of the field. Tsao et al. [21] used non-coding microsatellite loci as molecular tumour clocks in a number of human mutator phenotype colorectal tumours. Stochastic models of tumour growth and statistical inference were used to estimate ancestral features of the tumours, such as their age (defined as the time to loss of mismatch repair). Campbell et al. [4] used deep sequencing of a DNA region to characterise the phylogenetic relationships among clones within patients with B-cell chronic lymphocytic leukaemia. Siegmund et al. [18] used passenger mutations at particular CpG sites to infer aspects of the evolution of colorectal tumours in a number of patients, by examining the methylation patterns in different parts of each tumour.

The problem of comparing the molecular variation present in different parts of a tumour is akin to the following problem from population genetics. Suppose that R observers take samples of sizes $n_1, \ldots, n_R$ from a population, and record the molecular variation seen in each member of their sample. If the population were indeed homogeneous, it makes sense to ask about the relative amount of genetic variation seen in each sample. For example, how many genetic types are seen by all the observers, how many are seen by a single observer, and so on. Ewens et al. [8] discuss this problem in the case of $R = 2$ observers; the methodological contribution of the present paper addresses the case of multiple observers. The theory is used to study the spatial organization of the colorectal tumours studied in Siegmund et al. [18].

This paper is organized as follows. In Section 2 we describe the tumour data that form the motivation for our work. The Ewens Sampling Formula, which forms the basis for our modelling of the data, is described in Section 3, together with a simulation procedure for use in obtaining reference distributions for the statistics of interest. The procedure for testing whether the observers are homogeneous among themselves is illustrated in Section 4. The remainder of the paper is concerned with the large-sample asymptotics of the joint distributions of the allele counts from the different observers. While actual data should be evaluated with reference to the simulation procedure, the asymptotics serve to provide

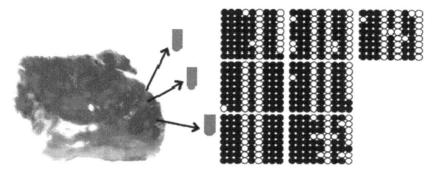

Figure 2.1 Left panel: sampling illustrated from three glands from one side of a colorectal tumour. Each gland contains 2,000–10,000 cells. Right panel: Methylation data from the BGN locus from 7 glands from the left side of Cancer 1 (CNC1, from [18]). 8 cells are sampled from each gland. Each row of 9 circles represents the methylation pattern in a cell. Solid circles denote methylated sites, open circles unmethylated. See Table 2.1 for further details.

theoretical guidelines, for instance with reference to the choice of possible statistics.

2 Colorectal cancer data

In this section we describe the colorectal cancer data that motivate the ensuing work. Yatabe et al. [24] describe an experimental procedure for sampling CpG DNA methylation patterns from cells. These methylation patterns change during cell division, due to random mutational events that result in switching an unmethylated site to a methylated one, or vice versa. The methylation patterns obtained from a particular locus may be represented as strings of binary outcomes, a 1 denoting a methylated site and a 0 an unmethylated one.

Siegmund et al. [18] studied 12 human colorectal tumours, each taken from male patients of known ages. Samples of cells were taken from 7 different glands from each of two sides of each tumour, and the methylation pattern at two neutral (passenger) CpG loci (BGN, 9 sites; and LOC, 14 sites; both are on the X chromosome) was measured in each of 8 cells from each gland. Figure 2.1 illustrates the sampling, and depicts the data from the left side of Cancer 1.

Data obtained from methylation patterns may be compared in several

Table 2.1 *Data for Cancer 1. 13 alleles were observed in the 7 samples. Columns labelled 1–7 give the distribution of the alleles observed in each sample, and column 8 shows the combined data. Data from cancer CNC1 in* [18].

i	1	2	3	4	5	6	7	8
n_i	8	8	8	8	8	8	8	56
K_i	4	2	1	3	3	5	5	13
$\hat{\theta}_i$	2.50	0.49	0.00	1.25	1.25	4.69	4.69	5.01
allele								
1	1			5	5	1	4	16
2	5					3		8
3	1							1
4	1					1		2
5		7	8		2			17
6		1						1
7				2				2
8				1	1		1	3
9						2		2
10						1		1
11							1	1
12							1	1
13							1	1

ways. We focus on the simplest method that considers whether or not cells have the same allele (that is, an identical pattern of 0s and 1s). Here we do not exploit information about the detailed structure of the methylation patterns, for which the reader is referred to [18]. In Table 2.1 we present the data from Cancer 1 shown in Figure 2.1 in a different way. The body of the table shows the numbers of cells of each allele (or type) in each of the 7 samples. The third row of the Table shows the numbers K_i of different alleles seen in each sample. In Table 2.2 we give a similar breakdown for data from the left side of Cancer 2.

The last column in Tables 2.1 and 2.2 gives the combined distribution of allelic variation at this locus in the two tumours. Qualitatively, the two tumours seem to have rather different behaviour: Cancer 1 has far fewer alleles than Cancer 2, and their allocation among the different samples is more homogeneous in Cancer 1 than in Cancer 2. In the next sections we develop some theory that allows us to analyse this variation more carefully.

Table 2.2 *Data for Cancer 2. 27 alleles were observed in the 7 samples. Columns labelled 1–7 give the distribution of the alleles observed in each sample, and column 8 shows the combined data. Data from cancer COC1 in* [18].

i	1	2	3	4	5	6	7	8
n_i	8	8	8	8	8	8	8	56
K_i	7	7	4	3	7	6	4	27
$\hat{\theta}_i$	23.11	23.11	2.50	1.25	23.11	9.23	2.50	19.88
allele								
1	1							1
2	1							1
3	2	1	2	1				6
4	1							1
5	1	1						2
6	1							1
7	1							1
8		1	4	4		3		12
9		1						1
10		2						2
11		1						1
12		1						1
13			1			1		2
14			1					1
15				3				3
16					1			1
17					2		1	3
18					1			1
19					1		1	2
20					1			1
21					1			1
22					1			1
23						1		1
24						1		1
25						1	4	5
26						1		1
27							2	2

3 The Ewens sampling formula

Our focus is on identifying whether the data are consistent with a uniformly mixing collection of tumour cells that are in approximate stasis, or are more typical of patterns of growth such as described in Siegmund et al. [17, 18, 19]. Whatever the model, the basic ingredients that must be specified include how the cells are related, the details of which depend on the demographic model used to describe the tumour evolution, and the mutation process that describes the methylation patterns. A review is provided in Siegmund et al. [19]. We use a simple null model in which the population of cells is assumed to have evolved for some time with an approximately constant, large size of N cells, the constancy of cell numbers mimicking stasis in tumour growth. The mutation model assumes that in each cell division there is probability u of a mutation resulting in a type that has not been seen before—admittedly a crude approximation to the nature of methylation mutations arising in our sample. The mutations are assumed to be neutral, a reasonable assumption given that the BGN gene is expressed in connective tissue but not in the epithelium. Thus our model is a classical one from population genetics, the so-called infinitely-many-neutral-alleles model.

Under this model the distribution of the types observed in the combined data (i.e., the allele counts derived from the right-most columns of data from Tables 2.1 and 2.2) has a distribution that depends on the parameter $\theta = 2Nu$. This distribution is known as the Ewens Sampling Formula [7], denoted by $\mathrm{ESF}(\theta)$, and may be described as follows. For a sample of n cells, we write $(C_1, C_2, \ldots, C_n)$ for the vector of counts given by

$$C_j = \text{number of types represented } j \text{ times in the sample,}$$

where $C_1 + 2C_2 + \cdots + nC_n = n$. For the Cancer 1 sample we have $n = 56$ and

$$C_1 = 6, C_2 = 3, C_3 = 1, C_8 = 1, C_{16} = 1, C_{17} = 1,$$

whereas for Cancer 2 we also have $n = 56$, but

$$C_1 = 17, C_2 = 5, C_3 = 2, C_5 = 1, C_6 = 1, C_{12} = 1.$$

The distribution $\mathrm{ESF}(\theta)$ is given by

$$\mathbf{P}[C_1 = c_1, \ldots, C_n = c_n] = \frac{n!}{\theta_{(n)}} \prod_{j=1}^{n} \left(\frac{\theta}{j}\right)^{c_j} \frac{1}{c_j!}, \qquad (3.1)$$

for $c_1 + 2c_2 + \cdots + nc_n = n$ and $\theta_{(n)} := \theta(\theta+1) \ldots (\theta+n-1)$. An explicit connection between mutations resulting in the ESF and the ancestral history of the individuals (cells) in the sample is provided by Kingman's coalescent [13, 12], and the connection with the infinite population limit is given in Kingman [11, 14].

We recall from [7] that $K = K_n := C_1 + \cdots + C_n$, the number of types in the sample, is a sufficient statistic for θ, and that the maximum-likelihood estimator of θ is the solution of the equation

$$K_n = \mathbf{E}_\theta(K_n) = \sum_{j=1}^{n} \frac{\theta}{\theta + j - 1}. \tag{3.2}$$

The conditional distribution of the counts $C_1, \ldots, C_n$ given K_n does not depend on θ, and thus may be used to assess the goodness-of-fit of the model.

3.1 The multi-observer ESF

So far, we have described the distribution of variation in the entire sample, rather than in each of the subsamples from the different glands separately. The joint law of the counts of different alleles seen in the R glands (that is, by the R observers) is precisely that obtained by taking a hypergeometric sample of sizes $n_1, n_2, \ldots, n_R$ from the n cells in the combined sample. It is a consequence of the consistency property of the ESF that the sample seen by each observer i has its own ESF, with parameters n_i and θ, $i = 1, 2, \ldots, R$. Tables 2.1 and 2.2 give the observed values for the two tumour examples.

We are interested in assessing the goodness-of-fit of the tumour data subsamples to our simple model of a homogeneous tumour in stasis. Because K_n is sufficient for θ in the combined sample, this can be performed by using the joint distribution of the counts seen by each observer, conditional on the value of K_n. To simulate from this distribution we use the Chinese Restaurant Process, as described in the next section.

3.2 The Chinese Restaurant Process

We use simulation to find the distribution of certain test statistics relating to the multiple observer data. To do this we exploit a simple way to simulate a sample of individuals (cells in our example) whose allele counts follow the ESF(θ). The method, known as the Chinese Restaurant Process (CRP), after Diaconis and Pitman [6], simulates individuals in a sample sequentially. The first individual is given type 1. The second individual is either a new type (labelled 2) with probability $\theta/(\theta + 1)$, or a copy of the type of individual 1, with probability $1/(\theta + 1)$. Suppose that $k - 1$ individuals have been assigned types. Individual k is assigned a new type (the lowest unused positive integer) with probability $\theta/(\theta + k - 1)$, or is assigned the type of one of individuals $1, 2, \ldots,$

$k - 1$ selected uniformly at random. Continuing until $k = n$ produces a sample of size n, and the joint distribution of the number of types represented once, twice, ... is indeed $\text{ESF}(\theta)$.

Once the sample of n individuals is generated, it is straightforward to subsample without replacement to obtain R samples, of sizes $n_1, \ldots, n_R$, in each of which the distribution of the allele counts follows the $\text{ESF}(\theta)$ of the appropriate size. This may be done sequentially, choosing n_1 without replacement to be the first sample, then n_2 from the remaining $n - n_1$ to form the second sample, and so on.

When samples of size n are required to have a given number of alleles, say $K_n = k$, this is most easily arranged by the rejection method: the CRP is run to produce an n-sample, and that run is rejected unless the correct value of k is observed. Since conditional on $K_n = k$ the distribution of the allele frequencies is independent of θ, we have freedom to choose θ, which may be taken as the MLE $\hat{\theta}$ determined in (3.2) to make the rejection probability as small as possible.

4 Analysis of the cancer data

We have noted that the combined data in the R-observer ESF have the $\text{ESF}(\theta)$ distribution with sample size $n = n_1 + \cdots + n_R$, while the ith observer's sample has $\text{ESF}(\theta)$ distribution with sample size n_i. Of course, these distributions are not independent. To test whether the combined data are consistent with the ESF, we may use a statistic suggested by Watterson [23], based on the distribution of the sample homozygosity

$$F = \sum_{j=1}^{n} C_j \left(\frac{j}{n} \right)^2$$

found after conditioning on the number of types seen in the combined sample. Each marginal sample may be tested in a similar way using the appropriate value of n.

Since our cancer data arise as the result of a spatial sampling scheme, it is natural to consider statistics that are aimed at testing whether the samples can be assumed homogeneous, that is, are described by the multi-observer ESF. Knowing the answer to this question would aid in understanding the dynamics of tumour evolution, which in turn has implications for understanding metastasis and response to therapy.

To assess this, we use as a simple illustration the sample variance of the numbers of types seen in each sample. The statistic may be written as

$$Q := \frac{1}{R-1} \sum_{i=1}^{R} (K_i - \bar{K})^2 = \frac{1}{R(R-1)} \sum_{1 \le i < j \le R} (K_i - K_j)^2, \quad (4.1)$$

the latter expression emphasizing its role as a measure of the average discrepancy between samples. In the next paragraphs, we discuss the structure of Cancers 1 and 2 using these statistics.

Cancer 1 We begin with a comparison of the data from the two sides of Cancer 1. In this case $n_1 = 56, n_2 = 56$ and the combined sample of $n = 112$ has $K_{112} = 16$ and $F = 0.237$. The 5th and 95th percentiles of the null distribution of F found by the conditional CRP simulation described in the last section are 0.108 and 0.277 respectively, suggesting no anomaly with the underlying ESF model. For the left side of the cancer (Table 2.1), $K_{56} = 13$ and $F = 0.209$, while for the right side (data not shown), $K_{56} = 10$ and $F = 0.293$. In both cases these observed values of F are consistent with the ESF. We then use the statistic Q to investigate whether the data from the 7 glands from the left side of the tumour are homogeneous. We observed $Q = 2.24$, and the null distribution of Q can also be found from the conditional CRP simulation. We obtained 5th and 95th percentiles of 0.29 and 2.48 respectively, supporting the conclusion of a homogeneous tumour.

Cancer 2 The comparison of the two sides of Cancer 2 is more interesting. Once more $n_1 = 56, n_2 = 56$ but the combined sample of $n = 112$ now has $K_{112} = 48$ and $F = 0.081$. The 99th percentile of the null distribution of F is 0.060, suggesting that the ESF model is not adequate to describe the combined data. At first glance the anomaly can be attributed to the data from the right side of the tumour (not shown here), for which $K_{56} = 29$ and $F = 0.105$, far exceeding the 99th percentile of 0.089. For the left side (Table 2.2), $F = 0.083$, just below the 95th percentile of 0.084. Thus the left side seems in aggregate to be adequately described by the ESF model. Further examination of the data from the 7 glands reveals a different story. From the third row of Table 2.2 we calculate $Q = 2.95$, far exceeding the estimated 99th percentile of 2.33. Thus a more detailed view of the way the mutations are shared among the glands shows that these data are indeed inconsistent with the homogeneity expected in the multi-observer ESF.

Of course, many other statistics could have been considered. A natural starting point for constructing them would be the numbers of alleles that are seen only by a specific subset A of the observers, where A ranges over the $2^R - 2$ non-empty proper subsets of the R observers. Such statistics form the basis of the results in Section 5.

Rejection of the null hypothesis of the uniformly mixing homogeneous tumour model can occur for many reasons, for example because of non-uniform mutation rates, different demography of cell growth, non-neutrality of the mutations (which might apply to the BGN locus if in fact it were expressed in tissue in the tumour), and unforeseen effects of the simple mutation model itself. Which of these hypotheses is most likely requires a far more detailed analysis of competing models, as for example outlined in [17, 18, 19].

5 Poisson approximation

In this section, we derive Poisson approximations to the joint distribution of the numbers of alleles that are seen only by specific subsets A of the observers. As mentioned above, functionals of these counts can be used as statistics to test for the homogeneity of (subgroups of) observers. Our approximations come together with bounds on the total variation distance between the actual and approximate distributions. We begin with the case of $R = 2$ observers, and with the statistic $K_1 - K_2$.

5.1 2 observers

We write $C := (C_1, C_2, \ldots)$, where $C_j = 0$ for $j > n$, and recall Watterson's result, that $(C_1, \ldots, C_n)$ are jointly distributed according to $\mathcal{L}(Z_1, Z_2, \ldots, Z_n \mid T_{0n}(Z) = n)$, where $(Z_j, j \geq 1)$ are independent with $Z_i \sim \mathrm{Po}\,(\theta/i)$, and

$$T_{rs}(c) \;=\; \sum_{j=r+1}^{s} j c_j, \qquad c \in \mathbb{Z}_+^{\infty}, \tag{5.1}$$

[22]. The sampled individuals can be labelled 1 or 2, according to which observer sampled them; under the above model, the n_1 1-labels and n_2 2-labels are distributed at random among the individuals, irrespective of their allelic type. Let K_r denote the number of distinct alleles observed by the r-th observer, $r = 1, 2$. Ewens et al. [8] observed that, in the case $n_1 = n_2$ and for large n, $(K_1 - K_2)/\log n$ is equivalent to the difference

in the estimates of the mutation rate made by the two observers. The same is asymptotically true also as n becomes large, if $n_1/n \sim p_1$ for any fixed p_1. This motivates us to look for a distributional approximation to the distribution of the difference $K_1 - K_2$.

Theorem 5.1   *For any n_1, n_2 and b,*

$$d_{TV}(\mathcal{L}(K_1 - K_2), \mathcal{L}(P_1 - P_2)) \leq \frac{kb}{n-1} + \frac{k'\rho^{b+1}}{(b+1)(1-\rho)},$$

for suitable constants k and k', where P_1 and P_2 are independent Poisson random variables having means $\theta \log\{1/(1-p_1)\}$ and $\theta \log\{1/(1-p_2)\}$ respectively, with $p_r := n_r/n$, and where $\rho = \max\{1 - p_1, 1 - p_2\}$. The choice $b = b_n = \lfloor \log n / \log(1/\rho) \rfloor$ gives a bound of order $O\left(\log n/(n \min\{p_1, p_2\})\right)$.

Proof   Group the individuals in the combined sample according to their allelic type, and let M_{js} denote the number of individuals that were observed by observer 1 in the s-th of the C_j groups of size j, the remaining $j - M_{js}$ being observed by observer 2. Define

$$S_j^1 := \sum_{s=1}^{C_j} I[M_{js} = j] \quad \text{and} \quad S_j^2 := \sum_{s=1}^{C_j} I[M_{js} = 0]$$

to be the numbers of j-groups observed only by observers 1 and 2, respectively. Then it follows that

$$K_1 - K_2 = S^1 - S^2,$$

where $S^r := \sum_{j=1}^{n} S_j^r$. The first step in the proof is to show that the effect of the large groups is relatively small.

Note that the probability that an allele which is present j times in the combined sample was *not* observed by observer 1 is

$$\prod_{i=0}^{j-1} \frac{n_1 - i}{n - i} \leq (1 - p_1)^j;$$

similarly, the probability that it was not observed by observer 2 is at most $(1 - p_2)^j$. Hence, conditional on C, the probability that any of the alleles present more than b times in the combined sample is seen by *only one* of the observers is at most

$$\mathbf{E}\left\{\left.\sum_{j=b+1}^{n}(S_j^1 + S_j^2)\right| C\right\} \leq \sum_{j=b+1}^{n} C_j\{(1 - p_1)^j + (1 - p_2)^j\}$$

$$\leq 2\sum_{j=b+1}^{n}\rho^j C_j,$$

whatever the value of b. Hence, writing $U_b := \sum_{j=1}^{b}(S_j^1 - S_j^2)$, we find that

$$\mathbf{P}[K_1 - K_2 \neq U_b] \leq 2\sum_{j=b+1}^{n}\rho^j\mathbf{E}C_j \leq \frac{2k_1\rho^{b+1}}{(b+1)(1-\rho)}, \tag{5.2}$$

where, by Watterson's formula [22] for the means of the component sizes, we can take $k_1 := (2\theta + e^{-1})$ if $n \geq 4(b+1)$ (and $k_1 := \theta$ if $\theta \geq 1$).

To approximate the distribution of U_b, note that, conditional on C, the number of 1-labels among the individuals in allele groups of at most b individuals has a hypergeometric distribution

$$\mathrm{HG}\,(T_{0b}(C); n_1; n),$$

where HG $(s; m; n)$ denotes the number of black balls obtained in s draws from an urn containing m black balls out of a total of n. By Theorem 3.1 of Holmes [10], we have

$$d_{\mathrm{TV}}(\mathrm{HG}\,(T_{0b}(C); n_1; n), \mathrm{Bi}\,(T_{0b}(C), p_1)) \leq \frac{T_{0b}(C) - 1}{n - 1}. \tag{5.3}$$

Hence, conditional on C, the joint distribution of labels among individuals differs in total variation from that obtained by independent Bernoulli random assignments, with label 1 having probability p_1 and label 2 probability p_2, by at most $(T_{0b}(C) - 1)/(n - 1)$.

Now, by Lemma 5.3 of Arratia et al. [2], we also have

$$d_{\mathrm{TV}}(\mathcal{L}(C_1, \ldots, C_b), \mathcal{L}(Z_1, \ldots, Z_b)) \leq \frac{c_\theta b}{n},$$

with $c_\theta \leq 4\theta(\theta + 1)/3$ if $n \geq 4b$. Hence, and from (5.3), it follows that

$$d_{\mathrm{TV}}(\mathcal{L}(C_1, \ldots, C_b; \{M_{js}, 1 \leq s \leq C_j, 1 \leq j \leq b\}),$$
$$\mathcal{L}(Z_1, \ldots, Z_b; \{N_{js}, 1 \leq s \leq Z_j, 1 \leq j \leq b\}))$$
$$\leq \frac{c_\theta b}{n} + \frac{\mathbf{E}(T_{0b}(C)) - 1}{n - 1}, \tag{5.4}$$

where $(N_{js}; s \geq 1, 1 \leq j \leq b)$ are independent of each other and of $Z_1,$

$\ldots, Z_b$, with $N_{js} \sim \mathrm{Bi}(j, p)$. But now the values of the N_{js}, $1 \leq s \leq Z_j$, $1 \leq j \leq b$, can be interpreted as the numbers of 1-labels assigned to each of Z_j groups of size j for each $1 \leq j \leq b$, again under independent Bernoulli random assignments, with label 1 having probability p_1 and label 2 probability p_2. Hence, since the Z_j are independent Poisson random variables, the counts

$$T_j^1 := \sum_{s=1}^{Z_j} I[N_{js} = j] \quad \text{and} \quad T_j^2 := \sum_{s=1}^{Z_j} I[N_{js} = 0]$$

are pairs of independent Poisson distributed random variables, with means $\theta j^{-1} p_1^j$ and $\theta j^{-1} p_2^j$, and are also independent of one another. Hence it follows that

$$V_b := \sum_{j=1}^{b} (T_j^1 - T_j^2) \sim P_{1b} - P_{2b}, \tag{5.5}$$

where P_{1b} and P_{2b} are independent Poisson random variables, with means $\theta \sum_{j=1}^{b} j^{-1} p_1^j$ and $\theta \sum_{j=1}^{b} j^{-1} p_2^j$, respectively. Comparing the definitions of U_b and V_b, and combining (5.4) and (5.5), it thus follows that

$$d_{\mathrm{TV}}(\mathcal{L}(U_b), \mathcal{L}(P_{1b} - P_{2b})) \leq \frac{(c_\theta + k_2)b}{n - 1}, \tag{5.6}$$

with $k_2 = 4\theta/3$ for $n \geq 4b$, once again by Watterson's formula [22].

With (5.2) and (5.6), the argument is all but complete; it simply suffices to observe that, much as in proving (5.2),

$$d_{\mathrm{TV}}(\mathcal{L}(P_1), \mathcal{L}(P_{1b})) + d_{\mathrm{TV}}(\mathcal{L}(P_2), \mathcal{L}(P_{2b})) \leq \frac{2\theta \rho^{b+1}}{(b+1)(1 - \rho)};$$

we take $k := 4 \vee (c_\theta + k_2)$ and $k' := 2(\theta + k_1)$.   □

5.2 *R* observers

The proof of Theorem 5.1 actually shows that the joint distribution of S^1 and S^2, the numbers of types seen respectively by observers 1 and 2 alone, is close to that of independent Poisson random variables P_1 and P_2. For $R \geq 3$ observers, we use a similar approach to derive an approximation to the joint distribution of the numbers of alleles seen by each proper subset A of the R observers.

Suppose that the r-th observer samples n_r individuals, $1 \leq r \leq R$, and set $n := \sum_{r=1}^{R} n_r$, $p_r := n_r/n$. Define the component frequencies in the combined sample as before, and set $M_{js} = m := (m_1, \ldots, m_R)$ if the

r-th observer sees m_r of the j individuals in the s-th of the C_j groups of size j. For any $\emptyset \neq A \subsetneq [R]$, where $[R] := \{1, 2, \ldots, R\}$, define

$$\mathcal{M}_{Aj} :=$$

$$\left\{ m \in \mathbb{Z}_+^R \colon \sum_{r=1}^{R} m_r = j, \ \{r \colon m_r \geq 1\} = A, \ \{r \colon m_r = 0\} = [R] \setminus A \right\},$$

and set

$$S_j^A := \sum_{s=1}^{C_j} I[M_{js} \in A].$$

Our interest lies now in approximating the joint distribution of the counts $(S^A, \emptyset \neq A \subsetneq [R])$, where $S^A := \sum_{j=1}^{n} S_j^A$. To do so, we need a set of independent Poisson random variables $(P^A, \emptyset \neq A \subsetneq [R])$, with $P^A \sim \mathrm{Po}\,(\lambda^A(\theta))$, where

$$\lambda_j^A(\theta) := \frac{\theta}{j}\,\mathrm{MN}\,(j; p_1, \ldots, p_R)\{\mathcal{M}_{Aj}\} \quad \text{and} \quad \lambda^A(\theta) := \sum_{j \geq 1} \lambda_j^A(\theta);$$

$$(5.7)$$

here, $\mathrm{MN}\,(j; p_1, \ldots, p_R)$ denotes the multinomial distribution with j trials and cell probabilities $p_1, \ldots, p_R$.

Theorem 5.2 *In the above setting, we have*

$$d_{TV}\left(\mathcal{L}((S^A, \emptyset \neq A \subsetneq [R])), \bigtimes_{\emptyset \neq A \subsetneq [R]} \mathrm{Po}\,(\lambda^A(\theta))\right)$$

$$\leq \frac{k_R b}{n} + \frac{k_R' \rho^{b+1}}{(b+1)(1-\rho)},$$

where $\rho := \max_{1 \leq r \leq R}(1 - p_r)$. *Again* $b = b_n = \lfloor \log n / \log(1/\rho) \rfloor$ *is a good choice.*

Proof The proof runs much as before. First, the bound

$$\mathbf{E}\left\{ \sum_{j=b+1}^{n} \sum_{\emptyset \neq A \subsetneq [R]} S_j^A \,\Big|\, C \right\} \leq \sum_{j=b+1}^{n} C_j \sum_{r=1}^{R}(1 - p_r)^j \leq R \sum_{j=b+1}^{n} \rho^j C_j$$

shows that

$$\mathbf{P}\left[\bigcup_{\emptyset \neq A \subsetneq [R]} \{S^A \neq S_{(b)}^A\} \right] \leq \frac{R k_1 \rho^{b+1}}{(b+1)(1-\rho)}, \qquad (5.8)$$

where $S^A_{(b)} := \sum_{j=1}^b S^A_j$. Then, by Theorem 4 of Diaconis and Freedman [5],

$$d_{TV}\left(\mathrm{HG}\left(T_{0b}(C); n_1, \ldots, n_R; n\right), \mathrm{MN}\left(T_{0b}(C); p_1, \ldots, p_R\right)\right) \leq \frac{RT_{0b}}{n},$$

from which it follows that

$$d_{TV}\Big(\mathcal{L}(C_1, \ldots, C_b; \{M_{js}, 1 \leq s \leq C_j, 1 \leq j \leq b\}),$$

$$\mathcal{L}(Z_1, \ldots, Z_b; \{N_{js}, 1 \leq s \leq Z_j, 1 \leq j \leq b\})\Big)$$

$$\leq \frac{c_\theta b}{n} + \frac{R\mathbf{E}(T_{0b}(C))}{n}, \tag{5.9}$$

where $(N_{js}; s \geq 1, 1 \leq j \leq b)$ are independent of each other and of $Z_1, \ldots, Z_b$, with $N_{js} \sim \mathrm{MN}(j; p_1, \ldots, p_R)$. Then the random variables

$$T^A_j := \sum_{s=1}^{Z_j} I[N_{js} \in A], \qquad \emptyset \neq A \subsetneq [R],$$

are independent and Poisson distributed, with means $\lambda^A_{(b)}(\theta) := \sum_{j=1}^b \lambda^A_j(\theta)$, and

$$d_{TV}\{\mathcal{L}(S^A_{(b)}, \emptyset \neq A \subsetneq [R]), \mathcal{L}(T^A_{(b)}, \emptyset \neq A \subsetneq [R])\} \leq \frac{k'_2 b}{n},$$

with $k'_2 := c_\theta + 4R\theta/3$, where

$$T^A_{(b)} := \sum_{j=1}^b T^A_j \sim \mathrm{Po}\left(\lambda^A_{(b)}(\theta)\right).$$

Finally, much as before,

$$d_{TV}\left(\mathcal{L}((T^A_{(b)}, \emptyset \neq A \subsetneq [R])), \underset{\emptyset \neq A \subsetneq [R]}{\times} \mathrm{Po}\left(\lambda^A(\theta)\right)\right) \leq \frac{R\theta\rho^{b+1}}{(b+1)(1-\rho)},$$

and we can take $k_R := 4 \vee (c_\theta + Rk_2)$ and $k'_R := R(\theta + k_1)$ in the theorem. $\qquad\square$

We note that the Poisson means $\lambda^A_j(\theta)$ appearing in (5.7) may be calculated using an inclusion-exclusion argument. For reasons of symmetry it is only necessary to compute $\lambda^A_j(\theta)$ for sets of the form $A = [r] = \{1, 2, \ldots, r\}$ for $r = 1, 2, \ldots, R - 1$. We obtain

$$\mathrm{MN}(j; p_1, \ldots, p_R)\{M_{[r]j}\} = \sum_{l=1}^r (-1)^{r-l} \sum_{J \subseteq [r], |J|=l} \left(\sum_{u \in J} p_u\right)^j, \tag{5.10}$$

from which the terms $\lambda^A(\theta)$ readily follow as

$$\lambda^{[r]}(\theta) = -\theta \sum_{l=1}^{r} (-1)^{r-l} \sum_{J \subseteq [r], |J|=l} \log \left(1 - \sum_{u \in J} p_u \right). \qquad (5.11)$$

5.3 The conditional distribution

In statistical applications, such as that discussed above, the value of θ is unknown, and has to be estimated. Defining

$$K_{st}(c) := \sum_{j=s+1}^{t} c_j,$$

the quantity $K_{0n}(C)$ is sufficient for θ, and the null distribution appropriate for testing model fit is then the conditional distribution

$$\mathcal{L}((S^A, \emptyset \neq A \subsetneq [R]) \mid K_{0n}(C) = k),$$

where k is the observed value of $K_{0n}(C)$. Hence we need to approximate this distribution as well. Because of sufficiency, the distribution no longer involves θ. However, for our approximation, we shall need to define means for the approximating Poisson random variables $P^A \sim \text{Po}(\lambda^A(\theta))$, as in (5.7), and these need a value of θ for their definition. We thus take $\lambda^A(\theta_k)$ for our approximation, for convenience with $\theta_k := k/\log n$; the MLE given in (3.2) could equally well have been used.

The proof again runs along the same lines. Supposing that the probabilities $p_1, \ldots, p_R$ are bounded away from 0, we can take $b := b_n := \lfloor \log n / \log(1/\rho) \rfloor$ in Theorem 5.2, and use (5.8) to show that it is enough to approximate $\mathcal{L}((S^A_{(b)}, \emptyset \neq A \subsetneq [R]) \mid K_{0n}(C) = k)$. Then, since the arguments conditional on the whole realization C remain the same when restricting C to the set $\{K_{0n}(C) = k\}$, it is enough to show that the distributions

$$\mathcal{L}(C_{[0,b]} \mid K_{0n}(C) = k) \quad \text{and} \quad \mathcal{L}_{\theta_k}(Z_{[0,b]})$$

are close enough, where $c_{[0,b]} := (c_1, \ldots, c_b)$, to conclude that the Poisson approximation of Theorem 5.2 with $\theta = \theta_k$ also holds conditionally on $\{K_{0n}(C) = k\}$. Note also that the event $\{K_{0n}(C) = k\}$ has probability at least as big as $c_1(\theta_k)k^{-1/2}$ for some positive function $c_1(\cdot)$, by (8.17) of Arratia et al. [2].

Defining $\lambda_{st}(\theta_k) := \sum_{j=s+1}^{t} j^{-1}\theta_k$, we can now prove the key lemma.

Lemma 5.3 *Fix any $\varepsilon, \eta > 0$. Suppose that n is large enough, so that $b + b^3 < n/2$. Then there is a constant κ such that, uniformly for $\varepsilon \leq k/\log n \leq 1/\varepsilon$, and for $c \in \mathbb{Z}_+^\infty$ with $K_{0b}(c) \leq \eta \log \log n$ and $T_{0b}(c) \leq b^{7/2}$,*

$$\left| \frac{\mathbf{P}[C_{[0,b]} = c_{[0,b]} \mid K_{0n}(C) = k]}{\mathbf{P}_{\theta_k}[Z_{[0,b]} = C_{[0,b]}]} - 1 \right| \leq \frac{\kappa \log \log n}{\log n}.$$

Proof Since $\mathcal{L}(C_1, \ldots, C_n) = \mathcal{L}(Z_1, \ldots, Z_n \mid T_{0n}(Z) = n)$, it follows that

$$\mathbf{P}[C_{[0,b]} = c_{[0,b]} \mid K_{0n}(C) = k]$$
$$= \frac{\mathbf{P}[K_{0n}(C) = k \mid C_{[0,b]} = c_{[0,b]}]\mathbf{P}[C_{[0,b]} = c_{[0,b]}]}{\mathbf{P}[K_{0n}(C) = k]}$$
$$= \frac{\mathbf{P}[K_{bn}(C) = k - K_{0b}(c) \mid T_{0b}(C) = T_{0b}(c)]\mathbf{P}[C_{[0,b]} = c_{[0,b]}]}{\mathbf{P}[K_{0n}(C) = k]}.$$

We now use results from §13.10 of Arratia et al. [2]. First, as on p. 323,

$$\mathbf{P}_{\theta_k}[K_{bn}(C) = k - K_{0b}(c) \mid T_{0b}(C) = T_{0b}(c)]$$
$$= \mathbf{P}_{\theta_k}[K_{bn}(Z) = k - K_{0b}(c) \mid T_{bn}(Z) = n - T_{0b}(c)],$$

and the estimate on p. 327 then gives

$$\mathbf{P}_{\theta_k}[K_{bn}(Z) = k - K_{0b}(c) \mid T_{bn}(Z) = n - T_{0b}(c)]$$
$$= \mathrm{Po}\,(\lambda_{bn}(\theta_k))\{k - K_{0b}(c) - 1\}\,\{1 + O((\log n)^{-1} \log \log n)\}, \quad (5.12)$$

uniformly in the chosen ranges of k, $T_{0b}(c)$ and $K_{0b}(c)$, because of the choice $\theta = \theta_k$. Then

$$\mathbf{P}_{\theta_k}[K_{0n}(C) = k] = \mathrm{Po}\,(\lambda_{0n}(\theta_k))\{k - 1\}\,\{1 + O((\log n)^{-1})\}, \quad (5.13)$$

again uniformly in k, $T_{0b}(c)$ and $K_{0b}(c)$, by Theorem 5.4 of Arratia et al. [1]. Finally,

$$\left| \frac{\mathbf{P}_{\theta_k}[C_{[0,b]} = c_{[0,b]}]}{\mathbf{P}_{\theta_k}[Z_{[0,b]} = c_{[0,b]}]} - 1 \right| = \left| \frac{\mathbf{P}_{\theta_k}[T_{bn}(Z) = n - T_{0b}(c)]}{\mathbf{P}_{\theta_k}[T_{bn}(Z) = n]} - 1 \right|$$
$$= O(n^{-1}b^{7/2}),$$

by (4.43), (4.45) and Example 9.4 of [2], if $b + b^3 < n/2$. The lemma now follows by considering the ratio of the Poisson probabilities in (5.12) and (5.13); note that $\lambda_{0n}(\theta_k) - \lambda_{bn}(\theta_k) = O(\log \log n)$. □

In order to deduce the main theorem of this section, we just need to bound the conditional probabilities of the events $\{K_{0b}(C) > \eta \log\log n\}$ and $\{T_{0b}(C) > b^{7/2}\}$, given $K_{0n}(C) = k$. For the first, note that

$$\mathbf{P}_{\theta_k}[K_{0b}(C) > \eta \log\log n] \leq \frac{c_{\theta_k} b}{n} + \mathbf{P}_{\theta_k}[K_{0b}(Z) > \eta \log\log n], \quad (5.14)$$

and that $K_{0b}(Z) \sim \text{Po}\left(\theta_k \sum_{j=1}^{b} j^{-1}\right)$ with mean of order $O(\log\log n)$. Hence there is an η large enough that

$$\mathbf{P}_{\theta_k}[K_{0b}(C) > \eta \log\log n] = O((\log n)^{-5/2}),$$

uniformly in the given range of k. Since also, from (5.13),

$$\mathbf{P}_{\theta_k}[K_{0n}(C) = k] \geq \eta'/\sqrt{\log n}$$

for some $\eta' > 0$, it follows immediately that

$$\mathbf{P}_{\theta_k}[K_{0b}(C) > \eta \log\log n \mid K_{0n}(C) = k] = O((\log n)^{-2}). \quad (5.15)$$

The second inequality is similar. We use the argument of (5.14) to reduce consideration to $\mathbf{P}_{\theta_k}[T_{0b}(Z) > b^{7/2}]$, and (4.44) of Arratia et al. [2] shows that

$$\mathbf{P}_{\theta_k}[T_{0b}(Z) > b^{7/2}] = O(b^{-5/2}) = O((\log n)^{-5/2});$$

the conclusion is now as for (5.15).

In view of these considerations, we have established the following theorem, justifying the Poisson approximation to the conditional distribution of the $(S^A, \emptyset \neq A \subsetneq [R])$, using the estimated value θ_k of θ as parameter.

Theorem 5.4 *For any $0 < \varepsilon < 1$, uniformly in $\varepsilon \leq k/\log n \leq 1/\varepsilon$, we have*

$$d_{TV}\left(\mathcal{L}((S^A, \emptyset \neq A \subsetneq [R]) \mid K_{0n}(C) = k), \underset{\emptyset \neq A \subsetneq [R]}{\times} \text{Po}\left(\lambda^A(\theta_k)\right)\right)$$

$$= O\left(\frac{\log\log n}{\log n}\right).$$

Note that the error bound is much larger for this approximation than those in the previous theorems. However, it is not unreasonable. From (5.8), the joint distribution of the S^A is almost entirely determined by that of $C_1, \ldots, C_b$. Now $\mathcal{L}(K_{0b}(C) \mid K_{0n}(C) = k)$ can be expected to be close to $\mathcal{L}(K_{0b}(Z) \mid K_{0n}(Z) = k)$, which is binomial $\text{Bi}(k, p_{b,n})$, where

$$p_{b,n} := \frac{\sum_{j=1}^{b} 1/j}{\sum_{j=1}^{n} 1/j} \approx \frac{\log b}{\log n} \approx \frac{\log\log n}{\log(1/\rho) \log n}.$$

On the other hand, from Lemma 5.3 of [2], the *unconditional* distribution of $K_{0b}(C)$ is very close to that of $K_{0b}(Z)$, a Poisson distribution. The total variation distance between the distributions Po (kp) and Bi (k, p) is of exact order p if kp is large (Theorem 2 of Barbour and Hall [3]). Since $p_{b,n} \asymp \log \log n / \log n$, an error of this order in Theorem 5.4 is thus in no way surprising.

We can now compute the mean μ of the approximation to the distribution of Q, as used in Section 4, obtained by using Theorem 5.4. We begin by noting that, using the theorem,

$$
K_r - K_s = \sum_{A:\ r \in A, s \notin A} S^A - \sum_{A:\ r \notin A, s \in A} S^A
$$

is close in distribution to

$$
\widehat{K}_{rs} - \widehat{K}_{sr} := \sum_{A:\ r \in A, s \notin A} P^A - \sum_{A:\ r \notin A, s \in A} P^A,
$$

where $P^A \sim \text{Po}\,(\lambda^A(\theta_k))$, $\emptyset \neq A \subsetneq [R]$, are independent. To compute the means

$$
\lambda_{rs} := \sum_{A:\ r \in A, s \notin A} \lambda^A(\theta_k) \quad \text{and} \quad \lambda_{sr} := \sum_{A:\ r \notin A, s \in A} \lambda^A(\theta_k)
$$

of $\widehat{K}_{rs}$ and $\widehat{K}_{sr}$, we note that

$$
\sum_{A:\ r \in A, s \notin A} \text{MN}\,(j; p_1, \ldots, p_R)\{\mathcal{M}_{Aj}\}
$$
$$
= (1 - p_s)^j \{1 - (1 - p_r/(1 - p_s))^j\}
$$
$$
= (1 - p_s)^j - (1 - p_r - p_s)^j,
$$

the probability under the multinomial scheme that the r-th cell is non-empty but the s-th cell is empty. Thus

$$
\lambda_{rs} = \sum_{j \geq 1} \frac{\theta_k}{j} \{(1 - p_s)^j - (1 - p_r - p_s)^j\} = \theta_k \log((p_r + p_s)/p_s),
$$

and $\lambda_{sr} = \theta_k \log((p_r + p_s)/p_r)$. Then, because $\widehat{K}_{rs}$ and $\widehat{K}_{sr}$ are independent and Poisson distributed,

$$
\mathbf{E}\{(\widehat{K}_{rs} - \widehat{K}_{sr})^2\} = (\lambda_{rs} - \lambda_{sr})^2 + \lambda_{rs} + \lambda_{sr}.
$$

This yields the formula

$$\mu := \frac{1}{R(R-1)} \sum_{1 \le r < s \le R} \left\{ \theta_k^2 \{\log(p_r/p_s)\}^2 + \theta_k \log\left(\frac{(p_r + p_s)^2}{p_r p_s} \right) \right\}.$$

(5.16)

In particular, if $p_r = 1/R$ for $1 \le r \le R$, then $\mu = \theta_k \log 2$, agreeing with the observation of Ewens et al. [8] in the case $R = 2$.

6 Conclusion

Our paper is about ancestral inference (albeit in a somatic cell setting rather than the typical population genetics one) and Poisson approximation. John Kingman has made fundamental and far-reaching contributions to both areas. It therefore gives us great pleasure to dedicate it to John on his birthday.

Acknowledgements ST acknowledges the support of the University of Cambridge, Cancer Research UK and Hutchison Whampoa Limited. ADB was supported in part by Schweizer Nationalfonds Projekt Nr. 20–117625/1.

References

[1] Arratia, R., Barbour, A. D., and Tavaré, S. 2000. The number of components in a logarithmic combinatorial structure. *Ann. Appl. Probab.*, **10**, 331–361.

[2] Arratia, R., Barbour, A. D., and Tavaré, S. 2003. *Logarithmic Combinatorial Structures: A Probabilistic Approach*. EMS Monogr. Math., vol. 1. Zürich: Eur. Math. Soc.

[3] Barbour, A. D., and Hall, P. G. 1984. On the rate of Poisson convergence. *Math. Proc. Cambridge Philos. Soc.*, **95**, 473–480.

[4] Campbell, P. J., Pleasance, E. D., Stephens, P. J., Dicks, E., Rance, R., Goodhead, I., Follows, G. A., Green, A. R., Futreal, P. A., and Stratton, M. R. 2008. Subclonal phylogenetic structures in cancer revealed by ultra-deep sequencing. *Proc. Natl. Acad. Sci. USA*, **105**, 13081–13086.

[5] Diaconis, P., and Freedman, D. 1980. Finite exchangeable sequences. *Ann. Probab.*, **8**, 745–764.

[6] Diaconis, P., and Pitman, J. 1986. *Permutations, Record Values and Random Measures*. Unpublished lecture notes, Statistics Department, University of California, Berkeley.

[7] Ewens, W. J. 1972. The sampling theory of selectively neutral alleles. *Theor. Population Biology*, **3**, 87–112.

[8] Ewens, W. J., RoyChoudhury, A., Lewontin, R. C., and Wiuf, C. 2007. Two variance results in population genetics theory. *Math. Popul. Stud.*, **14**, 1–18.

[9] Greenman, C., Wooster, R., Futreal, P. A., Stratton, M. R., and Easton, D. F. 2006. Statistical analysis of pathogenicity of somatic mutations in cancer. *Genetics*, **173**, 2187–2198.

[10] Holmes, S. 2004. Stein's method for birth and death chains. Pages 45–67 of: Diaconis, P., and Holmes, S. (eds), *Stein's Method: Expository Lectures and Applications*. IMS Lecture Notes Monogr. Ser., vol. 46. Beachwood, OH: Inst. Math. Statist.

[11] Kingman, J. F. C. 1982a. The coalescent. *Stochastic Process. Appl.*, **13**, 235–248.

[12] Kingman, J. F. C. 1982b. Exchangeability and the evolution of large populations. Pages 97–112 of: Koch, G., and Spizzichino, F. (eds), *Exchangeability in Probability and Statistics*. Amsterdam: North-Holland.

[13] Kingman, J. F. C. 1982c. On the genealogy of large populations. *J. Appl. Probab.*, **19A**, 27–43.

[14] Kingman, J. F. C. 1993. *Poisson Processes*. Oxford Studies in Probability, vol. 3. Oxford: Oxford University Press.

[15] Merlo, L. M. F., Pepper, J. W., Reid, B. J., and Maley, C. C. 2006. Cancer as an evolutionary and ecological process. *Nature Reviews Cancer*, **6**, 924–935.

[16] Shah, S. P., Morin, R. D., Khattra, J., Prentice, L., Pugh, T., Burleigh, A., Delaney, A., Gelmon, K., Guliany, R., Senz, J., Steidl, C., Holt, R. A., Jones, S., Sun, M., Leung, G., Moore, R., Severson, T., Taylor, G. A., Teschendorff, A. E., Tse, K., Turashvili, G., Varhol, R., Warren, R. L., Watson, P., Zhao, Y., Caldas, C., Huntsman, D., Hirst, M., Marra, M. A., and Aparicio, S. 2009. Mutational evolution in a lobular breast tumour profiled at single nucleotide resolution. *Nature*, **461**, 809–813.

[17] Siegmund, K. D., Marjoram, P., and Shibata, D. 2008. Modeling DNA methylation in a population of cancer cells. *Stat. Appl. Genet. Mol. Biol.*, **7**, Article 18.

[18] Siegmund, K. D., Marjoram, P., Woo, Y-J., Tavaré, S., and Shibata, D. 2009a. Inferring clonal expansion and cancer stem cell dynamics from dna methylation patterns in colorectal cancers. *Proc. Natl. Acad. Sci. USA*, **106**, 4828–4833.

[19] Siegmund, K. D., Marjoram, P., Tavaré, S., and Shibata, D. 2009b. Many colorectal cancers are "flat" clonal expansions. *Cell Cycle*, **8**, 2187–2193.

[20] Sjöblom, T., Jones, S., Wood, L. D., Parsons, D. W., Lin, J., Barber, T. D., Mandelker, D., Leary, R. J., Ptak, J., Silliman, N., Szabo, S., Buckhaults, P., Farrell, C., Meeh, P., Markowitz, S. D., Willis, J., Dawson, D., Willson, J. K. V., Gazdar, A. F., Hartigan, J., Wu, L., Liu, C., Parmigiani, G., Park, B. H., Bachman, K. E., Papadopoulos, N., Vogelstein, B., Kinzler, K. W., and Velculescu, V. E. 2006. The consensus coding sequences of human breast and colorectal cancers. *Science*, **314**, 268–274.

[21] Tsao, J. L., Yatabe, Y., Salovaara, R., Järvinen, H. J., Mecklin, J. P., Aaltonen, L. A., Tavaré, S., and Shibata, D. 2000. Genetic reconstruction of individual colorectal tumor histories. *Proc. Natl. Acad. Sci. USA*, **97**, 1236–1241.

[22] Watterson, G. A. 1974. The sampling theory of selectively neutral alleles. *Adv. in Appl. Probab.*, **6**, 463–488.

[23] Watterson, G. A. 1978. The homozygosity test of neutrality. *Genetics*, **88**, 405–417.

[24] Yatabe, Y., Tavaré, S., and Shibata, D. 2001. Investigating stem cells in human colon by using methylation patterns. *Proc. Natl. Acad. Sci. USA*, **98**, 10839–10844.

5

Branching out

J. D. Biggins[a]

Abstract

Results on the behaviour of the rightmost particle in the nth generation in the branching random walk are reviewed and the phenomenon of anomalous spreading speeds, noticed recently in related deterministic models, is considered. The relationship between such results and certain coupled reaction-diffusion equations is indicated.

AMS subject classification (MSC2010) 60J80

1 Introduction

I arrived at the University of Oxford in the autumn of 1973 for post-graduate study. My intention at that point was to work in Statistics. The first year of study was a mixture of taught courses and designated reading on three areas (Statistics, Probability, and Functional Analysis, in my case) in the ratio 2:1:1 and a dissertation on the main area. As part of the Probability component, I attended a graduate course that was an exposition, by its author, of the material in Hammersley (1974), which had grown out of his contribution to the discussion of John's invited paper on subadditive ergodic theory (Kingman, 1973). A key point of Hammersley's contribution was that the postulates used did not cover the time to the first birth in the nth generation in a Bellman–Harris process.[1] Hammersley (1974) showed, among other things, that these

---

a Department of Probability & Statistics, Hicks Building, University of Sheffield, Sheffield S3 7RH; J.Biggins@sheffield.ac.uk
[1] Subsequently, Liggett (1985) established the theorem under weaker postulates.

quantities did indeed exhibit the anticipated limit behaviour in probability. I decided not to be examined on this course, which was I believe a wise decision, but I was intrigued by the material. That interest turned out to be critical a few months later.

By the end of the academic year I had concluded that I wanted to pursue research in Probability rather than Statistics and asked to have John as supervisor. He agreed. Some time later we met and he asked me whether I had any particular interests already—I mentioned Hammersley's lectures. When I met him he was in the middle of preparing something (which I could see, but not read upside down). He had what seemed to be a pile of written pages, a part written page and a pile of blank paper. There was nothing else on the desk. A few days later a photocopy of a handwritten version of Kingman (1975), essentially identical to the published version, appeared in my pigeon-hole with the annotation "the multitype version is an obvious problem"—I am sure this document was what he was writing when I saw him. (Like all reminiscences, this what I recall, but it is not necessarily what happened.) This set me going. For the next two years, it was a privilege to have John as my thesis supervisor. He supplied exactly what I needed at the time: an initial sense of direction, a strong encouragement to independence, an occasional nudge on the tiller about what did or did not seem tractable, the discipline of explaining orally what I had done, and a ready source on what was known, and where to look for it. However, though important, none of these get to the heart of the matter, which is that I am particularly grateful to have had that period of contact with, and opportunity to appreciate first-hand, such a gifted mathematician.

Kingman (1975) considered the problem Hammersley had raised in its own right, rather than as an example of, and adjunct to, the general theory of subadditive processes. Here, I will say something about some recent significant developments on the first-birth problem. I will also go back to my beginnings, by outlining something new about the multitype version that concerns the phenomenon of 'anomalous spreading speeds', which was noted in a related context in Weinberger et al. (2007). Certain martingales were deployed in Kingman (1975). These have been a fruitful topic in their own right, and have probably received more attention since then than the first-birth problem itself (see Alsmeyer and Iksanov (2009) for a recent nice contribution on when these martingales are integrable). However, those developments will be ignored here.

2 The basic model

The branching random walk (BRW) starts with a single particle located at the origin. This particle produces daughter particles, which are scattered in $\mathbb{R}$, to give the first generation. These first generation particles produce their own daughter particles similarly to give the second generation, and so on. Formally, each family is described by the collection of points in $\mathbb{R}$ giving the positions of the daughters relative to the parent. Multiple points are allowed, so that in a family there may be several daughter particles born in the same place. As usual in branching processes, the nth generation particles reproduce independently of each other. The process is assumed supercritical, so that the expected family size exceeds one (but need not be finite—indeed even the family size itself need not be finite). Let $\mathbf{P}$ and $\mathbf{E}$ be the probability and expectation for this process and let Z be the generic reproduction process of points in $\mathbb{R}$. Thus, $\mathbf{E}Z$ is the intensity measure of Z and $Z(\mathbb{R})$ is the family size, which will also be written as N. The assumption that the process is supercritical becomes that $\mathbf{E}Z(\mathbb{R}) = \mathbf{E}N > 1$. To avoid burdening the description with qualifications about the survival set, let $\mathbf{P}(N = 0) = 0$, so that the process survives almost surely.

The model includes several others. One is when each daughter receives an independent displacement, another is when all daughters receive the same displacement, with the distribution of the displacement being independent of family size in both cases. These will be called the BRW with *independent* and *common* displacements respectively. Obviously, in both of these any line of descent follows a trajectory of a random walk. (It is possible to consider an intermediate case, where displacements have these properties conditional on family size, but that is not often done.) Since family size and displacements are independent, these two processes can be coupled in a way that shows that results for one will readily yield results for the other. In a common displacement BRW imagine each particle occupying the (common) position of its family. Then the process becomes an independent displacement BRW, with a random origin given by the displacement of the first family, and its nth generation occupies the same positions as the $(n+1)$th generation in the original common displacement BRW. Really this just treats each family as a single particle.

In a different direction, the points of Z can be confined to $(0, \infty)$ and interpreted as the mother's age at the birth of that daughter: the framework adopted in Kingman (1975). Then the process is the general branching process associated with the names of Ryan, Crump, Mode and Jagers. Finally, when all daughters receive the same positive displacement with a distribution independent of family size the process is the Bellman–Harris branching process: the framework adopted in Hammersley (1974).

There are other 'traditions', which consider the BRW but introduce and describe it rather differently and usually with other problems in focus. There is a long tradition phrased in terms of 'multiplicative cascades' (see for example Liu (2000) and the references there) and a rather shorter one phrased in terms of 'weighted branching' (see for example Alsmeyer and Rösler (2006) and the references there). The model has arisen in one form or another in a variety of areas. The most obvious is as a model for a population spreading through an homogeneous habitat. It has also arisen in modelling random fractals (Peyrière, 2000) commonly in the language of multiplicative cascades, in the theoretical study of algorithms (Mahmoud, 1992), in a problem in group theory (Abért and Virág, 2005) and as an ersatz for both lattice-based models of spin glasses in physics (Koukiou, 1997) and a number theory problem (Lagarias and Weiss, 1992).

3 Spreading out: old results

Let $Z^{(n)}$ be the positions occupied by the nth generation and $B^{(n)}$ its rightmost point, so that

$$B^{(n)} = \sup\{z : z \text{ a point of } Z^{(n)}\}.$$

One can equally well consider the leftmost particle, and the earliest studies did that. Reflection of the whole process around the origin shows the two are equivalent: all discussion here will be expressed in terms of the rightmost particle. The first result, stated in a moment, concerns $B^{(n)}/n$ converging to a constant, Γ, which can reasonably be interpreted as the speed of spread in the positive direction.

A critical role in the theory is played by the Laplace transform of the intensity measure $\mathbf{E}Z$: let $\kappa(\phi) = \log \int e^{\phi z} \mathbf{E}Z(dz)$ for $\phi \geq 0$ and $\kappa(\phi) = \infty$ for $\phi < 0$. It is easy to see that when this is finite for some $\phi > 0$ the intensity measures of Z and $Z^{(n)}$ are finite on bounded sets, and decay exponentially in their right tail. The behaviour of the leftmost particle is governed by the behaviour of the transform for negative values of its argument. The definition of κ discards these, which simplifies later formulations by automatically keeping attention on the right tail. In order to give one of the key formulae for Γ and for later explanation, let κ^* be the Fenchel dual of κ, which is the convex function given by

$$\kappa^*(a) = \sup_{\theta}\{\theta a - \kappa(\theta)\}. \tag{3.1}$$

This is sufficient notation to give the first result.

Theorem 3.1 *When there is a $\phi > 0$ such that*

$$\kappa(\phi) < \infty, \tag{3.2}$$

there is a constant Γ such that

$$\frac{B^{(n)}}{n} \to \Gamma \quad a.s. \tag{3.3}$$

and $\Gamma = \sup\{a : \kappa^*(a) < 0\} = \inf\{\kappa(\theta)/\theta : \theta\}$.

This result was proved for the common BRW with only negative displacements with convergence in probability in Hammersley (1974, Theorem 2). It was proved in Kingman (1975, Theorem 5) for Z concentrated on $(-\infty, 0)$ and with $0 < \kappa(\phi) < \infty$ instead of (3.2). The result stated above is contained in Biggins (1976a, Theorem 4), which covers the irreducible multitype case also, of which more later. The second of the formulae for Γ is certainly well-known but cannot be found in the papers mentioned—I am not sure where it first occurs. It is not hard to establish from the first one using the definition and properties of κ^*.

The developments described here draw on features of transform theory, to give properties of κ, and of convexity theory, to give properties of κ^* and the speed Γ. There are many presentations of, and notations for, these, tailored to the particular problem under consideration. In this review, results will simply be asserted. The first of these provides a context for the next theorem and aids interpretation of $\sup\{a : \kappa^*(a) < 0\}$ in the previous one. It is that when κ is finite somewhere on $(0, \infty)$, κ^* is an increasing, convex function, which is continuous from the left, with minimum value $-\kappa(0) = -\log \mathbf{E}N$, which is less than zero.

A slight change in focus derives Theorem 3.1 from the asymptotics of the numbers of particles in suitable half-infinite intervals. As part of the derivation of this the asymptotics of the expected numbers are obtained. Specifically, it is shown that when (3.2) holds

$$n^{-1}\log\left(\mathbf{E}Z^{(n)}[na, \infty)\right) \to -\kappa^*(a)$$

(except, possibly, at one a). The trivial observation that when the expectation of integer-valued variables decays geometrically the variables themselves must ultimately be zero implies that $\log Z^{(n)}[na, \infty)$ is ultimately infinite on $\{a : \kappa^*(a) > 0\}$. This motivates introducing a notation for sweeping positive values of κ^*, and later other functions, to infinity and so we let

$$f^\circ(a) = \begin{cases} f(a) & \text{when } f(a) \le 0 \\ \infty & \text{when } f(a) > 0 \end{cases} \qquad (3.4)$$

and $\kappa^\circledast = (\kappa^*)^\circ$. The next result can be construed as saying that in crude asymptotic terms this is the only way actual numbers differ from their expectation.

Theorem 3.2   *When* (3.2) *holds*,

$$\frac{1}{n} \log \left(Z^{(n)}[na, \infty) \right) \to -\kappa^\circledast(a) \quad a.s., \qquad (3.5)$$

for all $a \ne \Gamma$.

From this result, which is Biggins (1977a, Theorem 2), and the properties of $\kappa^\circledast$, Theorem 3.1 follows directly.

A closely related continuous-time model arises when the temporal development is a Markov branching process (Bellman–Harris with exponential lifetimes) or even a Yule process (binary splitting too) and movement is Brownian, giving binary branching Brownian motion. The process starts with a single particle at the origin, which then moves with a Brownian motion with variance parameter V. This particle splits in two at rate λ, and the two particles continue, independently, in the same way from the splitting point. (Any discrete skeleton of this process is a branching random walk.)

Now, let $B^{(t)}$ be the position of the rightmost particle at time t. Then $u(x, t) = \mathbf{P}(B^{(t)} \le x)$ satisfies the (Fisher/Kolmogorov–Petrovski–Piscounov) equation

$$\frac{\partial u}{\partial t} = V \frac{1}{2} \frac{\partial^2 u}{\partial x^2} - \lambda u(1 - u), \qquad (3.6)$$

which is easy to see informally by conditioning on what happens in $[0, \delta t]$. The deep studies of Bramson (1978a, 1983) show, among other things, that (with $V = \lambda = 1$) $B^{(t)}$ converges in distribution when centred on its median and that median is (to $O(1)$)

$$\sqrt{2}t - \frac{1}{\sqrt{2}} \left(\frac{3}{2} \log t \right),$$

which implies that $\Gamma = \sqrt{2}$ here. For the skeleton at integer times, $\kappa(\theta) = \theta^2/2 + 1$ for $\theta \geq 0$, and using Theorem 3.1 on this confirms that $\Gamma = \sqrt{2}$. Furthermore, for later reference, note that $\theta\Gamma - \kappa(\theta) = 0$ when $\theta = \sqrt{2}$.

Theorem 3.1 is for discrete time, counted by generation. There are corresponding results for continuous time, where the reproduction is now governed by a random collection of points in time and space ($\mathbb{R}^+ \times \mathbb{R}$). The first component gives the mother's age at the birth of this daughter and the second that daughter's position relative to her mother. Then the development in time of the process is that of a general branching process rather than the Galton–Watson development that underpins Theorem 3.1. This extension is discussed in Biggins (1995) and Biggins (1997). In it particles may also move during their lifetime and then branching Brownian motion becomes a (very) special case. Furthermore, there are also natural versions of Theorems 3.1 and 3.2 when particle positions are in $\mathbb{R}^d$ rather than $\mathbb{R}$—see Biggins (1995, §4.2) and references there.

4 Spreading out: first refinements

Obviously rate-of-convergence questions follow on from (3.3). An aside in Biggins (1977b, p33) noted that, typically, $B^{(n)} - n\Gamma$ goes to $-\infty$. The following result on this is from Biggins (1998, Theorem 3), and much of it is contained also in Liu (1998, Lemma 7.2). When $\mathbf{P}(Z(\Gamma, \infty) > 0) > 0$, so displacements greater than Γ are possible, and (3.2) holds, there is a finite $\vartheta > 0$ with $\vartheta\Gamma - \kappa(\vartheta) = 0$. Thus the condition here, which will recur in later theorems, is not restrictive.

Theorem 4.1   *If there is a finite $\vartheta > 0$ with $\vartheta\Gamma - \kappa(\vartheta) = 0$, then*

$$B^{(n)} - n\Gamma \to -\infty \quad a.s., \tag{4.1}$$

and the condition is also necessary when $\mathbf{P}(Z(\Gamma, \infty) > 0) > 0$.

The theorem leaves some loose ends when $\mathbf{P}(Z(\Gamma, \infty) = 0) = 1$. Then $B^{(n)} - n\Gamma$ is a decreasing sequence, and so it does have a limit, but whether (4.1) holds or not is really the explosion (i.e. regularity) problem for the general branching process: whether, with a point z from Z corresponding to a birth time of $\Gamma - z$, there can be an infinite number of births in a finite time. This is known to be complex—see Grey (1974) for example. In the simpler cases it is properties of $Z(\{\Gamma\})$, the number of daughters displaced by exactly Γ, that matters.

If $Z(\{\Gamma\})$ is the family size of a surviving branching process (so either $\mathbf{E}Z(\{\Gamma\}) > 1$ or $\mathbf{P}(Z(\{\Gamma\}) = 1) = 1$) it is easy to show that $(B^{(n)} - n\Gamma)$ has a finite limit—so (4.1) fails—using embedded surviving processes resulting from focusing on daughters displaced by Γ: see Biggins (1976b, Proposition II.5.2) or Dekking and Host (1991, Theorem 1). In a similar vein, with extra conditions, Addario-Berry and Reed (2009, Theorem 4) show $\mathbf{E}(B^{(n)} - n\Gamma)$ is bounded.

Suppose now that (3.2) holds. When $\mathbf{P}(Z(a, \infty) = 0) = 1$, simple properties of transforms imply that $\theta a - \kappa(\theta) \uparrow - \log \mathbf{E}Z(\{a\})$ as $\theta \uparrow \infty$. Then, when $\mathbf{E}Z(\{a\}) < 1$ a little convexity theory shows that $\Gamma < a$ and that there is a finite ϑ with $\vartheta\Gamma - \kappa(\vartheta) = 0$, so that Theorem 4.1 applies. This leaves the case where (3.2) holds, $\mathbf{P}(Z(\Gamma, \infty) = 0) = 1$ and $\mathbf{E}Z(\{\Gamma\}) = 1$ but $\mathbf{P}(Z(\{\Gamma\}) = 1) < 1$, which is sometimes called, misleadingly in my opinion, the *critical* branching random walk because the process of daughters displaced by exactly Γ from their parent forms a critical Galton–Watson process. For this case, Bramson (1978b, Theorem 1) and Dekking and Host (1991, §9) show that (4.1) holds under extra conditions including that displacements lie in a lattice, and that the convergence is at rate $\log \log n$. Bramson (1978b, Theorem 2) also gives conditions under which (4.1) fails.

5 Spreading out: recent refinements

The challenge to derive analogues for the branching random walk of the fine results for branching Brownian motion has been open for a long time. Progress was made in McDiarmid (1995) and, very recently, a nice result has been given in Hu and Shi (2009, Theorem 1.2), under reasonably mild conditions. Here is its translation into the current notation. It shows that the numerical identifications noted in the branching Brownian motion case in §3 are general.

Theorem 5.1 *Suppose that there is a $\vartheta > 0$ with $\vartheta\Gamma - \kappa(\vartheta) = 0$, and that, for some $\epsilon > 0$, $\mathbf{E}(N^{1+\epsilon}) < \infty$, $\kappa(\vartheta + \epsilon) < \infty$ and $\int e^{-\epsilon z}\mathbf{E}Z(dz) < \infty$. Then*

$$-\frac{3}{2} = \liminf_n \frac{\vartheta(B^{(n)} - n\Gamma)}{\log n} < \limsup_n \frac{\vartheta(B^{(n)} - n\Gamma)}{\log n} = -\frac{1}{2} \quad a.s.$$

and

$$\frac{\vartheta(B^{(n)} - n\Gamma)}{\log n} \to -\frac{3}{2} \quad \text{in probability.}$$

Good progress has also been made on the tightness of the distributions of $B^{(n)}$ when centred suitably. Here is a recent result from Bramson and Zeitouni (2009, Theorem 1.1).

Theorem 5.2 *Suppose the BRW has independent or common displacements according to the random variable X. Suppose also that for some $\epsilon > 0$, $\mathbf{E}(N^{1+\epsilon}) < \infty$ and that for some $\psi > 0$ and $y_0 > 0$*

$$\mathbf{P}(X > x + y) \le e^{-\psi y}\mathbf{P}(X > x) \qquad \forall x > 0, y > y_0. \tag{5.1}$$

Then the distributions of $\{B^{(n)}\}$ are tight when centred on their medians.

It is worth noting that (5.1) ensures that (3.2) holds for all $\phi \in [0, \psi)$. There are other results too—in particular, McDiarmid (1995, Theorem 1) and Dekking and Host (1991, §3) both give tightness results for the (general) BRW, but with Z concentrated on a half-line. Though rather old for this section, Dekking and Host (1991, Theorem 2) is worth recording here: the authors assume the BRW is concentrated on a lattice, but they do not use that in the proof of this theorem. To state it, let $\widetilde{D}$ be the second largest point in Z when $N \ge 2$ and the only point otherwise.

Theorem 5.3 *If the points of Z are confined to $(-\infty, 0]$ and $\mathbf{E}\widetilde{D}$ is finite, then $\mathbf{E}B^{(n)}$ is finite and the distributions of $\{B^{(n)}\}$ are tight when centred on their expectations.*

The condition that $\mathbf{E}\widetilde{D}$ is finite holds when $\int e^{\phi z}\mathbf{E}Z(dz)$ is finite in a neighbourhood of the origin, which is contained within the conditions in Theorem 5.1. In another recent study Addario-Berry and Reed (2009, Theorem 3) give the following result, which gives tightness and also estimates the centring.

Theorem 5.4 *Suppose that there is a $\vartheta > 0$ with $\vartheta \Gamma - \kappa(\vartheta) = 0$, and that, for some $\epsilon > 0$, $\kappa(\vartheta + \epsilon) < \infty$ and $\int e^{-\epsilon z}\mathbf{E}Z(dz) < \infty$. Suppose also that the BRW has a finite maximum family size and independent displacements. Then*

$$\mathbf{E}B^{(n)} = n\Gamma - \frac{3}{2\vartheta}\log n + O(1),$$

and there are $C > 0$ and $\delta > 0$ such that

$$\mathbf{P}\left(|B^{(n)} - \mathbf{E}B^{(n)}| > x\right) \le Ce^{-\delta x} \qquad \forall x.$$

The conditions in the first sentence here have been stated in a way that keeps them close to those in Theorem 5.1 rather than specialising them for independent displacements. Now, moving from tightness to convergence in distribution—which cannot be expected to hold without a non-lattice assumption—the following result, which has quite restrictive conditions, is taken from Bachmann (2000, Theorem 1).

Theorem 5.5   *Suppose that the BRW has* $\mathbf{E}N < \infty$ *and independent displacements according to a random variable with density function f where* $-\log f$ *is convex. Then the variables* $B^{(n)}$ *converge in distribution when centred on medians.*

It is not hard to use the coupling mentioned in §2 to see that Theorems 5.4 and 5.5 imply that these two results also hold for common displacements.

6   Deterministic theory

There is another, deterministic, stream of work concerned with modelling the spatial spread of populations in a homogeneous habitat, and closely linked to the study of reaction-diffusion equations like (3.6). The main presentation is Weinberger (1982), with a formulation that has much in common with that adopted in Hammersley (1974). Here the description of the framework is pared-down. This sketch draws heavily on Weinberger (2002), specialised to the homogeneous (i.e. aperiodic) case and one spatial dimension. The aim is to say enough to make certain connections with the BRW.

Let $u^{(n)}(x)$ be the density of the population (or the gene frequency, in an alternative interpretation) at time n and position $x \in \mathbb{R}$. This is a discrete-time theory, so there is an updating operator Q satisfying $u^{(n+1)} = Q(u^{(n)})$. More formally, let $\mathcal{F}$ be the non-negative continuous functions on $\mathbb{R}$ bounded by β. Then Q maps $\mathcal{F}$ into itself and $u^{(n)} = Q^{(n)}(u^{(0)})$, where $u^{(0)}$ is the initial density and $Q^{(n)}$ is the nth iterate of Q. The operator is to satisfy the following restrictions. The constant functions at 0 and at β are both fixed points of Q. For any function $u \in \mathcal{F}$ that is not zero everywhere, $Q^{(n)}(u) \to \beta$, and $Q(\alpha) \geq \alpha$ for non-zero constant functions in $\mathcal{F}$. (Of course, without the spatial component, this is all reminiscent of the basic properties of the generating function of the family-size.) The operator Q is order-preserving, in that if $u \leq v$ then $Q(u) \leq Q(v)$, so increasing the population anywhere never has deleterious effects in the future; it is also translation-invariant, because the habitat is homogeneous, and suitably continuous. Finally, every sequence $u_m \in \mathcal{F}$ contains a subsequence $u_{m(i)}$ such that $Q(u_{m(i)})$ converges uniformly on compacts. Such a Q can be obtained by taking the development of a reaction-diffusion equation for a time τ. Then $Q^{(n)}$ gives the development to time $n\tau$, and the results for this discrete formulation transfer to such equations.

Specialising Weinberger (2002, Theorem 2.1), there is a spreading speed Γ in the following sense. If $u^{(0)}(x) = 0$ for $x \geq L$ and $u^{(0)}(x) \geq \delta > 0$ for all $x \leq K$, then for any $\epsilon > 0$

$$\sup_{x \geq n(\Gamma + \epsilon)} |u^{(n)}(x)| \to 0 \quad \text{and} \quad \sup_{x \leq n(\Gamma - \epsilon)} |u^{(n)}(x) - \beta| \to 0. \tag{6.1}$$

In some cases the spreading speed can be computed through linearisation —see Weinberger (2002, Corollary 2.1) and Lui (1989a, Corollary to Theorem 3.5)—in that the speed is the same as that obtained by replacing Q by a truncation of its linearisation at the zero function. So $Q(u) = Mu$ for small u and $Q(u)$ is replaced by $\min\{\omega, Mu\}$, where ω is a constant, positive function with $M\omega > \omega$. The linear functional $Mu(y)$ must be represented as an integral with respect to some measure, and so, using the translation invariance of M, there is a measure μ such that

$$Mu(y) = \int u(y - z)\,\mu(dz). \tag{6.2}$$

Let $\tilde{\kappa}(\theta) = \log \int e^{\theta z} \mu(dz)$. Then the results show that the speed Γ in (6.1) is given by

$$\Gamma = \inf_{\theta > 0} \frac{\tilde{\kappa}(\theta)}{\theta}. \tag{6.3}$$

Formally, this is one of the formulae for the speed in Theorem 3.1. In fact, the two frameworks can be linked, as indicated next.

In the BRW, suppose the generic reproduction process Z has points $\{z_i\}$. Define Q by

$$Q(u(x)) = 1 - \mathbf{E}\left[1 - \prod_i u(x - z_i)\right].$$

This has the general form described above with $\beta = 1$. On taking $u^{(0)}(x) = \mathbf{P}(B^{(0)} > x)$ (i.e. Heaviside initial data) it is easily established by induction that $u^{(n)}(x) = \mathbf{P}(B^{(n)} > x)$. This is in essence the same as the observation that the distribution of the rightmost particle in branching Brownian motion satisfies the differential equation (3.6). The idea is explored in the spatial spread of the 'deterministic simple epidemic' in Mollison and Daniels (1993), a continuous-time model which, like branching Brownian motion, has BRW as its discrete skeleton. Now Theorem 3.1 implies that (6.1) holds, and that, for Q obtained in this way, the speed is indeed given by the (truncated) linear approximation. The other theorems about $B^{(n)}$ also translate into results about such Q. For example, Theorem 5.5 gives conditions for $u^{(n)}$ when centred suitably to converge to a fixed (travelling wave) profile.

7 The multitype case

Particles now have types drawn from a finite set, $\mathcal{S}$, and their reproduction is defined by random points in $\mathcal{S} \times \mathbb{R}$. The distribution of these points depends on the parent's type. The first component gives the daughter's type and the second component gives the daughter's position, relative to the parent's. As previously, Z is the generic reproduction process, but now let Z_σ be the points (in $\mathbb{R}$) corresponding to those of type σ; $Z^{(n)}$ and Z_σ^n are defined similarly. Let $\mathbf{P}_\nu$ and $\mathbf{E}_\nu$ be the probability and expectation associated with reproduction from an initial ancestor with type $\nu \in \mathcal{S}$. Let $B_\sigma^{(n)}$ be the rightmost particle of type σ in the nth generation, and let $B^{(n)}$ be the rightmost of these, which is consistent with the one-type notation.

The type space can be classified, using the relationship 'can have a descendant of this type', or, equivalently, using the non-negative expected family-size matrix, $\mathbf{E}_\nu Z_\sigma(\mathbb{R})$. Two types are in the same class when each can have a descendant of the other type in some generation. When there is a single class the family-size matrix is *irreducible* and the process is similarly described. When the expected family-size matrix is *aperiodic* (i.e. primitive) the process is also called aperiodic, and it is supercritical when this matrix has Perron–Frobenius (i.e. non-negative and of maximum modulus) eigenvalue greater than one. Again, to avoid qualifications about the survival set, assume extinction is impossible from the starting type used.

For $\theta \geq 0$, let $\exp(\kappa(\theta))$ be the Perron–Frobenius eigenvalue of the matrix of transforms $\int e^{\theta z} \mathbf{E}_\nu Z_\sigma(dz)$, and let $\kappa(\theta) = \infty$ for $\theta < 0$. If there is just one type, this definition agrees with that of κ at the start of §3. The following result, which is Biggins (1976a, Theorem 4), has been mentioned already.

Theorem 7.1 *Theorem 3.1 holds for any initial type in a supercritical irreducible BRW.*

The simplest multitype version of Theorem 3.2 is the following, which is proved in Biggins (2009). When $\sigma = \nu$ it is a special case of results indicated in Biggins (1997, §4.1).

Theorem 7.2 *For a supercritical aperiodic BRW for which* (3.2) *holds,*

$$\frac{1}{n} \log \left(Z_\sigma^{(n)}[na, \infty) \right) \to -\kappa^\bullet(a) \quad a.s.\text{-}\mathbf{P}_\nu \tag{7.1}$$

for $a \neq \sup\{a : \kappa^*(a) < 0\} = \Gamma$, *and*

$$\frac{B_\sigma^{(n)}}{n} \to \Gamma \quad a.s.\text{-}\mathbf{P}_\nu.$$

Again there is a deterministic theory, following the pattern described in §6 and discussed in Lui (1989a,b), which can be related to Theorem 7.1. Recent developments in that area raise some interesting questions that are the subject of the next two sections.

8 Anomalous spreading

In the multitype version of the deterministic context of §6, recent papers (Weinberger et al., 2002, 2007; Lewis et al., 2002; Li et al., 2005) have considered what happens when the type space is reducible. Rather than set out the framework in its generality, the simplest possible case, the reducible two-type case, will be considered here, for the principal issue can be illustrated through it. The two types will be ν and η. Now, the vector-valued non-negative function $u^{(n)}$ gives the population density of two species—the two types, ν and η—at $x \in \mathbb{R}$ at time n, and Q models growth, interaction and migration, as the populations develop in discrete time. The programme is the same as that indicated in §6, that is to investigate the existence of spreading speeds and when these speeds can be obtained from the truncated linear approximation.

In this case the approximating linear operator, generalising that given in (6.2), is

$$(Mu(y))_\eta = \int u_\eta(y - z)\, \mu_{\eta\eta}(dz),$$

$$(Mu(y))_\nu = \int u_\nu(y - z)\, \mu_{\nu\nu}(dz) + \int u_\eta(y - z)\, \mu_{\nu\eta}(dz).$$

Simplifying even further, assume there is no spatial spread associated with the 'interaction' term here, so that $\int u_\eta(y - z)\, \mu_{\nu\eta}(dz) = cu_\eta(y)$

for some $c > 0$. The absence of $\mu_{\eta\nu}$ in the first of these makes the linear approximation reducible. The first equation is really just for the type η and so will have the speed that corresponds to $\mu_{\eta\eta}$, given through its transform by (6.3), and written Γ_η. In the second, on ignoring the interaction term, it is plausible that the speed must be at least that of type ν alone, which corresponds to $\mu_{\nu\nu}$ and is written Γ_ν. However, it can also have the speed of u_η from the 'interaction' term. It is claimed in Weinberger et al. (2002, Lemma 2.3) that when Q is replaced by the approximating operator $\min\{\omega, Mu\}$ this does behave as just outlined, with the corresponding formulae for the speeds: thus that of η is Γ_η and that for ν is $\max\{\Gamma_\eta, \Gamma_\nu\}$. However, in Weinberger et al. (2007) a flaw in the argument is noted, and an example is given where the speed of ν in the truncated linear approximation can be faster than this, the anomalous spreading speed of their title, though the actual speed is not identified. The relevance of the phenomenon to a biological example is explored in Weinberger et al. (2007, §5).

As in §6, the BRW provides some particular examples of Q that fall within the general scope of the deterministic theory. Specifically, suppose the generic reproduction process Z has points $\{\sigma_i, z_i\} \in \mathcal{S} \times \mathbb{R}$. Now let Q, which operates on vector functions indexed by the type space $\mathcal{S}$, be defined by

$$Q\left(u(x)\right)_\nu = 1 - \mathbf{E}_\nu\left[1 - \prod_i u_{\sigma_i}(x - z_i)\right].$$

Then, just as in the one-type case, when $u_\nu^{(0)}(x) = \mathbf{P}_\nu(B^{(0)} > x)$ induction establishes that $u_\nu^{(n)}(x) = \mathbf{P}_\nu\left(B^{(n)} > x\right)$. It is perhaps worth noting that in the BRW the index ν is the starting type, whereas it is the 'current' type in Weinberger et al. (2007). However, this makes no formal difference.

Thus, the anomalous spreading phenomenon should be manifest in the BRW, and, given the more restrictive framework, it should be possible to pin down the actual speed there, and hence for the corresponding Q with Heaviside initial data. This is indeed possible. Here the discussion stays with the simplifications already used in looking at the deterministic results.

Consider a two-type BRW in which each type ν always produces at least one daughter of type ν, on average produces more than one, and can produces daughters of type η—but type η never produce daughters of type ν. Also for $\theta \geq 0$ let

$$\kappa_\nu(\theta) = \log \int e^{\theta z} \mathbf{E}_\nu Z_\nu(dz) \quad \text{and} \quad \kappa_\eta(\theta) = \log \int e^{\theta z} \mathbf{E}_\eta Z_\eta(dz)$$

and let these be infinite for $\theta < 0$. Thus Theorem 3.2 applies to type ν considered alone to show that

$$\frac{1}{n} \log \left(Z_\nu^{(n)}[na, \infty) \right) \to -\kappa_\nu^{\circledast}(a) \quad \text{a.s.-}\mathbf{P}_\nu.$$

It turns out that this estimation of numbers is critical in establishing the speed for type η. It is possible for the growth in numbers of type ν, through the numbers of type η they produce, to increase the speed of type η from that of a population without type ν. This is most obvious if type η is subcritical, so that any line of descent from a type η is finite, for the only way they can then spread is through the 'forcing' from type ν. However, if in addition the dispersal distribution at reproduction for η has a much heavier tail than that for ν it is now possible for type η to spread faster than type ν.

For any two functions f and g, let $\mathfrak{C}[f, g]$ be the greatest (lower semi-continuous) convex function beneath both of them. The following result is a very special case of those proved in Biggins (2009). The formula given in the next result for the speed $\Gamma^\dagger$ is the same as that given in Weinberger et al. (2007, Proposition 4.1) as the upper bound on the speed of the truncated linear approximation.

Theorem 8.1 *Suppose that* $\max\{\kappa_\nu(\phi_\nu), \kappa_\eta(\phi_\eta)\}$ *is finite for some* $\phi_\eta \geq \phi_\nu > 0$ *and that*

$$\int e^{\theta z} \mathbf{E}_\nu Z_\eta(dz) < \infty \quad \forall \theta \geq 0. \tag{8.1}$$

Let $r = \mathfrak{C}[\kappa_\nu^{\circledast}, \kappa_\eta^*]^\circ$. *Then*

$$\frac{1}{n} \log \left(Z_\eta^{(n)}[na, \infty) \right) \to -r(a) \quad \text{a.s.-}\mathbf{P}_\nu, \tag{8.2}$$

for $a \neq \sup\{a : r(a) < 0\} = \Gamma^\dagger$, *and*

$$\frac{B_\eta^{(n)}}{n} \to \Gamma^\dagger. \quad \text{a.s.-}\mathbf{P}_\nu. \tag{8.3}$$

Furthermore,

$$\Gamma^\dagger = \inf_{0 < \varphi \leq \theta} \max \left\{ \frac{\kappa_\nu(\varphi)}{\varphi}, \frac{\kappa_\eta(\theta)}{\theta} \right\}. \tag{8.4}$$

From this result it is possible to see how $\Gamma^\dagger$ can be anomalous. Suppose that $r(\Gamma^\dagger) = 0$, so that $\Gamma^\dagger$ is the speed, and that r is strictly below both $\kappa_\nu^\bullet$ and κ_η^* at $\Gamma^\dagger$. This will occur when the minimum of the two convex functions κ_ν^* and κ_η^* is not convex at $\Gamma^\dagger$, and then the largest convex function below both will be linear there. In these circumstances, $\kappa_\nu^*(\Gamma^\dagger) > 0$, which implies that $\kappa_\nu^\bullet(\Gamma^\dagger) = \infty$, and $\kappa_\eta^*(\Gamma^\dagger) > 0$. Thus $\Gamma^\dagger$ will be strictly greater than both Γ_ν and Γ_η, giving a 'super-speed'— Figure 8.1 illustrates a case that will soon be described fully where Γ_ν and Γ_η are equal and $\Gamma^\dagger$ exceeds them. Otherwise, that is when $\Gamma^\dagger$ is not in a linear portion of r, $\Gamma^\dagger$ is just the maximum of Γ_ν and Γ_η.

The example in Weinberger et al. (2007) that illustrated anomalous speed was derived from coupled reaction-diffusion equations. When there is a branching interpretation, which it must be said will be the exception not the rule, the actual speed can be identified through Theorem 8.1 and its generalisations. This will now be illustrated with an example. Suppose type η particles form a binary branching Brownian motion, with variance parameter and splitting rate both one. Suppose type ν particles form a branching Brownian motion, but with variance parameter V, splitting rate λ and, on splitting, type ν particles produce a (random) family of particles of both types. There are $1 + N_\nu$ of type ν and N_η of type η, so that the family always contains at least one daughter of type ν; the corresponding bivariate probability generating function is $\mathbf{E}a^{1+N_\nu}b^{N_\eta} = af(a, b)$. Let $v(x, t) = \mathbf{P}_\eta(B^{(t)} \le x)$ and $w(x, t) = \mathbf{P}_\nu(B^{(t)} \le x)$. These satisfy

$$\frac{\partial v}{\partial t} = \frac{1}{2}\frac{\partial^2 v}{\partial x^2} - v(1 - v),$$
$$\frac{\partial w}{\partial t} = V\frac{1}{2}\frac{\partial^2 w}{\partial x^2} - \lambda w(1 - f(w, v)).$$

Here, when the initial ancestor is of type ν and at the origin the initial data are $w(x, 0) = 1$ for $x \ge 0$ and 0 otherwise and $v(x, 0) \equiv 0$. Note that, by a simple change of variable, these can be rewritten as equations in $\mathbf{P}_\eta(B^{(t)} > x)$ and $\mathbf{P}_\nu(B^{(t)} > x)$ where the differential parts are unchanged, but the other terms look rather different.

Now suppose that $af(a, b) = a^2(1 - p + pb)$, so that a type ν particle always splits into two type ν and with probability p also produces one type η. Looking at the discrete skeleton at integer times, $\kappa_\nu(\theta) = V\theta^2/2 + \lambda$

for $\theta \geq 0$, giving

$$\kappa_\nu^*(a) = \begin{cases} -\lambda & a < 0 \\ -\lambda + \dfrac{1}{2}\dfrac{a^2}{V} & a \geq 0 \end{cases}$$

and speed $(2V\lambda)^{1/2}$, obtained by solving $\kappa_\nu^*(a) = 0$. The formulae for κ_η^* are just the special case with $V = \lambda = 1$. Now, for convenience, take $V = \lambda^{-1}$, so that both types, considered alone, have the same speed. Then, sweeping positive values to infinity,

$$\kappa_\nu^\bullet(a) = \begin{cases} -\lambda & a < 0, \\ -\lambda\left(1 - \dfrac{a^2}{2}\right) & a \in [0, 2^{1/2}], \\ \infty & a > 2^{1/2}. \end{cases}$$

Now $\mathfrak{C}[\kappa_\nu^\bullet, \kappa_\eta^*]$ is the largest convex function below this and κ_η^*. When $\lambda = 3$ these three functions are drawn in Figure 8.1.

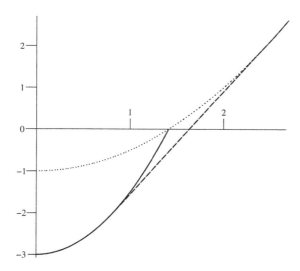

Key: solid — $\kappa_\nu^\bullet$ (with $\lambda = 3$); dotted — κ_η^*; dashed — $\mathfrak{C}[\kappa_\nu^\bullet, \kappa_\eta^*]$.
Speeds: $\Gamma_\nu = \Gamma_\eta = \sqrt{2} = 1.41\ldots$; $\Gamma^\dagger = 4/\sqrt{6} = 1.63\ldots$.

Figure 8.1 Illustration of how anomalous speed arises.

The point where each of them meets the horizontal axis gives the value of speed for that function. Thus, $\Gamma^\dagger$ exceeds the other two, which are both $\sqrt{2}$. Here $\Gamma^\dagger = 4/\sqrt{6}$. In general, for $\lambda > 1$, it is $(1+\lambda)/\sqrt{2\lambda}$, which can be made arbitrarily large by increasing λ sufficiently.

9 Discussion of anomalous spreading

The critical function in Theorem 8.1 is $r = \mathfrak{C}[\kappa_\nu^\circledast, \kappa_\eta^*]^\circ$. Here is how it arises. The function $\kappa_\nu^\circledast$ describes the growth in numbers and spread of the type ν. Conditional on these, $\mathfrak{C}[\kappa_\nu^\circledast, \kappa_\eta^*]$ describes the growth and spread in expectation of those of type η. To see why this might be so, take a b with $\kappa_\nu^\circledast(b) < 0$ so that (3.5) describes the exponential growth of $Z_\nu^{(m)}[mb, \infty)$: there are roughly $\exp(-m\kappa_\nu^\circledast(b))$ such particles in generation m. Suppose now, for simplicity, that each of these produces a single particle of type η at the parent's position. As noted just before Theorem 3.2, the expected numbers of type η particles in generation r and in $[rc, \infty)$ descended from a single type η at the origin is roughly $\exp(-r\kappa_\eta^*(c))$. Take $\lambda \in (0,1)$ with $m = \lambda n$ and $r = (1 - \lambda)n$. Then, conditional on the development of the first m generations, the expectation of the numbers of type η in generation n and to the right of $mb + rc = n(\lambda b + (1 - \lambda)c)$ will be (roughly) at least $\exp(-n(\lambda\kappa_\nu^\circledast(b)+(1-\lambda)\kappa_\eta^*(c)))$. As b, c and λ vary with $\lambda b+(1-\lambda)c = a$, the least value for $\lambda\kappa_\nu^\circledast(b) + (1 - \lambda)\kappa_\eta^*(c)$ is given by $\mathfrak{C}[\kappa_\nu^\circledast, \kappa_\eta^*](a)$. There is some more work to do to show that this lower bound on the conditional expected numbers is also an upper bound—it is here that (8.1) comes into play. Finally, as indicated just before Theorem 3.2, this corresponds to actual numbers only when negative, so the positive values of this convex minorant are swept to infinity.

When the speed is anomalous, this indicative description of how $r = \mathfrak{C}[\kappa_\nu^\circledast, \kappa_\eta^*]^\circ$ arises makes plausible the following description of lines of descent with speed near $\Gamma^\dagger$. They will arise as a 'dog-leg', with the first portion of the trajectory, which is a fixed proportion of the whole, being a line of descent of type η with a speed less than Γ_η. The remainder is a line of descent of type ν, with a speed faster than Γ_μ (and also than $\Gamma^\dagger$).

Without the truncation, the linear operator approximating (near $u \equiv 1$) a Q associated with a BRW describes the development of its expected numbers, and so it is tempting to define the speed using this, by looking at when expected numbers start to decay. In the irreducible case,

Theorem 7.2 has an analogue for expected numbers, that

$$\frac{1}{n}\log\left(\mathbf{E}_\nu Z_\sigma^{(n)}[na,\infty)\right) \to -\kappa^*(a),$$

and so here the speed can indeed be found by looking at when expected numbers start to decay. In contrast, in the set up in Theorem 8.1

$$\frac{1}{n}\log\left(\mathbf{E}_\nu Z_\eta^{(n)}[na,\infty)\right) \to -\mathfrak{C}[\kappa_\nu^*,\kappa_\eta^*](a),$$

and the limit here can be lower than $\mathfrak{C}[\kappa_\nu^\circledast,\kappa_\eta^*]$—the distinction between the functions is whether or not positive values are swept to infinity in the first argument. Hence the speed computed by simply asking when expectations start to decay can be too large. In Figure 8.1, $\mathfrak{C}[\kappa_\nu^\circledast,\kappa_\eta^*]$ is the same as $\mathfrak{C}[\kappa_\nu^*,\kappa_\eta^*]$, but it is easy to see, reversing the roles of κ_ν^* and κ_η^*, that $\mathfrak{C}[\kappa_\eta^\circledast,\kappa_\nu^*]$ is the same as κ_ν^*. Thus if η could produce ν, rather than the other way round, expectations would still give the speed $\Gamma^\dagger$ but the true speed would be $\Gamma_\nu (= \Gamma_\eta)$.

The general case, with many classes, introduces a number of additional challenges (mathematical as well as notational). It is discussed in Biggins (2009). The matrix of transforms now has irreducible blocks on its diagonal, corresponding to the classes, and their Perron–Frobenius eigenvalues supply the κ for each class, as would be anticipated from §7. Here a flavour of some of the other complications. The rather strong condition (8.1) means that the spatial distribution of type η daughters to a type ν mother is irrelevant to the form of the result. If convergence is assumed only for some $\theta > 0$ rather than all this need not remain true. One part of the challenge is to describe when these 'off-diagonal' terms remain irrelevant; another is to say what happens when they are not. If there are various routes through the classes from the initial type to the one of interest these possibilities must be combined: in these circumstances, the function r in (8.2) need not be convex (though it will be increasing). It turns out that the formula for $\Gamma^\dagger$, which seems as if it might be particular to the case of two classes, extends fully—not only in the sense that there is a version that involves more classes, but also in the sense that the speed can usually be obtained as the maximum of that obtained using (8.4) for all pairs of classes where the first can have descendants in the second (though the line of descent may have to go through other classes on the way).

References

Abért, M., and Virág, B. 2005. Dimension and randomness in groups acting on rooted trees. *J. Amer. Math. Soc.*, **18**(1), 157–192.

Addario-Berry, L., and Reed, B. 2009. Minima in branching random walks. *Ann. Probab.*, **37**(3), 1044–1079.

Alsmeyer, G., and Iksanov, A. 2009. A log-type moment result for perpetuities and its application to martingales in supercritical branching random walks. *Electron. J. Probab.*, **14**(10), 289–312.

Alsmeyer, G., and Rösler, U. 2006. A stochastic fixed point equation related to weighted branching with deterministic weights. *Electron. J. Probab.*, **11**(2), 27–56.

Bachmann, M. 2000. Limit theorems for the minimal position in a branching random walk with independent logconcave displacements. *Adv. in Appl. Probab.*, **32**(1), 159–176.

Biggins, J. D. 1976a. The first- and last-birth problems for a multitype age-dependent branching process. *Adv. in Appl. Probab.*, **8**(3), 446–459.

Biggins, J. D. 1976b. *Asymptotic Properties of the Branching Random Walk.* D.Phil thesis, University of Oxford.

Biggins, J. D. 1977a. Chernoff's theorem in the branching random walk. *J. Appl. Probab.*, **14**(3), 630–636.

Biggins, J. D. 1977b. Martingale convergence in the branching random walk. *J. Appl. Probab.*, **14**(1), 25–37.

Biggins, J. D. 1995. The growth and spread of the general branching random walk. *Ann. Appl. Probab.*, **5**(4), 1008–1024.

Biggins, J. D. 1997. How fast does a general branching random walk spread? Pages 19–39 of: Athreya, K. B., and Jagers, P. (eds), *Classical and Modern Branching Processes (Minneapolis, MN, 1994)*. IMA Vol. Math. Appl., vol. 84. New York: Springer-Verlag.

Biggins, J. D. 1998. Lindley-type equations in the branching random walk. *Stochastic Process. Appl.*, **75**(1), 105–133.

Biggins, J. D. 2009. *Spreading Speeds in Reducible Multitype Branching Random Walk.* Submitted.

Bramson, M. D. 1978a. Maximal displacement of branching Brownian motion. *Comm. Pure Appl. Math.*, **31**(5), 531–581.

Bramson, M. D. 1978b. Minimal displacement of branching random walk. *Z. Wahrscheinlichkeitstheorie verw. Gebiete*, **45**(2), 89–108.

Bramson, M. D. 1983. Convergence of solutions of the Kolmogorov equation to travelling waves. *Mem. Amer. Math. Soc.*, **44**(285), iv+190.

Bramson, M. D., and Zeitouni, O. 2009. Tightness for a family of recursion equations. *Ann. Probab.*, **37**(2), 615–653.

Dekking, F. M., and Host, B. 1991. Limit distributions for minimal displacement of branching random walks. *Probab. Theory Related Fields*, **90**(3), 403–426.

Grey, D. R. 1974. Explosiveness of age-dependent branching processes. *Z. Wahrscheinlichkeitstheorie verw. Gebiete*, **28**, 129–137.

Hammersley, J. M. 1974. Postulates for subadditive processes. *Ann. Probab.*, **2**, 652–680.

Hu, Y., and Shi, Z. 2009. Minimal position and critical martingale convergence in branching random walk and directed polymers on disordered trees. *Ann. Probab.*, **37**(2), 742–789.

Kingman, J. F. C. 1973. Subadditive ergodic theory. *Ann. Probab.*, **1**, 883–909. With discussion and response.

Kingman, J. F. C. 1975. The first birth problem for an age-dependent branching process. *Ann. Probab.*, **3**(5), 790–801.

Koukiou, F. 1997. Directed polymers in random media and spin glass models on trees. Pages 171–179 of: Athreya, K. B., and Jagers, P. (eds), *Classical and Modern Branching Processes (Minneapolis, MN, 1994)*. IMA Vol. Math. Appl., vol. 84. New York: Springer-Verlag.

Lagarias, J. C., and Weiss, A. 1992. The $3x+1$ problem: two stochastic models. *Ann. Appl. Probab.*, **2**(1), 229–261.

Lewis, M. A., Li, B., and Weinberger, H. F. 2002. Spreading speed and linear determinacy for two-species competition models. *J. Math. Biol.*, **45**(3), 219–233.

Li, B., Weinberger, H. F., and Lewis, M. A. 2005. Spreading speeds as slowest wave speeds for cooperative systems. *Math. Biosci.*, **196**(1), 82–98.

Liggett, T. M. 1985. An improved subadditive ergodic theorem. *Ann. Probab.*, **13**(4), 1279–1285.

Liu, Q. 1998. Fixed points of a generalized smoothing transformation and applications to the branching random walk. *Adv. in Appl. Probab.*, **30**(1), 85–112.

Liu, Q. 2000. On generalized multiplicative cascades. *Stochastic Process. Appl.*, **86**(2), 263–286.

Lui, R. 1989a. Biological growth and spread modeled by systems of recursions, I: mathematical theory. *Math. Biosci.*, **93**(2), 269–295.

Lui, R. 1989b. Biological growth and spread modeled by systems of recursions, II: biological theory. *Math. Biosci.*, **93**(2), 297–311.

Mahmoud, H. M. 1992. *Evolution of Random Search Trees*. Wiley-Intersci. Ser. Discrete Math. Optim. New York: John Wiley & Sons.

McDiarmid, C. 1995. Minimal positions in a branching random walk. *Ann. Appl. Probab.*, **5**(1), 128–139.

Mollison, D., and Daniels, H. 1993. The "deterministic simple epidemic" unmasked. *Math. Biosci.*, **117**(1-2), 147–153.

Peyrière, J. 2000. Recent results on Mandelbrot multiplicative cascades. Pages 147–159 of: Bandt, C., Graf, S., and Zähle, M. (eds), *Fractal Geometry and Stochastics, II (Greifswald/Koserow, 1998)*. Progr. Probab., vol. 46. Basel: Birkhäuser.

Weinberger, H. F. 1982. Long-time behavior of a class of biological models. *SIAM J. Math. Anal.*, **13**(3), 353–396.

Weinberger, H. F. 2002. On spreading speeds and traveling waves for growth and migration models in a periodic habitat. *J. Math. Biol.*, **45**(6), 511–548.

Weinberger, H. F., Lewis, M. A., and Li, B. 2002. Analysis of linear determinacy for spread in cooperative models. *J. Math. Biol.*, **45**(3), 183–218.

Weinberger, H. F., Lewis, M. A., and Li, B. 2007. Anomalous spreading speeds of cooperative recursion systems. *J. Math. Biol.*, **55**(2), 207–222.

6

Kingman, category and combinatorics

N. H. Bingham[a] and A. J. Ostaszewski[b c]

Abstract

Kingman's Theorem on skeleton limits—passing from limits as $n \to \infty$ along nh ($n \in \mathbb{N}$) for enough $h > 0$ to limits as $t \to \infty$ for $t \in \mathbb{R}$—is generalized to a Baire/measurable setting via a topological approach. We explore its affinity with a combinatorial theorem due to Kestelman and to Borwein and Ditor, and another due to Bergelson, Hindman and Weiss. As applications, a theory of 'rational' skeletons akin to Kingman's integer skeletons, and more appropriate to a measurable setting, is developed, and two combinatorial results in the spirit of van der Waerden's celebrated theorem on arithmetic progressions are given.

Keywords Baire property, bitopology, complete metrizability, density topology, discrete skeleton, essential contiguity, generic property, infinite combinatorics, measurable function, Ramsey theory, Steinhaus theory

AMS subject classification (MSC2010) 26A03

1 Introduction

The background to the theme of the title is Feller's theory of *recurrent events*. This goes back to Feller in 1949 [F1], and received its first

[a] Mathematics Department, Imperial College London, London SW7 2AZ; n.bingham@ic.ac.uk, nick.bingham@btinternet.com
[b] Mathematics Department, London School of Economics, Houghton Street, London WC2A 2AE; a.j.ostaszewski@lse.ac.uk
[c] This paper was completed on the final day of the life of Zofia Ostaszewska, the second author's mother, and is dedicated to her memory.

textbook synthesis in [F2] (see e.g. [GS] for a recent treatment). One is interested in something ('it', let us say for now—we can proceed informally here, referring to the above for details) that happens (by default, or by fiat) at time 0, may or may not happen at discrete times $n = 1$, 2, ..., and is such that its happening 'resets the clock', so that if one treats this random time as a new time-origin, the subsequent history is a probabilistic replica of the original situation. Motivating examples include return to the origin in a simple random walk (or coin-tossing game); attaining a new maximum in a simple random walk; returning to the initial state i in a (discrete-time) Markov chain. Writing u_n for the probability that 'it' happens at time n (so $u_0 = 1$), one calls $u = (u_n)$ a *renewal sequence*. Writing f_n for the probability that 'it' happens *for the first time* at $n > 1$ ($f_0 := 0$), $f = (f_n)$, the generating functions U, F of u, f satisfy the *Feller relation* $U(s) = 1/(1 - F(s))$.

It is always worth a moment when teaching stochastic processes to ask the class whether time is discrete or continuous. It is both, but which aspect is uppermost depends on how we measure, or experience, time— whether our watch is digital or has a sweep second hand, one might say. In continuous time, one encounters analogues of Feller's theory above in various probabilistic contexts—e.g., the server in an $M/G/1$ queue being idle. In the early 1960s, the Feller theory, queueing theory (in the phase triggered by Kendall's work, [Ken1]) and John Kingman were all young. Kingman found himself drawn to the task of creating a continuous-time version of Feller's theory of recurrent events (see [King4] for his reminiscences of this time)—a task he triumphantly accomplished in his theory of *regenerative phenomena*, for which his book [King3] remains the standard source. Here the role of the renewal sequence is played by the Kingman p-function, where $p(t)$ is the probability that the regenerative phenomenon Φ occurs at time $t \geq 0$.

A continuous-time theory contains within itself infinitely many versions of a discrete-time theory. For each fixed $h > 0$, one obtains from a Kingman regenerative phenomenon Φ with p-function $p(t)$ a Feller recurrent event (or regenerative phenomenon in discrete time, as one would say nowadays), Φ_h say, with renewal sequence $u_n(h) = p(nh)$—called the discrete *skeleton* of Φ for time-step h—the h-skeleton, say.

While one can pass from continuous to discrete time by taking skeletons, it is less clear how to proceed in the opposite direction—how to combine discrete-time information for various time-steps h to obtain continuous-time information. A wealth of information was available in the discrete-time case—for example, limit theorems for Markov chain transition probabilities. It was tempting to seek to use such information to study corresponding questions in continuous time, as was done in [King1]. There, Kingman made novel use of the Baire category theorem, to extend a result of Croft [Cro], making use of a lemma attributed both to Golomb and Gould and to Anderson and Fine (see [NW] for both).

While in the above we have limits at infinity through integer multiples nx, we shall also be concerned with limits through positive rational multiples qx (we shall always use the notations $\lim_n$ and $\lim_q$ for these). There are at least three settings in which such rational limits are probabilistically relevant:

(i) *Infinitely divisible p-functions.* The Kingman p-functions form a semigroup under pointwise multiplication (if p_i come from Φ_i with Φ_1, Φ_2 independent, $p := p_1 p_2$ comes from $\Phi := \Phi_1 \cap \Phi_2$, in an obvious notation). The arithmetic of this semigroup has been studied in detail by Kendall [Ken2].

(ii) *Embeddability of infinitely divisible laws.* If a probability law is infinitely divisible, one can define its qth convolution power for any positive rational q. The question then arises as to whether one can embed these rational powers into a continuous semigroup of real powers. This is the question of *embeddability*, studied at length (see e.g. [Hey, Ch. III]) in connection with the Lévy–Khintchine formula on locally compact groups.

(iii) *Embeddability of branching processes.* While for simple branching processes both space (individuals) and time (generations) are discrete, it makes sense in considering e.g. the biomass of large populations to work with branching processes where space and/or time may be continuous. While one usually goes from the discrete to the continuous setting by taking limits, embedding is sometimes possible; see e.g. Karlin and McGregor [KMcG], Bingham [Bin].

In addition, (i) led Kendall [Ken2, Th. 16] to study *sequential regular variation* (see e.g. [BGT, §1.9]). The interplay between the continuous and sequential aspects of regular variation, and between measurable and Baire aspects, led us to our recent theory of *topological regular variation* (see e.g. [BOst2] and our other recent papers), our motivation here.

In Section 2 we discuss the relation between Kingman's Theorem and the Kestelman–Borwein–Ditor Theorem (KBD), introducing definitions and summarizing background results which we need (including the density topology). In Section 3 we generalize to a Baire/measurable setting the Kingman Theorem (originally stated for open sets). Our (bi-)topological approach (borrowed from [BOst7]) allows the two cases to be treated as one that, by specialization, yields either case. (This is facilitated by the density topology.) The theorem is applied in Section 4 to establish a theory of 'rational' skeletons parallel to Kingman's integer skeletons. In Section 5 we offer a new proof of KBD in a 'consecutive' format suited to proving in Section 6 combinatorial results in the spirit of van der Waerden's celebrated theorem on arithmetic progressions. Again a bitopological (actually 'bi-metric') approach allows unification of the Baire/measurable cases. Our work in Section 6 is based on a close reading of [BHW], our debt to which is clear.

2 Preliminaries

In this section we motivate and define notions of contiguity. Then we gather classical results from topology and measure theory (complete metrizability and the density topology). We begin by recalling the following result, in which the expression 'for generically all t' means for all t except in a meagre or null set according to context. We use the terms Baire set/function to mean a set/function with the Baire property. Evidently, the interesting cases are with T Baire non-meagre/measurable non-null. The result in this form is due in the measure case to Borwein and Ditor [BoDi], but was already known much earlier albeit in somewhat weaker form by Kestelman [Kes, Th. 3], and rediscovered by Kemperman [Kem] and later by Trautner [Trau] (see [BGT, p. xix and footnote p. 10]). We note a cognate result in [HJ, Th. 2.3.7].

Theorem KBD (Kestelman–Borwein–Ditor Thorem, KBD) *Let* $\{z_n\} \to 0$ *be a null sequence of reals. If T is Baire/Lebesgue-measurable, then for generically all $t \in T$ there is an infinite set $\mathbb{M}_t$ such that*

$$\{t + z_m : m \in \mathbb{M}_t\} \subseteq T.$$

We give a new unified proof of the measure and Baire cases in Section 5, based on 'almost-complete metrizability' (= *almost complete + complete metrizability*, see below) and a *Generic Dichotomy* given in Section 3; earlier unification was achieved through a bitopological approach (as here to the Kingman Theorem) in [BOst7]. This result is a theorem about additive infinite combinatorics. It is of fundamental and unifying importance in contexts where additive structure is key; its varied applications include proofs of classical results such as Ostrowski's Theorem on the continuity of Baire/Lebesgue convex (and so additive) functions (cf. [BOst8]), a plethora of results in the theory of subadditive functions (cf. [BOst1, BOst4]), the Steinhaus Theorem on distances [BOst8, BOst6] and the Uniform Convergence Theorem of regular variation [BOst3]. Its generalizations to normed groups may be used to prove the Uniform Boundedness Theorem (see [Ost]). Recently it has found applications to additive combinatorics in the area of Ramsey Theory (for which see [GRS, HS]), best visualized in the language of colour: one seeks monochromatic structures in finitely coloured situations. Two examples are included in Section 6.

The KBD theorem is about shift-embedding of subsequences of a null sequence $\{z_n\}$ into a *single* set T with an assumption of *regularity* (Baire/measurable). Our generalizations in Section 3 of a theorem of Kingman's have been motivated by the wish to establish 'multiple embedding' versions of KBD: we seek conditions on a sequence $\{z_n\}$ and a *family* of sets $\{T_k\}_{k \in \omega}$ which together guarantee that *one* shift embeds (different) subsequences of $\{z_n\}$ into *all* members of the family.

Evidently, if $t + z_n$ lies in several sets infinitely often, then the sets in question have a common limit point, a sense in which they are contiguous at t. Thus *contiguity conditions* are one goal, the other two being *regularity conditions* on the family, and *admissibility conditions* on the null sequences.

We view the original Kingman Theorem as studying contiguity at infinity, so that divergent sequences z_n (i.e. with $z_n \to +\infty$) there replace the null sequences of KBD. The theorem uses *openness* as a regularity condition on the family, *cofinality* at infinity (e.g. *unboundedness* on the right) as the simplest contiguity condition at infinity, and

$$\frac{z_{n+1}}{z_n} \to 1 \ \text{(multiplicative form)}, \quad z_{n+1} - z_n \to 0 \ \text{(additive form)} \quad (*)$$

as the admissibility condition on the divergent sequence z_n ($(*)$ follows from regular variation by Weissman's Lemma, [BGT, Lemma 1.9.6]). Taken together, these three guarantee multiple embedding (at infinity).

One can switch from $\pm\infty$ to 0 by an inversion $x \to 1/x$, and thence to any τ by a shift $y \to y + \tau$. *Openness* remains the *regularity* condition, a property of *density at zero* becomes the analogous *admissibility* condition on null sequences, and cofinality (or accumulation) at τ the contiguity condition. The transformed theorem then asserts that for admissible null sequences ζ_n there exists a scalar σ such that the sequence

$\sigma\zeta_n + \tau$ has subsequences in all the open sets T_k provided these all accumulate at τ.

In the next section, we will replace Kingman's regularity condition of openness by the Baire property, or alternatively measurability, to obtain two versions of Kingman's theorem—one for measure and one for category. We develop the regularity theme bitopologically, working with two topologies, so as to deduce the measure case from the Baire case by switching from the Euclidean to the density topology.

Definitions and notation (Essential contiguity conditions) We use the notation $B_r(x) := \{y : |x - y| < r\}$ and $\omega := \{0, 1, 2, \dots\}$. Likewise for $a \in A \subseteq \mathbb{R}$ and metric $\rho = \rho_A$ on A, $B_r^\rho(a) := \{y \in A : \rho(a,y) < r\}$ and cl_A denotes closure in A. For S given, put $S^{>m} = S \backslash B_m(0)$. $\mathbb{R}_+$ denotes the (strictly) positive reals. When we regard $\mathbb{R}_+$ as a multiplicative group, we write

$$A \cdot B := \{ab : a \in A,\ b \in B\}, \qquad A^{-1} := \{a^{-1} : a \in A\},$$

for A, B subsets of $\mathbb{R}_+$.

Call a Baire set S *essentially unbounded* if for each $m \in \mathbb{N}$ the set $S^{>m}$ is non-meagre. This may be interpreted in the sense of the metric (Euclidean) topology, or as we see later in the measure sense by recourse to the density topology. To distinguish the two, we will qualify the term by referring to the category/metric or the measure sense.

Say that a set $S \subset \mathbb{R}_+$ *accumulates essentially* at 0 if S^{-1} is essentially unbounded. (In [BHW] such sets are called measurably/Baire *large* at 0.) Say that $S \subset \mathbb{R}_+$ *accumulates essentially* at t if $(S - t) \cap \mathbb{R}_+$ accumulates essentially at 0.

We turn now to some topological notions. Recall (see e.g. [Eng, 4.3.23 and 24]) that a metric space A is *completely metrizable* iff it is a $\mathcal{G}_\delta$ subset of its completion (i.e. $A = \bigcap_{n \in \omega} G_n$ with each G_n open in the completion of A), in which case it has an equivalent metric under which it is complete. Thus a $\mathcal{G}_\delta$ subset A of the line has a metric $\rho = \rho_A$, equivalent to the Euclidean metric, under which it is complete. (So for each $a \in A$ and $\varepsilon > 0$ there exists $\delta > 0$ such that $B_\delta(a) \subseteq B_\varepsilon^\rho(a)$, which enables the construction of sequences with ρ-limit guaranteed to be in A.)

This motivates the definition below, which allows us to capture a feature of measure-category duality: both exhibit $\mathcal{G}_\delta$ *inner-regularity*, modulo sets which we are prepared to neglect. (The definition here takes advantage of the fact that $\mathbb{R}$ is complete; for the general metric group context see [BOst6, Section 5].)

Definition   Call $A \subset \mathbb{R}$ *almost complete* (in category/measure) if

(i) there is a meagre set N such that $A\backslash N$ is a $\mathcal{G}_\delta$, or

(ii) for each $\varepsilon > 0$ there is a measurable set N with $|N| < \varepsilon$ and $A\backslash N$ a $\mathcal{G}_\delta$.

Thus A almost complete is Baire resp. measurable. A bounded non-null measurable subset A is almost complete: for each $\varepsilon > 0$ there is a compact (so $\mathcal{G}_\delta$) subset K with $|A\backslash K| < \varepsilon$, so we may take $N = A\backslash K$. Likewise a Baire non-meagre set is almost complete—this is in effect a restatement of Baire's Theorem:

Theorem B (Baire's Theorem—almost completeness of Baire sets)   *For $A \subseteq \mathbb{R}$ Baire non-meagre there is a meagre set M such that $A\backslash M$ is completely metrizable.*

Proof   For $A \subseteq \mathbb{R}$ Baire non-meagre we have $A = (U\backslash M_0) \cup M_1$ with M_i meagre and U a non-empty open set. Now $M_0 = \bigcup_{n\in\omega} N_n$ with N_n nowhere dense; the closure $F_n := \bar{N}_n$ is also nowhere dense (and the complement $E_n = \mathbb{R}\backslash F_n$ is dense, open). The set $M_0' = \bigcup_{n\in\omega} F_n$ is also meagre, so $A_0 := U\backslash M_0' = \bigcap_{n\in\omega} U \cap E_n \subseteq A$. Taking $G_n := U \cap E_n$, we see that A_0 is completely metrizable.   □

The tool whereby we interpret measurable functions as Baire functions is *refinement* of the usual metric (Euclidean) topology of the line $\mathbb{R}$ to a non-metric one: the *density topology* (see e.g. [Kech, LMZ, CLO]). Recall that for T measurable, t is a (metric) density point of T if $\lim_{\delta\to 0} |T \cap I_\delta(t)|/\delta = 1$, where $I_\delta(t) = (t - \delta/2, t + \delta/2)$. By the Lebesgue Density Theorem almost all points of T are density points ([Hal, Section 61], [Oxt, Th. 3.20], or [Goff]). A set U is *d*-open (density-open = open in the density topology d) if (it is measurable and) each of its points is a density point of U. We mention five properties:

(i) The density topology is finer than (contains) the Euclidean topology [Kech, 17.47(ii)].

(ii) A set is Baire in the density topology iff it is (Lebesgue) measurable [Kech, 17.47(iv)].

(iii) A Baire set is meagre in the density topology iff it is null [Kech, 17.47(iii)]. So (since a countable union of null sets is null), the conclusion of the Baire theorem holds for the line under d.

(iv) $(\mathbb{R}, d)$ is a *Baire* space, i.e. the conclusion of the Baire theorem holds (cf. [Eng, 3.9]).

(v) A function is d-continuous iff it is approximately continuous in Denjoy's sense ([Den]; [LMZ, pp. 1, 149]).

The reader unfamiliar with the density topology may find it helpful to recall Littlewood's Three Principles ([Lit, Ch. 4], [Roy, Section 3.6, p. 72]): general situations are 'nearly' the easy situations—i.e. are easy situations modulo small sets. Theorem 3.0 below is in this spirit. We refer now to Littlewood's Second Principle, of a measurable function being continuous on nearly all of its domain, in a form suited to our d-topology context.

Theorem L (Lusin's Theorem; cf. [Hal, end of Section 55]) *For $f : \mathbb{R}_+ \to \mathbb{R}$ measurable, there is a density-open set S which is almost all of $\mathbb{R}_+$ and an increasing decomposition $S := \bigcup_m S_m$ into density-open sets S_m such that each $f|S_m$ is continuous in the usual sense.*

Proof By a theorem of Lusin (see e.g. [Hal]), there is an increasing sequence of (non-null) compact sets K_n ($n = 1, 2, \ldots$) covering almost all of $\mathbb{R}_+$ with the function f restricted to K_n continuous on K_n. Let S_n comprise the density points of K_n, so S_n is density-open, is almost all of K_n (by the Lebesgue Density Theorem) and f is continuous on S_n. Put $S := \bigcup_m S_m$; then S is almost all of $\mathbb{R}_+$, is density-open and $f|S_m$ is continuous for each m. □

Two results, Theorem S below and Theorem 3.1 in the next section, depend on the following consequence of Steinhaus's theorem concerning the existence of interior points of $A \cdot B^{-1}$ ([St], cf. [BGT, Th. 1.1.1]) for A, B measurable non-null. The first is in multiplicative form a sharper version of Sierpiński's result that any two non-null measurable sets realize a rational distance.

Lemma S (Multiplicative Sierpiński Lemma; [Sier]) *For a, b density points of their respective measurable sets A, B in $\mathbb{R}_+$ and for $n = 1, 2, \ldots$, there exist positive rationals q_n and points a_n, b_n converging to a, b through A, B respectively such that $b_n = q_n a_n$.*

Proof For $n = 1, 2, \ldots$ and the consecutive values $\varepsilon = 1/n$ the sets $B_\varepsilon(a) \cap A$ and $B_\varepsilon(b) \cap B$ are measurable non-null, so by Steinhaus's theorem the set $[B \cap B_\varepsilon(b)] \cdot [A \cap B_\varepsilon(a)]^{-1}$ contains interior points and so in particular a rational point q_n. Thus for some $a_n \in B_\varepsilon(a) \cap A$

and $b_n \in B_\varepsilon(b) \cap B$ we have $q_n = b_n a_n^{-1}$, and as $|a - a_n| < 1/n$ and $|b - b_n| < 1/n$, $a_n \to a$, $b_n \to b$. □

Remarks 1. For the purposes of Theorem 3.2 below, we observe that q_n may be selected arbitrarily large, for fixed a, by taking b sufficiently large (since $q_n \to ba^{-1}$).

2. The Lemma addresses d-open sets but also holds in the metric topology (the proof is similar but simpler), and so may be restated bitopologically (from the viewpoint of [BOst7]) as follows.

Theorem S (Sierpiński, [Sier]) *For $\mathbb{R}_+$ with either the Euclidean or the density topology, if a, b are respectively in the open sets A, B, then for $n = 1, 2, \ldots$ there exist positive rationals q_n and points a_n, b_n converging metrically to a, b through A, B respectively such that*

$$b_n = q_n a_n.$$

3 A bitopological Kingman theorem

We begin by simplifying essential unboundedness modulo null/meagre sets.

Theorem 3.0 *In $\mathbb{R}_+$ with the Euclidean or density topology, for S Baire/measurable and essentially unbounded there exists an open/density-open unbounded G and meagre/null M with $G \backslash M \subset S$.*

Proof Choose integers m_n inductively with $m_0 = 0$ and $m_{n+1} > m_n$ the least integer such that $(m_n, m_{n+1}) \cap S$ is non-meagre; for given m_n the integer m_{n+1} is well-defined, as otherwise for each $m > m_n$ we would have $(m_n, m) \cap S$ meagre, and so also

$$(m_n, \infty) \cap S = \bigcup_{m > m_n} (m_n, m) \cap S \quad \text{meagre,}$$

contradicting S essentially unbounded. Now, as $(m_n, m_{n+1}) \cap S$ is Baire/measurable, we may choose G_n open/density-open and M_n, M_n' meagre subsets of (m_n, m_{n+1}) such that

$$((m_n, m_{n+1}) \cap S) \cup M_n = G_n \cup M_n'.$$

Hence G_n is non-empty. Put $G := \bigcup_n G_n$ and $M := \bigcup_n M_n$. Then M is meagre and G is open unbounded and, since $M \cap (m_n, m_{n+1}) = M_n$

and $G \cap (m_n, m_{n+1}) = G_n$,

$$G \backslash M = \bigcup_n G_n \backslash M = \bigcup_n G_n \backslash M_n \subset \bigcup_n (m_n, m_{n+1}) \cap S = S,$$

as asserted.   □

Definition (Weakly Archimedean property—an admissibility condition)   Let $\mathbb{I}$ be $\mathbb{N}$ or $\mathbb{Q}$, with the ordering induced from the reals. Our purpose will be to take limits through subsets J of $\mathbb{I}$ which are *unbounded* on the right (more briefly: unbounded). According as $\mathbb{I}$ is $\mathbb{N}$ or $\mathbb{Q}$, we will write $n \to \infty$ or $q \to \infty$. Denote by X the line with either the metric or density topology and say that a family $\{h_i : i \in \mathbb{I}\}$ of self-homeomorphisms of the topological space X is *weakly Archimedean* if for each non-empty open set V in X and any $j \in \mathbb{I}$ the open set

$$U_j(V) := \bigcup_{i \geq j} h_i(V)$$

meets every essentially unbounded set in X.

Theorem 3.1 (implicit in [BGT, Th. 1.9.1(i)])   *In the multiplicative group of positive reals $\mathbb{R}_+^*$ with Euclidean topology, the functions $h_n(x) = d_n x$ for $n = 1, 2, \ldots$, are homeomorphisms and $\{h_n : n \in \mathbb{N}\}$ is weakly Archimedean, if d_n is divergent and the multiplicative form of $(*)$ holds. For any interval $J = (a, b)$ with $0 < a < b$ and any m,*

$$U_m(J) := \bigcup_{n \geq m} d_n J$$

contains an infinite half-line, and so meets every unbounded open set. Similarly this is the case in the additive group of reals $\mathbb{R}$ with $h_n(x) = d_n + x$ and $U_m(J) = \bigcup_{n \geq m} (d_n + J)$.

Proof   For given $\varepsilon > 0$ and all large enough n, $1 - \varepsilon < d_n/d_{n+1} < 1 + \varepsilon$. Write $x := (a+b)/2 \in J$. For ε small enough $a < x(1-\varepsilon) < x(1+\varepsilon) < b$, and then $a < x d_n/d_{n+1} < b$, so $x d_n \in d_{n+1} J$, and so $d_n J$ meets $d_{n+1} J$. Thus for large enough n consecutive $d_n J$ overlap; as $d_n \to \infty$, their union is thus a half-line.   □

Remark   Some such condition as $(*)$ is necessary, otherwise the set $U_m(J)$ risks missing an unbounded sequence of open intervals. For an indirect example, see the remark in [BGT] after Th. 1.9.2 and G. E. H. Reuter's elegant counterexample to a corollary of Kingman's Theorem, a break-down caused by the absence of our condition. For a direct example, note that if $d_n = r^n \log n$ with $r > 1$ and $J = (0, 1)$, then $d_n + J$ and

$d_{n+1} + J$ miss the interval $(1 + r^n \log n, r^{n+1} \log(1+n))$ and the omitted intervals have union an unbounded open set; to see that the omitted intervals are non-degenerate note that their lengths are unbounded:

$$r^{n+1} \log(1 + n) - r^n \log n - 1 \to \infty.$$

Theorem 3.1 does not extend to the real line under the density topology; the homeomorphisms $h_n(x) = nx$ are no longer weakly Archimedean, as we demonstrate by an example in Theorem 4.6. We are thus led to an alternative approach:

Theorem 3.2 *In the multiplicative group of reals $\mathbb{R}_+^*$ with the density topology, the family of homeomorphisms $\{h_q : q \in \mathbb{Q}_+\}$ defined by $h_q(x) := qx$, where $\mathbb{Q}_+$ has its natural order, is weakly Archimedean. For any density-open set A and any $j \in \mathbb{Q}_+$,*

$$U_j(A) := \bigcup_{q \geq j,\, q \in \mathbb{Q}_+} qA$$

contains almost all of an infinite half-line, and so meets every unbounded density-open set.

Proof Let B be Baire and essentially unbounded in the d-topology. Then B is measurable and essentially unbounded in the sense of measure. From Theorem 3.0, we may assume that B is density-open. Let A be non-empty density-open. Fix $a \in A$ and $j \in \mathbb{Q}_+$. Since B is unbounded, we may choose $b \in B$ such that $b > ja$. By Theorem S there is a $q \in \mathbb{Q}_+$ with $j < q < ba^{-1}$ such that $qa' = b'$, with $a' \in A$ and $b' \in B$. Thus

$$U_j(A) \cap B \supseteq h_q(A) \cap B = qA \cap B \neq \varnothing,$$

as required.

If $U_j(A)$ fails to contain almost all of any infinite half-line, then its complement $B := \mathbb{R}_+ \backslash U_j(A)$ is essentially unbounded in the sense of measure and so, as above, must meet $U_j(A)$, a contradiction. $\qquad \square$

Remarks 1. For A the set of irrationals in $(0, 1)$ the set $U_j(A)$ is again a set of irrationals which contains almost all, but not all, of an infinite half-line. Our result is thus best possible.

2. Note that $(*)$ is relevant to the distinction between integer and rational skeletons; see the prime-divisor example on p. 53 of [BGT]. Theorem 3.2 holds with $\mathbb{Q}_+$ replaced by any countable *dense* subset of $\mathbb{R}_+^*$, although later we use the fact that $\mathbb{Q}_+$ is closed under

multiplication. There is an affinity here with the use of a dense 'skeleton set' in the Heiberg–Seneta Theorem, Th. 1.4.3 of [BGT], and its extension Th. 3.2.5 therein.

Kingman's Theorem below, like KBD, is a generic assertion about embedding into target sets. We address first the source of this genericity: a property inheritable by supersets either holds generically or fails outright. This is now made precise.

Definition   Recall that X denotes $\mathbb{R}_+$ with Euclidean or density topology. Denote by $\mathcal{B}a(X)$, or just $\mathcal{B}a$, the Baire sets of the space X, and recall these form a σ-algebra. Say that a correspondence $F : \mathcal{B}a \to \mathcal{B}a$ is *monotonic* if $F(S) \subseteq F(T)$ for $S \subseteq T$.

The nub is the following simple result, which we call the Generic Dichotomy Principle.

Theorem 3.3 (Generic Dichotomy Principle)   *For $F : \mathcal{B}a \to \mathcal{B}a$ monotonic: either*

(i)   *there is a non-meagre $S \in \mathcal{B}a$ with $S \cap F(S) = \varnothing$, or*
(ii) *for every non-meagre $T \in \mathcal{B}a$, $T \cap F(T)$ is quasi all of T.*

Equivalently: the existence condition that $S \cap F(S) \neq \varnothing$ should hold for all non-meagre $S \in \mathcal{B}a$ implies the genericity condition that, for each non-meagre $T \in \mathcal{B}a$, $T \cap F(T)$ is quasi all of T.

Proof   Suppose that (i) fails. Then $S \cap F(S) \neq \varnothing$ for every non-meagre $S \in \mathcal{B}a$. We show that (ii) holds. Suppose otherwise; thus for some T non-meagre in $\mathcal{B}a$, the set $T \cap F(T)$ is not almost all of T. Then the set $U := T \backslash F(T) \subseteq T$ is non-meagre (it is in $\mathcal{B}a$ as T and $F(T)$ are) and so

$$\varnothing \neq U \cap F(U) \qquad (S \cap F(S) \neq \varnothing \text{ for every non-meagre } S)$$
$$\subseteq U \cap F(T) \qquad (U \subseteq T \text{ and } F \text{ monotonic}).$$

But as $U := T \backslash F(T)$, $U \cap F(T) = \varnothing$, a contradiction.

The final assertion simply rephrases the dichotomy as an implication. $\square$

The following corollary transfers the onus of verifying the existence condition of Theorem 3.3 to topological completeness.

Theorem 3.4 (Generic Completeness Principle)   *For $F : \mathcal{B}a \to \mathcal{B}a$ monotonic, if $W \cap F(W) \neq \varnothing$ for all non-meagre $W \in \mathcal{G}_\delta$ then, for each non-meagre $T \in \mathcal{B}a$, $T \cap F(T)$ is quasi all of T.*

That is, either

(i) *there is a non-meagre $S \in \mathcal{G}_\delta$ with $S \cap F(S) = \varnothing$, or*
(ii) *for every non-meagre $T \in \mathcal{B}a$, $T \cap F(T)$ is quasi all of T.*

Proof From Theorem B, for S non-meagre in $\mathcal{B}a$ there is a non-meagre $W \subseteq S$ with $W \in \mathcal{G}_\delta$. So $W \cap F(W) \neq \varnothing$ and thus $\varnothing \neq W \cap F(W) \subseteq S \cap F(S)$, by monotonicity. By Theorem 3.3 for every non-meagre $T \in \mathcal{B}a$, $T \cap F(T)$ is quasi all of T. □

Remarks In regard to the role of $\mathcal{G}_\delta$ sets, we note *Solecki's analytic dichotomy theorem* (reformulating and generalizing a specific instance discovered by Petruska, [Pet]) as follows. For $\mathcal{I}$ a family of closed sets (in any Polish space), let $\mathcal{I}_{\text{ext}}$ denote the sets covered by a countable union of sets in $\mathcal{I}$. Then, for A an analytic set, either $A \in \mathcal{I}_{\text{ext}}$, or A contains a $\mathcal{G}_\delta$ set not in $\mathcal{I}_{\text{ext}}$. See [Sol1], where a number of classical theorems, asserting that a 'large' analytic set contains a 'large' compact subset, are deduced, and also [Sol2] for further applications of dichotomy. A superficially similar, but more distant result, is *Kuratowski's Dichotomy*—([Kur-B], [Kur-1], [McSh, Cor. 1]): suppose a set H of auto-homeomorphisms acts transitively on a space X, and $Z \subseteq X$ is Baire and has the property that for each $h \in H$

$$Z = h(Z) \text{ or } Z \cap h(Z) = \varnothing,$$

i.e. under each $h \in H$, either Z is invariant or Z and its image are disjoint. Then, either H is meagre or it is clopen.

Examples Here are four examples of monotonic correspondences. The first two relate to standard results. The following two are canonical for the current paper as they relate to KBD and to Kingman's Theorem in its original form. Each correspondence F below gives rise to a correspondence $\Phi(A) := F(A) \cap A$ which is a lower or upper density and arises in the theory of *lifting* [IT1, IT2] and category measures [Oxt, Th. 22.4], and so gives rise to a fine topology on the real line. See also [LMZ, Section 6F] on lifting topologies.

1. Here we apply Theorem 3.3 to the real line with the density topology, in which the meagre sets are the null sets. Let $\mathcal{B}$ denote a countable basis of Euclidean open neighbourhoods. For any set T and $0 < \alpha < 1$ put

$$\mathcal{B}_\alpha(T) := \{I \in \mathcal{B} : |I \cap T| > \alpha |I|\},$$

which is countable, and

$$F(T) := \bigcap_{\alpha \in \mathbb{Q} \cap (0,1)} \bigcup \{I : I \in \mathcal{B}_\alpha(T)\}.$$

Thus F is increasing in T, $F(T)$ is measurable (even if T is not) and $x \in F(T)$ iff x is a density point of T. If T is measurable, the set of points x in T for which $x \in I \in \mathcal{B}$ implies that $|I \cap T| < \alpha |I|$ is null (see [Oxt, Th. 3.20]). Hence any non-null measurable set contains a density point. It follows that almost all points of a measurable set T are density points. This is the Lebesgue Density Theorem ([Oxt, Th. 3.20], or [Kucz, Section 3.5]).

2. In [PWW, Th. 2] a category analogue of the Lebesgue Density Theorem is established. This follows more simply from our Theorem 3.3.

3. For KBD, let $z_n \to 0$ and put $F(T) := \bigcap_{n \in \omega} \bigcup_{m > n} (T - z_m)$. Thus $F(T) \in \mathcal{B}a$ for $T \in \mathcal{B}a$ and F is monotonic. Here $t \in F(T)$ iff there is an infinite $\mathbb{M}_t$ such that $\{t + z_m : m \in \mathbb{M}_t\} \subseteq T$. The Generic Dichotomy Principle asserts that once we have proved (for which see Theorem 5.3 below) that an arbitrary non-meagre set T contains a 'translator', i.e. an element t which shift-embeds a subsequence z_m into T, then quasi all elements of T are translators.

4. For $z_n = n$ and $\{S_k\}$ a family of unbounded open sets (in the Euclidean sense), put $F(T) := T \cap \bigcap_{k \in \omega} \bigcap_{n \in \omega} \bigcup_{m > n} (S_k - z_m)$. Thus $F(T) \in \mathcal{B}a$ for $T \in \mathcal{B}a$ and F is monotonic. Here $t \in F(T)$ iff $t \in T$ and for each $k \in \omega$, there is an infinite $\mathbb{M}_t^k$ such that $\{t + z_m : m \in \mathbb{M}_t^k\} \subseteq S_k$. In Kingman's version of his theorem, as stated, we know only that $F(V)$ is non-empty for any non-empty open set V; but in Theorem 3.5 below we adjust his argument to show that $F(T)$ is non-empty for arbitrary non-meagre sets $T \in \mathcal{B}a$, hence that quasi all members of T are in $F(T)$, and in particular that this is so for $T = \mathbb{R}_+$.

Theorem 3.5 (Bitopological Kingman Theorem—[King1, Th. 1], [King2], where $\mathbb{I} = \mathbb{N}$) *If $X = \mathbb{R}_+$ under the Euclidean or density topology,*

(i) $\{h_i : i \in \mathbb{I}\}$ *is a countable, linearly ordered, weakly Archimedean family of self-homeomorphisms of X, and*

(ii) $\{S_k : k = 1, 2, \ldots\}$ *are essentially unbounded Baire sets,*

then for quasi all $\eta \in X$ and all $k \in \mathbb{N}$ there exists an unbounded subset

$\mathbb{J}_\eta^k$ *of* $\mathbb{I}$ *with*

$$\{h_j(\eta) : j \in \mathbb{J}_\eta^k\} \subset S_k.$$

Equivalently, if (i) *and*

(i)′ $\{A_k : k = 1, 2, \ldots\}$ *are Baire and all accumulate essentially at* 0,

then for quasi all η *and every* $k = 1$, 2, ... *there exists* $\mathbb{J}_\eta^k$ *unbounded with*

$$\{h_j(\eta)^{-1} : j \in \mathbb{J}_\eta^k\} \subset A_k.$$

Proof We will apply Theorem 3.3 (Generic Dichotomy), so consider an arbitrary non-meagre Baire set T. We may assume without loss of generality that $T = V \backslash M$ with V non-empty open and M meagre. For each $k = 1$, 2, ... choose G_k open and N_k and N_k' meagre such that $S_k \cup N_k' = G_k \cup N_k$. Put $N := M \cup \bigcup_{n,k} h_n^{-1}(N_k')$; then N is meagre (as h_n, and so h_n^{-1}, is a homeomorphism).

As S_k is essentially unbounded, G_k is unbounded (otherwise, for some m, $G_k \subset (-m, m)$, and so $S_k \cap (m, \infty) \subset N_k$ is meagre). Define the open sets $G_{jk} := \bigcup_{i \geq j} h_i^{-1}(G_k)$. We first show that each G_{jk} is dense. Suppose, for some j, k, there is a non-empty open set V such that $V \cap G_{jk} = \varnothing$. Then for all $i \geq j$,

$$V \cap h_i^{-1}(G_k) = \varnothing; \qquad G_k \cap h_i(V) = \varnothing.$$

So $G_k \cap \bigcup_{i \geq j} h_i(V) = \varnothing$, i.e., for U^j the open set $U^j := \bigcup_{i \geq j} h_i(V)$ we have $G_k \cap U^j = \varnothing$. But as G_k is unbounded, this contradicts $\{h_i\}$ being a weakly Archimedean family.

Thus the open set G_{jk} is dense (meets every non-empty open set); so, as $\mathbb{I}$ is countable, the $\mathcal{G}_\delta$ set

$$H := \bigcap_{k=1}^\infty \bigcap_{j \in \mathbb{I}} G_{jk}$$

is dense (as X is a Baire space). So as V is a non-empty open subset we may choose $\eta \in (H \cap V) \backslash N$. (Otherwise $N \cup (X \backslash H)$ and hence V is of first category.) Thus $\eta \in T$ and for all $k = 1$, 2, ...

$$\eta \in V \cap \bigcap_{j \in \mathbb{I}} \bigcup_{i \geq j} h_i^{-1}(G_k) \text{ and } \eta \notin N. \tag{eta}$$

For all m, as $h_m(\eta) \notin h_m(N)$ we have for all m, k that $h_m(\eta) \notin N_k'$. Using (eta), for each k select an unbounded $\mathbb{J}_\eta^k$ such that for $j \in \mathbb{J}_\eta^k$,

$\eta \in h_j^{-1}(G_k)$; for such j we have $\eta \in h_j^{-1}(S_k)$. That is, for some $\eta \in T$ we have

$$\{h_j(\eta) : j \in \mathbb{J}_\eta^k\} \subset S_k.$$

Now

$$F(T) := T \cap \bigcap_{k=1}^{\infty} \bigcap_{j \in \mathbb{I}} \bigcup_{i \geq j} h_i^{-1}(G_k)$$

takes Baire sets to Baire sets and is monotonic. Moreover, $\eta \in F(T)$ iff $\eta \in T$ and for each k there is an unbounded $\mathbb{J}_\eta^k$ with $\{h_j(\eta) : j \in \mathbb{J}_\eta^k\} \subset S_k$. We have just shown that $T \cap F(T) \neq \varnothing$ for T arbitrary non-meagre, so the Generic Dichotomy Principle implies that $X \cap F(X)$ is quasi all of X, i.e. for quasi all η in X and each k there is an unbounded $\mathbb{J}_\eta^k$ with $\{h_j(\eta) : j \in \mathbb{J}_\eta^k\} \subset S_k$. $\square$

Working in either the density or the Euclidean topology, we obtain the following conclusions.

Theorem 3.6C (Kingman Theorem for Category)   *If $\{S_k : k = 1, 2, \dots\}$ are Baire and essentially unbounded in the category sense, then for quasi all η and each $k \in \mathbb{N}$ there exists an unbounded subset $\mathbb{J}_\eta^k$ of $\mathbb{N}$ with*

$$\{n\eta : n \in \mathbb{J}_\eta^k\} \subset S_k.$$

In particular this is so if the sets S_k are open.

Theorem 3.6M (Kingman Theorem for Measure)   *If $\{S_k : k = 1, 2, \dots\}$ are measurable and essentially unbounded in the measure sense, then for almost all η and each $k \in \mathbb{N}$ there exists an unbounded subset $\mathbb{J}_\eta^k$ of $\mathbb{Q}_+$ with*

$$\{q\eta : q \in \mathbb{J}_\eta^k\} \subset S_k.$$

In the corollary below $\mathbb{J}_t^k$ refers to unbounded subsets of $\mathbb{N}$ or $\mathbb{Q}_+$ according to the category/measure context. It specializes down to a KBD result for a single set T when $T_k \equiv T$, but it falls short of KBD in view of the extra admissibility assumption and the factor σ (the latter an artefact of the multiplicative setting).

Corollary   *For $\{T_k : k \in \omega\}$ Baire/measurable and $z_n \to 0$ admissible, for generically all $t \in \mathbb{R}$ there exist σ_t and unbounded $\mathbb{J}_t^k$ such that for $k = 1, 2, \dots$*

$$t \in T_k \implies \{t + \sigma_t z_m : m \in \mathbb{J}_t^k\} \subset T_k.$$

Proof For T Baire/measurable, let $N = N(T)$ be the set of points $t \in T$ that are not points of essential accumulation of T; then $t \in N$ if for some $n = n(t)$ the set $T \cap B_{1/n}(t)$ is meagre/null. As $\mathbb{R}$ with the Euclidean topology is (hereditarily) second-countable, it is hereditarily Lindelöf (see [Eng, Th. 3.8.1] or [Dug, Th. 8.6.3]), so for some countable $S \subset N$

$$N \subset \bigcup_{t \in S} T \cap B_{1/n(t)}(t),$$

and so N is meagre/null. Thus the set N_k of points $t \in T_k$ such that $T_k - t$ does not accumulate essentially at 0 is meagre/null, as is $N = \bigcup_k N_k$. For $t \notin N$, put $\Omega_t := \{k \in \omega : T_k - t \text{ accumulates essentially at } 0\}$. Applying Kingman's Theorem to the sets $\{T_k - t : k \in \Omega_t\}$ and the sequence $z_n \to 0$, there exist σ_t and unbounded $\mathbb{J}_t^k$ such that for $k \in \Omega_t$

$$\{\sigma_t z_m : m \in \mathbb{J}_t^k\} \subset T_k - t, \text{ i.e. } \{t + \sigma_t z_m : m \in \mathbb{J}_t^k\} \subset T_k.$$

Thus for $t \notin N$, so for generically all t, there exist σ_t and unbounded $\mathbb{J}_t^k$ such that for $k = 1, 2, \ldots$

$$t \in T_k \implies \{t + \sigma_t z_m : m \in \mathbb{J}_t^k\} \subset T_k. \qquad \square$$

4 Applications—rational skeletons

In [King1] Kingman's applications were concerned mostly with limiting behaviour of continuous functions, studied by means of h-skeletons defined by

$$L_{\mathbb{N}}(h) := \lim_{n \to \infty} f(nh),$$

assumed to exist for all h in some interval I. This works for Baire functions; but in our further generalization to measurable functions $f : \mathbb{R}_+ \to \mathbb{R}$, we are led to study limits $L_{\mathbb{Q}}(h) := \lim_{q \to \infty} f(qh)$, taken through the rationals. Using the decomposition

$$q := n(q) + r(q), \qquad n(q) \in \omega, \ r(q) \in [0, 1),$$

the limit $L_{\mathbb{Q}}(h)$ may be reduced to, and so also computed as, $L_{\mathbb{N}}(h)$ (provided we admit perturbations on h, making the assumption of convergence here more demanding)—see Theorem 4.5 below.

Theorem 4.1 (Conversion of sequential to continuous limits at infinity—cf. [King1, Cor. 2 to Th. 1])   *For $f : \mathbb{R}_+ \to \mathbb{R}$ measurable* *and V a non-empty, density-open set (in particular, an open interval), if*

$$\lim_{q\to\infty} f(qx) = 0, \text{ for each } x \in V,$$

then

$$\lim_{t\to\infty} f(t) = 0.$$

The category version holds, mutatis mutandis, with f Baire and V open.

Proof Suppose not; choose $c > 0$ with $\limsup_{t\to\infty} |f(t)| > c > 0$. By Theorem L there is a density-open set S which is almost all of $\mathbb{R}_+$ and an increasing decomposition $S := \bigcup_m S_m$ such that each $f|S_m$ is continuous. Put $B = \{s \in S : |f(s)| > c\}$. For any $M > 0$, there is an $s^* \in S_m$ for some m with $s^* > M$ such that $|f(s^*)| > c$. Then by continuity of $f|S_m$, for some $\delta > 0$ we have $|f(s)| > c$ for $s \in B_\delta(s^*) \cap S_m$. Thus B is essentially unbounded. By Theorem 3.6M there exists $v \in V$ such that $qv \in B$, for unboundedly many $q \in \mathbb{Q}_+$; but, for such a v, we have $\lim_{q\to\infty} f(qv) \neq 0$, a contradiction. □

We will need the following result, which is of independent interest. The Baire case is implicit in [King1, Th. 2]. A related Baire category result is in [HJ, Th. 2.3.7] (with $G = \mathbb{R}_+$ and $T = \mathbb{Q}$ there).

Theorem 4.2 (Constancy of rationally invariant functions) *If for $f : \mathbb{R}_+ \to \mathbb{R}$ measurable*

$$f(qx) = f(x), \text{ for } q \in \mathbb{Q}_+ \text{ and almost all } x \in \mathbb{R}_+,$$

then $f(x)$ takes a constant value almost everywhere. The category version holds, mutatis mutandis, with f Baire.

Proof (for the measure case) Again by Theorem L, there is a density-open set S which is almost all of $\mathbb{R}_+$ and an increasing decomposition $S := \bigcup_m S_m$ such that each $f|S_m$ is continuous. We may assume without loss of generality that $f(qx) = f(x)$, for $q \in \mathbb{Q}_+$ and all $x \in S$. We claim that on S the function f is constant. Indeed, by Theorem S if a, b are in S_m, then, since they are density points of S_m, there are a_n, b_n in S_m and q_n in $\mathbb{Q}_+$ such that $b_n = q_n a_n$ with $a_n \to a$ and $b_n \to b$, as $n \to \infty$ (in the metric sense). Hence, since $f(q_n a_n) = f(a_n)$, relative continuity gives

$$f(b) = \lim_{n\to\infty} f(b_n) = \lim_{n\to\infty} f(q_n a_n) = \lim_{n\to\infty} f(a_n) = f(a).$$

The Baire case is similar, but simpler. □

Of course, if $f(x)$ is the indicator function $1_{\mathbb{Q}}(x)$, which is measurable/Baire, then $f(x)$ is constant almost everywhere, but not constant, so the result in either setting is best possible.

Theorem 4.3 (Uniqueness of limits—cf. [King1, Th. 2])   *For f : $\mathbb{R}_+ \to \mathbb{R}$ measurable, suppose that for each $x > 0$ the limit*

$$L(x) := \lim_{q \to \infty} f(qx)$$

exists and is finite on a density-open set V (in particular an interval). Then $L(x)$ takes a constant value a.e. in $\mathbb{R}_+$, L say, and

$$\lim_{t \to \infty} f(t) = L.$$

The category version holds, mutatis mutandis, with f Baire and V open.

Proof Since $\mathbb{Q}_+$ is countable, the function $L(x)$ is measurable/Baire. Note that for $q \in \mathbb{Q}_+$ one has $L(qx) = L(x)$, so if L is defined for $x \in V$, then L is defined for $x \in qV$ for each $q \in \mathbb{Q}_+$, so by Theorem 3.2 for almost all x in some half-infinite interval and thus for almost all $x \in \mathbb{R}_+$. The result now follows from Theorem 4.2. As for the final conclusion, replacing $f(x)$ by $f(x) - L$, we may suppose that $L = 0$, and so may apply Theorem 4.1. □

We now extend an argument in [King1]. Recall that f is *essentially bounded* on S if $\operatorname{ess\,sup}_S f < \infty$.

Definition Call f *essentially bounded at infinity* if for some M $\operatorname{ess\,sup}_{(n,\infty)} f \leq M$, for all large enough n, i.e. $\limsup_{n\to\infty} [\operatorname{ess\,sup}_{(n,\infty)} f] < \infty$.

Theorem 4.4 (Essential boundedness theorem—cf. [King1, Cor. 3 to Th. 1])   *For V non-empty density-open and $f : \mathbb{R}_+ \to \mathbb{R}$ measurable, suppose that for each $x \in V$*

$$\sup\{f(qx) : q \in \mathbb{Q}_+\} < \infty.$$

Then $f(t)$ is essentially bounded at infinity. The category version holds, mutatis mutandis, with f Baire and V open.

Proof Suppose not. Then for each $n = 1, 2, \ldots$ there exists $m \in \mathbb{N}$, arbitrarily large, such that $\operatorname{ess\,sup}_{(m,\infty)} f > n$. We proceed inductively. Suppose that $m(n)$ has been defined so that $\operatorname{ess\,sup}_{(m(n),\infty)} f > n$. Choose $m(n+1) > m(n)$ so that

$$G_n := \{t : m(n) < t < m(n+1) \text{ and } |f(t)| > n\}$$

is non-null (otherwise, off a null set, we would have $|f(t)| \leq n$ for all $t > m(n)$, making n an essential bound of f on (m, ∞), for each $m > m(n)$, contradicting the assumed essential unboundedness). Each G_n is measurable non-null, so defining

$$G := \bigcup_n G_n$$

yields G essentially unbounded. So there exists $v \in V$ such that $qv \in G$, for an unbounded set of $q \in \mathbb{Q}_+$. Since each set G_n is bounded, the set $\{qv : q \in \mathbb{Q}_+\}$ meets infinitely many of the disjoint sets G_n, and so

$$\sup\{|f(qv)| : q \in \mathbb{Q}_+\} = \infty,$$

contradicting our assumption.   □

We close with the promised comparison of $L_\mathbb{Q}(h)$ with $L_\mathbb{N}(h)$ (of course, if $L_\mathbb{Q}(h)$ exists, then so does $L_\mathbb{N}(h)$ and they are equal). We use the decomposition $q = n(q) + r(q)$, with $n(q) \in \mathbb{N}$ and $r(q) \in [0, 1) \cap \mathbb{Q}$.

Theorem 4.5 (Perturbed skeletons)   *For $f : \mathbb{R}_+ \to \mathbb{R}$ measurable, the limit $L_\mathbb{Q}(h)$ exists for all h in the non-empty interval $I = (0, b)$ with $b > 0$ iff for all h in I the limit*

$$\lim_n f(n(h + z_n))$$

exists, for every null sequence z_n with

$$r_n := n(z_n/h) \in [0, 1) \cap \mathbb{Q}.$$

Furthermore, if either limit exists on I then it exists a.e. on $\mathbb{R}_+$, and then both limits are equal a.e. to $L_\mathbb{N}(h)$.

If $I = (a, b)$ with $0 < a < b$, the assertion holds far enough to the right.

Proof   First we prove the asserted equivalence.

Suppose the limit $L_\mathbb{Q}(h)$ exists for all h in the interval I. Then given z_n as above, take $q_n := n + r_n$, $r_n = n(z_n/h)$; then $n(q_n) = n$, $r(q_n) = r_n$, and

$$q_n h = n(h + z_n).$$

So the following limit exists:

$$\lim_n f(n(h + z_n)) = \lim_n f(q_n h) = L_\mathbb{Q}(h).$$

For the converse, take z_n as above (so $z_n = r_n h/n$ with $r_n \in [0,1) \cap \mathbb{Q}$); our assumption is that $L(h, \{r_n\})$ exists for all $h \in I$, where

$$L(h, \{r_n\}) := \lim_n f(n(h + z_n)).$$

Write $\partial L(\{r_n\})$ for the 'domain of L'—the set of h for which this limit exists. Thus $I \subseteq \partial L(\{r_n\})$. Let $q_n \to \infty$ be arbitrary in $\mathbb{Q}_+$. So $q_n = n(q_n) + r(q_n)$ and, writing $r_n := r(q_n)$ and $z_n := r_n h/n(q_n)$, we find

$$q_n h = n(q_n)[h + r_n h/n(q_n)] = n(q_n)[h + z_n],$$

and

$$r_n = n(z_n/h) \in [0,1) \cap \mathbb{Q}.$$

By assumption, $\lim_n f(q_n h) = L(h, \{r_n\})$ exists for $h \in I$. Restricting from $\{n\}$ to $\{pn\}$, the limit $\lim_n f(np(h + z_n))$ exists for each $p \in \mathbb{N}$, and

$$L(h, \{r_n\}) = \lim_n f(np(h + z_n)) = \lim_n f(n(h' + z'_n)),$$

where $z'_n = pz_n$ and $h' = ph$. So

$$L(h, \{r_n\}) = L(ph, \{r_n\}),$$

as

$$n(z'_n/h') = n(z_n/h) = r_n \in [0,1) \cap \mathbb{Q}.$$

That is, $L(ph, \{r_n\})$ exists for p a positive integer, whenever $L(h, \{r_n\})$ exists, and equals $L(h, \{r_n\})$. As $h/p \in I = (0,b)$ for $h \in I$ and $L(h, \{r_n\}) = L(p(h/p), \{r_n\}) = L(h/p, \{r_n\})$, $L(rh, \{r_n\})$ exists whenever r is a positive rational. So the domain ∂L of L includes all intervals of the form rI for positive rational r, and so includes the whole of $\mathbb{R}_+$. Moreover, $L(\cdot, \{r_n\})$ is rationally invariant. But, since f is measurable, L is a measurable function on $I \times \mathbb{Q}_+^\omega$. Now $\mathbb{Q}_+$ can be identified with $\mathbb{N} \times \mathbb{N}$, and hence $\mathbb{Q}_+^\omega$ can be identified with $\mathbb{N}^\omega$. This in turn may be identified with the irrationals $\mathcal{I}$ (see e.g. [JR, p. 9]). So L is measurable on $\mathbb{R}_+ \times \mathcal{I}$ and so, by Theorem 4.2, $L(\cdot, \{r_n\})$ is almost everywhere constant.

This proves the equivalence asserted. For the final assertion, the argument in the last paragraph shows that, given the assumptions, $L_\mathbb{Q}(h)$ exists for a.e. positive h; from here the a.e. equality is immediate. This completes the case $I = (0,b)$. For $I = (a,b)$ with $a > 0$, $\bigcup_{p \in \mathbb{N}} pI$ contains some half-line $[c, \infty)$ by Theorem 3.1. □

The following example, due to R. O. Davies, clarifies why use of the natural numbers and hence also of discrete skeletons $L_{\mathbb{N}}(h)$ is inadequate in the measure setting. (We thank Roy Davies for this contribution.)

Theorem 4.6 *The open set*

$$G := \bigcup_{m=1}^{\infty} (m - 2^{-(m+2)}, m)$$

is disjoint for each $n = 1, 2, \ldots$ from the dilation nF of the non-null closed set F defined by

$$F := \left[\frac{1}{2}, 1\right] \setminus \left(\bigcup_{m=1}^{\infty} \bigcup_{n=m}^{2m-1} \left(\frac{m}{n} - \frac{1}{n2^{m+2}}, \frac{m}{n} \right) \right).$$

Proof Suppose not. Put $z_m := 2^{-(m+2)}$ and

$$E := \bigcup_{m=1}^{\infty} \bigcup_{n=m}^{2m-1} \left(\frac{m}{n} - \frac{1}{n}z_m, \frac{m}{n} \right).$$

Then for some n, there are $f \in F$ and $g \in G$ such that $nf = g$. So for some $m = 1, 2, \ldots$

$$m - z_m < nf = g < m, \text{ i.e. } \frac{m}{n} - \frac{z_m}{n} < f < \frac{m}{n}.$$

But as $1/2 \le f \le 1$, we have $n/2 \le m$ and

$$\frac{m}{n} - \frac{1}{n}z_m < 1, \text{ i.e. } m - z_m < n.$$

Thus $m \le n \le 2m$, yielding the contradiction that $f \notin F$. Put

$$a_m := \sum_{n=m}^{2m-1} \frac{1}{n}, \text{ so that } \frac{1}{2} \le a_m \le 1.$$

Then

$$|E| = \sum_{m=1}^{\infty} a_m z_m \le \sum_{m=1}^{\infty} 2^{-(m+2)} = \frac{1}{4},$$

and $|E| \ge 1/8$. Hence the complementary set F has measure at least $1/4$. $\square$

5 KBD in van der Waerden style

Fix p. Let z_n be a null sequence. We prove a generalization of KBD inspired by the van der Waerden theorem on arithmetic progressions (see Section 6). For this we need the notation

$$t + \bar{z}_{pm} = t + z_{pm+1}, t + z_{pm+1}, \ldots, t + z_{pm+p}$$

as an abbreviation for a block of consecutive terms of the null sequence all shifted by t. Our unified proof, based on the $\mathcal{G}_\delta$-inner regularity common to measure and category noted in Section 2, is inspired by a technique in [BHW].

Theorem 5.1 (Kestelman–Borwein–Ditor Theorem—consecutive form; [BOst5]) *Let $\{z_n\} \to 0$ be a null sequence of reals. If T is Baire/Lebesgue-measurable and $p \in \mathbb{N}$, then for generically all $t \in T$ there is an infinite set $\mathbb{M}_t$ such that*

$$\{t + \bar{z}_{pm} : m \in \mathbb{M}_t\} := \{t + z_{pm+1}, t + z_{pm+1}, \ldots, t + z_{pm+p} : m \in \mathbb{M}_t\} \subseteq T.$$

This will follow from the two results below, both important in their own right. The first and its corollary address displacements of open sets in the density and the Euclidean topologies; it is mentioned in passing in a note added in proof (p. 32) in Kemperman [Kem, Th. 2.1, p. 30], for which we give an alternative proof. The second parallels an elegant result for the measure case treated in [BHW].

Theorem K (Displacements Lemma—Kemperman's Theorem; [Kem, Th. 2.1] with $B_i = E$, $a_i = t$) *If E is non-null Borel, then $f(x) := |E \cap (E + x)|$ is continuous at $x = 0$, and so for some $\varepsilon = \varepsilon(E) > 0$*

$$E \cap (E + x) \text{ is non-null, for } |x| < \varepsilon.$$

More generally, $f(x_1, \ldots, x_p) := |(E + x_1) \cap \cdots \cap (E + x_p)|$ is continuous at $x = (0, \ldots, 0)$, and so for some $\varepsilon = \varepsilon_p(E) > 0$

$$(E + x_1) \cap \cdots \cap (E + x_p) \text{ is non-null, for } |x_i| < \varepsilon \quad (i = 1, \ldots, p).$$

Proof 1 (After [BHW]; cf. e.g. [BOst6, Th. 6.2 and 7.5].) Let t be a density point of E. Choose $\varepsilon > 0$ such that

$$|E \cap B_\varepsilon(t)| > \frac{3}{4}|B_\varepsilon(0)|.$$

Now $|B_\varepsilon(t)\backslash B_\varepsilon(t + x)| \le (1/4)|B_\varepsilon(t + x)|$ for $x \in B_{\varepsilon/2}(0)$, so

$$|E \cap B_\varepsilon(t + x)| > \frac{1}{2}|B_\varepsilon(0))|.$$

By invariance of Lebesgue measure we have

$$|(E + x) \cap B_\varepsilon(t + x)| > \frac{3}{4}|B_\varepsilon(0))|.$$

But, again by invariance, as $B_\varepsilon(t) + x = B_\varepsilon(0) + t + x$ this set has measure $|B_\varepsilon(0)|$. Using $|A_1 \cup A_2| = |A_1| + |A_2| - |A_1 \cap A_2|$ with $A_1 := E \cap B_\varepsilon(t + x)$ and $A_2 := (E + x) \cap B_\varepsilon(t + x)$ now yields

$$|E \cap (E + x)| \ge |E \cap (E + x) \cap (B_\varepsilon(t) + x)| > \frac{5}{4}|B_\varepsilon(0)| - |B_\varepsilon(0)| > 0.$$

Hence, for $x \in B_{\varepsilon/2}(0)$, we have $|E \cap (E + x)| > 0$.

For the p-fold form we need some notation. Let t again denote a density point of E and $x = (x_1, \ldots, x_n)$ a vector of variables. Set $A_j := B(t) \cap E \cap (E + x_j)$ for $1 \le j \le n$. For each multi-index $\mathbf{i} = (i(1), \ldots, i(d))$ with $0 < d < n$, put

$$f_{\mathbf{i}}(x) := |A_{i(1)} \cap \cdots \cap A_{i(d)}|;$$
$$f_n(x) := |A_1 \cap \cdots \cap A_n|, \qquad f_0 = |B(t) \cap E|.$$

We have already shown that the functions $f_j(x) = |B(t) \cap E \cap (E + x_j)|$ are continuous at 0. Now argue inductively: suppose that, for $\mathbf{i}$ of length less than n, the functions $f_{\mathbf{i}}$ are continuous at $(0, \ldots, 0)$. Then for given $\varepsilon > 0$, there exists $\delta > 0$ such that for $\|x\| < \delta$ and each such index $\mathbf{i}$ we have

$$-\varepsilon < f_{\mathbf{i}}(x) - f_0 < \varepsilon,$$

where $f_0 = |B(t) \cap E|$. Noting that

$$\bigcup_{i=1}^n A_i \subset B(t) \cap E,$$

and using upper or lower approximations, according to the signs in the inclusion-exclusion identity

$$\left|\bigcup_{i=1}^n A_i\right| = \sum_i |A_i| - \sum_{i<j} |A_i \cap A_j| + \cdots + (-1)^{n-1}\left|\bigcap_i A_i\right|,$$

one may compute linear functions $L(\varepsilon)$, $R(\varepsilon)$ such that

$$L(\varepsilon) < f_n(x) - f_0 < R(\varepsilon).$$

Indeed, taking $x_i = 0$ in the identity, both sides collapse to the value f_0. Continuity follows. □

Proof 2 Apply instead Theorem 61.A of [Hal, Ch. XII, p. 266] to establish the base case, and then proceed inductively as before. □

Corollary *Theorem K holds for non-meagre Baire sets E in place of Borel sets in the form:*

for each p in $\mathbb{N}$ there exists $\varepsilon = \varepsilon_p(E) > 0$ such that

$$(E + x_1) \cap \cdots \cap (E + x_p) \text{ is non-meagre, for } |x_i| < \varepsilon \quad (i = 1, \ldots, p).$$

Proof A non-meagre Baire set differs from an open set by a meagre set. □

We will now prove Theorem 5.1 using the Generic Completeness Principle (Theorem 3.4); this amounts to proceeding in two steps. To motivate the proof strategy, note that the embedding property is upward-hereditary (i.e. monotonic in the sense of Section 3): if T includes a subsequence of z_n by a shift t in T, then so does any superset of T. We first consider a non-meagre $\mathcal{G}_\delta$/non-null closed set T, just as in [BHW], modified to admit the consecutive format. We next deduce the theorem by appeal to $\mathcal{G}_\delta$ inner-regularity of category/measure and Generic Dichotomy. (The subset E of exceptional shifts can only be meagre/null.)

Theorem 5.2 (Generalized BHW Lemma—Existence of sequence embedding; cf. [BHW, Lemma 2.2]) *For T Baire non-meagre /measurable non-null and a null sequence $z_n \to 0$, there exist $t \in T$ and an infinite $\mathbb{M}_t$ such that*

$$\{t + \bar{z}_{pm} : m \in \mathbb{M}_t\} \subseteq T.$$

Proof The conclusion of the theorem is inherited by supersets (is upward hereditary), so without loss of generality we may assume that T is Baire non-meagre/measurable non-null and completely metrizable, say under a metric $\rho = \rho_T$. (For T measurable non-null we may pass down to a compact non-null subset, and for T Baire non-meagre we simply take away a meagre set to leave a Baire non-meagre $\mathcal{G}_\delta$ subset.) Since this is an equivalent metric, for each $a \in T$ and $\varepsilon > 0$ there exists $\delta = \delta(\varepsilon) > 0$ such that $B_\delta(a) \subseteq B_\varepsilon^\rho(a)$. Thus, by taking $\varepsilon = 2^{-n-1}$ the δ-ball $B_\delta(a)$ has ρ-diameter less than 2^{-n}.

Working inductively in steps of length p, we define subsets of T (of possible translators) B_{pm+i} of ρ-diameter less than 2^{-m} for $i = 1, \ldots, p$ as follows. With $m = 0$, we take $B_0 = T$. Given $n = pm$ and B_n open in T, choose N such that $|z_k| < \min\{\frac{1}{2}|x_n|, \varepsilon_p(B_n)\}$, for all $k > N$. For $i = 1, \ldots, p$, let $x_{n-1+i} = z_{N+i} \in Z$; then by Theorem K or its Corollary $B_n \cap (B_n - x_n) \cap \cdots \cap (B_n - x_{n+p})$ is non-empty (and open). We may now choose a non-empty subset B_{n+i} of T which is open in A with ρ-diameter less than 2^{-m-1} such that $\mathrm{cl}_T\, B_{n+i} \subset B_n \cap (B_n - x_n) \cap \cdots \cap (B_n - x_{n+i}) \subseteq B_{n+i-1}$. By completeness, the intersection $\bigcap_{n \in \mathbb{N}} B_n$ is non-empty. Let

$$t \in \bigcap_{n \in \mathbb{N}} B_n \subset T.$$

Now $t + x_n \in B_n \subset T$, as $t \in B_{n+1}$, for each n. Hence $\mathbb{M}_t := \{m : z_{mp+1} = x_n$ for some $n \in \mathbb{N}\}$ is infinite. Moreover, if $z_{pm+1} = x_n$ then $z_{pm+2} = x_{n+1}, \ldots, z_{pm+p} = x_{n+p-1}$ and so

$$\{t + \bar{z}_{pm} : m \in \mathbb{M}_t\} \subseteq T. \qquad \square$$

We now apply Theorem 3.3 (Generic Dichotomy) to extend Theorem 5.2 from an existence to a genericity statement, thus completing the proof of Theorem 5.1.

Theorem 5.3 (Genericity of sequence embedding) *For T Baire/ measurable and $z_n \to 0$, for generically all $t \in T$ there exists an infinite $\mathbb{M}_t$ such that*

$$\{t + \bar{z}_{pm} : m \in \mathbb{M}_t\} \subseteq T.$$

Hence, if $Z \subseteq \mathbb{R}_+$ accumulates at 0 (has an accumulation point there), then for some $t \in T$ the set $Z \cap (T - t)$ accumulates at 0 (along Z). Such a t may be found in any open set with which T has non-null intersection.

Proof Working as usual in X, the correspondence

$$F(T) := \bigcap_{n \in \omega} \bigcup_{m > n} [(T - z_{pm+1}) \cap \cdots \cap (T - z_{pm+p})]$$

takes Baire sets to Baire sets and is monotonic. Here $t \in F(T)$ iff there exists an infinite $\mathbb{M}_t$ such that $\{t + \bar{z}_{pm} : m \in \mathbb{M}_t\} \subseteq T$. By Theorem 5.2 $F(T) \cap T \neq \varnothing$, for T Baire non-meagre, so we may appeal to Generic Dichotomy (Th. 3.3) to deduce that $F(T) \cap T$ is quasi all of T (cf. Example 1 of Section 3).

With the main assertion proved, let $Z \subseteq X$ accumulate at 0 and suppose that z_n in Z converges to 0. Take $p = 1$. Then, for some $t \in T$, there is an infinite $\mathbb{M}_t$ such that $\{t + z_m : n \in \mathbb{M}_t\} \subseteq T$. Thus $\{z_m : n \in \mathbb{M}_t\} \subseteq Z \cap (T - t)$ has 0 as a joint accumulation point. $\qquad \square$

The preceding argument identifies only that $Z \cap (T - t)$ has a point of simple, rather than essential, contiguity. More in fact is true, as we show in Theorem 5.4 below.

Notation Omitting the superscript if context allows, denote by $\mathcal{M}_0^{Ba}$ resp. $\mathcal{M}_0^{Leb}$ the family of Baire/Lebesgue-measurable sets which accumulate essentially at 0.

The following is a strengthened version of the two results in Lemma 2.4 (a) and (b) of [BHW] (embraced by (iii) below).

Theorem 5.4 (Shifted-filter property of $\mathcal{M}_0$) *Let A be Baire non-meagre/measurable non-null, $B \in \mathcal{M}_0^{Ba/Leb}$.*

(i) *If $(A - t) \cap B$ accumulates (simply) at 0, then $(A - t) \cap B \in \mathcal{M}_0$.*

(ii) *For $A, B \in \mathcal{M}_0$, and generically all $t \in A$, the set $(A-t) \cap B \in \mathcal{M}_0$.*

(iii) *For $B \in \mathcal{M}_0$ and t such that $(B - t) \cap B$ accumulates (simply) at 0, the set $(B - t) \cap B$ accumulates essentially at 0.*

Proof We will prove (i) separately for the two cases (a) Baire (b) measure. From KBD (i) implies (ii), while (i) specializes to (iii) by taking $A = B$.

(a) Baire case. Assume that A is Baire non-meagre and that B accumulates essentially at 0.

Suppose that $A \cup N_1 = U \backslash N_0$ with U open, non-empty, and N_0 and N_1 meagre. Put $M = N_0 \cup N_1$ and fix $t \in A \backslash M$, so that t is quasi-any point in A; put $M_t^- := M \cup (M - t)$, which is meagre. As $U \backslash M \subset A$, note that by translation $(U - t) \backslash (M - t) \subset A - t$.

Let $\varepsilon > 0$. Without loss of generality $B_\varepsilon(0) \subset U - t$. By the assumption on B, $B \cap B_\varepsilon(0)$ is non-meagre, and thus so is $[B \cap B_\varepsilon(0)] \backslash M_t^-$. But the latter set is included in $B \cap (A - t)$; indeed

$$[B \cap B_\varepsilon(0)] \backslash M_t^- \subset [B \cap (u - t)] \backslash (M - t) = B_\varepsilon(0) \cap B \cap (A - t).$$

As ε was arbitrary, $B \cap (A - t)$ accumulates essentially at 0.

(b) Measure case. Let A, B be non-null Borel, with B accumulating essentially at e. Without loss of generality both are density-open (all points are density points). By KBD, $(A - t) \cap B$ accumulates (simply) at e for almost all $t \in A$. Fix such a t.

Let $\varepsilon > 0$ be given. Pick $x \in (A - t) \cap B$ with $|x| < \varepsilon/2$ (possible since $B \cap (A - t)$ accumulates at e). As x and $x - t$ are density points of B and A (resp.) pick $\delta < \varepsilon/2$ such that

$$|B \cap B_\delta(x)| > \frac{3}{4}|B_\delta(x)| = \frac{3}{4}|B_\delta(e)|$$

and

$$|A \cap B_\delta(x - t)| > \frac{3}{4}|B_\delta(x - t)| = \frac{3}{4}|B_\delta(e)|,$$

which is equivalent to

$$|(A - t) \cap B_\delta(x)| > \frac{3}{4}|B_\delta(e)|.$$

By inclusion-exclusion as before, with $A_1 := (A - t) \cap B_\delta(x)$ and $A_2 := B \cap B_\delta(x)$,

$$|(A - t) \cap B \cap B_\delta(x)| > \frac{3}{2}|B_\delta(e)| - |B_\delta(e)| > 0.$$

But $|x| < \varepsilon/2 < \varepsilon - \delta$ so $|x| + \delta < \varepsilon$, and thus $B_\delta(x) \subseteq B_\varepsilon(e)$, hence

$$|(A - t) \cap B \cap B_\varepsilon(e)| > 0.$$

As $\varepsilon > 0$ was arbitrary, $(A - t) \cap B$ is measurably large at e.   □

6 Applications: additive combinatorics

Recall van der Waerden's theorem [vdW] of 1927, that in any finite colouring of the natural numbers, one colour contains arbitrarily long arithmetic progressions. This is one of Khinchin's three pearls of number theory [Kh, Ch. 1]. It has had enormous impact, for instance in Ramsey theory ([Ram1]; [GRS], [HS, Ch. 18]) and additive combinatorics ([TV]; [HS, Ch. 14]).

An earlier theorem of the same type, but for finite partitions of the *reals* into measurable cells, is immediately implied by the theorem of Ruziewicz [Ruz] in 1925, quoted below. We deduce its category and measure forms from the consecutive form of the KBD Theorem. The Baire case is new.

Theorem R (Ruziewicz's Theorem [Ruz]; cf. [Kem] after Lemma 2.1 for the measure case) *Given* p *positive real numbers* $k_1, \ldots, k_p$ *and any Baire non-meagre/measurable non-null set* T, *there exist* d *and points* $x_0 < x_1 < \cdots < x_p$ *in* T *such that*

$$x_i - x_{i-1} = k_i d, \qquad i = 1, \ldots, p.$$

Proof Given $k_1, \ldots, k_p$, define a null sequence by the condition $z_{pm+i} = (k_1 + \cdots + k_i)2^{-m}$ $(i = 1, \ldots, p)$. Then there are $t \in T$ and m such that

$$\{t + z_{mp+1}, \ldots, t + z_{mp+p}\} \subseteq T.$$

Taking $d = 2^{-m}$, $x_0 = t$ and for $i = 1, \ldots, p$

$$x_i = t + z_{mp+i} = t + (k_1 + \cdots + k_i)d,$$

we have $x_0 < x_1 < \cdots < x_p$ and

$$x_{i+1} - x_i = k_i d. \qquad \qquad \square$$

Remarks 1. If each $k_i = 1$ above, then the sequence $x_0, \ldots, x_p$ is an arithmetic progression of arbitrarily small step d (which we can take as 2^{-m} with m arbitrarily large) and arbitrarily large length p. So if $\mathbb{R}$ is partitioned into a finite number of Baire/measurable cells, one cell T is necessarily non-meagre/non-null, and contains arbitrarily long arithmetic progressions of arbitrarily short step. This is similar to the van der Waerden theorem.

2. By referring to the continuity properties of the functions f_i in Theorem K, Kemperman strengthens the Ruziewicz result in the measure case, by establishing the existence of an upper bound for d, which depends on p and T only.

We now use almost completeness and the shifted-filter property (Th. 5.2) to prove the following.

Theorem BHW [BHW, Th. 2.6 and 2.7] *For a Baire/measurable set A which accumulates essentially at 0, there exists in A a sequence of reals $\{t_n\}$ such that $\sigma_F(t) := \sum_{i \in F} t_i \in A$ for every $F \subseteq \omega$.*

Proof As in Theorem 5.2 the conclusion is upward hereditary, so without loss of generality we may assume that A is completely metrizable (for A measurable non-null we may pass down to a compact non-null subset accumulating essentially at 0, and for A Baire non-meagre we simply take away a meagre set to leave a Baire non-null $\mathcal{G}_\delta$ subset). Let $\rho = \rho_A$ be a complete metric equivalent to the Euclidean metric. Denote by ρ-diam the ρ diameter of a set.

Referring to the shifted-filter property of $\mathcal{M}_e^{Ba}$ or $\mathcal{M}_e^{Leb}$, we inductively choose decreasing sets $A_n \subseteq A$ and points $t_n \in A_n$. Assume inductively that

(i) $(A_n - t_n)$ accumulates at 0,

(ii) $\sigma_F = \sum_{i \in F} t_i \in A_{\max F}$, for any finite set of indices $F \subseteq \{0, 1, \ldots, n\}$, and

(iii) $\rho\text{-diam}(\sigma_F) \leq 2^{-n}$ for all finite $F \subseteq \{0, 1, \ldots, n\}$.

By Theorem 5.4,

$$A_{n+1} := A_n \cap (A_n - t_n) \text{ accumulates essentially at } 0.$$

Let $\delta_n \in (0, t_n)$ be arbitrary (to be chosen later). By above, we may pick

$$t_{n+1} \in A_{n+1} \cap (0, \delta_n/2) \text{ such that } (A_{n+1} - t_{n+1}) \text{ accumulates at } 0.$$

Thus t_n is chosen inductively with $t_{n+1} \in A_{n+1} \cap (A_{n+1} - t_{n+1})$ and $\sum_{i \in I} t_i$ convergent for any I. Also

$$\sum_{i=n+1}^{\infty} t_i \leq t_{n+1} \sum_{i=n+1}^{\infty} 2^{-i} = \delta_n 2^{-n} < \delta_n.$$

Evidently $t_1 \in A_1$. As $A_n \subset A_{n+1} \subset A_n - t_n$, we see that, as $t_1 + \cdots + t_n \in A_n$, we have $t_1 + \cdots + t_{n+1} \in A_{n+1}$. More generally, $\sigma_F = \sum_{i \in F} t_i \in A_{\max F}$ for any finite set of indices $F \subseteq \{0, 1, \ldots, n+1\}$. For $\varepsilon = 2^{-n-1}$ there exists $\delta = \delta(\varepsilon) > 0$ small enough such that for all finite $F \subseteq \{0, 1, \ldots, n+1\}$

$$B_\delta(\sigma_F) \subseteq B_\varepsilon^\rho(\sigma_F).$$

Taking $\delta_n < \delta(2^{-n-1})$ in the inductive step above implies that, for any infinite set I, the sequence $\sigma_{I \cap \{0, \ldots, n\}}$ is Cauchy under ρ, and so $\sigma_I \in A$. □

Remark (Generalizations) Much of the material here (which extends immediately from additive to multiplicative formats) can be taken over to the more general contexts of $\mathbb{R}^d$ and beyond—to normed groups (including Banach spaces), for which see [BOst6]. We choose to restrict here to the line—Kingman's setting—for simplicity, and in view of Mark Kac's dictum: No theory can be better than its best example.

Postscript It is no surprise that putting a really good theorem and a really good mathematician together may lead to far-reaching consequences. We hope that John Kingman will enjoy seeing his early work on category still influential forty-five years later. The link with combinatorics is much more recent, and still pleases and surprises us—as we hope it will him, and our readers.

Acknowledgments It is a pleasure to thank the referee for a close reading of the paper, Roy Davies for his helpful insights (Th. 4.4), and Dona Strauss for discussions on the corpus of combinatorial work established by her, Neil Hindman and their collaborators. We salute her 75th birthday.

References

[BHW] Bergelson, V., Hindman, N., and Weiss, B. 1997. All-sums sets in (0,1]—category and measure. *Mathematika*, **44**, 61–87.

[Bin] Bingham, N. H. 1988. On the limit of a supercritical branching process. *A Celebration of Applied Probability, J. Appl. Probab.*, **25A**, 215–228.

[BGT] Bingham, N. H., Goldie, C. M., and Teugels, J. L. 1989. *Regular Variation*, 2nd ed. (1st ed. 1987). Encyclopedia Math. Appl., vol. 27. Cambridge: Cambridge Univ. Press.

[BOst1] Bingham, N. H., and Ostaszewski, A. J. 2008. Generic subadditive functions. *Proc. Amer. Math. Soc.*, **136**, 4257–4266.

[BOst2] Bingham, N. H., and Ostaszewski, A. J. 2009 Beyond Lebesgue and Baire: generic regular variation. *Colloq. Math.*, **116**, 119–138.

[BOst3] Bingham, N. H., and Ostaszewski, A. J. 2009. Infinite combinatorics and the foundations of regular variation. *J. Math. Anal. Appl.*, **360**, 518–529.

[BOst4] Bingham, N. H., and Ostaszewski, A. J. 2009. New automatic properties: subadditivity, convexity, uniformity. *Aequationes Math.*, **78**, 257–270.

[BOst5] Bingham, N. H., and Ostaszewski, A. J. 2009. Infinite combinatorics in function spaces: category methods. *Publ. Inst. Math. (Beograd) (N.S.)*, **86(100)**, 55–73.

[BOst6] Bingham, N. H., and Ostaszewski, A. J. Normed versus topological groups: dichotomy and duality. *Dissertationes Math. (Rozprawy Mat.)*, to appear.

[BOst7] Bingham, N. H., and Ostaszewski, A. J. Beyond Lebesgue and Baire II: bitopology and measure-category duality. *Colloq. Math.*, to appear.

[BOst8] Bingham, N. H., and Ostaszewski, A. J. 2007. *Infinite Combinatorics and the Theorems of Steinhaus and Ostrowski*. Preprint.

[BoDi] Borwein, D., and Ditor, S. Z. 1978. Translates of sequences in sets of positive measure. *Canad. Math. Bull.*, **21**, 497–498.

[CLO] Ciesielski, K., Larson, L., and Ostaszewski, K. 1994. $\mathcal{I}$-density continuous functions. *Mem. Amer. Math. Soc.*, **107**, no. 515.

[Cro] Croft, H. T. 1957. A question of limits. *Eureka*, **20**, 11–13.

[Den] Denjoy, A. 1915. Sur les fonctions dérivées sommable. *Bull. Soc. Math. France*, **43**, 161–248.

[Dug]   Dugundji, J. 1966. *Topology*. Boston: Allyn and Bacon.

[Eng]   Engelking, R. 1989. *General Topology*. Berlin: Heldermann Verlag.

[F1]   Feller, W. 1949. Fluctuation theory of recurrent events. *Trans. Amer. Math. Soc.*, **67**, 98–119.

[F2]   Feller, W. 1968. *An Introduction to Probability Theory and its Applications*, vol. I, 3rd ed. (1st ed. 1950, 2nd ed. 1957). New York: John Wiley & Sons.

[Goff]   Goffman, C. 1950. On Lebesgue's density theorem. *Proc. Amer. Math. Soc.*, **1**, 384–388.

[GRS]   Graham, R. L., Rothschild, B. L., and Spencer, J. H. 1990. *Ramsey Theory*, 2nd ed. (1st ed. 1980). New York: John Wiley & Sons.

[GS]   Grimmett, G. R., and Stirzaker, D. R. 2001. *Probability and Random Processes*, 3rd ed. (1st ed. 1982, 2nd ed. 1992). Oxford: Oxford Univ. Press.

[Hal]   Halmos, P. R. 1974. *Measure Theory*. Grad. Texts in Math., vol. 18. New York: Springer-Verlag (1st ed. Van Nostrand, 1950).

[Hey]   Heyer, H. 1977. *Probability Measures on Locally Compact Groups*. Ergeb. Math. Grenzgeb., vol. 94. Berlin: Springer-Verlag.

[HS]   Hindman, N., and Strauss, D. 1998. *Algebra in the Stone-Čech Compactification. Theory and Applications*. De Gruyter Exp. Math., vol. 27. Berlin: Walter de Gruyter.

[HJ]   Hoffmann-Jørgensen, J. Automatic continuity. Section 3 of [THJ].

[IT1]   Ionescu Tulcea, A., and Ionescu Tulcea, C. 1961. On the lifting property I. *J. Math. Anal. Appl.*, **3**, 537–546.

[IT2]   Ionescu Tulcea, A., and Ionescu Tulcea, C. 1969. *Topics in the Theory of Lifting*. Ergeb. Math. Grenzgeb., vol. 48. Berlin: Springer-Verlag.

[JR]   Jayne, J., and Rogers, C. A. Analytic Sets. Part 1 of [Rog].

[KMcG]   Karlin, S., and McGregor, J. 1968. Embeddability of discrete time simple branching processes into continuous time branching processes. *Trans. Amer. Math. Soc.*, **132**, 115–136.

[Kech]   Kechris, A. S. 1995. *Classical Descriptive Set Theory*. Grad. Texts in Math., vol. 156. New York: Springer-Verlag.

[Kem]   Kemperman, J. H. B. 1957. A general functional equation. *Trans. Amer. Math. Soc.*, **86**, 28–56.

[Ken1]   Kendall, D. G. 1951. Some problems in the theory of queues. *J. Roy. Statist. Soc. Ser. B*, **13**, 151–173; discussion: 173–185.

[Ken2]   Kendall, D. G. 1968. Delphic semi-groups, infinitely divisible regenerative phenomena, and the arithmetic of p-functions. *Z. Wahrscheinlichkeitstheorie verw. Geb.*, **9**, 163–195 (reprinted in §2.2 of: Kendall, D. G., and Harding, E. F. (eds), 1973. *Stochastic Analysis*. London: John Wiley & Sons).

[Kes]   Kestelman, H. 1947. The convergent sequences belonging to a set. *J. Lond. Math. Soc.*, **22**, 130–136.

[Kh]   Khinchin, A. Ya. 1998/1952. *Three Pearls of Number Theory* (tr. 2nd Russian ed. 1948, 1st ed. 1947). New York: Dover/Baltimore: Graylock Press.

[King1] Kingman, J. F. C. 1963. Ergodic properties of continuous-time Markov processes and their discrete skeletons. *Proc. Lond.Math. Soc. (3)*, **13**, 593–604.

[King2] Kingman, J. F. C. 1964. A note on limits of continuous functions. *Quart. J. Math. Oxford Ser. (2)*, **15**, 279–282.

[King3] Kingman, J. F. C. 1972. *Regenerative Phenomena*. New York: John Wiley & Sons.

[King4] Kingman, J. F. C. 2010. A fragment of autobiography, 1957–67. In: Bingham, N. H., and Goldie, C. M. (eds), *Probability and Mathematical Genetics: Papers in Honour of Sir John Kingman*. London Math. Soc. Lecture Note Ser. Cambridge: Cambridge Univ. Press.

[Kucz] Kuczma, M. 1985. *An Introduction to the Theory of Functional Equations and Inequalities. Cauchy's Functional Equation and Jensen's Inequality*. Warsaw: PWN.

[Kur-B] Kuratowski, C. 1933. Sur la propriété de Baire dans les groupes métriques. *Studia Math.*, **4**, 38–40.

[Kur-1] Kuratowski, K. 1966. *Topology*, vol. I. Warsaw: PWN.

[Lit] Littlewood, J. E. 1944. *Lectures on the Theory of Functions*. Oxford: Oxford Univ. Press.

[LMZ] Lukeš, J., Malý, J., and Zajíček, L. 1986. *Fine Topology Methods in Real Analysis and Potential Theory*. Lecture Notes in Math. vol. 1189. New York: Springer-Verlag.

[McSh] McShane, E. J. 1950. Images of sets satisfying the condition of Baire. *Ann. of Math.*, **51**, 380–386.

[NW] Newman, D. J., Weissblum, W. E., Golomb, M., Gould, S. H., Anderson, R. D., and Fine, N. J. 1955. Advanced Problems and Solutions: Solutions: 4605. *Amer. Math. Monthly*, **62**, 738.

[Ost] Ostaszewski, A. J. 2010. Regular variation, topological dynamics, and the uniform boundedness theorem. *Topology Proc.*, **36**, 1–32.

[Oxt] Oxtoby, J. C. 1980. *Measure and Category*, 2nd ed. (1st ed. 1971). Grad. Texts in Math., vol. 2. New York: Springer-Verlag.

[Pet] Petruska, Gy. 1992–93. On Borel sets with small cover. *Real Anal. Exchange*, **18**(2), 330–338.

[PWW] Poreda, W., Wagner-Bojakowska, E., and Wilczyński, W. 1985. A category analogue of the density topology. *Fund. Math.*, **125**, 167–173.

[Ram1] Ramsey, F. P. 1930. On a problem in formal logic. *Proc. Lond. Math. Soc. (2)*, **30**, 264–286 (reprinted as Ch. III in [Ram2]).

[Ram2] Ramsey, F. P. 1931. *The Foundations of Mathematics*, (ed. Braithwaite, R. B.). London: Routledge & Kegan Paul.

[Rog] Rogers, C. A., Jayne, J., Dellacherie, C., Topsøe, F., Hoffmann-Jørgensen, J., Martin, D. A., Kechris, A. S., and Stone, A. H. 1980. *Analytic Sets*. New York: Academic Press.

[Roy] Royden, H. L. 1988. *Real Analysis*, 3rd ed. New York: Macmillan.

[Ruz] Ruziewicz, St. 1925. Contribution à l'étude des ensembles de distances de points. *Fund. Math.*, **7**, 141–143.

[Sier] Sierpiński, W. 1920. Sur l'équation fonctionelle $f(x+y) = f(x)+f(y)$. *Fund. Math.*, **1**, 116–122 (reprinted in *Oeuvres choisis,* vol. 2, 331–336, Warsaw: PWN, 1975).

[Sol1] Solecki, S. 1994. Covering analytic sets by families of closed sets. *J. Symbolic Logic*, **59**, 1022–1031.

[Sol2] Solecki, S. 1999. Analytic ideals and their applications, *Ann. Pure Appl. Logic*, **99** 51–72.

[St]   Steinhaus, H. 1920. Sur les distances des points de mesure positive. *Fund. Math.*, **1**, 93–104.

[TV]   Tao, T. and Vu, V. N. 2006. *Additive Combinatorics.* Cambridge: Cambridge Univ. Press.

[THJ]   Topsøe, F., and Hoffmann-Jørgensen, J. Analytic spaces and their applications. Part 3 of [Rog].

[Trau] Trautner, R. 1987. A covering principle in real analysis. *Quart. J. Math. Oxford Ser. (2)*, **38**, 127–130.

[vdW]   Waerden, B. van der. 1927. Beweis einer Baudetschen Vermutung. *Nieuw Arch. Wiskd.*, **19**, 212–216.

7

Long-range dependence in a Cox process directed by an alternating renewal process

D. J. Daley[a]

Abstract

A Cox process N directed by a stationary random measure ξ has second moment $\operatorname{var} N(0, t] = \operatorname{E}[\xi(0, t]] + \operatorname{var} \xi(0, t]$, where by stationarity $\operatorname{E}[\xi(0, t]] = (\text{const.})t = \operatorname{E}[N(0, t]]$, so long-range dependence (LRD) properties of N coincide with LRD properties of the random measure ξ. When $\xi(A) = \int_A I(u) \, du$ is determined by the indicator process of (say) the ON periods of an alternating renewal process, the random measure is LRD if and only if the generic cycle durations $Y = X_0 + X_1$ of the process have infinite second moment, where X_j ($j = 0, 1$) denote independent generic lifetimes of the OFF and ON phases. Then the Cox process has the same Hurst index as the renewal process N_{cyc} of cycle durations, and as $t \to \infty$,

$$\operatorname{var} \xi(0, t] \sim 2\lambda^2 \left[\varpi^2 \int_0^t \mathcal{F}_0(u) \, du + (1 - \varpi)^2 \int_0^t \mathcal{F}_1(u) \, du \right],$$

where $\varpi = \lambda \operatorname{E}(X_1)$, $1/\lambda = \operatorname{E}(Y) = \operatorname{E}(X_0 + X_1)$, and $\mathcal{F}_j(t) = \int_0^\infty \min(u, t)[1 - F_j(u)] \, du$, whereas $\operatorname{var} N_{\text{cyc}}(0, t] \sim 2\lambda^3 \int_0^t [\mathcal{F}_0(u) + \mathcal{F}_1(u)] \, du$, independent of ϖ. An example is given of lifetimes for which $\mathcal{F}_0(t)/\mathcal{F}_1(t)$ does not have a limit as $t \to \infty$ and hence for which $\operatorname{var} \xi(0, t] \not\sim (\text{const.}) \operatorname{var} N_{\text{cyc}}(0, t]$.

Keywords alternating renewal process, long-range dependence, regenerative phenomenon, renewal process

a Department of Mathematics & Statistics, University of Melbourne, Parkville, VIC 3010, Australia; dndaley@gmail.com

AMS subject classification (MSC2010) Primary 60K05, Secondary 60K15

0 Preamble

John Kingman was my first research supervisor on my coming to Cambridge in October 1963. This was the time that he had not long finished his first lengthy study of regenerative phenomena (Kingman, 1964a), and he suggested that I investigate what can be said about the product $h_1 h_2$ of two renewal density functions (r.d.f.s) (the question arises by analogy with the product of two p-functions, known on probabilistic grounds to be another p-function). It seems appropriate therefore that this contribution, to a Festschrift that celebrates both John's seventieth birthday and decades of his deep contributions to the academic enterprise, should be concerned with a problem from renewal theory. The upshot (Daley, 1965) of his suggestion in 1963 is only a partial answer, just as there is a question related to this paper which remains open (see the Postlude below); one stimulus for my interest in this last matter is the subadditivity of the integral $r(x) \equiv \int_0^x p(u) \, du$ of any standard p-function, "a curious fact ... of which I know no application" (Kingman, 1972, p. 100).

1 Introduction

This paper characterizes long-range dependence (LRD) in those stationary Cox processes that are driven by a stationary alternating renewal process (ARP) being in a particular phase. Renewal and alternating renewal processes in which at least one of the underlying lifetime distributions has finite first moment but infinite second moment have been favourite components of models giving rise to LRD and other heavy-tail phenomena, yet the ARPs themselves seem to have escaped full analysis beyond their pragmatic treatment in Cox (1962) and a lengthy examination of a particular case in a paper by Heath *et al.* (1998) that avoids the term 'alternating renewal process': Heath *et al.* ultimately assume that the ON periods have a distribution with a regularly varying tail and appeal to work of Frenk (1987), whereas we use a more powerful and simpler real-variable result of Sgibnev (1981) in a setting of more general ARPs. By using Sgibnev's work, which should be better known, we describe exact asymptotics of the LRD phenomenon in ARPs, and thereby characterize LRD Cox processes driven by a stationary ARP.

We start by recalling that a stationary random measure $\xi(\cdot)$ on the Borel sets of the real line $\mathbb{R}$ with boundedly finite second moment is LRD when

$$\limsup_{t \to \infty} \frac{\operatorname{var} \xi(0, t]}{t} = \infty; \tag{1.1}$$

this definition includes Daley and Vesilo's (1997) definition of LRD stationary point processes $N(\cdot)$ on $\mathbb{R}$ with boundedly finite second moment. Further, the Hurst index H of such a random measure is defined by

$$H = \inf \left\{ h : \limsup_{t \to \infty} \frac{\operatorname{var} \xi(0, t]}{t^{2h}} < \infty \right\}, \tag{1.2}$$

so $\frac{1}{2} \leq H \leq 1$.

Now the first and second moment properties of a Cox process $N(\cdot)$ driven by a random measure ξ are determined by the first and second moment properties of ξ, namely, for bounded Borel sets A of the real line $\mathbb{R}$,

$$\mathrm{E}[N(A)] = \mathrm{E}[\xi(A)], \qquad \operatorname{var} N(A) = \mathrm{E}[\xi(A)] + \operatorname{var} \xi(A) \tag{1.3}$$

(see e.g. Proposition 6.2.II of Daley and Vere-Jones (2003)). Trivially, we then have the following result in which infinite limit suprema (cf. (1.1)) identify LRD point processes and random measures.

Proposition 1.1 *For a stationary Cox process $N(\cdot)$ driven by a stationary random measure ξ with boundedly finite second moment measure, $N(\cdot)$ is LRD if and only if $\xi(\cdot)$ is LRD; they then have the same Hurst index.*

The particular interest of this paper is the study of such LRD Cox processes when the random measure is determined by the ON phases of a stationary alternating renewal process:

$$\xi(A) = \int_A I(u) \, \mathrm{d}u \qquad \text{say,} \tag{1.4}$$

where the stationary indicator process $I(t) = 1$ when at time t a stationary ARP is in the ON phase, $= 0$ otherwise. From stationarity, $\mathrm{E}[\xi(A)] = (\text{const.})\ell(A)$, where ℓ denotes Lebesgue measure. Writing $\xi_0(A) = \int_A [1 - I(u)] \, \mathrm{d}u$ for the random measure determined by the OFF phases of this ARP, clearly $\xi(0, t] + \xi_0(0, t] = t$, so

$$- \operatorname{cov}\big(\xi(0, t], \, \xi_0(0, t]\big) = \operatorname{var} \xi(0, t] = \operatorname{var} \xi_0(0, t], \tag{1.5}$$

from which the following is immediately evident.

Proposition 1.2   *The variance functions of the two Cox processes driven by the ON and OFF phases respectively of a stationary alternating renewal process differ at most by a multiple of Lebesgue measure, and thus have the same Hurst index.*

Indeed, in this result, we could as easily replace the 'phases of a stationary alternating renewal process' by a 'stationary measurable indicator process and its complement' (i.e. by $I(\cdot)$ and $1 - I(\cdot)$) and the statement would remain true.

In Section 2 we introduce our notation and recall results for both stationary renewal and stationary alternating renewal processes, and deduce the main analytical properties we use including first-order moments. In Section 3 we consider second-order moments in more detail and prove our main result. Section 4 describes an example of a random measure driven by an ARP for which the variance functions of the two phases have the same Hurst index but are not asymptotically proportional as holds for simpler examples, thereby demonstrating that the detail of our main result is not vacuous.

2 Stationary renewal and alternating renewal processes

For a stationary renewal process whose generic lifetime r.v. Y has d.f. G with finite mean $\mathrm{E}(Y) = 1/\lambda$ and renewal function $U = \sum_{n=0}^{\infty} G^{n*}$, so that $U(t) \sim \lambda t$ by the elementary renewal theorem, it is a standard result that when G is a nonarithmetic d.f. and $\mathrm{E}(Y^2) < \infty$,

$$U(t) - \lambda t \to \tfrac{1}{2}\lambda^2 \mathrm{E}(Y^2) = \lambda^2 \int_0^\infty u\overline{G}(u)\,\mathrm{d}u \qquad (t \to \infty), \qquad (2.1)$$

where $\overline{G}(u) = 1 - G(u)$ denotes the tail of G (e.g. Feller, 1971, Section XI.4). When $\mathrm{E}(Y^2) = \infty$ and with no arithmeticity requirement, Sgibnev (1981) showed that then in place of (2.1),

$$U(t) - \lambda t \sim \lambda^2 \int_0^\infty \min(u,t)\,\overline{G}(u)\,\mathrm{d}u \equiv \lambda^2 \mathcal{G}(t) \qquad (t \to \infty) \qquad (2.2)$$

(there is a brief textbook account in Daley and Vere-Jones (2003), Exercise 4.4.5). For a stationary renewal process with this lifetime d.f., assuming that the process is orderly (so that $G(0+) = 0$), its variance function is given by

$$\operatorname{var} N(0, t] = \lambda \int_0^t \left(2[U(u) - \lambda u] - 1 \right) du \qquad (2.3)$$

(e.g. *op. cit.* p. 62), and therefore

$$\operatorname{var} N(0, t] \sim 2\lambda^3 \int_0^t \mathcal{G}(u)\, du \qquad (t \to \infty). \qquad (2.4)$$

When G has moment index

$$\kappa_G = \inf \left\{ k \colon \int_0^\infty u^k\, G(du) = \infty \right\}, \quad = \liminf_{u \to \infty} \frac{-\log \overline{G}(u)}{\log u} \qquad (2.5)$$

(see Baltrūnas, Daley and Klüppelberg (2003) for the second equality here), so $1 \le \kappa_G \le 2$ under our assumptions on G, the stationary renewal process we are considering is LRD and has Hurst index $\frac{1}{2}(3 - \kappa_G)$, and

$$U(t) - \lambda t = o(t^{2+\epsilon-\kappa_G}) \qquad \text{for every } \epsilon > 0, \qquad (2.6)$$

with $U(t) - \lambda t = O(t^{2-\kappa_G})$ when $\int_0^\infty u^{\kappa_G} G(du) < \infty$ [see Daley (1999) for proof; this proof can be streamlined by using Sgibnev's more specific result quoted at (2.2), because the fact that $\mathcal{G}(t) = O(t^{2-\beta})$ when $\mathrm{E}(Y^\beta) < \infty$ and $\beta < 2$ follows from

$$\frac{1}{t^{2-\beta}} \int_0^\infty \min(u, t)\overline{G}(u)\, du \le \int_0^\infty \min(u^{\beta-1}, t^{\beta-1})\overline{G}(u)\, du$$

$$\le \int_0^\infty u^{\beta-1}\overline{G}(u)\, du = \frac{\mathrm{E}(Y^\beta)}{\beta} < \infty].$$

For the alternating renewal processes (ARPs) we consider, assume that the independent generic lifetime r.v.s X_j of its phases have finite first moments and d.f.s F_j, $j = 0, 1$ for the OFF and ON phases respectively. Write $G = F_0 * F_1$ for the d.f. of a generic cycle time $X_0 + X_1 = Y$ say, and $1/\lambda = \mathrm{E}(Y) = \int_0^\infty \overline{G}(u)\, du$ for its mean. Define also $\varpi = \lambda\mathrm{E}(X_1)$. The point process $N_{\mathrm{cyc}}(\cdot)$ consisting of the successive epochs where e.g. OFF phases start is a renewal process with lifetime d.f. G. Below we use U to denote its renewal process, and write H_{cyc} for its Hurst index when $\mathrm{E}(Y^2) = \infty$, in which case the asymptotic behaviour $U(t) \sim \lambda t$ $(t \to \infty)$ of the elementary renewal theorem is supplemented by (2.2) above.

Now consider a Cox process N directed by the random measure ξ at (1.4). Since $\xi(0, t] \approx \sum_{i=1}^{N_{\mathrm{cyc}}(0,t]} X_{1i}$, where X_{1i} denote the lifetimes of the ON phases in the interval $(0, t]$, an obvious conjecture is that the random measure ξ should have asymptotic behaviour for its variance proportional to that of var $N_{\mathrm{cyc}}(0,t]$ at (2.2), and that the Hurst index H as at (1.2) should coincide with H_{cyc}. We ultimately show that this is broadly correct (see part (b) of Theorem 2.1) under the supplementary Condition A concerning a ratio that plays a bridging role between the LRD ARP $N_{\mathrm{cyc}}(\cdot)$ and the lifetime d.f.s for its component phases. Without this condition the more detailed conclusion in part (a) of the theorem holds.

Condition A   The ratio

$$\frac{\mathcal{F}_1(t)}{\mathcal{F}_0(t)} = \frac{\int_0^\infty \min(u,t)\,\overline{F}_1(u)\,\mathrm{d}u}{\int_0^\infty \min(u,t)\,\overline{F}_0(u)\,\mathrm{d}u} \tag{2.7}$$

converges as $t \to \infty$ to a possibly infinite limit γ.

Theorem 2.1   *Let $\xi(\cdot)$ denote a stationary random measure as at (1.4), being determined by the ON phases of an alternating renewal process whose phases have nonzero lifetimes with finite means but whose cycle times have infinite second moment. Then $\xi(\cdot)$ is long-range dependent and has Hurst index*

$$H_\xi = \inf\left\{h: \limsup_{t\to\infty} \frac{\operatorname{var}\xi(0,t]}{t^{2h}} < \infty\right\} = \tfrac{1}{2}(3 - \kappa_G) \tag{2.8}$$

that coincides with the Hurst index H_{cyc} of $N_{\mathrm{cyc}}(\cdot)$.

(a) *As $t \to \infty$,*

$$\lambda^{-2}\operatorname{var}\xi(0,t] \sim 2\varpi^2 \int_0^t \mathcal{F}_0(u)\,\mathrm{d}u + 2(1-\varpi)^2 \int_0^t \mathcal{F}_1(u)\,\mathrm{d}u. \tag{2.9}$$

(b) *When Condition A is satisfied so the limit γ of (2.7) exists,*

$$\begin{aligned}
\frac{\operatorname{var}\xi(0,t]}{\lambda^2 \operatorname{var}N_{\mathrm{cyc}}(0,t]} &\to \frac{\varpi^2 + \gamma(1-\varpi)^2}{1+\gamma} \\
&= \frac{\lambda^2\big([\mathrm{E}(X_1)]^2 + \gamma[\mathrm{E}(X_0)]^2\big)}{1+\gamma} \quad (t \to \infty).
\end{aligned} \tag{2.10}$$

On the other hand, when the cycle times have finite second moment $\mathrm{E}(Y^2)$,

$$\begin{aligned}
\operatorname{var}\xi(0,t] &= \lambda^2\big(\varpi^2\mathrm{E}(X_0^2) + (1-\varpi)^2\mathrm{E}(X_1^2)\big)t + O(1) \\
&= \lambda^4\big([\mathrm{E}(X_1)]^2\mathrm{E}(X_0^2) + [\mathrm{E}(X_0)]^2\mathrm{E}(X_1^2)\big)t + O(1). \tag{2.11}
\end{aligned}$$

Inspection of the formulae in the three cases shows them to be consistent *modulo* the different assumptions of finiteness.

When we establish these second order properties in the next section, we use certain probabilities and first moment properties that we develop here. Start by defining for $t > 0$

$$p_{01}(t) = \Pr\{I(t) = 1 \mid \text{OFF phase starts at } 0\}, \qquad (2.12a)$$

$$p_{11}(t) = \Pr\{I(t) = 1 \mid \text{ON phase starts at } 0\}, \qquad (2.12b)$$

$$\pi_{11}(t) = \Pr\{I(0) = I(t) = 1\}, \qquad (2.12c)$$

and the associated first moment functions

$$M_0(t) = \mathrm{E}[\xi(0,t] \mid \text{OFF phase starts at } 0]$$

$$= \mathrm{E}\left[\int_0^t I(u)\,\mathrm{d}u \,\middle|\, \text{OFF phase starts at } 0\right] = \int_0^t p_{01}(u)\,\mathrm{d}u,$$

$$(2.13a)$$

$$M_1(t) = \mathrm{E}[\xi(0,t] \mid \text{ON phase starts at } 0] = \int_0^t p_{11}(u)\,\mathrm{d}u. \qquad (2.13b)$$

Elementary renewal arguments justify the relations

$$M_0(t) = \int_0^t M_1(t-u)\,F_0(\mathrm{d}u), \qquad (2.14a)$$

$$M_1(t) = t\overline{F}_1(t) + \int_0^t \left(u + M_0(t-u)\right) F_1(\mathrm{d}u)$$

$$= \int_0^t \overline{F}_1(u)\,\mathrm{d}u + \int_0^t M_1(t-v)\,G(\mathrm{d}v). \qquad (2.14b)$$

Equation (2.14b) shows that $M_1(\cdot)$ satisfies a generalized renewal equation, and is therefore expressible as

$$M_1(t) = \int_0^t U(\mathrm{d}u) \int_0^{t-u} \overline{F}_1(v)\,\mathrm{d}v = \int_0^t U(t-v)\overline{F}_1(v)\,\mathrm{d}v. \qquad (2.15)$$

Now appeal to the elementary renewal theorem and apply dominated convergence to conclude that

$$\frac{M_1(t)}{t} = \int_0^t \frac{U(t-v)}{t}\,\overline{F}_1(v)\,\mathrm{d}v \to \lambda\mathrm{E}(X_1) = \varpi \qquad (t \to \infty). \quad (2.16)$$

From (2.15) we can also write

$$M_1(t) = \int_0^t [U(t-v) - \lambda(t-v)]\overline{F}_1(v)\,\mathrm{d}v + \lambda \int_0^t (t-v)\overline{F}_1(v)\,\mathrm{d}v,$$

where the last term equals $\lambda t E(X_1) - \lambda \int_0^\infty \min(v, t) \overline{F}_1(v)\, dv = \varpi t - \lambda \mathcal{F}_1(t)$, so

$$M_1(t) - \varpi t = \int_0^t [U(t-v) - \lambda(t-v)]\overline{F}_1(v)\, dv - \lambda \mathcal{F}_1(v). \qquad (2.17)$$

If now $E(Y^2) < \infty$ and G is nonarithmetic, then as $t \to \infty$,

$$M_1(t) - \varpi t \to \tfrac{1}{2}\lambda^2 E(X_1)E(Y^2) - \tfrac{1}{2}\lambda E(X_1^2)$$
$$= \tfrac{1}{2}\lambda\big[\varpi E(X_0^2) + 2\varpi E(X_0)E(X_1) - (1-\varpi)E(X_1^2),\big] \qquad (2.18a)$$

while when $E(Y^2) = \infty$, with $\mathcal{G}(\cdot)$ as in (2.2),

$$M_1(t) - \varpi t \sim E(X_1)\lambda^2 \mathcal{G}(t) - \lambda \mathcal{F}_1(t)$$
$$= \lambda\big(\varpi \mathcal{G}(t) - \mathcal{F}_1(t)\big) \qquad (2.18b)$$
$$\sim \lambda\big(\varpi \mathcal{F}_0(t) - (1-\varpi)\mathcal{F}_1(t)\big);$$

this last relation comes from (2.21) below and the assumption that $E(Y^2) = \infty$ so as $t \to \infty$, $\mathcal{G}(t) \to \infty$ and $\mathcal{F}_j(t) \to \infty$ for at least one $j = 0, 1$. Condition A may now be called into play to yield a ratio form of this asymptotic result:

$$\frac{M_1(t) - \varpi t}{\lambda \mathcal{G}(t)} = \int_0^t \frac{U(t-v) - \lambda(t-v)}{\lambda \mathcal{G}(t)}\, \overline{F}_1(v)\, dv - \frac{\mathcal{F}_1(t)}{\mathcal{G}(t)}. \qquad (2.19)$$

On the right-hand side, as $t \to \infty$, the ratio in the integral converges for fixed v to λ by Sgibnev's result at (2.2), and therefore, since $\mathcal{G}(t)$ is monotonic in t, the whole integral converges to $\lambda E(X_1) = \varpi$ by Lemma A. From the first inequality in (2.21), the last term in (2.19) is bounded by 0 and 1 for all t. We therefore have the condition in (a) of Proposition 2.2 below for the left-hand side of (2.19) to converge as $t \to \infty$.

Part (b) of Proposition 2.2 follows from the relation, stemming from (2.14a), that

$$M_0(t) - \varpi t = \int_0^t [M_1(t-u) - \varpi(t-u)]\, F_0(du) - \varpi \int_0^t \overline{F}_0(u)\, du. \quad (2.20)$$

Proposition 2.2 (a) *When $\mathcal{G}(t) \to \infty$ as $t \to \infty$, the limit $\lim_{t \to \infty} [M_1(t) - \varpi t]/\lambda \mathcal{G}(t)$ exists if and only if Condition A is satisfied, and the limit then equals $[\varpi - \gamma(1-\varpi)]/(1+\gamma)$.*

(b) *When the limit in (a) exists, so also does $\lim_{t \to \infty} [M_0(t) - \varpi t]/\lambda \mathcal{G}(t)$ and the limits are the same.*

By substituting from (2.17) into (2.20), we obtain the alternative form

$$M_0(t) - \varpi t = \int_0^t [U(v) - \lambda v] \, [\overline{G}(t - v) - \overline{F}_0(t - v)] \, dv$$

$$- \lambda \int_0^t \mathcal{F}_1(t - u) \, F_0(du) - \varpi \int_0^t \overline{F}_0(v) \, dv \qquad (2.20')$$

$$\sim \lambda \varpi \mathcal{G}(t) - \lambda \mathcal{F}_1(t) + O(1).$$

The term involving $\overline{G}$ in (2.20') equals $t - \lambda \int_0^t (t - v) \overline{G}(v) \, dv = \lambda \mathcal{G}(t)$, but we do not see what simplification may come from this.

The following result is elementary but worth recording here for reference (it has already been used above).

Lemma A *When the nonnegative function $f(\cdot)$ is integrable over $\mathbb{R}_+$ and the measurable boundedly finite positive function $g(\cdot)$ satisfies $g(t)/g(t + u) \to 1$ $(t \to \infty)$ for every fixed u, $\int_0^t g(t - u) f(u) \, du \sim g(t) \int_0^\infty f(u) \, du$ for $t \to \infty$.*

Part of Theorem 2.1 exploits a decomposition of $\mathcal{G}(t)$ at (2.2) in terms of $\mathcal{F}_j(t)$ $(j = 0, 1)$, as in Lemma B.

Lemma B *The integrated tail functions $\mathcal{G}(\cdot)$ and $\mathcal{F}_j(\cdot)$ $(j = 0, 1)$, where $G = F_0 * F_1$, satisfy*

$$0 \leq \mathcal{G}(t) - \mathcal{F}_0(t) - \mathcal{F}_1(t) \leq \mathrm{E}(X_0) \mathrm{E}(X_1) \qquad (\text{all } 0 < t < \infty). \quad (2.21)$$

Proof Starting from $\overline{G}(u) = \overline{F}_0(u) + \int_0^u \overline{F}_1(u - v) \, F_0(dv)$, we have

$$\int_0^\infty \min(u, t)[\overline{G}(u) - \overline{F}_0(u)] \, du$$

$$= \int_0^\infty \min(u, t) \, du \int_0^u \overline{F}_1(u - v) F_0(dv)$$

$$= \int_0^\infty F_0(dv) \int_v^\infty \min(u, t) \overline{F}_1(u - v) \, du$$

$$= \int_0^\infty F_0(dv) \int_0^\infty \min(u + v, t) \overline{F}_1(u) \, du,$$

so

$$\mathcal{G}(t) - \mathcal{F}_0(t) - \mathcal{F}_1(t)$$

$$= \int_0^\infty \min(u, t)[\overline{G}(u) - \overline{F}_0(u) - \overline{F}_1(u)] \, du$$

$$= \int_0^\infty F_0(dv) \int_0^\infty [\min(u + v, t) - \min(u, t)] \overline{F}_1(u) \, du,$$

which is both ≥ 0 and $\leq \int_0^\infty v F_0(dv) \int_0^\infty \overline{F}_1(u)\, du = \mathrm{E}(X_0)\mathrm{E}(X_1)$ as asserted in (2.21).   □

In a companion paper (Daley *et al.*, 2007) we have used the extension of (2.21), proved by induction, to the tail function $\mathcal{F}^{n*}$ of F^{n*} for $n = 2$, 3, ... as

$$0 \leq \mathcal{F}^{n*}(t) - n\mathcal{F}(t) \leq \tfrac{1}{2}n(n-1)[\mathrm{E}(X)]^2. \tag{2.22}$$

3 Second moments

Equation (1.4) defines a stationary random measure $\xi(\cdot)$ which has boundedly finite second moment because its density $I(\cdot)$ is bounded, with $\mathrm{E}[I(t)] = \lambda\mathrm{E}(X_1) = \varpi$ so that

$$\mathrm{E}\big(\xi(0,t]\big) = \varpi t \qquad (t \geq 0). \tag{3.1}$$

Below we write $\lambda_j = 1/\mathrm{E}(X_j)$ for $j = 0, 1$, and suppose that $\mathrm{E}(Y^2) = \infty$. We are interested in

$$\begin{aligned}
\mathrm{var}\,\xi(0,t] &= \mathrm{E}\bigg(\int_0^t [I(u) - \varpi]\, du \int_0^t I(v)\, dv \bigg) \\
&= 2\int_0^t (t-v)\mathrm{E}\big(I(0)[I(v) - \varpi]\big) dv,
\end{aligned} \tag{3.2}$$

where the latter expression arises from stationarity and shows that $\mathrm{var}\,\xi(0,t]$ is differentiable with derivative

$$\begin{aligned}
2\int_0^t \mathrm{E}\big([I(0) - \varpi]I(u)\big) du &= 2\int_0^t \mathrm{E}\big(I(0)I(u)\big)\, du - 2\varpi^2 t \\
&= 2\int_0^t [\pi_{11}(u) - \varpi^2]\, du.
\end{aligned} \tag{3.3}$$

Then when the right-hand side here $\to \infty$ for $t \to \infty$, it follows that

$$\mathrm{var}\,\xi(0,t] \sim 2\int_0^t du \int_0^u [\pi_{11}(v) - \varpi^2]\, dv.$$

At time 0 the stationary alternating renewal process is in an ON phase with probability ϖ, conditional on which the time x backwards to the most recent start of the phase has density $\lambda_1 \overline{F}_1(x)$. By also tracing the ARP forward to the start of the next OFF phase (if there is such in

$(0, t))$, we deduce that the second-order joint probability $\pi_{11}(\cdot)$ satisfies

$$\pi_{11}(t)$$

$$= \varpi \int_0^\infty \left(\frac{\overline{F}_1(x+t)}{\overline{F}_1(x)} + \int_x^{x+t} \frac{F_1(dv)}{\overline{F}_1(x)} p_{01}\big(t - (v - x)\big) \right) \lambda_1 \overline{F}_1(x)\,dx$$

$$= \lambda \int_0^\infty \left(\overline{F}_1(x+t) + \int_x^{x+t} p_{01}(t + x - v) F_1(dv) \right) dx. \qquad (3.4)$$

Thus,

$$\int_0^t \pi_{11}(u)\,du = \lambda \int_0^t du \int_0^\infty \overline{F}_1(x+u)\,dx$$

$$+ \lambda \int_0^t du \int_0^\infty dx \int_x^{x+u} p_{01}(u + x - v) F_1(dv).$$

After manipulation the first term on the right-hand side here becomes $\lambda \int_0^\infty \min(x, t)\, \overline{F}_1(x)\,dx = \lambda \mathcal{F}_1(t)$, while the other term equals

$$\lambda \int_0^t du \int_0^\infty F_1(dv) \int_{(v-u)_+}^v p_{01}(u + x - v)\,dx$$

$$= \lambda \int_0^t du \int_0^\infty [M_0(u) - M_0((u - v)_+)] F_1(dv)$$

$$= \lambda \int_0^t M_0(u)\,du - \lambda \int_0^t du \int_0^u M_0(u - v) F_1(dv)$$

$$= \lambda \int_0^t M_0(u)\,du - \lambda \int_0^t M_0(u) F_1(t - u)\,du$$

$$= \lambda \int_0^t M_0(t - u)\, \overline{F}_1(u)\,du$$

$$= \lambda \int_0^t [M_0(t - u) - \varpi(t - u)]\, \overline{F}_1(u)\,du + \lambda \varpi \int_0^t (t - u) \overline{F}_1(u)\,du.$$

Now $\varpi^2 t = \lambda \varpi t \big(\int_0^t + \int_t^\infty \big) \overline{F}_1(x)\,dx$, so

$$\int_0^t [\pi_{11}(u) - \varpi^2]\,du$$

$$= \lambda(1 - \varpi) \int_0^\infty \min(x, t)\, \overline{F}_1(x)\,dx$$

$$+ \lambda \int_0^t [M_0(t - u) - \varpi(t - u)]\, \overline{F}_1(u)\,du$$

$$= \lambda(1 - \varpi)\mathcal{F}_1(t)$$

$$+ \lambda \int_0^t \overline{F}_1(t - u) \, du \left[\int_0^u [M_1(u - v) \right.$$

$$\left. - \varpi(u - v)] \, F_0(du) - \varpi \int_0^u \overline{F}_0(v) \, dv \right]$$

$$= \lambda(1 - \varpi)\mathcal{F}_1(t) + \lambda \int_0^t [M_1(w) - \varpi w] \left[\overline{G}(t - w) - \overline{F}_0(t - w) \right] dw$$

$$- \lambda \varpi \int_0^t \overline{F}_1(t - u) \, du \int_0^u \overline{F}_0(v) \, dv.$$

The term involving $M_1(\cdot)$ is asymptotically like $\lambda[\varpi \mathcal{F}_0(t) - (1 - \varpi)\mathcal{F}_1(t)]$ (see below (2.17)), and the last term $\to \lambda\varpi E(X_1)E(X_0) = \varpi^2 E(X_0)$ as $t \to \infty$, so when $E(Y^2) = \infty$,

$$\int_0^t [\pi_{11}(u) - \varpi^2] \, du \sim \lambda^2[\varpi^2 \mathcal{F}_0(t) + (1 - \varpi)^2 \mathcal{F}_1(t)] \qquad (t \to \infty), \quad (3.5)$$

which is monotonic. By inspection, the right-hand side $\to \infty$ as $t \to \infty$. Using (3.3) and integrating this derivative of $\operatorname{var}\xi(0, t]$, (2.9) follows.

From (2.3) and (2.20), $\operatorname{var} N_{\text{cyc}}(0, t]$ has derivative that $\sim \lambda^3 \mathcal{G}(t)$, so when Condition A holds, (2.10) follows from (2.9). Similarly, (2.11) follows from formulae for $\operatorname{var}\xi(0, t]$.

Notice that we cannot refine the term $O(1)$ in (2.11) without imposing some condition like at least one of F_j being nonarithmetic: consideration of the variance function of a stationary deterministic point process justifies this assertion.

It is straightforward to check that the rate of growth of $\mathcal{G}$ is the smaller of the rates of $\mathcal{F}_j$, and hence the Hurst index of $\xi(\cdot)$ (see (1.2)) is equal to the Hurst index of $N_{\text{cyc}}(\cdot)$.

Theorem 2.1 is established.

4 An alternating renewal process not satisfying Condition A

Let the r.v. X_0 have the atomic distribution F_0 that has masses $p_n = C_0/x_n^{3/2}$ at points $x_n = \exp(\gamma^{2n})$ for some finite $\gamma > 1$, $n = 1, 2,$ ... and C_0 a normalizing constant. Then $E(X_0)$ is finite but $E(X_0^2) = \sum_n C_0 x_n^{1/2} = \infty$. Since $x_n/x_{n+1} = \exp(-\gamma^{2n}(\gamma^2 - 1))$, we have both $x_n^{1/2}/(x_n^{1/2} + \cdots + x_1^{1/2}) \to 1$ and $x_n^{-1/2}/(x_n^{-1/2} + x_{n+1}^{-1/2} + \cdots) \to 1$, so for $\mathcal{F}_0$ as below (2.17),

$$\mathcal{F}_0(t) = \int_0^t \tfrac{1}{2}u^2 \, F_0(\mathrm{d}u) + t\mathrm{E}(X_0-t)_+ = C_0\left(\tfrac{1}{2}\sum_{x_n \leq t} x_n^{1/2} + t\sum_{x_n > t} \frac{x_n - t}{x_n^{3/2}}\right),$$

(4.1)

and $\mathcal{F}_0(x_n) \sim \tfrac{1}{2}C_0 x_n^{1/2}$ for $n \to \infty$.

Let the r.v. X_1 also be atomic but with atoms of mass $C_1/y_n^{3/2}$ at points $y_n = \exp(\gamma^{2n+1})$, with γ as for F_0, so $y_{n-1} < x_n < y_n$. Then $\mathcal{F}_0(y_n) \sim \tfrac{1}{2}C_0 x_n^{1/2}$ as for $\mathcal{F}_0(x_n)$. For $\mathcal{F}_1$, the dominant terms in $\mathcal{F}_1(x_n)$ and $\mathcal{F}_1(y_n)$ are $\tfrac{1}{2}C_1 y_{n-1}^{1/2}$ and $\tfrac{1}{2}C_1 y_n^{1/2}$ respectively. Thus, for $n \to \infty$,

$$\frac{\mathcal{F}_1(x_n)}{\mathcal{F}_0(x_n)} \approx \exp[\tfrac{1}{2}(\gamma^{2n-1} - \gamma^{2n})] \to 0$$

but

$$\frac{\mathcal{F}_1(y_n)}{\mathcal{F}_0(y_n)} \approx \exp[\tfrac{1}{2}(\gamma^{2n+1} - \gamma^{2n})] \to \infty.$$

It follows that for such F_0 and F_1 the ratio of Condition A oscillates ever more widely as t increases. Hence, the distinction between parts (a) and (b) of Theorem 2.1 is non-vacuous.

This counter-example has wider implications, namely, it can be adapted to show that if Y_n is an irreducible stationary aperiodic discrete-time Markov chain on countable state space that is long-range dependent in the sense that the variance of the number of visits to any state at the time epochs $1, \ldots, N$ is $> O(N)$ for large N (and there is then a uniform asymptotic rate applicable as shown in Sgibnev (1996) and, using a different technique of proof, in Carpio and Daley (2007)), the plausible relation that $\mathrm{var}\left(y_{Y_1} + \cdots + y_{Y_N}\right)$ should have the same asymptotic behaviour for any functional y_{Y_n} with finite variance—here y_i is an arbitrary real-valued function on the state space subject to this moment condition—cannot be established in such generality.

5 Postlude

For an alternating renewal process in which the lifetimes of the ON phases are exponentially distributed the function $p_{11}(\cdot)$ at (2.12b) is in fact a p-function of moderately general form (e.g. Kingman, 1972, pp. 37–39). Two aspects of studying its long-range dependence behaviour are pertinent. First, from Kingman (1964b) it follows that the stationary Cox process N_{Cox} driven by these ON phases is in fact a stationary renewal process, for which long-range dependence behaviour is described in Daley (1999). Second, it is appropriate to describe the associated regenerative phenomenon as long-range dependent when the generic lifetime of the OFF phases of the alternating renewal process, of finite first moment, has infinite second moment; for this case the limit γ of (2.7) in Condition A is zero.

Finally, Appendix B of the companion paper (Daley *et al.*, 2007), after noting that the renewal function U of a renewal process is subadditive in the sense that

$$U(x+y) \le U(x) + U(y) \qquad (x, y \ge 0), \tag{5.1}$$

asserts that the analogous expectation function for a Markov renewal process is subadditive also, but this is false as shown by an alternating renewal process (ARP) with generic lifetimes $X_0 = \epsilon$ and $X_1 = 2 - \epsilon$ for some ϵ in $(0, \frac{1}{2})$. For such an ARP,

$$U_{\mathrm{ARP}}(x) = \begin{cases} 1\frac{1}{2} & (\epsilon < x < 2 - \epsilon), \\ 3\frac{1}{2} & (2 + \epsilon < x < 4 - \epsilon), \end{cases}$$

so for $1 + \frac{1}{2}\epsilon < x < 2 - \epsilon$, $U_{\mathrm{ARP}}(2x) = 3\frac{1}{2} > 3 = 2U_{\mathrm{ARP}}(x)$, hence (5.1) does not hold universally. What is true here is that $V(x) = \frac{1}{2} + U_{\mathrm{ARP}}(x)$ is subadditive.

This counter example is easily extended to a Markov renewal process (MRP) on d states through which the chain proceeds in cyclic order, and whose lifetimes are 'small' for $d - 1$ states and sufficiently large for the remaining state so as to give a total cycle time equal to d. The analogous expectation function $U_{\mathrm{MRP}}(x) = \frac{1}{2}(d+1)$ for x around $\frac{1}{2}d$, and the function $V(x) = \frac{1}{2}(d-1) + U_{\mathrm{MRP}}(x)$ is subadditive. More generally, there is a finite positive constant c such that $c + U_{\mathrm{MRP}}(x)$ is subadditive as at (5.1), for every irreducible MRP on d states. The example just outlined suggests that the least possible universal constant equals $\frac{1}{2}(d-1)$, but we have no proof at present.

For a p-function, $1 + r(x)$ (cf. the Preamble above) is the renewal function of an orderly point process for which subadditivity at (5.1) holds, though as noted on p. 100 of Kingman (1964a) the smaller function $r(x)$ is subadditive, stemming from the lack of memory property of the exponential function and zero rate (for N_{Cox}) of the OFF state.

Acknowledgement This work was done first using facilities as an Honorary Visiting Fellow in the Centre for Mathematics and its Applications at the Australian National University, and subsequently as a Professorial Associate in the School of Mathematics and Statistics at the University of Melbourne.

References

Baltrūnas, A., Daley, D. J., and Klüppelberg, C. 2004. Tail behaviour of the busy period of a GI/G/1 queue with subexponential service times. *Stochastic Process Appl.*, **111**, 237–258.

Carpio, K. J. E., and Daley, D. J. 2007. Long-range dependence of Markov chains in discrete time on countable state space. *J. Appl. Probab.*, **44**, 1047–1055.

Cox, D. R. 1962. *Renewal Theory*. London: Methuen.

Daley, D. J. 1965. On a class of renewal functions. *Proc. Cambridge Philos. Soc.*, **61**, 519–526.

Daley, D. J. 1999. The Hurst index of long-range dependent renewal processes. *Ann. Probab.*, **27**, 2035–2041.

Daley, D. J. 2001. The moment index of minima. *Probability, Statistics and Seismology* (Festschrift for D. Vere-Jones), *J. Appl. Probab.*, **38A**, 33–36.

Daley, D. J., Rolski, T., and Vesilo, R. 2007. Long-range dependence in a Cox process directed by a Markov renewal process. *Statistics and Applied Probability: a Tribute to Jeffrey J. Hunter, J. Appl. Math. Decis. Sci.*, Article ID 83852.

Daley, D. J., and Vere-Jones, D. 2003. *An Introduction to the Theory of Point Processes, vol. I: Elementary Theory and Methods*, 2nd ed. New York: Springer-Verlag.

Daley, D. J., and Vesilo, R. 1997. Long range dependence of point processes, with queueing examples. *Stochastic Process Appl.*, **70**, 265–282.

Feller, W. 1971. *An Introduction to Probability Theory and its Applications*, vol. II, 2nd ed. New York: John Wiley & Sons.

Frenk, J. B. G. 1987. *On Banach Algebras, Renewal Measures and Regenerative Phenomena*. CWI Tracts, vol. 38. Amsterdam: Centrum voor Wiskunde en Informatica.

Heath, D., Resnick, S. I., and Samorodnitsky, G. 1998. Heavy tails and long range dependence in ON/OFF processes and associated fluid models. *Math. Oper. Res.*, **23**, 145–165.

Kingman, J. F. C. 1964a. The stochastic theory of regenerative events. *Z. Wahrscheinlichkeitstheorie verw. Geb.*, **2**, 180–224.

Kingman, J. F. C. 1964b. On doubly stochastic Poisson processes. *Proc. Cambridge Philos. Soc.*, **60**, 923–930.

Kingman, J. F. C. 1972. *Regenerative Phenomena*. London: John Wiley & Sons.

Sgibnev, M. S. 1981. On the renewal theorem in the case of infinite variance. *Sibirsk. Mat. Zh.*, **22**(5), 178–189. Translation in *Siberian Math. J.*, **22**, 787–796.

Sgibnev, M. S. 1996. An infinite variance solidarity theorem for Markov renewal functions. *J. Appl. Probab.*, **33**, 434–438.

8

Kernel methods and minimum contrast estimators for empirical deconvolution

Aurore Delaigle[a] and Peter Hall[b]

Abstract

We survey classical kernel methods for providing nonparametric solutions to problems involving measurement error. In particular we outline kernel-based methodology in this setting, and discuss its basic properties. Then we point to close connections that exist between kernel methods and much newer approaches based on minimum contrast techniques. The connections are through use of the sinc kernel for kernel-based inference. This 'infinite order' kernel is not often used explicitly for kernel-based deconvolution, although it has received attention in more conventional problems where measurement error is not an issue. We show that in a comparison between kernel methods for density deconvolution, and their counterparts based on minimum contrast, the two approaches give identical results on a grid which becomes increasingly fine as the bandwidth decreases. In consequence, the main numerical differences between these two techniques are arguably the result of different approaches to choosing smoothing parameters.

Keywords bandwidth, inverse problems, kernel estimators, local linear methods, local polynomial methods, minimum contrast methods, nonparametric curve estimation, nonparametric density estimation, nonparametric regression, penalised contrast methods, rate of convergence, sinc kernel, statistical smoothing

AMS subject classification (MSC2010) 62G08, 62G05

[a] Department of Mathematics and Statistics, The University of Melbourne, Parkville, VIC 3010, Australia; A.Delaigle@ms.unimelb.edu.au
[b] Department of Mathematics and Statistics, The University of Melbourne, Parkville, VIC 3010, Australia; halpstat@ms.unimelb.edu.au

1 Introduction

1.1 Summary

Our aim in this paper is to give a brief survey of kernel methods for solving problems involving measurement error, for example problems involving density deconvolution or regression with errors in variables, and to relate these 'classical' methods (they are now about twenty years old) to new approaches based on minimum contrast methods. Section 1.1 motivates the treatment of problems involving errors in variables, and section 1.2 describes conventional kernel methods for problems where the extent of measurement error is so small as to be ignorable. Section 2.1 shows how those standard techniques can be modified to take account of measurement errors, and section 2.2 outlines theoretical properties of the resulting estimators.

In section 3 we show how kernel methods for dealing with measurement error are related to new techniques based on minimum contrast ideas. For this purpose, in section 3.1 we specialise the work in section 2 to the case of the sinc kernel. That kernel choice is not widely used for density deconvolution, although it has previously been studied in that context by Stefanski and Carroll (1990), Diggle and Hall (1993), Barry and Diggle (1995), Butucea (2004), Meister (2004) and Butucea and Tsybakov (2007a,b). Section 3.2 outlines some of the properties that are known of sinc kernel estimators, and section 3 points to the very close connection between that approach and minimum contrast, or penalised contrast, methods.

1.2 Errors in variables

Measurement errors arise commonly in practice, although only in a minority of statistical analyses is a special effort made to accommodate them. Often they are minor, and ignoring them makes little difference, but in some problems they are important and significant, and we neglect them at our peril.

Areas of application of deconvolution, and regression with measurement error, include the analysis of seismological data (e.g. Kragh and Laws, 2006), financial analysis (e.g. Bonhomme and Robin, 2008), disease epidemiology (e.g. Brookmeyer and Gail, 1994, Chapter 8), and nutrition.

The latter topic is of particular interest today, for example in connection with errors-in-variables problems for data gathered in food frequency questionnaires (FFQs), or dietary questionnaires for epidemiological studies (DQESs). Formally, an FFQ is 'A method of dietary assessment in which subjects are asked to recall how frequently certain foods were consumed during a specified period of time,' according to the Nutrition Glossary of the European Food Information Council. An FFQ seeks detailed information about the nature and quantity of food eaten by the person filling in the form, and often includes a query such as, "How many of the above servings are from fast food outlets (McDonalds, Taco Bell, etc.)?" (Stanford University, 1994). This may seem a simple question to answer, but nutritionists interested in our consumption of fat generally find that the quantity of fast food that people admit to eating is biased downwards from its true value. The significant concerns in Western society about fat intake, and about where we purchase our oleaginous food, apparently influences our truthfulness when we are asked probing questions about our eating habits.

Examples of the use of statistical deconvolution in this area include the work of Stefanski and Carroll (1990) and Delaigle and Gijbels (2004b), who address nonparametric density deconvolution from measurement-error data, obtained from FFQs during the second National Health and Nutrition Examination Survey (1976–1980); Carroll *et al.* (1997), who discuss design and analysis aspects of linear measurement-error models when data come from FFQs; Carroll *et al.* (2006), who use measurement-error models, and deconvolution methods, to develop marginal mixed measurement-error models for each nutrient in a nutrition study, again when FFQs are used to supply the data; and Staudenmayer *et al.* (2008), who employ a dataset from nutritional epidemiology to illustrate the use of techniques for nonparametric density deconvolution. See Carroll *et al.* (2006, p. 7) for further discussion of applications to data on nutrition.

How might we correct for errors in variables? One approach is to use methods based on deconvolution, as follows. Let us write Q for the quantity of fast food that a person admits to eating, in a food frequency questionnaire; let Q_0 denote the actual amount of fast food; and put $R = Q/Q_0$. We expect that the distribution of R will be skewed towards values greater than 1, and we might even have an idea of the shape of the distribution responsible for this effect, i.e. the distribution of $\log R$. Indeed, we typically work with the logarithm of the formula $Q = Q_0 R$, and in that context, writing $W = \log Q$, $X = \log Q_0$ and $U = \log R$, the equation defining the variables of interest is:

$$W = X + U . \tag{1.1}$$

We have data on W, and from that we wish to estimate the distribution of X, i.e. the distribution of the logarithm of fast-food consumption.

It can readily be seen that this problem is generally not solvable unless the distribution of U, and the joint distribution of X and U, are known. In practice we usually take X and U to be independent, and undertake empirical deconvolution (i.e. estimation of the distribution, or density, of X from data on W) for several candidates for the distribution of U. If we are able to make repeated measurements of X, in particular to gather data on $W^{(j)} = X + U^{(j)}$ for $1 \leq j \leq m$, say, then we have an opportunity to estimate the distribution of U as well.

It is generally reasonable to assume that X, $U^{(1)}$, ..., $U^{(M)}$ are independent random variables. The distribution of U can be estimated whenever $m \geq 2$ and the distribution is uniquely determined by $|\phi_U|^2$, where ϕ_U denotes the characteristic function of U. The simplest example of this type is arguably that where U has a symmetric distribution for which the characteristic function does not vanish on the real line. One example of repeated measurements in the case $m = 2$ is that where a food frequency questionnaire asks at one point how many times we visited a fast food outlet, and on a distant page, how many hamburgers or servings of fried chicken we have purchased.

The model at (1.1) is simple and interesting, but in examples from nutrition science, and in many other problems, we generally wish to estimate the response to an explanatory variable, rather than the distribution of the explanatory variable. Therefore the proper context for our food frequency questionnaire example is really regression, not distribution or density estimation. In regression with errors in variables we observe data pairs (W, Y), where

$$W = X + U, \quad Y = g(X) + V, \tag{1.2}$$

$g(x) = E(Y \mid X = x)$, and the random variable V, denoting an experimental error, has zero mean. In this case the standard regression problem is altered on account of errors that are incurred when measuring the value of the explanatory variable. In (1.2) the variables U, V and X are assumed to be independent.

The measurement error U, appearing in (1.1) and (1.2), can be interpreted as the result of a 'laboratory error' in determining the 'dose' X which is applied to the subject. For example, a laboratory technician might use the dose X in an experiment, but in attempting to determine the dose after the experiment they might commit an error U, with the result that the actual dose is recorded as $X + U$ instead of X. Another

way of modelling the effect of measurement error is to reverse the roles
of X and W, so that we observe (W, Y) generated as

$$X = W + U, \quad Y = g(X) + V. \tag{1.3}$$

Here a precise dose W is specified, but when measuring it prior to the
experiment our technician commits an error U, with the result that
the actual dose is $W + U$. In (1.3) it assumed that U, V and W are
independent.

The measurement error model (1.2) is standard. The alternative model
(1.3) is believed to be much less common, although in some circumstances it is difficult to determine which of (1.2) and (1.3) is the more
appropriate. The model at (1.3) was first suggested by Berkson (1950),
for whom it is named.

1.3 Kernel methods

If the measurement error U were very small then we could estimate the
density f of X, and the function g in the model (1.2), using standard
kernel methods. For example, given data $X_1, \ldots, X_n$ on X we could
take

$$\hat{f}(x) = \frac{1}{nh} \sum_{i=1}^{n} K\left(\frac{x - X_i}{h}\right) \tag{1.4}$$

to be our estimator of $f(x)$. Here K is a kernel function and h, a positive
quantity, is a bandwidth. Likewise, given data $(X_1, Y_1), \ldots, (X_n, Y_n)$ on
(X, Y) we could take

$$\hat{g}(x) = \frac{\sum_i Y_i K\{(x - X_i)/h\}}{\sum_i K\{(x - X_i)/h\}} \tag{1.5}$$

to be our estimator of $g(x)$, where g is as in the model at (1.2).

The estimator at (1.4) is a standard kernel density estimator, and is
itself a probability density if we take K to be a density. It is consistent
under particularly weak conditions, for example if f is continuous and
$h \to 0$ and $nh \to \infty$ as n increases. Density estimation is discussed at
length by Silverman (1986) and Scott (1992). The estimator $\hat{g}$, which we
generally also compute by taking K to be a probability density, is often
referred to as the 'local constant' or Nadaraya–Watson estimator of g.
The first of these names follows from the fact that $\hat{g}(x)$ is the result of

fitting a constant to the data by local least squares:

$$\hat{g}(x) = \operatorname*{argmin}_{c} \sum_{i=1}^{n} (Y_i - c)^2 \, K\left(\frac{x - X_i}{h}\right). \tag{1.6}$$

The estimator $\hat{g}$ is also consistent under mild conditions, for example if the variance of the error, V, in (1.2) is finite, if f and g are continuous, if $f > 0$ at the point x where we wish to estimate g, and if $h \to 0$ and $nh \to \infty$ as n increases. General kernel methods are discussed by Wand and Jones (1995), and statistical smoothing is addressed by Simonoff (1996).

Local constant estimators have the advantage of being relatively robust against uneven spacings in the sequence $X_1, \ldots, X_n$. For example, the ratio at (1.5) never equals a nonzero number divided by zero. However, local constant estimators are particularly susceptible to boundary bias. In particular, if the density of X is supported and bounded away from zero on a compact interval, then $\hat{g}$, defined by (1.5) or (1.6), is generally inconsistent at the endpoints of that interval. Issues of this type have motivated the use of local polynomial estimators, which are defined by $\hat{g}(x) = \hat{c}_0(x)$ where, in a generalisation of (1.6),

$$(\hat{c}_0(x), \ldots, \hat{c}_p(x)) = \operatorname*{argmin}_{(c_0, \ldots, c_p)} \sum_{i=1}^{n} \left\{ Y_i - \sum_{j=0}^{p} c_j \, (x - X_i)^j \right\}^2 K\left(\frac{x - X_i}{h}\right). \tag{1.7}$$

See, for example, Fan and Gijbels (1996). In (1.7), p denotes the degree of the locally fitted polynomial. The estimator $\hat{g}(x) = \hat{c}_0(x)$, defined by (1.7), is also consistent under the conditions given earlier for the estimator defined by (1.5) and (1.6).

In the particular case $p = 1$ we obtain a local-linear estimator of $g(x)$:

$$\hat{g}(x) = \frac{S_2(x) \, T_0(x) - S_1(x) \, T_1(x)}{S_0(x) \, S_2(x) - S_1(x)^2}, \tag{1.8}$$

where

$$
\begin{aligned}
S_r(x) &= \frac{1}{nh} \sum_{i=1}^{n} \left(\frac{x - X_i}{h}\right)^r K\left(\frac{x - X_i}{h}\right), \\
T_r(x) &= \frac{1}{nh} \sum_{i=1}^{n} Y_i \left(\frac{x - X_i}{h}\right)^r K\left(\frac{x - X_i}{h}\right),
\end{aligned} \tag{1.9}
$$

h denotes a bandwidth and K is a kernel function.

Estimators of all these types can be quickly extended to cases where errors in variables are present, for example as in the models at (1.1) and (1.2), simply by altering the kernel function K so that it acts to cancel out the influence of the errors. We shall give details in section 2. Section 3 will discuss recently introduced methodology which, from some viewpoints looks quite different from, but is actually almost identical to, kernel methods.

2 Methodology and theory

2.1 Definitions of estimators

We first discuss a generalisation of the estimator at (1.4) to the case where there are errors in the observations of X_i, as per the model at (1.1). In particular, we assume that we observe data $W_1, \ldots, W_n$ which are independent and identically distributed as $W = X + U$, where X and U are independent and the distribution of U has known characteristic function ϕ_U which does not vanish anywhere on the real line. Let K be a kernel function, write $\phi_K = \int e^{itx} K(x)\, dx$ for the associated Fourier transform, and define

$$K_U(x) = \frac{1}{2\pi} \int e^{-itx} \frac{\phi_K(t)}{\phi_U(t/h)}\, dt. \qquad (2.1)$$

Then, to construct an estimator $\hat{f}$ of the density $f = f_X$ of X, when all we observe are the contaminated data $W_1, \ldots, W_n$, we simply replace K by K_U, and X_i by W_i, in the definition of $\hat{f}$ at (1.4), obtaining the estimator

$$\hat{f}_{\text{decon}}(x) = \frac{1}{nh} \sum_{i=1}^{n} K_U\left(\frac{x - W_i}{h}\right). \qquad (2.2)$$

Here the subscript 'decon' signifies that $\hat{f}_{\text{decon}}$ involves empirical deconvolution. The adjustment to the kernel takes care of the measurement error, and results in consistency in a wide variety of settings. Likewise, if data pairs $(W_1, Y_1), \ldots, (W_n, Y_n)$ are generated under the model at (1.2) then, to construct the local constant estimator at (1.5), or the local linear estimator defined by (1.8) and (1.9), all we do is replace each X_i by W_i, and K by K_U. Other local polynomial estimators can be calculated using a similar rule, replacing $h^{-r}(x - X_i)^r K\{(x - X_i)/h\}$ in S_r and T_r by $K_{U,r}\{(x - W_i)/h\}$, where

$$K_{U,r}(x) = \frac{1}{2\pi i^r} \int e^{-itx} \frac{\phi_K^{(r)}(t)}{\phi_U(t/h)}\, dt.$$

The estimator at (2.2) dates from work of Carroll and Hall (1988) and

Stefanski and Carroll (1990). Deconvolution-kernel regression estimators in the local-constant case were developed by Fan and Truong (1993), and extended to the general local polynomial setting by Delaigle *et al.* (2009).

The kernel K_U is deliberately constructed to be the function whose Fourier transform is $\phi_K/\phi_U(\cdot/h)$. This adjustment permits cancellation of the influence of errors in variables, as discussed at the end of section 1.3. To simplify calculations, for example computation of the integral in (1.2), we generally choose K not to be a density function but to be a smooth, symmetric function for which ϕ_K vanishes outside a compact interval. The commonly-used candidates for ϕ_K are proportional to functions that are used for K, rather than ϕ_K, in the case of regular kernel estimation discussed in section 1.3. For example, kernels K for which $\phi_K(t) = (1 - |t|^r)^s$ for $|t| \leq 1$, and $\phi_K(t) = 0$ otherwise, are common; here r and s are integers. Taking $r = 2s = 2$, $r = s = 2$ and $r = \frac{2}{3} s = 2$ corresponds to the Fourier inverses of the biweight, quartic and triweight kernels, respectively. Taking $s = 0$ gives the inverse of the uniform kernel, i.e. the sinc kernel, which we shall meet again in section 3. Further information about kernel choice is given by Delaigle and Hall (2006).

These kernels, and others, have the property that $\phi_K(t) = 1$ when $t = 0$, thereby guaranteeing that $\int K = 1$. The latter condition ensures that the density estimator, defined at (2.2) and constructed using this kernel, integrates to 1. (However, the estimator defined by (2.2) will generally take negative values at some points x.) The normalisation property is not so important when the kernel is used to construct regression estimators, where the effects of multiplying K by a constant factor cancel from the 'deconvolution' versions of formulae (1.5) and (1.8), and likewise vanish for all deconvolution-kernel estimators based on local polynomial methods.

Note that, as long as ϕ_K and ϕ_U are supported either on the whole real line or on a symmetric compact domain, the kernel K_U, defined by (2.1), and its generalised form $K_{U,r}$, are real-valued. Indeed, using properties of the complex conjugate of Fourier transforms of real-valued functions, and the change of variable $u = -t$, we have, using the notation $\overline{a}(t)$ for the complex conjugate of a complex-valued function a of a real variable t,

$$\overline{K}_{U,r}(x) = (-1)^{-r} \frac{1}{2\pi i^r} \int e^{itx} \frac{\overline{\phi_K^{(r)}(t)}}{\overline{\phi_U}(t/h)} \, dt$$

$$= (-1)^{-r} \frac{1}{2\pi i^r} \int e^{itx} \frac{(-1)^{-r} \phi_K^{(r)}(-t)}{\phi_U(-t/h)} \, dt$$

$$= \frac{1}{2\pi i^r} \int e^{-iux} \frac{\phi_K^{(r)}(u)}{\phi_U(u/h)} \, du = K_{U,r}(x).$$

In practice it is almost always the case that the distribution of U is symmetric, and in the discussion of variance in section 2.2, below, we shall make this assumption. We shall also suppose that K is symmetric, again a condition which holds almost invariably in practice.

The estimators discussed above were based on the assumption that the characteristic function ϕ_U of the errors in variables is known. This enabled us to compute the deconvolution kernel K_U at (2.1). In cases where the distribution of U is not known, but can be estimated from replicated data (see section 1.2), we can replace ϕ_U by an estimator of it and, perhaps after a little regularisation, compute an empirical version of K_U. This can give good results, in both theory and practice. In particular, in many cases the resulting estimator of the density of X, or the regression mean g, can be shown to have the same first-order properties as estimators computed under the assumption that the distribution of U is known. Details are given by Delaigle *et al.* (2008).

Methods for choosing the smoothing parameter, h, in the estimators discussed above have been proposed by Hesse (1999), Delaigle and Gijbels (2004a,b) and Delaigle and Hall (2008).

2.2 Bias and variance

The expected value of the estimator at (2.2) equals

$$E\{\hat{f}_{\text{decon}}(x)\} = \frac{1}{2\pi h} \int E\left[e^{-it\{x-W\}/h}\right] \frac{\phi_K(t)}{\phi_U(t/h)} \, dt$$

$$= \frac{1}{2\pi} \int e^{-itx} \frac{\phi_K(ht)}{\phi_U(t)} \, \phi_X(t) \, \phi_U(t) \, dt$$

$$= \frac{1}{2\pi} \int e^{-itx} \phi_K(ht) \, \phi_X(t) \, dt = \frac{1}{h} \int K(u/h) \, f(x-u) \, du$$

$$= E\{\hat{f}(x)\}, \tag{2.3}$$

where the first equality uses the definition of K_U, and the fourth equality uses Plancherel's identity. Therefore the deconvolution estimator $\hat{f}_{\text{decon}}(x)$, calculated from data contaminated by measurement errors, has exactly the same mean, and therefore the same bias, as $\hat{f}(x)$, which would be computed using values of X_i observed without measurement error. This confirms that using the deconvolution kernel estimator does indeed allow for cancellation of measurement errors, at least in terms of their presence in the mean.

Of course, variance is a different matter. Since $\hat{f}_{\mathrm{decon}}(x)$ equals a sum of independent random variables then

$$
\mathrm{var}\{\hat{f}_{\mathrm{decon}}(x)\}
$$
$$
= \left(nh^2\right)^{-1} \mathrm{var}\left\{K_U\left(\frac{x-W}{h}\right)\right\}
$$
$$
\sim (nh)^{-1} f_W(x) \int K_U^2 = \frac{f_W(x)}{2\pi nh} \int \phi_K(t)^2 \, |\phi_U(t/h)|^{-2} \, dt\,. \quad (2.4)
$$

(Here the relation $\sim$ means that the ratio of the left- and right-hand sides converges to 1 as $h \to 0$.) Thus it can be seen that the variance of $\hat{f}_{\mathrm{decon}}(x)$ depends intimately on tail behaviour of the characteristic function ϕ_U of the measurement-error distribution.

If ϕ_K vanishes outside a compact set, which, as we noted in section 2.1, is generally the case, and if $|\phi_U|$ is asymptotic to a positive regularly varying function ψ (see Bingham *et al.*, 1989), in the sense that $|\phi_U(t)| \asymp \psi(t)$ (meaning that the ratio of both sides is bounded away from zero and infinity as $t \to \infty$), then the integral on the right-hand side of (2.3) is bounded between two constant multiples of $\psi(1/h)^{-2}$ as $h \to 0$. Therefore by (2.4), provided that $f_W(x) > 0$,

$$
\mathrm{var}\{\hat{f}_{\mathrm{decon}}(x)\} \asymp (nh)^{-1} \, \psi(1/h)^{-2} \quad (2.5)
$$

as n increases and h decreases. Recall that we are assuming that f_U and K are both symmetric functions.

If the density f of X has two bounded and continuous derivatives, and if K is bounded and symmetric and satisfies $\int x^2 \, |K(x)| \, dx < \infty$, then the bias of $\hat{f}_{\mathrm{decon}}$ can be found from (2.3), using elementary calculus and arguments familiar in the case of standard kernel estimators:

$$
\mathrm{bias}(x) = E\{\hat{f}_{\mathrm{decon}}(x)\} - f(x) = E\{\hat{f}(x)\} - f(x)
$$
$$
= \int K(u) \, \{f(x-hu) - f(x)\} \, du = \tfrac{1}{2} h^2 \, \kappa \, f''(x) + o\!\left(h^2\right) \quad (2.6)
$$

as $h \to 0$, where $\kappa = \int x^2 \, K(x) \, dx$. Therefore, provided that $f''(x) \neq 0$, the bias of the conventional kernel estimator $\hat{f}(x)$ is exactly of size h^2 as $h \to 0$. Combining this property, (2.3) and (2.5) we deduce a relatively concise asymptotic formula for the mean squared error of $\hat{f}_{\mathrm{decon}}(x)$:

$$
E\{\hat{f}_{\mathrm{decon}}(x) - f(x)\}^2 \asymp h^4 + (nh)^{-1} \, \psi(1/h)^{-2}\,. \quad (2.7)
$$

For a given error distribution we can work out the behaviour of $\psi(1/h)$ as $h \to 0$, and then from (2.7) we can calculate the optimal bandwidth and determine the exact rate of convergence of $\hat{f}_{\mathrm{decon}}(x)$ to $f(x)$, in mean square. In many instances this rate is optimal, in a minimax sense; see, for example, Fan (1991). It is also generally optimal in the case of the errors-in-variables regression estimators discussed in section 2.1, based on deconvolution-kernel versions of local polynomial estimators. See Fan and Truong (1993).

Therefore, despite their almost naive simplicity, deconvolution-kernel estimators of densities and regression functions have features that can hardly be bettered by more complex, alternative approaches. The results derived in the previous paragraph, and their counterparts in the regression case, imply that the estimators are limited by the extent to which they can recover from the data. (This is reflected in the fact that the rate of decay of the tails of ϕ_U drives the results on convergence rates.) However, the fact that the estimators are nevertheless optimal, in terms of their rates of convergence, implies that this restriction is inherent to the problem, not just to the estimators; no other estimators would have a better convergence rate, at least not uniformly in a class of problems.

3 Relationship to minimum contrast methods

3.1 Deconvolution kernel estimators based on the sinc kernel

The sinc, or Fourier integral, kernel is given by

$$L(x) = \begin{cases} (\pi x)^{-1} \sin(\pi x) & \text{if } x \neq 0 \\ 1 & \text{if } x = 0 \,. \end{cases} \tag{3.1}$$

Its Fourier transform, defined as a Riemann integral, is the 'boxcar function', $\phi_L(t) = 1$ if $|t| \leq 1$ and $\phi_L(t) = 0$ otherwise. In particular, ϕ_L vanishes outside a compact set, which property, as we noted in section 2.1, aids computation. The version of K_U, at (2.1), for the sinc kernel is

$$L_U(x) = \frac{1}{2\pi} \int_{-1}^{1} e^{-itx} \, \phi_U(t/h)^{-1} \, dt = \frac{1}{\pi} \int_0^1 \cos(tx) \, \phi_U(t/h)^{-1} \, dt \,,$$

where the second identity holds if the distribution of U is symmetric and has no zeros on the real line.

The kernel L is sometimes said to be of 'infinite order', in the sense that if a is any function with an infinite number of bounded, integrable derivatives then

$$\int \left[\int \{a(x+hu) - a(x)\} L(u)\, du \right]^2 dx = O(h^r) \qquad (3.2)$$

as $h \downarrow 0$, for all $r > 0$. If K were of finite order then (3.2) would hold only for a finite range of values of r, no matter how many derivatives the function a enjoyed. For example, if K were a symmetric function for which $\int u^2 K(u)\, du \neq 0$, and if we were to replace L in (3.2) by K, then (3.2) would hold only for $r \leq 4$, not for all r. In this case we would say that K was of second order, because

$$\int \{a(x+hu) - a(x)\} K(u)\, du = O(h^2)\,.$$

If we take a to be the density, f, of the random variable X, and take K in the definition of $\hat{f}$ at (1.4) to be the sinc kernel, L, then (3.2) equals the integral of the squared bias of $\hat{f}$. Therefore, in the case of a very smooth density, the 'infinite order' property of the sinc kernel ensures particularly small bias, in an average sense.

Properties of conventional kernel density estimators, but founded on the sinc kernel, for data without measurement errors, have been studied by, for example, Davis (1975, 1977). Glad *et al.* (1999) have provided a good survey of properties of sinc kernel methods for density estimation, and have argued that those estimators have received an unfairly bad press. Despite criticism of sinc kernel estimators (see e.g. Politis and Romano, 1999), the approach is "more accurate for quite moderate values of the sample size, has better asymptotics in non-smooth cases (the density to be estimated has only first derivative), [and] is more convenient for bandwidth selection etc" than its conventional competitors, suggest Glad *et al.* (1999).

The property of greater accuracy is borne out in both theoretical and numerical studies, and derives from the infinite-order property noted above. Indeed, if f is very smooth then the low level of average squared bias can be exploited to produce an estimator $\hat{f}$ with particularly low mean squared error, in fact of order n^{-1} in some cases. The most easily seen disadvantage of sinc-kernel density estimators is their tendency to suffer from spurious oscillations, inherited from the infinite number of oscillations of the kernel itself.

These properties can be expected to carry over to density and regression estimators based on contaminated data, when we use the sinc kernel. To give a little detail in the case of density estimation from data contaminated by measurement errors, we note that if the density f of X is infinitely differentiable, but we observe only the contaminated data $W_1, \ldots, W_n$ distributed as W, generated as at (1.1); if we use the density estimator at (1.4), but computed using $K = L$, the sinc kernel; and if $|\phi_U(t)| \geq C(1 + |t|)^{-\alpha}$ for constants C, $\alpha > 0$; then, in view of (2.3), (2.4) and (3.2), we have for all $r > 0$,

$$
\int \{\hat{f}_{\text{decon}}(x) - f(x)\}^2 \, dx
$$
$$
= \int \left\{ E\hat{f}(x) - f(x) \right\}^2 + (nh^2)^{-1} \int \text{var} \left\{ L_U \left(\frac{x - W}{h} \right) \right\} dx
$$
$$
\leq \int \left[\int \{f(x + hu) - f(x)\} L(u) \, du \right]^2 dx + (nh)^{-1} \int L_U^2
$$
$$
= O \left\{ h^r + (nh)^{-1} \int_{-1}^{1} |\phi_U(t/h)|^{-2} \, dt \right\}
$$
$$
= O \left\{ h^r + \left(nh^{2\alpha+1} \right)^{-1} \right\}. \tag{3.3}
$$

It follows that, if f has infinitely many integrable derivatives and if the tails of $\phi_U(t)$ decrease at no faster than a polynomial rate as $|t| \to \infty$, then the bandwidth h can be chosen so that the mean integrated squared error of a deconvolution kernel estimator of f, using the sinc kernel, converges at rate $O(n^{\epsilon-1})$ for any given $\epsilon > 0$.

This very fast rate of convergence contrasts with that which occurs if the kernel K is of only finite order. For example, if K is a second-order kernel, in which case (3.2) holds only for $r \leq 4$ when L is replaced by K, the argument at (3.3) gives:

$$
\int \{\hat{f}_{\text{decon}}(x) - f(x)\}^2 \, dx = O \left\{ h^4 + \left(nh^{2\alpha+1} \right)^{-1} \right\}.
$$

The fastest rate of convergence of the right-hand side to zero is attained with $h = n^{-1/(2\alpha+5)}$, giving

$$
\int \{\hat{f}_{\text{decon}}(x) - f(x)\}^2 \, dx = O \left(n^{-4/(2\alpha+5)} \right).
$$

In fact, this is generally the best rate of convergence of mean integrated squared error that can be obtained using a second-order kernel when the characteristic function ϕ_U decreases like $|t|^{-\alpha}$ in the tails, even if the density f is exceptionally smooth. Nevertheless, second-order kernels are often preferred to the sinc kernel in practice, since they do not suffer from the unwanted oscillations that afflict estimators based on the sinc kernel.

3.2 Minimum contrast estimators, and their relationship to deconvolution kernel estimators

In the context of the measurement error model at (1.1), Comte *et al.* (2007) suggested an interesting minimum contrast estimator of the density f of X. Their approach has applications in a variety of other settings (see Comte *et al.*, 2006, 2008; Comte and Taupin, 2007), including to the regression model at (1.2), and the conclusions we shall draw below apply in these cases too. Therefore, for the sake of brevity we shall treat only the density deconvolution problem.

To describe the minimum contrast estimator in that setting, define

$$\hat{a}_{k\ell} = \frac{1}{2\pi n} \sum_{j=1}^{n} \int \exp(it\, W_j)\, \phi_{L_{k\ell}}(t)\, \phi_U(t)^{-1}\, dt \,,$$

where $\phi_{L_{k\ell}}$ denotes the Fourier transform of the function $L_{k\ell}$ defined by $L_{k\ell}(x) = \ell^{1/2}\, L(\ell\, x - k)$, k is an integer and $\ell > 0$. In this notation the minimum contrast nonparametric density estimator is

$$\tilde{f}(x) = \sum_{k=-k_0}^{k_0} \hat{a}_{k\ell}\, L_{k\ell}(x)\,.$$

There are two tuning parameters, k_0 and ℓ. Comte *et al.* (2007) suggest choosing ℓ to minimise a penalisation criterion.

The resulting minimum contrast estimator is called a penalised contrast density estimator. The penalisation criterion suggested by Comte *et al.* (2007) for choosing ℓ is related to cross-validation, although its exact form, which involves the choice of additional terms and multiplicative constants, is based on simulation experiments. It is clear on inspecting the definition of $\tilde{f}$ that ℓ plays a role similar to that of the inverse of bandwidth in a conventional deconvolution kernel estimator. In particular, ℓ should diverge to infinity with n. Comte *et al.* (2007) suggest taking $k_0 = 2^m - 1$, where $m \geq \log_2(n + 1)$ is an integer. In numerical experiments they use $m = 8$, which gives good performance in the cases they consider. More generally, k_0/ℓ should diverge to infinity as sample size increases.

The minimum contrast density estimator of Comte *et al.* (2007) is actually very close to the standard deconvolution kernel density estimator at (1.4), where in the latter we use the sinc kernel at (3.1). Indeed, as the theorem below shows, the two estimators are exactly equal on a grid, which becomes finer as the bandwidth, h, for the sinc kernel density estimator decreases. However, this relationship holds only for values of x for which $|x| \leq k_0/\ell$; for larger values of $|x|$ on the grid, $\tilde{f}(x)$ vanishes. (This property is one of the manifestations of the fact that, as noted earlier, k and ℓ generally should be chosen to depend on sample size in such a manner that $k_0/\ell \to \infty$ as $n \to \infty$.)

Theorem *Let $\hat{f}_{\mathrm{decon}}$ denote the deconvolution kernel density estimator at (1.4), constructed using the sinc kernel and employing the bandwidth $h = \ell^{-1}$. Then, for any point $x = hk$ with k an integer, we have*

$$\tilde{f}(x) = \begin{cases} \hat{f}_{\mathrm{decon}}(x) & \text{if } |x| \leq k_0/\ell \\ 0 & \text{if } |x| > k_0/\ell . \end{cases}$$

A proof of the theorem will be given in section 3.3. Between grid points the estimator $\tilde{f}$ is a nonstandard interpolation of values of the kernel estimator $\hat{f}_{\mathrm{decon}}$. Note that, if we take $h = \ell^{-1}$, the weights $L(\ell x - k) = L\{(x - hk)/h\}$ used in the interpolation decrease quickly as k moves further from x/h, and, except for small k, neighbour weights are close in magnitude but differ in sign. (Here L is the sinc kernel defined at (3.1).) In effect, the interpolation is based on rather few values $\hat{f}_{\mathrm{decon}}(k/\ell)$ corresponding to those k for which k is close to x/h.

In practice the two estimators are almost indistinguishable. For example, Figure 3.1 compares them using the bandwidth that minimises the integrated squared difference between the true density and the estimator, for one generated sample in the case where X is normal N(0, 1), U is Laplace with $\mathrm{var}(U)/\mathrm{var}(X) = 0.1$, and $n = 100$ or $n = 1000$. In the left graphs the two estimators can hardly be distinguished. The right graphs show magnifications of these estimators for $x \in [-\frac{1}{2}, 0]$. Here it can be seen more clearly that the minimum contrast estimator is an approximation of the deconvolution kernel estimator, and is exactly equal to the latter at $x = 0$.

These results highlight the fact that the differences in performance between the two estimators derive more from different tuning parameter choices than from anything else. In their comparison, Comte *et al.* (2007) used a minimum contrast estimator with the sinc kernel L and a bandwidth chosen by penalisation, whereas for the deconvolution kernel estimator they employed a conventional second-order kernel K and a different bandwidth-choice procedure. Against the background of

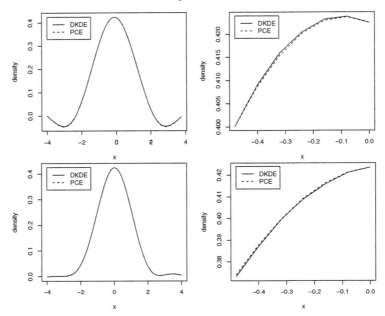

Figure 3.1 Deconvolution kernel density estimator (DKDE) and minimum contrast estimator (PCE) for a particular sample of size
$n = 100$ (upper panels) or $n = 1000$ (lower panels) in the case
$\text{var}(U)/\text{var}(X) = 0.1$. Right panels show magnifications of the estimates for $x \in [-0.5, 0]$ in the respective upper panels.

the theoretical analysis in section 3.1, the different kernel choices (and
different ways of choosing smoothing parameters) explain the differences
observed between the penalised contrast density estimator and the deconvolution kernel density estimator based on a second-order kernel.

3.3 Proof of Theorem

Note that $\phi_{L_{k\ell}}(t) = \ell^{-1/2} \exp(itk/\ell)\,\phi_L(t/\ell)$ and

$$\hat{a}_{k\ell} = \frac{1}{2n\pi\ell^{1/2}} \sum_{j=1}^{n} \int_{-\ell\pi}^{\ell\pi} \exp\left\{ - it\left(k\,\ell^{-1} - W_j\right)\right\} \frac{\phi_L(t/\ell)}{\pi_U(t)}\,dt\,.$$

Therefore,

$$\tilde{f}(x)$$

$$= \frac{1}{2n\pi} \sum_{k=-k_0}^{k_0} L(\ell x - k) \sum_{j=1}^{n} \int_{-\ell\pi}^{\ell\pi} \exp\left\{ -it\left(k\ell^{-1} - W_j\right) \right\} \frac{\phi_L(t/\ell)}{\pi_U(t)} \, dt$$

$$= \sum_{k=-k_0}^{k_0} L(\ell x - k) \, \hat{f}_{\text{decon}}(k/\ell) \, . \tag{3.4}$$

If r is a nonzero integer then $L(r) = 0$. Therefore, if $x = kh = s/\ell$ for an integer s then $L(\ell x - k) = 0$ whenever $k \neq s$, and $L(\ell x - k) = 1$ if $k = s$. Hence, (3.4) implies that $\tilde{f}(x) = \hat{f}_{\text{decon}}(x)$ if $|k| \leq k_0$, and $\tilde{f}(x) = 0$ otherwise.

References

Barry, J., and Diggle, P. J. 1995. Choosing the smoothing parameter in a Fourier approach to nonparametric deconvolution of a density estimate. *J. Nonparametr. Stat.*, **4**, 223–232.

Berkson, J. 1950. Are there two regression problems? *J. Amer. Statist. Assoc.*, **45**, 164–180.

Bingham, N. H., Goldie, C. M., and Teugels, J. L. 1989. *Regular Variation*, revised ed. Encyclopedia Math. Appl., vol. 27. Cambridge: Cambridge Univ. Press.

Bonhomme, S., and Robin, J.-M. 2008. *Generalized Nonparametric Deconvolution with an Application to Earnings Dynamics*. University College London, Centre for Microdata Methods & Practice working paper 3/08; http://www.cemmap.ac.uk/wps/cwp308.pdf.

Brookmeyer, R., and Gail, M. H. 1994. *AIDS Epidemiology: a Quantitative Approach*. Oxford: Oxford Univ. Press.

Butucea, C. 2004. Deconvolution of supersmooth densities with smooth noise. *Canad. J. Statist.*, **32**, 181–192.

Butucea, C., and Tsybakov, A. B. 2007a. Sharp optimality for density deconvolution with dominating bias, I. *Theory Probab. Appl.*, **52**, 111–128.

Butucea, C., and Tsybakov, A. B. 2007b. Sharp optimality for density deconvolution with dominating bias, II. *Theory Probab. Appl.*, **52**, 336–349.

Carroll, R. J., and Hall, P. 1988. Optimal rates of convergence for deconvolving a density. *J. Amer. Statist. Assoc.*, **83**, 1184–1186.

Carroll, R. J., Freedman, L. S., and Pee, D. 1997. Design aspects of calibration studies in nutrition, with analysis of missing data in linear measurement error models. *Biometrics*, **53**, 1440–1457.

Carroll, R. J., Midthune, D., Freedman, L. S., and Kipnis, V. 2006. Seemingly unrelated measurement error models, with application to nutritional epidemiology. *Biometrics*, **62**, 75–84.

Comte, F., Rozenholc, Y., and Taupin, M.-L. 2006. Penalized contrast estimator for adaptive density deconvolution. *Canad. J. Statist.*, **34**, 431–452.

Comte, F., Rozenholc, Y., and Taupin, M.-L. 2007. Finite sample penalization in adaptive density deconvolution. *J. Stat. Comput. Simul.*, **77**, 977–1000.

Comte, F., Rozenholc, Y., and Taupin, M.-L. 2008. Adaptive density estimation for general ARCH models. *Econometric Theory*, **24**, 1628–1662.

Comte, F., and Taupin, M.-L. 2007. Nonparametric estimation of the regression function in an errors-in-variables model. *Statist. Sinica*, **17**, 1065–1090.

Davis, K. B. 1975. Mean square error properties of density estimates. *Ann. Statist.*, **3**, 1025–1030.

Davis, K. B. 1977. Mean integrated square error properties of density estimates. *Ann. Statist.*, **5**, 530–535.

Delaigle, A., Fan, J., and Carroll, R. J. 2009. A design-adaptive local polynomial estimator for the errors-in-variables problem. *J. Amer. Statist. Assoc.*, **104(485)**, 348–359.

Delaigle, A., and Gijbels, A. 2004a. Bootstrap bandwidth selection in kernel density estimation from a contaminated sample. *Ann. Inst. Statist. Math.*, **56**, 19–47.

Delaigle, A., and Gijbels, A. 2004b. Practical bandwidth selection in deconvolution kernel density estimation. *Comput. Statist. Data Anal.*, **45**, 249–267.

Delaigle, A., and Hall, P. 2006. On the optimal kernel choice for deconvolution. *Statist. Probab. Lett.*, **76**, 1594–1602.

Delaigle, A., and Hall, P. 2008. Using SIMEX for smoothing-parameter choice in errors-in-variables problems. *J. Amer. Statist. Assoc.*, **103**, 280–287.

Delaigle, A., Hall, P., and Meister, A. 2008. On deconvolution with repeated measurements. *Ann. Statist.*, **36**, 665–685.

Diggle, P., and Hall, P. 1993. A Fourier approach to nonparametric deconvolution of a density estimate. *J. Roy. Statist. Soc. Ser. B*, **55**, 523–531.

Fan, J. 1991. On the optimal rates of convergence for nonparametric deconvolution problems. *Ann. Statist.*, **19**, 1257–1272.

Fan, J., and Gijbels, I. 1996. *Local Polynomial Modelling and its Applications*. London: Chapman and Hall.

Fan, J., and Truong, Y. K. 1993. Nonparametric regression with errors in variables. *Ann. Statist.*, **21**, 1900–1925.

Glad, I. K., Hjort, N. L., and Ushakov, N. 1999. *Density Estimation Using the Sinc Kernel*. Department of Mathematical Sciences, Norwegian University of Science & Technology, Trondheim, Statistics Preprint No. 2/2007; http://www.math.ntnu.no/preprint/statistics/2007/S2-2007.pdf.

Hesse, C. 1999. Data-driven deconvolution. *J. Nonparametr. Stat.*, **10**, 343–373.

Kragh, E., and Laws, R. 2006. Rough seas and statistical deconvolution. *Geophysical Prospecting*, **54**, 475–485.

Meister, A. 2004. On the effect of misspecifying the error density in a deconvolution problem. *Canad. J. Statist.*, **32**, 439–449.

Politis, D. N., and Romano, J. P. 1999. Multivariate density estimation with general flat-top kernels of infinite order. *J. Multivariate Anal.*, **68**, 1–25.

Scott, D. W. 1992. *Multivariate Density Estimation. Theory, Practice, and Visualization.* New York: John Wiley & Sons.

Silverman, B. W. 1986. *Density Estimation for Statistics and Data Analysis.* London: Chapman and Hall.

Simonoff, J. S. 1996. *Smoothing Methods in Statistics.* New York: Springer-Verlag.

Stanford University. 1994. *Food Frequency Questionnaire #1 2 3 4.* Available at `http://www.permanente.net/homepage/kaiser/pdf/6116.pdf`.

Staudenmayer, J., Ruppert, D., and Buonaccorsi, J. P. 2008. Density estimation in the presence of heteroscedastic measurement error. *J. Amer. Statist. Assoc.*, **103**, 726–736.

Stefanski, L., and Carroll, R. J. 1990. Deconvoluting kernel density estimators. *Statistics*, **2**, 169–184.

Wand, M. P., and Jones, M. C. 1995. *Kernel Smoothing.* London: Chapman and Hall.

9

The coalescent and its descendants

Peter Donnelly[a] and Stephen Leslie[b]

Abstract

The coalescent revolutionised theoretical population genetics, simplifying, or making possible for the first time, many analyses, proofs, and
derivations, and offering crucial insights about the way in which the
structure of data in samples from populations depends on the demographic history of the population. However statistical inference under
the coalescent model is extremely challenging, effectively because no explicit expressions are available for key sampling probabilities. This led
initially to approximation of these probabilities by ingenious application
of modern computationally-intensive statistical methods. A key breakthrough occurred when Li and Stephens introduced a different model,
similar in spirit to the coalescent, for which efficient calculations are
feasible. In turn, the Li and Stephens model has changed statistical inference for the wealth of data now available which documents molecular
genetic variation within populations. We briefly review the coalescent
and associated measure-valued diffusions, describe the Li and Stephens
model, and introduce and apply a generalisation of it for inference of
population structure in the presence of linkage disequilibrium.

AMS subject classification (MSC2010) 60J70, 62M05, 92D10

[a] Wellcome Trust Centre for Human Genetics, University of Oxford, Roosevelt
Drive, Oxford OX3 7BN, and Department of Statistics, 1 South Parks Road,
Oxford OX1 3TG; donnelly@stats.ox.ac.uk

[b] Department of Statistics, University of Oxford, 1 South Parks Road, Oxford
OX1 3TG; leslie@stats.ox.ac.uk

1 Introduction

John Kingman made a number of incisive and elegant contributions to modelling in the field of genetics, several of which are described elsewhere in this volume. But it is probably the coalescent, or 'Kingman coalescent' as it is often known, which has had the greatest impact. Several authors independently developed related ideas around the same time [20], [53], [30] but it was Kingman's description and formulation, together with his proofs of the key robustness results [34], [33], [32] which had the greatest impact in the mathematical genetics community.

More than 25 years later the coalescent remains central to much of population genetics, with book-level treatments of the subject now available [54], [28], [55]. As others have noted, population genetics as a field was theory rich and data poor for much of its history. Over the last five years this has changed beyond recognition. The development and wide application of high-throughput experimental techniques for assaying molecular genetic variation means that scientists are now awash with data. Enticing as this seems, it turns out that the coalescent model cannot be fully utilised for analysis of these data—it is simply not computationally feasible to do so. Instead, a closely related model, due to Li and Stephens, has proved to be computationally tractable and reliable as a basis for inference for modern genetics data. Alternative approaches are based on approximate inference under the coalescent.

Our purpose here is to give a sense of these developments, before describing and applying an extension of the Li and Stephens model to populations with geographical structure. We do not attempt an extensive review.

The initial historical presentation is necessarily somewhat technical in nature, and provides some explanation of the theoretical developments leading up to the Li and Stephens model. Readers interested in just the Li and Stephens model and its application may begin at Section 4 as the presentation of this material does not heavily depend on the previous sections.

Before doing so, one of us (PD) will indulge in a brief personal reminiscence. John Kingman was my doctoral supervisor. More accurately, he acted as my supervisor for a year, before leaving Oxford to Chair the then UK Science and Engineering Research Council (I have always hoped the decision to change career direction was unrelated to his supervisory experiences). During the year in question, Kingman wrote his three seminal coalescent papers. A photocopy of one of the manuscripts, in John's immaculate handwriting, unblemished by corrections or second thoughts as to wording, still survives in my filing cabinet.

There is a certain irony to the fact that although the coalescent was the unifying theme of much of my academic work for the following 20 years, it formed no part of my work under Kingman's supervision. John's strategy with new research students, or at least with this one, was to direct them to the journals in the library, note that some of the papers therein contained unsolved problems, and suggest that he would be happy to offer advice or suggestions if one were stuck in solving one of these. This was daunting, and the attempts were largely unsuccessful. It was only as Kingman was leaving Oxford that he passed on copies of the coalescent manuscripts, and, embarrassingly, it was some years before I saw the connection between the coalescent and aspects of my doctoral work on interacting particle systems, then under Dominic Welsh's supervision.

2 The coalescent and the Fleming–Viot process

To set the scene, we briefly outline the context in which the coalescent arises, and then describe the coalescent itself along with the corresponding process forward in time, the so-called Fleming–Viot measure-valued diffusion. We aim here only to give a brief flavour of the two processes rather than a detailed exposition.

The most basic, and oldest, models in population genetics are finite Markov chains which describe the way in which the genetic composition of the population changes over time. In most cases, these models are not tractable, and interest moves to their limiting behaviour as the population size grows large, under suitable re-scalings of time. When examined forward in time, this leads to a family of measure-valued diffusions, called Fleming–Viot processes. In a complementary, and for many purposes more powerful, approach one can instead look backwards in time, and focus on the genealogical tree relating sampled chromosomes. In the large population limit, these (random) trees converge to a particular process called the coalescent.

We start with the simplest setting in which individuals are *haploid*; that is they carry a single copy of their genetic material which is inherited from a single parent. Many organisms, including humans, are *diploid*, with each cell carrying two copies of the individual's DNA— these copies being inherited one from each of the individual's two parents. It turns out that the haploid models described below also apply to diploid organisms provided one is interested in modelling the evolution of small contiguous segments of DNA—for many purposes we can ignore the fact that in diploid organisms these small segments of DNA occur in pairs in individuals, and instead model the population of individual chromosomes, or more precisely of small segments of them taken from the same genomic region, in each individual. In what follows, we will respect this particular perspective and refer to the haploid 'individuals' in the population we are modelling as 'chromosomes'.

One simple discrete model for population demography is the Wright–Fisher model. Consider a population of fixed size N chromosomes which evolves in discrete generations. The random mechanism for forming the next generation is as follows: each chromosome in the next generation chooses a chromosome in the current generation (uniformly at random) and copies it, with the choices made by different chromosomes being independent. An equivalent description is that each chromosome in the current generation gives rise to a random number of copies in the next generation, with the joint distribution of these offspring numbers being symmetric multinomial.

In addition to modelling the demography of a population, a population genetics model needs to say something about the genetic types carried by the chromosomes in the population, and the way in which these change (probabilistically) when being copied from parental to offspring chromosomes. Formally, this involves specifying a set, E, of possible types (usually, if unimaginatively, called the *type space*), and a matrix of transition probabilities Γ whose i, jth entry, γ_{ij}, specifies for each $i, j \in E$, the probability that an offspring chromosome will be of type j when the parental chromosome is of type i. The generality has real advantages: different choices of type space E can be used for modelling different kinds of genetic information. In most genetic contexts, offspring are extremely likely to have the same type as their parent, with changes to this type, referred to as *mutations*, being extremely rare.

Under an assumption of genetic neutrality, all variants in a population are equally fit and are thus equally likely to be transmitted. This assumption allows a crucial simplification: the random process describing demography is independent of the genetic types carried by the individuals in the population. In this case, one can first generate the demography of the population using, say, the Wright–Fisher model, and then independently superimpose the genetic type for each chromosome, and the details of the (stochastic) mutation process which may change types. The separation of demography from genetic types lies at the heart of the simplification offered by the coalescent: the coalescent models the parts of the demography relevant to the population at the current time; information about genetic types can be added independently. The extent to which the neutrality assumption applies is rather controversial in general, and for humans in particular, but it seems likely that it provides a reasonable description for many parts of the genome.

The Wright–Fisher model may also be extended to allow for more realistic demographic effects, including variation in population size, and geographical spatial structure in the population (so that offspring chromosomes are more likely to be located near to their parents). We will not describe these here. Somewhat surprisingly, it transpires that the simple model described above, (constant population size, random mating, and neutrality—the so-called 'standard neutral' model), or rather its large population limit, captures many of the important features of the evolution of human and other populations. There is an aphorism in statistics that "all models are false, but some are useful". The standard neutral model has proved to be extremely useful.

In a Wright–Fisher, or any other, model, we could describe the genetic composition of the population at any point in time by giving a list of the genetic types currently present, and the proportion of the population currently of each type. Such a description corresponds to giving a probability measure on the set E of possible types. It is sometimes helpful to think of this measure as the distribution of the type of an individual picked at random from the population. Note that summarising the population composition in this way at a particular time point involves an assumption of exchangeability across individuals: it is only the types present, and the numbers of individuals of each type, in a particular population which matter, with information about precisely which individuals carry particular types not being relevant. In this framework, when we add details of the mutation process to the Wright–Fisher model, by specifying E and Γ, we obtain a discrete time (probability-) measure-valued Markov process. As N becomes large a suitable rescaling of the process converges to a diffusion limit: time is measured in units of N generations, and mutation probabilities, the off-diagonal entries of the matrix Γ above, are scaled as N^{-1}. For general genetic systems, the limit is naturally formulated as a measure-valued process, called the Fleming–Viot diffusion. The classical so-called Wright–Fisher diffusion is a one-dimensional diffusion on $[0,1]$ which arises when there are only two possible genetic types and one tracks the population frequency of one of the types. This is a special case of the Fleming–Viot diffusion, in which we can identify the value of the classical diffusion, $p \in [0, 1]$, with a probability measure on a set with just two elements. The beauty of the more general, measure-valued, formulation is that it allows much more complicated genetic types, which could track DNA sequences, or more exotically even keep track of the time since particular mutations arose in the population.

The Fleming–Viot process can thus be thought of as an approximation to a large population evolving according to the Wright–Fisher model. As noted, for the Wright–Fisher model, time is measured in units of N generations in this approximation (and the approximation applies when mutation probabilities are of order N^{-1}). In fact the Fleming–Viot process arises as the limit of a wide range of demographic models (and we refer to such models as being within the domain of attraction of the Fleming–Viot process), although the appropriate time scaling can differ between models. (See, for example, [14].) For background, including explicit formulations of the claims made above, see [10], [11] [13], [14], [15].

Donnelly and Kurtz [10], [11] give a discrete construction of the Fleming–Viot process. As a consequence, the process can actually be thought of as describing the evolution of a hypothetically infinite population, and it explicitly includes the demography of that population. Exchangeability figured prominently in Kingman's work in genetics. It provides a linking thread here: the Donnelly–Kurtz construction embeds population models for each finite population size N in an infinite exchangeable sequence. The value taken by the Fleming–Viot diffusion at a particular time point is just the de Finetti representing measure for the infinite exchangeable sequence. Given the value of the measure, the types of individuals in the population are independent and identically distributed according to that measure.

The coalescent arises by looking backwards in time. Consider again the discrete Wright–Fisher model. If we consider two different chromosomes in the current generation, they will share an ancestor in the previous generation with probability $1/N$. If not, they retain distinct ancestries, and will share an ancestor in the generation before that with probability $1/N$. The number of generations until they share an ancestor is thus geometrically distributed with success probability $1/N$ and mean N. In the limit for large N, with time measured in units of N generations, this geometric random variable converges to an exponential random variable with mean 1.

More generally, if we consider k chromosomes, then for fixed k and large N, they will descend from k distinct ancestors in the previous generation with probability

$$1 - \binom{k}{2}\frac{1}{N} + O(N^{-2}).$$

Exactly two will share a common ancestor in the previous generation with probability $\binom{k}{2}\frac{1}{N} + O(N^{-2})$, and more than a single pair will share a common ancestor with probability $O(N^{-2})$. In the limit as $N \to \infty$, with time measured in units of N generations, the time until any of the k share an ancestor will be exponentially distributed with mean $1/\binom{k}{2}$, after which time a randomly chosen pair of chromosomes will share an ancestor.

Thus, in the large population limit, with time measured in units of N generations, the genealogical history of a sample of size n may be described by a random binary tree. The tree initially has n branches, for a period of time T_n, after which a pair of branches (chosen uniformly at random, independently of all other events) will join, or coalesce. More generally, the times T_k, $k = n$, $n-1$, ..., 2 for which the tree has k branches are independent exponential random variables with

$$\mathrm{E}(T_k) = \binom{k}{2}^{-1},$$

after which a pair of branches (chosen uniformly at random independently of all other events) will join, or coalesce. The resulting random tree is called the n-coalescent, or often just the coalescent. Note that we have described the coalescent as a random tree. Kingman's original papers elegantly formulated the n-coalescent as a stochastic process on the set of equivalence relations on $\{1, 2, \dots, n\}$. The two formulations are equivalent. We view the tree description as more intuitive.

In a natural sense the tree describes the important part of the genealogical history of the sample, in terms of their genetic composition. It captures their shared ancestry, due to the demographic process. As noted above, a key observation is that in neutral models the distribution of this ancestry is independent of the genetic types which happen to be carried by the individuals in the population. Probabilistically, one can thus sample the coalescent tree and then superimpose genetic types. For example, at stationarity, first choose a type for the most recent common ancestor of the population (the type at the root of the coalescent tree) according to the stationary distribution of the mutation process, and then track types forward through the tree from the common ancestor, where they will possibly be changed by mutation.

The preceding recipe gives a simple means of simulating the genetic types of a sample of size n from the population. Note that this is an early example of what has more recently come to be termed 'exact simulation': a finite amount of simulation producing a sample with the exact distribution given by the stationary distribution of a Markov process. In addition, it is much more computationally efficient than simulating the entire population forward in time for a long period and then taking a sample from it. Finally, it reveals the complex structure of the distribution of genetics models at stationarity—the types of each of the sampled chromosomes are (positively) correlated, exactly because of their shared ancestral history.

We motivated the coalescent from the Wright–Fisher model, but the same limiting genealogical tree arises for any of the large class of demographic models in the domain of attraction of the Fleming–Viot diffusion. See Kingman [34] for an elegant formulation and proof of this kind of robustness result. Moreover, the ways in which the tree shape changes under different demographic scenarios (e.g. changes in population size or geographical population structure) is well understood [55], [28].

The discrete construction of the Fleming–Viot process described above actually embeds the coalescent and the forward diffusion in the same framework, so that one can think of the coalescent as describing the genealogy of a sample from the diffusion.

There is even a natural limit, as $n \to \infty$, of the n-coalescents, introduced and studied by Kingman [32]. This can be thought of as the limit of the genealogy of the whole population, or as the genealogy of the infinite population described by the Fleming–Viot process (though this perspective was not available when Kingman introduced the process). The analysis underlying the relevant limiting results for this population-genealogical process is much more technical than that outlined above for the fixed-sample-size case [8], [9], [10]. It is easiest to describe this tree from the root, representing the common ancestor of the population, forward to the tips, each of which represents an individual alive at the reference time. The tree has k branches for a random period of time T_k, after which a branch, chosen uniformly at random, independently for each k, splits to form two branches. The times T_k, $k = 2, 3, \ldots$, are independent exponential random variables, and independent of the topology of the tree, with

$$\mathrm{E}(T_k) = \binom{k}{2}^{-1}.$$

Write

$$T = \sum_{k=1}^{\infty} T_k$$

for the total depth of the tree, or equivalently for the time back until the population first has a common ancestor. Note that T is a.s. finite. In fact $\mathrm{E}(T) = 2$.

To date, we have focussed on models for a small segment of DNA. For larger segments, in diploid populations, one has to allow for the process of *recombination*. Consider a particular human chromosome inherited from a parent. The parent will have two (slightly different) copies of this chromosome. Think of the process which produces the chromosome to be passed on to the offspring as starting on one of the two chromosomes in the parent and copying from it along the chromosome. Occasionally, and for our purposes randomly, the copying process will 'cross over' to the other chromosome in the parent, and then copy from that, perhaps later jumping back and copying from the original chromosome, and so on. The chromosome passed on to the offspring will thus be made up as a mosaic of the two chromosomes in the parent. The crossings over are referred to as recombination events. In practice, these recombination events are relatively rare along the chromosome: for example in humans, there will typically be only a few recombination events per chromosome.

The formulation described above can be extended to allow for recombination. In the coalescent framework, the consequence is that in going backwards in time, different parts of a chromosome may be inherited from different chromosomes in the previous generation. One way of conceptualising this is to imagine each position in the DNA as having its own coalescent tree, tracing the ancestry of the DNA in that position. This coalescent tree, marginally, will have the same distribution as the coalescent. As one moves along the DNA sequence, these trees for different positions are highly positively correlated. In fact, two neighbouring positions will have the same tree iff there is no recombination event between those positions, on a lineage leading to the current sample, since their joint most recent common ancestor. If there is such a recombination, the trees for the two positions will be identical back to that point, but (in general) different before it. The correlation structure between the trees for different positions is complex. For example, when regarded as a process on trees as one moves along the sequence, it is not Markov. Nonetheless it is straightforward to simulate from the relevant joint distribution of trees, and hence of sampled sequences. The trees for each position can be embedded in a more general probabilistic object (this time a graph rather than a tree) called the ancestral recombination graph [21], [22].

3 Inference under the coalescent

The coalescent has revolutionised the way we think about and analyse population genetics models, and changed the way we simulate from these models. There are several important reasons for this. One is the separation, for neutral models, of demography from the effects of mutation. This means that many of the properties of samples taken from genetics models follow from properties of the coalescent tree. A second reason is that the coalescent has a simple, indeed beautiful, structure, which is amenable to calculation. Most questions of interest in population genetics can be rephrased in terms of the coalescent, and the coalescent is a fundamentally simpler process than the traditional forwards-in-time models.

The body of work outlined above has an applied probability flavour; some of it more applied (for example solving genetics questions of interest), and some more pure (for example the links with measure-valued diffusions). Historically, much of it occurred in the 10–15 years after Kingman's coalescent papers, but even today 'coalescent methods' as they have become known in population genetics, are central to the analysis of genetics models.

If an applied probability perspective prevailed over the first 10–15 years of the coalescent's existence, the last 10–15 years have seen a parallel interest in statistical questions. Since the early 1990s there has been a steady growth in data documenting molecular genetic variation in samples taken from real populations. Over recent years this has become a deluge, especially for humans. Instead of trying to study probabilistic properties of the coalescent, the statistical perspective assumes that some data come from a coalescent model, and asks how to do statistical inference for parameters in the model, or comparisons between models (for example arising from different demographic histories for the population).

There have been two broad approaches. One has been to attempt to use all the information in the data, by basing inference (in either a frequentist or Bayesian framework) on the likelihood under the model: the probability, regarded as a function of parameters of interest, of observing the configuration actually observed in the sample. This is the thread we will follow below. Full-likelihood inference under the coalescent turns out to be a difficult problem. A second approach has been to summarise the information in the data via a small set of summary statistics, and then to base inference on these statistics. One particular, Bayesian, version of this approach has come to be called *approximate Bayesian computation* (ABC): one approximates the full posterior distribution of parameters of interest conditional on the data by their conditional distribution given just the summary statistics [2].

Full-likelihood inference under the coalescent is not straightforward, for a simple reason. Although the coalescent enjoys many nice properties, and lends itself to many calculations, no explicit expressions are available for the required likelihoods. There is one exception to this, namely settings in which the mutation probabilities, γ_{ij}, that a chromosome is of type j when its parent is of type i, depend only on j. This so-called *parent-independent mutation* is unrealistic for most modern data, notwithstanding the fact that any two-allele model (that is, when the type space E consists of only two elements) can be written in this form. For parent-independent mutation models, the likelihood is multinomial.

In the absence of explicit expressions for the likelihood, indirect approaches, typically relying on sophisticated computational methods, were developed. Griffiths and Tavaré were the pioneers [23], [24], [25], [26]. They devised an ingenious computational approach whereby the likelihood was expressed as a functional of a Markov chain arising from systems of recursive equations for probabilities of interest. Felsenstein [19] later showed the Griffiths–Tavaré (GT) approach to be a particular implementation of importance sampling. In contrast to the GT approach, Felsenstein and colleagues developed Markov chain Monte Carlo (MCMC) methods for evaluating coalescent likelihoods. These were not without challenges: the space which MCMC methods explored was effectively that of coalescent trees, and thus extremely high-dimensional, and assessment of mixing and convergence could be fraught. As subsequent authors pointed out, failure of the Markov chains to mix properly resulted in poor approximations to the likelihood [18].

Donnelly and Stephens [49] adopted an importance sampling approach. They reasoned that if the GT approach was implicitly doing importance sampling, under a particular proposal distribution which arose automatically, then it might be possible to improve performance by explicitly choosing the proposal distribution. In particular, they noted that the optimal proposal distribution was closely related to a particular conditional probability under the coalescent, namely the probability that an additional, $n + 1$st sampled chromosome will have type j conditional on the observed types in a sample of n chromosomes from the population. This conditional probability under the coalescent does not seem to be available explicitly (except under the unrealistic parent-independent mutation assumption)—indeed an explicit expression for this probability leads naturally to one for the required likelihoods, and conversely.

Donnelly and Stephens exploited the structure of the discrete representation of the Fleming–Viot diffusion to approximate the key conditional probabilities. In effect, in the Donnelly–Kurtz process they fixed the types on the first n levels and ran the level $n+1$ process. This led naturally to an approximation to the conditional distribution of the $n + 1$st sampled chromosome given the types of the first n chromosomes, which in turn leads naturally to importance sampling proposal distributions. As had been hoped, importance sampling under this family of proposal distributions was considerably more efficient than under the GT scheme [49].

There has been considerably more activity in the area of inference under the coalescent over the last 10 years. We will not pursue this here, as our narrative will take a different path. Interested readers are referred to [54] and [48].

4 The Li and Stephens model

As we have noted, statistical inference under the coalescent is hard. From our perspective, a key breakthrough came earlier this decade from Li and Stephens [36]. Their idea was very simple, and it turns out to have had massive impact. Li and Stephens argued that instead of trying to do inference under the coalescent one should appreciate that the coalescent is itself only an approximation to reality, and that one might instead do inference under a model which shares many of the nice properties of the coalescent but also enjoys the additional property that full likelihood inference is straightforward.

Li and Stephens changed the model. Inference then became a tractable problem. What matters is how good these inferences are for real data sets. Although not obvious in advance, it turns out that for a very wide range of questions, inference under the Li and Stephens model works well in practice.

A forerunner to the Li and Stephens approach arose in connection with the problem of estimating haplotype phase from genotype data. Stephens, Smith, and Donnelly [52] introduced an algorithm, PHASE, in which the conditional distribution underpinning the Donnelly–Stephens importance-sampling proposal distribution was used directly in a pseudo Gibbs sampler. PHASE has been widely used, and even today provides one of the most accurate methods for computational recovery of haplotype phase. (Several recent approaches aim to speed up computations to allow phasing of genome-wide data sets, typically at a slight cost in accuracy e.g. Beagle [5], [4]; FastPhase [46]; and IMPUTE 2 [29]. See [39] for a review of some of these methods.)

We now describe the Li and Stephens model. For most modern data sets it is natural to do so in the context of 'SNPs'. A *SNP*, or single nucleotide polymorphism, is a position in the DNA sequence which is known to vary across chromosomes. At the overwhelming majority of SNPs there will be exactly two variants present in a population, and we assume this here. For ease, we will often code the variants as 0 and 1. To simplify the description of the model we assume haplotype data are available. This is equivalent to knowing the types at each SNP separately along each of the two chromosomes in a diploid individual. (Most experimental methods provide only the unordered pair of types on the two chromosomes at each SNP, without giving the additional information as to which variant is on which chromosome. As noted above, there are good statistical methods for estimating the separate haplotypes from these genotype data.)

It is convenient, and for many purposes most helpful, to describe the Li and Stephens model via the conditional probabilities it induces, and in particular by specifying the probability distribution for the $n + 1$st sampled chromosome given the types of the first n chromosomes. This in turn can be described by a recipe for simulating from this conditional distribution. (We return below to a probabilistic aside on this perspective.)

In effect, the Li and Stephens model simulates the $n+1$st chromosome as an imperfect mosaic of the first n chromosomes. To simulate the $n+1$st chromosome, first pick one of the existing n chromosomes at random. At the first SNP copy the type from the chosen chromosome, but with random 'error' in a way we will describe below. With high probability (specified below) the second SNP will be probabilistically copied from the same chromosome. Alternatively, the chromosome for copying at the second SNP will be re-chosen, uniformly and independently. Having simulated the type on the $n + 1$st chromosome at the kth SNP, the $k + 1$st SNP will be copied from the same chromosome as the kth SNP with high probability, and otherwise copied from a chromosome chosen independently, and uniformly at random, from the first n. It remains to specify the probabilistic copying mechanism: with high probability at a particular SNP, the value of the $n + 1$st chromosome will be the same as that on the chromosome being copied, otherwise it will have the opposite type.

The connection with the coalescent comes from the following. Consider the position of the first SNP on the $n + 1$st chromosome. Ignoring coalescences amongst the first n sampled chromosomes at this position, the ancestry of the $n + 1$st sampled chromosome coalesces with exactly one of the lineages leading to the n sampled chromosomes. Ignoring mutation, the type of the $n + 1$st chromosome at this position will be the same as the type on the chromosome with which its ancestry coalesces. To incorporate mutation one allows mis-copying of the ancestral type. This mis-copying is an oversimplification of the effect of mutation on the coalescent tree at this position. Now, moving along the $n + 1$st chromosome, there will be a segment, up to the first recombination event in the relevant history, which shares the same ancestry (and so is copied from the same one of the sampled chromosomes). The effect of recombination is to follow different ancestral chromosomes and this is mimicked in the Li and Stephens approach by choosing a different chromosome from which to copy. The probabilities of this change will depend on the recombination rates between the SNPs, and in a coalescent also on n, because coalescence of the lineage of the $n + 1$st chromosome to one of the other lineages happens faster for larger n.

We now describe the model more formally. Suppose that n chromosomes (each of which can be thought of as a haplotype) have been sampled from the population, where the jth haplotype has the SNP information at l SNPs, $c^j = \{c_1^j, c_2^j, \ldots, c_l^j\}$. Let us call this set of chromosomes C. Now suppose an additional chromosome i has been sampled and has SNP information $h^i = \{h_1^i, h_2^i, \ldots, h_l^i\}$. We seek to determine the probability of sampling this chromosome, based on its SNP haplotype and the SNP haplotypes of the previously sampled chromosomes C. The model takes as input fine-scale estimates of recombination rates in the region: $r = \{r_0, r_1, \ldots, r_l\}$ where $r_{j+1} - r_j$ is the average rate of crossover per unit physical distance per meiosis between sites j and $j + 1$ times the physical distance between them. We set $r_0 = 0$. We obtain this map from elsewhere (for example [31]) rather than estimating it for ourselves[1]. Note that the SNPs (and the map) are ordered by the position of the SNP (or map point) on the chromosome (for convenience we refer to the first SNP position as the leftmost position and the lth SNP position as the rightmost). We define the per-locus recombination probability $\rho_s = 1 - \exp(-4N_e(r_{s+1} - r_s)/n)$ and then define transition probabilities for a Markov chain on $\{1, 2, \ldots, n\}$ from state j (indicating that it is the jth haplotype of those that have been previously sampled that is 'parental') at position s to state k at position $s + 1$:

$$q(j_s, k_{s+1}) = \begin{cases} 1 - \rho_s + \rho_s/n, & j = k, \\ \rho_s/n, & j \neq k, \end{cases} \tag{4.1}$$

where N_e is the so-called effective population size, a familiar quantity in population genetics models. Equation (4.1) is related to the fact that recombination events occur along the sequence as a Poisson process. Here we use the Poisson rate $4N_e(r_{s+1} - r_s)/n$. Given the rate, the probability that there is no recombination between sites s and $s + 1$ is $\exp(-4N_e(r_{s+1} - r_s)/n) = 1 - \rho_s$. The probability of at least one recombination between sites s and $s+1$ is thus ρ_s. In this case the model has the assumption that it is equally likely that the recombination occurs with any of the n sampled haplotypes. In particular, as ρ_s incorporates the probability that multiple recombinations occur between sites s and $s + 1$, the first case in Equation (4.1) includes a ρ_s term to allow for the possibility that the same haplotype is parental at each site s and $s + 1$ even when one or more recombinations have occurred.

We define the copying probabilities in terms of the 'population mutation rate' for the given sample size (θ, defined below), another familiar quantity from population genetics. The mismatch (or not) between the SNP allele of the jth 'parent' chromosome at SNP s, c_s^j, and the SNP allele of the ith additional 'daughter' chromosome, h_s^i, is defined as

$$e(h_s^i, c_s^j) = \begin{cases} \frac{n}{n+\theta} + \frac{1}{2}\frac{\theta}{n+\theta}, & h_s^i = c_s^j, \\ \frac{1}{2}\frac{\theta}{n+\theta}, & h_s^i \neq c_s^j. \end{cases} \tag{4.2}$$

Notice that as $\theta \to \infty$ the alleles 0 and 1 at any given site become

---

[1] In fact, Li and Stephens developed their model precisely for estimating the genetic map, but for our purposes we wish to utilize the model for inference in other settings and thus we utilize a known genetic map.

equally likely. Equation (4.2) is motivated by a similar coalescent argument to that used for the transition probabilities above. In this case the probability of no mutations occurring at the site s is $n/(n+\theta)$ and thus the probability of at least one mutation occurring at s is $\theta/(n+\theta)$. It is possible to allow for the population mutation rate to vary sitewise if it is necessary to account for known variable mutation rates.

The particular form of the transition and copying probabilities in the model follow from the informal coalescent arguments given four paragraphs above ([36]). We noted that the recombination probabilities typically come from available estimates. It turns out that the accuracy of these estimates can be important in applications of the model. In contrast, such applications are generally not especially sensitive to the exact value of θ used. Thus, for the mutation probabilities, we follow Li and Stephens and set

$$\theta = \left(\sum_{z=1}^{n-1} \frac{1}{z} \right)^{-1}. \tag{4.3}$$

We can view the Li and Stephens process as defining a path through the previously sampled sequences C. This is illustrated in Figure 4.1.

A key feature of the conditional probabilities just described for the Li and Stephens model is that they take the form of a hidden Markov model (HMM) [12]. The path through the sampled chromosomes is a Markov chain on the set $\{1, 2, \ldots, n\}$ which indexes these chromosomes: the value of the chain at a particular SNP specifies which chromosome is being copied at that SNP to produce the new chromosome. In the language of HMMs, the *transition probabilities* specify the probability of either continuing to copy the same chromosome or choosing another chromosome to copy, and the *emission probabilities* specify the probability of observing a given value at the SNP on the new chromosome given its type on the chromosome being copied. The latter have the simple form that with high probability the new chromosome will just copy the type from the chromosome being copied, with the remaining probability being for a switch to the opposite type from that on the chromosome being copied. Viewed this way, the transition probabilities relate to the recombination process and the emission probabilities to the mutation process. The reason that the HMM structure is crucial is that very efficient algorithms are available for calculations in this context. For example, under the Li and Stephens model, given values for all the SNPs on the $n+$1st chromosome, and good computational resources, one can

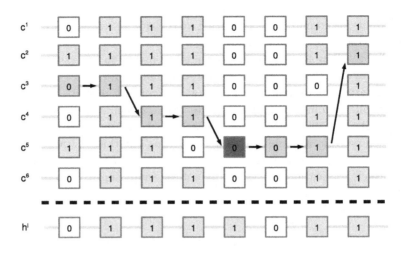

Figure 4.1 A pictorial representation of the Li and Stephens model for the simulation of a new haplotype. Here we have sampled sequences $c^1, \ldots, c^6$ and seek to simulate a new haplotype h^i. The model first simulates a 'path' through the sampled sequences, indicated in the figure by the arrows, which show which of the c^i is copied at each SNP locus. Recombination with another parental sequence is indicated by changing the sequence being copied (for example, between the second and third loci). The dark box on the path at the fifth locus indicates that a mutation occurred in the copying of that locus.

calculate the conditional probability of each possible path through the first n chromosomes, or of the maximum likelihood path (see [12] for details).

Not only is the Li and Stephens model tractable, in a way that the coalescent is not, but it turns out that its use for inference in real populations has been very successful. Examples include inference of haplotype phase [46], [51], [37], [38]; inference of fine-scale recombination rates [36]; imputation of unobserved genotype data [40], [47]; and imputation of classical HLA types from SNP data [35]. It seems that the model captures enough of the features of real data to provide a good framework for inference. Because inference under the coalescent is impossible, it is not known whether this would have properties which are better or worse than those under Li and Stephens, though we would not expect major differences.

We have specified the Li and Stephens model via its conditional probabilities, and these are what is crucial in most applications. But there is a probabilistic curiosity which is worth mentioning. One could calculate the probability, under the model, for a particular configuration of n chromosomes via these conditional probabilities: for a particular ordering, this would simply be the product of the marginal probability for the first chromosome, the Li and Stephens conditional probability for the second chromosome given the first, the Li and Stephens conditional probability for the third given the first and second, and so on. In fact, Li and Stephens themselves referred to their model as the product of approximate conditionals, or PAC, likelihood. The probabilistic curiosity is that in this formulation the likelihood, or equivalently sampling probabilities, in general depend on the order in which the chromosomes are considered. One could solve the problem by averaging over all possible orders, or approximately solve it by averaging over many orders, the approach adopted by Li and Stephens. But for most applications, including the one we describe below, it is the conditional distributions which matter, either in their own right (e.g. [35]) or for use in what resembles a Gibbs sampler. This latter approach gives rise to another curiosity: one can write down, and implement, an algorithm using the Li and Stephens conditionals as if it were a Gibbs sampler, even though the conditionals do not obviously correspond to a proper joint distribution. These algorithms (of which PHASE was perhaps the first) often perform extremely well in practice, notwithstanding the gap in their probabilistic pedigree, an observation which might warrant further theoretical investigation.

5 Application: modelling population structure

In this section we describe an extension of the Li and Stephens model appropriate for geographically structured populations, and then show how inference under the model performs well on real data.

Real populations often consist of genetically distinct subpopulations, and there has been considerable interest in the development of statistical methods which detect such subpopulations, and assign sampled individuals to them, on the basis of population genetics data [3], [44], [43], [45], [16], [6], [7]. Understanding population structure is of interest in conservation biology, human anthropology, human disease genetics, and forensic science. It is important to detect hidden population structure for disease mapping or studies of gene flow, where failure to detect such structure may result in misleading inferences. Population structure is common amongst organisms, and is usually caused by subpopulations forming due to geographical subdivision. It results in genetic differentiation—frequencies of variants, (called *alleles* in genetics) that differ between subpopulations. This may be due to natural selection in different environments, genetic drift (stochastic fluctuations in population composition) in distinct subpopulations, or chance differences in the genetic make up of the founders of subpopulations [27].

Model-based approaches to detecting and understanding population structure have been very successful [45], [16], [6], [7]. Broadly, one specifies a statistical model describing data from different subpopulations and then performs inference under the model. These models are examples of the statistical class of *mixture models*, in which observations are modelled as coming from a (typically unknown) number of distinct classes. In our context the classes consist of the subpopulations, and what is required is a model for the composition of each, and the way in which these are related. Model-based approaches to understanding population structure have several natural advantages. Results are readily interpretable, and, at least for Bayesian inference procedures, they provide a coherent assessment of the uncertainty associated with the assignment of individuals to subpopulations, and the assessment of the number of populations in the sample.

Where the data consist of SNPs taken from distinct regions of the genome it is natural to model these as being independent of each other, within subpopulations, and it remains only to model the frequency of the alleles at each SNP in each subpopulation, and the joint distribution of these across subpopulations. This approach was taken by Pritchard, Stephens and Donnelly [45] and implemented in the program STRUCTURE which has been successfully used in a variety of applications.

In some contexts the population in question arose from the mixing, or admixing as it is known in genetics, of distinct populations at some time in the relatively recent past. African-American populations are an example of admixture, in this case between Caucasian and African populations. Such admixture is well known to lead to correlations between SNPs over moderately large scales (say 5–10 million bases) across chromosomes, known as admixture linkage disequilibrium. Falush, Stephens and Pritchard [16] adapted the model underlying STRUCTURE to incorporate these correlations, with the new model also having been widely applied [16], [17].

SNPs which are close to each other on the chromosome exhibit correlations, known in genetics as *linkage disequilibrium*, due to their shared ancestral history. Both the coalescent with recombination and the Li and Stephens model explicitly capture these correlations for a single randomly-mating population. To employ a model-based approach to population structure for nearby SNPs with linkage disequilibrium, one needs to model these correlations within and between subpopulations. We introduce such a model below as a natural extension of the Li and Stephens model.

For simplicity throughout, we describe our model assuming the haplotypes of the sampled individuals are known, so we assume that the phase of the data is known, either directly or by being inferred using a phasing method such as [52] or [46]. Given the accuracy of statistical phasing methods, this is a reasonable assumption in most cases, but as we note below, the assumption can be dropped, at some computational cost.

Suppose we have DNA sequence data (SNPs) for N haploid individuals (or N phased haplotypes) sampled from K populations at L loci. Call these data $H = \{h_1, \ldots, h_N\}$, where the index i represents individual i. Define the population assignment of individual i to be the discrete random variable Z_i which can take values in $\{1, \ldots, K\}$. In contrast to the previous methods, however, we make no assumption that the SNPs are independent (or merely loosely dependent as is the case for [16]) i.e. we explicitly deal with linkage disequilibrium. In order to model linkage disequilibrium, by applying the model of Li and Stephens [36], we require a genetic map of the region covering the L loci. We assume we have such a map, obtained from other sources (e.g. the program LDhat [41], [42], [1] or from applying the Li and Stephens method for estimating recombination rates [36]). In specifying the model we assume we know the value of K.

We wish to allocate sequences to populations based on the data H. As is natural in mixture models, we do so in a Bayesian framework, via Gibbs sampling (or to be more precise, as we note below, via pseudo-Gibbs sampling) over the unobserved allocation variables, the Z_i. To do so we need to specify the conditional distribution of Z_i given the data H and the values of the allocation variables Z for individuals other than individual i.

We specify the prior distribution on the Z_i to be uniform, i.e.

$$\mathbf{P}(Z_i = 1) = \mathbf{P}(Z_i = 2) = \cdots = \mathbf{P}(Z_i = K) = \frac{1}{K}$$

for every $i = 1, \ldots, N$. It is a simple matter to include an informative prior if required. Furthermore, we assume that in the absence of any data h_i, Z_i is independent of all of the Z_j and h_j for $i \neq j$.

We first informally describe the model in the special case in which the ancestry of any chromosome only ever involves chromosomes in its own population. In effect this means that there has been no migration between subpopulations over the time back to the common ancestors of sampled chromosomes. In this special case, we can think of calculating the conditional distribution as follows. (We describe the calculation of the conditional distribution of Z_1 given H and the other Z_i.) Suppose we know the current population assignments in our sampler of all of the haplotypes. We wish to update the assignment of haplotype 1 based on the assignments of all of the other haplotypes. First, we remove haplotype 1 from the population it is currently assigned to. Then, we calculate the Li and Stephens conditional probability that h_1 would be the next sampled haplotype from each of the K populations. We normalize these probabilities by dividing by their sum and draw the new population assignment of haplotype 1 from the multinomial distribution with these normalized probabilities as the population weights.

We now introduce the general version of our new model, which extends the simple Li and Stephens models above to incorporate migration or recent admixture, and use this as the basis for our method for classifying individuals into populations. Importantly, the model reduces to the Li and Stephens model in certain limiting cases. The method uses SNP haplotype data and explicitly models correlations between SNP loci using the model and estimates of recombination rates between the loci. We test our method on a sample of individuals from three continents and show that it performs well when classifying individuals to these continental populations.

We extend the Li and Stephens [36] model to incorporate admixture or inter-population migration by adding another parameter to the model, represented by α, which models the extent of shared ancestry between populations. This parameter attempts to capture the contribution to an individual's ancestry which is derived from a population other than its own. It does not have a natural interpretation in terms of the dynamics of the populations forward in time. In this sense it has more in common with parameters in statistical models than in probability models. This extended model is perhaps best understood by considering the 'copying path' representation of the Li and Stephens model (see Figure 4.1). For simplicity we present the extended model for the case of two populations (i.e. $K = 2$) with a fixed value of α. We discuss further generalizations after the initial presentation of the model.

As in the special case above, the model allows calculation of the probability that a particular haplotype is sampled from a given population. Normalising these probabilities then gives the probabilities for resampling the allocation variables.

The extension of Li and Stephens relates to recombination, or in the sense of Figure 4.1, to the step when there is a potential change to the chromosome being copied. When such a potential change occurs in the simple model a new chromosome is chosen to be copied uniformly at random (including the chromosome currently being copied). In the extension described above, this remains the case, but with the new chromosome allowed to be chosen only from the population being considered. Our generalisation allows the possibility that the copying process could switch to a chromosome in a different population, and the parameter α controls the relative probability of this.

We now give the details of the model we use. Suppose that n_1 and n_2 (where $n_1 + n_2 = n$) chromosomes have been sampled from populations 1 and 2 respectively, where the jth haplotype in the total population has the SNP information at l SNPs, $c^j = \{c_1^j, c_2^j, \ldots, c_l^j\}$. Let us call this set of chromosomes $C = C_1 \cup C_2$, where C_1 are the chromosomes sampled from Population 1, and C_2 from Population 2. Now suppose an additional chromosome i has been sampled and has SNP information $h^i = \{h_1^i, h_2^i, \ldots, h_l^i\}$. Without loss of generality we seek to determine the probability of sampling this chromosome from Population 1, based on its SNP haplotype and the SNP haplotypes of the previously sampled chromosomes C. We define the per locus recombination probability $\rho_s = 1 - \exp(-4N_e(r_{s+1} - r_s)/(n_1 + \alpha_1 n_2))$ and then define the transition probabilities from state j (indicating that it is the jth haplotype of those that have been previously sampled that is 'parental') at position

s to state k at position $s + 1$:

$$
q(j_s, k_{s+1}) = \begin{cases}
1 - \rho_s + \frac{\rho_s}{n_1 + \alpha_1 n_2}, & j = k, \ j \in C_1, \\
\frac{\rho_s}{n_1 + \alpha_1 n_2}, & j \neq k, \ j \in C_1, \ k \in C_1, \\
\alpha_1 \frac{\rho_s}{n_1 + \alpha_1 n_2}, & j \neq k, \ j \in C_1, \ k \in C_2, \\
1 - \rho_s + \alpha_1 \frac{\rho_s}{n_1 + \alpha_1 n_2}, & j = k, \ j \in C_2, \\
\frac{\rho_s}{n_1 + \alpha_1 n_2}, & j \neq k, \ j \in C_2, \ k \in C_1, \\
\alpha_1 \frac{\rho_s}{n_1 + \alpha_1 n_2}, & j \neq k, \ j \in C_2, \ k \in C_2,
\end{cases}
\tag{5.1}
$$

where N_e is the effective population size. Notice that when $\alpha_1 = 0$ the transition probabilities so defined reduce to the Li and Stephens transition probabilities for a single sample from Population 1. Also note that when $\alpha_1 = 1$ we have effectively combined the two samples into a single sample from one population and again we obtain the Li and Stephens transition probabilities for this case.

We define the emission probabilities in terms of the 'population mutation rate' for our given sample size, where in this case our 'sample size' is $n_1 + \alpha_1 n_2$,

$$
\theta = \left(\sum_{z=1}^{n_1-1} \frac{1}{z} + \sum_{z=n_1}^{n_1+n_2-1} \frac{\alpha_1}{z} \right)^{-1}.
\tag{5.2}
$$

Again, we obtain the desired Li and Stephens population mutation rates for the $\alpha_1 = 0$ and $\alpha_1 = 1$ cases. We then define the mismatch (or not) between the SNP allele of the jth 'parent' chromosome at SNP s, c_s^j, and the SNP allele of the ith additional 'daughter' chromosome, h_s^i:

$$
e(h_s^i, c_s^j) = \begin{cases}
\frac{n_1 + \alpha_1 n_2}{n_1 + \alpha_1 n_2 + \theta} + \frac{1}{2} \frac{\theta}{n_1 + \alpha_1 n_2 + \theta}, & h_s^i = c_s^j, \\
\frac{1}{2} \frac{\theta}{n_1 + \alpha_1 n_2 + \theta}, & h_s^i \neq c_s^j.
\end{cases}
\tag{5.3}
$$

As before, notice that these emission probabilities reduce to the analogous Li and Stephens emission probabilities for the cases $\alpha_1 = 0$ and $\alpha_1 = 1$.

As was the case for a single population, we can view this process as defining a path through the previously sampled sequences C. This is illustrated in Figure 5.1.

Using this model we proceed as before. To calculate the conditional probability of observing the additional haplotype we sum over all possible paths through the potential parental chromosomes C. We use the forward algorithm which gives a computationally efficient means of performing the required summation [12]. For each of the n previously

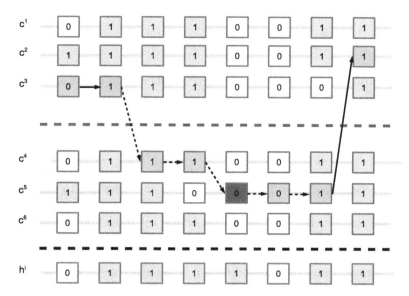

Figure 5.1 A pictorial representation of the calculation for a single path using the new model. Here we have sampled sequences $c^1, \ldots,$ c^6 from two populations, represented by the first three rows and the second three rows in the figure, respectively. We seek to calculate the probability of sampling h^i (seventh row) from the first population by summing the probabilities obtained over all possible paths. We illustrate this with a single path. The arrows indicate the 'path' taken through the sampled sequences, indicating which of the c^i has the 'parental' type at a given SNP locus. Recombination with another parental sequence is indicated by changing the sequence being copied (for example, between the second and third loci). The dark box on the path at the fifth locus indicates that a mutation occurred in the copying of that locus. A dashed arrow indicates that the copying at the next position is taking place with a sequence not in the population we are interested in (in this case the first population) and thus the *recombination term* in the model is scaled by the α parameter for these terms.

sampled chromosomes, we initialise the forward algorithm:

$$
f_1^j = \begin{cases} \frac{1}{n_1 + \alpha_1 n_2} \times e(h_1^i, c_1^j), & j \in C_1, \\ \frac{\alpha_1}{n_1 + \alpha_1 n_2} \times e(h_1^i, c_1^j), & j \in C_2. \end{cases} \tag{5.4}
$$

The forward algorithm moves along the sequence such that at each SNP s, where $1 < s \leq l$,

$$f_s^j = e(h_s^i, c_s^j) \sum_{k=1}^{n} f_{s-1}^k \times q(k_{s-1}, j_s). \tag{5.5}$$

The probability of observing the SNP configuration of the additional chromosome is given by

$$\hat{\pi}(h^i | C, \theta, \rho) = \sum_{j=1}^{n} f_l^j. \tag{5.6}$$

A simple extension for $K > 2$ treats the current population of interest as one population and all other populations are combined to form the second population, thus reducing this case to the $K = 2$ case. This is what we consider here. To further generalize the model one may define a matrix of α parameters for each ordered pair of populations. Inference for the number of populations represented in the sample (K) may be performed using the method of [45] or by other methods. We focus here only on the problem of classifying individuals where the number of subpopulations is known.

As noted above, it is a slight abuse of terminology to refer to the process just described as Gibbs sampling. In this case, we cannot be certain that our approximations do in fact correspond to a proper joint probability. Nonetheless, there are successful precedents for this use of 'pseudo-Gibbs' sampling, notably the program PHASE [52], [50] which has proved successful for inferring haplotypes from genotypes. Furthermore, our method is successful in practice. Given these caveats, we continue our abuse of terminology throughout.

It is also possible to set the proportion of the sample coming from each population as a parameter and update this value at each step. This would allow us to deal easily with populations of different sizes in a straightforward manner, although we have not implemented this extension. Extending the model to incorporate genotypes rather than haplotypes is also straightforward in principle, for example by analogy with [40].

Our model applies when SNPs are sampled from the same small chromosomal region. In practice, it is more likely that data of this type will be available for a number of different, separated, chromosomal regions. In this context we apply the model separately to each region and combine probabilities across regions multiplicatively. We implement this by setting $\rho_s = 1$ in Equation (5.1) for the transition from the last SNP in one region to the first SNP in the next region (the order of the regions is immaterial).

We tested the method on phased data available from the Phase II HapMap [31]. In particular the data consisted of SNP haplotypes from samples from four populations: Yoruba ancestry from Ibadan in Nigeria (YRI); European ancestry from Utah (CEU); Han Chinese ancestry from Beijing (CHB); and Japanese ancestry from Tokyo (JPT). We use the SNP haplotype data available from the HapMap and the recombination map estimated in that study from the combined populations. The YRI and CEU HapMap samples are taken from 30 parent offspring trios in each case. Of these we used the data from the parents only, giving 60 individuals (120 haplotypes) in each population. The Chinese and Japanese samples derive from 45 unrelated individuals (90 haplotypes) in each case. Following common practice, we combine the Chinese and Japanese samples into a single 'analysis panel', which we denote by CHB+JPT. Thus our sample consists of a total of 420 haplotypes deriving from three groups.

From these data we selected 50 independent regions each of 50 kilobases in length (i.e. regions not in linkage disequilibrium with each other). Within these regions we selected only those SNPs that are variable in all three populations and have a minor allele frequency of at least 0.05. A summary of the data is given in Table 5.1.

We then applied our new method to these data. In all cases we set the effective population size for our model, N_e, to 15,000 and set $\alpha = \alpha_1 = \alpha_2$ (this may be thought of as the simple situation when there is an equal amount of gene flow in each direction).

As a first test of using the new model we decided to test the sensitivity to the number of regions used and the number of SNPs selected across those regions, where we specify the correct number of populations in the sample ($K = 3$) and set $\alpha = 0.1$. In our experiments we ignore the information as to which haplotypes come from which populations—this is what we aim to infer, and we can then compare inferences with the truth. We denote by r the number of independent regions chosen, where $r \in \{5, 10, 20, 30, 40, 50\}$, and by s the number of SNPs chosen in the selected regions, where $s \in \{10, 20, 50, 80, 100\}$. For every pair of values r and s we independently at random selected r regions and then selected s SNPs from the selected region to be included in the analysis. We did this 20 times for each pair of values r and s and tested the method on each of the resulting samples from our data.

Table 5.1 *Summary of the data: The data consist of HapMap samples for 420 haplotypes (120 CEU, 180 CHB+JPT, 120 YRI) sampled from 50 independent regions of ~50kb each, taken across all autosomes. SNPs with a minor allele frequency of less than 0.05 in any of the population groups have been excluded. For each region the number of SNPs segregating in all populations is shown, as well as the region size, which is the distance between the first and last SNP in the region.*

Chromosome Number	Region ID	Number of SNPs	Region Size (bp)	Chromosome Number	Region ID	Number of SNPs	Region Size (bp)	Chromosome Number	Region ID	Number of SNPs	Region Size (bp)
1	0	6	38600	1	22	68	47223	1	44	6	44916
2	1	42	48726	2	23	56	49044	2	45	24	47963
3	2	20	46100	3	24	42	48586	3	46	16	46999
4	3	49	49031	4	25	38	47651	4	47	32	46961
5	4	82	43511	5	26	56	44973	5	48	31	49698
6	5	53	49377	6	27	139	49617	6	49	25	48619
7	6	37	48387	7	28	34	47905				
8	7	86	49212	8	29	35	47465				
9	8	55	49700	9	30	34	49603				
10	9	59	49620	10	31	23	49827				
11	10	29	46888	11	32	25	49406				
12	11	76	49762	12	33	63	45207				
13	12	30	48056	13	34	53	49803				
14	13	23	49648	14	35	26	45723				
15	14	33	47289	15	36	26	44137				
16	15	85	47207	16	37	30	44109				
17	16	38	47062	17	38	42	48621				
18	17	61	48001	18	39	38	49694				
19	18	33	48359	19	40	38	47725				
20	19	44	48119	20	41	80	44893				
21	20	45	49195	21	42	77	47874				
22	21	1	0	22	43	8	23443				

In each run we used a burn-in of 200 iterations and then kept the following 1,000 iterations. We ran several analyses to confirm that these values are sufficient for convergence. Haplotypes were assigned to the cluster in which they spent the majority of iterations (after the burn-in). To check whether label-switching was occurring we kept the pairwise assignments matrix for each run. As with [45] and [16], label-switching was not observed and thus the clusters are well-defined. In order to assess performance, clusters were labelled by the population from which the majority of their assigned samples were derived. As we shall see, given the accuracy of the method, this does not result in any ambiguity. Performance was measured as the proportion of haplotypes assigned to the correct population, where we measured these proportions for each sample population individually, and also for the combined sample. We show the average performance over the 20 runs for each pair of values r and s in Figure 5.2.

Examination of Figure 5.2 reveals some insights about the performance of the method. As would be expected, across all populations, for each fixed value of the number of regions selected (r), average performance increases with the number of SNPs used. Thus, the more SNP information we have, the more accurate are our population allocations. In general, increasing the number of independent regions used has less effect on the accuracy of the population assignments, although accuracy does increase with increasing the number of regions used. We conclude that applying our method to data sets comprising at least 80 SNPs derived from at least 10 independent regions will give high accuracy ($\geq 95\%$) with good precision (the standard deviation over all runs observed in this case was less than 2%).

We then tested the effect of varying the α parameter, for α in the range $(0, 0.3)$. In this case we used a single set of SNPs for testing. We used a set of 50 SNPs derived from each of 10 regions which had given average accuracy over previous tests (approximately 90% of individuals classified correctly in the initial tests). We selected this set as it represents a typical possible application of the method and also gives reasonable scope to examine the effect of α both in decreasing and increasing accuracy. For each value of α we ran the method 10 times with randomly chosen initial assignments, using a burn-in of 200 iterations and retaining 1,000 iterations after the burn-in. We observe that for a given value of α up to 0.2 the sampler converges to virtually the same result over all ten runs for all population groups. For α in this range, performance varies across populations but remains consistently above 90% of total chromosomes assigned correctly. The best overall performance is observed when $\alpha = 0.01$. For values of α greater than 0.2 the method no longer converges and the average accuracy of assignments is reduced. The number of individuals classified correctly for various values of α in the range 0 to 0.2 differs by only a small amount, so it is

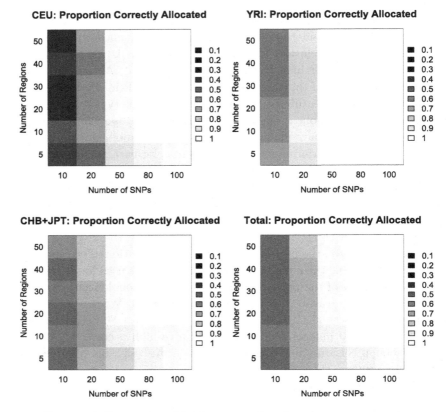

Figure 5.2 Proportion of haplotypes assigned to the correct population. In this application of our method the number of populations is
set to the correct number ($K = 3$), only SNPs with minor allele frequency (MAF) > 0.05 are included in the analysis, and $\alpha = 0.1$. Each
entry at position (r, s) in the four charts relates to a random selection of r regions out of the 50 regions in the data, with s SNPs then
randomly selected for each region. Values shown are averaged over
20 runs of the method in each case, with a burn-in of 200 iterations
and 1000 samples kept after the burn-in. Sequences are allocated to
the cluster to which they are assigned for the majority of the iterations after the burn-in. The top left chart shows the proportion of
CEU haplotypes assigned correctly to the CEU cluster. The second,
third and fourth charts show the equivalent proportions for the YRI,
CHB+JPT and Total Population respectively.

difficult to draw too many conclusions from the limited experiments we
have performed. Close examination of the posterior probabilities for all

individuals in the analyses may provide fruitful insights, although this is left to future work.

In conclusion, the method performed well and, provided sufficient SNPs from sufficient independent regions are available for the haplotypes that are to be classified, gives better than 95% accuracy for classifying sequences into these continental populations. The method converges rapidly: in most cases investigated, a burn-in of 200 iterations with a further 1,000 iterations retained for analysis was seen to be sufficient, but we would advocate the use of more iterations than this. It is feasible to apply the method to data sets of the scale considered here.

Note that there are natural, more sophisticated, approaches to handling the uncertainty associated with assignment of individuals to populations. Long-run proportions of time spent in different assignments naturally lead to posterior probabilities for assignment. These could be carried through into subsequent analyses, or thresholded at some (high) value, so that population assignments are made only when these posterior probabilities are close to unity.

Acknowledgement PD acknowledges support from the Royal Society.

References

[1] Auton, A., and McVean, G. 2007. Recombination rate estimation in the presence of hotspots. *Genome Research*, **17**(8), 1219.

[2] Beaumont, M. A., Zhang, W., and Balding, D. J. 2002. Approximate Bayesian computation in population genetics. *Genetics*, **162**(4), 2025–2035.

[3] Bowcock, A. M., Ruiz-Linares, A., Tomfohrde, J., Minch, E., Kidd, J. R., and Cavalli-Sforza, L. L. 1994. High resolution of human evolutionary trees with polymorphic microsatellites. *Nature*, **368**(6470), 455–457.

[4] Browning, B. L., and Browning, S. R. 2009. A unified approach to genotype imputation and haplotype-phase inference for large data sets of trios and unrelated individuals. *American J. Human Genetics*, **84**(2), 210–223.

[5] Browning, S. R., and Browning, B. L. 2007. Rapid and accurate haplotype phasing and missing-data inference for whole-genome association studies by use of localized haplotype clustering. *American J. Human Genetics*, **81**(5), 1084–1097.

[6] Corander, J., Waldmann, P., Marttinen, P., and Sillanpaa, M. J. 2004. BAPS 2: enhanced possibilities for the analysis of genetic population structure. *Bioinformatics*, **20**(15), 2363–2369.

[7] Dawson, K. J., and Belkhir, K. 2001. A Bayesian approach to the identification of panmictic populations and the assignment of individuals. *Genetical Research*, **78**(1), 59–77.

[8] Donnelly, P. 1991. Weak convergence to a Markov chain with an entrance boundary: ancestral processes in population genetics. *Ann. Probab.*, 1102–1117.

[9] Donnelly, P., and Joyce, P. 1992. Weak convergence of population genealogical processes to the coalescent with ages. *Ann. Probab.*, 322–341.

[10] Donnelly, P., and Kurtz, T. G. 1996. A countable representation of the Fleming-Viot measure-valued diffusion. *Ann. Probab.*, **24**(2), 698–742.

[11] Donnelly, P., and Kurtz, T. G. 1999. Particle representations for measure-valued population models. *Ann. Probab.*, **27**(1), 166–205.

[12] Durbin, R., Eddy, S. R., Krogh, A., and Mitchison, G. 1998. *Biological Sequence Analysis: Probabilistic Models of Proteins and Nucleic Acids.* Cambridge: Cambridge Univ. Press.

[13] Ethier, S. N., and Kurtz, T. G. 1986. *Markov Processes: Characterization and Convergence.* Wiley Ser. Probab. Math. Stat. New York: John Wiley & Sons.

[14] Ethier, S. N., and Kurtz, T. G. 1993. Fleming-Viot processes in population genetics. *SIAM J. Control Optim.*, **31**(2), 345–386.

[15] Ewens, W. J. 2004. *Mathematical Population Genetics, vol. I: Theoretical Introduction.* 2nd edn. Interdiscip. Appl. Math., no. 27. New York: Springer-Verlag.

[16] Falush, D., Stephens, M., and Pritchard, J. K. 2003a. Inference of population structure using multilocus genotype data: linked loci and correlated allele frequencies. *Genetics*, **164**, 1567–1587.

[17] Falush, D., Wirth, T., Linz, B., Pritchard, J. K., Stephens, M., Kidd, M., Blaser, M. J., Graham, D. Y., Vacher, S., Perez-Perez, G. I., Yamaoka, Y., Mégraud, F., Otto, K., Reichard, U., Katzowitsch, E., Wang, X., Achtman, M., and Suerbaum, S. 2003b. Traces of human migrations in Helicobacter pylori populations. *Science*, **299**(5612), 1582.

[18] Fearnhead, P., and Donnelly, P. 2002. Approximate likelihood methods for estimating local recombination rates. *J. R. Stat. Soc. Ser. B Stat. Methodol.*, **64**, 657–680.

[19] Felsenstein, J., Kuhner, M. K., Yamato, J., and Beerli, P. 1999. Likelihoods on coalescents: a Monte Carlo sampling approach to inferring parameters from population samples of molecular data. Pages 163–185 of: Seillier-Moiseiwitsch, F. (ed), *Statistics in Molecular Biology and Genetics.* IMS Lecture Notes Monogr. Ser., vol. 33. Hayward, CA: Inst. Math. Statist. Selected Proceedings of the Joint AMS-IMS-SIAM Summer Conference, 1997.

[20] Griffiths, R. C. 1980. Lines of descent in the diffusion approximation of neutral Wright–Fisher models. *Theor. Population Biology*, **17**(1), 37–50.

[21] Griffiths, R. C., and Marjoram, P. 1996. Ancestral inference from samples of DNA sequences with recombination. *J. Comput. Biol.*, **3**, 479–502.

[22] Griffiths, R. C., and Marjoram, P. 1997. An ancestral recombination graph. Pages 257–270 of: Donnelly, P., and Tavaré, S. (eds), *Progress in Population Genetics and Human Evolution*. IMA Vol. Math. Appl., vol. 87. Berlin: Springer-Verlag.

[23] Griffiths, R. C., and Tavaré, S. 1994a. Ancestral inference in population genetics. *Statist. Sci.*, 307–319.

[24] Griffiths, R. C., and Tavaré, S. 1994b. Sampling theory for neutral alleles in a varying environment. *Phil. Trans. R. Soc. Lond. Ser. B Biol. Sci.*, **344**(1310), 403–410.

[25] Griffiths, R. C., and Tavaré, S. 1994c. Simulating probability distributions in the coalescent. *Theor. Population Biology*, **46**(2), 131–159.

[26] Griffiths, R. C., and Tavaré, S. 1999. The ages of mutations in gene trees. *Ann. Appl. Probab.*, **9**(3), 567–590.

[27] Hartl, D. L., and Clark, A. G. 1997. *Principles of Population Genetics*. 3rd edn. Sunderland, MA: Sinauer Associates.

[28] Hein, J., Schierup, M. H., and Wiuf, C. 2005. *Gene Genealogies, Variation and Evolution: a Primer in Coalescent Theory*. Oxford: Oxford Univ. Press.

[29] Howie, B. N., Donnelly, P., and Marchini, J. 2009. A flexible and accurate genotype imputation method for the next generation of genome-wide association studies. *PLoS Genetics*, **5**(6).

[30] Hudson, R. R. 1983. Properties of a neutral allele model with intragenic recombination. *Theor. Population Biology*, **23**(2), 183–201.

[31] International HapMap Consortium. 2007. A second generation human haplotype map of over 3.1 million SNPs. *Nature*, **449**, 851–861.

[32] Kingman, J. F. C. 1982a. The coalescent. *Stochastic Process. Appl.*, **13**(3), 235–248.

[33] Kingman, J. F. C. 1982b. Exchangeability and the evolution of large populations. Pages 97–112 of: Koch, G., and Spizzichino, F. (eds), *Exchangeability in Probability and Statistics*. Amsterdam: Elsevier. Proceedings of the International Conference on Exchangeability in Probability and Statistics, Rome, 6th-9th April, 1981, in honour of Professor Bruno de Finetti.

[34] Kingman, J. F. C. 1982c. On the genealogy of large populations. Pages 27–43 of: Gani, J., and Hannan, E. J. (eds), *Essays in Statistical Science: Papers in Honour of P. A. P. Moran*. J. Appl. Probab., Special Volume 19A. Sheffield: Applied Probability Trust.

[35] Leslie, S., Donnelly, P., and McVean, G. 2008. A statistical method for predicting classical HLA alleles from SNP data. *American J. Human Genetics*, **82**(1), 48–56.

[36] Li, N., and Stephens, M. 2003. Modelling linkage disequilibrium and identifying recombination hotspots using single-nucleotide polymorphism data. *Genetics*, **165**, 2213–2233.

[37] Li, Y., Ding, J., and Abecasis, G. R. 2006. *Mach1.0: Rapid Haplotype Reconstruction and Missing Genotype Inference*. Paper presented at the Annual Meeting of the American Society of Human Genetics, 9–13

October 2006, New Orleans, LA. Abstract 2290, http://www.ashg.org/genetics/ashg06s/index.shtml.

[38] Li, Y., Willer, C. J., Ding, J., Scheet, P., and Abecasis, G. R. 2007. *In Silico Genotyping for Genome-Wide Association Studies*. Paper presented at the Annual Meeting of the American Society of Human Genetics, 23–27 October 2007, San Diego, CA. Abstract 2071, http://www.ashg.org/genetics/ashg07s/index.shtml.

[39] Marchini, J., Cutler, D., Patterson, N., Stephens, M., Eskin, E., Halperin, E., Lin, S., Qin, Z. S., Munro, H. M., Abecasis, G. R., and Donnelly, P., for the International HapMap Consortium. 2006. A comparison of phasing algorithms for trios and unrelated individuals. *American J. Human Genetics*, **78**(3), 437–450.

[40] Marchini, J., Howie, B., Myers, S., McVean, G., and Donnelly, P. 2007. A new multipoint method for genome-wide association studies by imputation of genotypes. *Nature Genetics*, **39**, 906–913.

[41] McVean, G. A. T., Awadalla, P., and Fearnhead, P. 2002. A coalescent-based method for detecting and estimating recombination from gene sequences. *Genetics*, **160**(3), 1231–1241.

[42] McVean, G. A. T., Myers, S. R., Hunt, S., Deloukas, P., Bentley, D. R., and Donnelly, P. 2004. The fine-scale structure of recombination rate variation in the human genome. *Science*, **304**(5670), 581–584.

[43] Patterson, N., Price, A. L., and Reich, D. 2006. Population structure and eigenanalysis. *PLoS Genetics*, **2**(12), e190.

[44] Price, A. L., Patterson, N. J., Plenge, R. M., Weinblatt, M. E., Shadick, N. A., and Reich, D. 2006. Principal components analysis corrects for stratification in genome-wide association studies. *Nature Genetics*, **38**(8), 904–909.

[45] Pritchard, J. K., Stephens, M., and Donnelly, P. 2000. Inference of population structure using multilocus genotype data. *Genetics*, **155**, 945–959.

[46] Scheet, P., and Stephens, M. 2006. A fast and flexible statistical model for large-scale population genotype data: applications to inferring missing genotypes and haplotypic phase. *American J. Human Genetics*, **78**(4), 629–644.

[47] Servin, B., and Stephens, M. 2007. Imputation-based analysis of association studies: candidate regions and quantitative traits. *PLoS Genetics*, **3**(7), e114.

[48] Stephens, M. 2007. Inference under the coalescent. Chap. 26, pages 878–908 of: Balding, D., Bishop, M., and Cannings, C. (eds), *Handbook of Statistical Genetics*, 3rd edn., vol. 2. Chichester: Wiley-Interscience.

[49] Stephens, M., and Donnelly, P. 2000. Inference in molecular population genetics. *J. R. Stat. Soc. Ser. B Stat. Methodol.*, **62**, 605–655.

[50] Stephens, M., and Donnelly, P. 2003. A comparison of Bayesian methods for haplotype reconstruction from population genotype data. *American J. Human Genetics*, **73**, 1162–1169.

[51] Stephens, M., and Scheet, P. 2005. Accounting for decay of linkage disequilibrium in haplotype inference and missing-data imputation. *American J. Human Genetics*, **76**(3), 449–462.

[52] Stephens, M., Smith, N. J., and Donnelly, P. 2001. A new statistical method for haplotype reconstruction from population data. *American J. Human Genetics*, **68**, 978–989.

[53] Tajima, F. 1983. Evolutionary relationship of DNA sequences in finite populations. *Genetics*, **105**(2), 437–460.

[54] Tavaré, S. 2004. Ancestral inference in population genetics. Pages 1–188 of: Picard, J. (ed), *École d'Été de Probabilités de Saint-Flour XXXI—2001*. Lecture Notes in Math., vol. 1837. New York: Springer-Verlag.

[55] Wakeley, J. 2009. *Coalescent Theory: an Introduction*. Greenwood Village, CO: Roberts & Co.

10

Kingman and mathematical population genetics

Warren J. Ewens[a]* and Geoffrey A. Watterson[b]

Abstract

Mathematical population genetics is only one of Kingman's many research interests. Nevertheless, his contribution to this field has been crucial, and moved it in several important new directions. Here we outline some aspects of his work which have had a major influence on population genetics theory.

AMS subject classification (MSC2010) 92D25

1 Introduction

In the early years of the previous century, the main aim of population genetics theory was to validate the Darwinian theory of evolution, using the Mendelian hereditary mechanism as the vehicle for determining how the characteristics of any daughter generation depended on the corresponding characteristics of the parental generation. By the 1960s, however, that aim had been achieved, and the theory largely moved in a new, retrospective and statistical, direction.

This happened because, at that time, data on the genetic constitution of a population, or at least on a sample of individuals from that population, started to become available. What could be inferred about the past history of the population leading to these data? Retrospective

[a] 324 Leidy Laboratories, Department of Biology, University of Pennsylvania, Philadelphia, PA 19104, USA; wewens@sas.upenn.edu

[b] 15 Brewer Road, Brighton East, Victoria 3187, Australia; geoffreywmailbox-monashfriends@yahoo.com.au

* Corresponding author

questions of this type include: "How do we estimate the time at which mitochondrial Eve, the woman whose mitochondrial DNA is the most recent ancestor of the mitochondrial DNA currently carried in the human population, lived? How can contemporary genetic data be used to track the 'Out of Africa' migration? How do we detect signatures of past selective events in our contemporary genomes?" Kingman's famous coalescent theory became a central vehicle for addressing questions such as these. The very success of coalescent theory has, however, tended to obscure Kingman's other contributions to population genetics theory. In this note we review his various contributions to that theory, showing how coalescent theory arose, perhaps naturally, from his earlier contributions.

2 Background

Kingman attended lectures in genetics at Cambridge in about 1960, and his earliest contributions to population genetics date from 1961. It was well known at that time that in a randomly mating population for which the fitness of any individual depended on his genetic make-up at a single gene locus, the mean fitness of the population increased from one generation to the next, or at least remained constant, if only two possible alleles, or gene types, often labelled A_1 and A_2, were possible at that gene locus. However, it was well known that more than two alleles could arise at some loci (witness the ABO blood group system, admitting three possible alleles, A, B and O). Showing that in this case the mean population fitness is non-decreasing in time under random mating is far less easy to prove. This was conjectured by Mandel and Hughes (1958) and proved in the 'symmetric' case by Scheuer and Mandel (1959) and Mulholland and Smith (1959), and more generally by Atkinson *et al.* (1960) and (very generally) Kingman, (1961a,b). Despite this success, Kingman then focused his research in areas quite different from genetics for the next fifteen years. The aim of this paper is to document some of his work following his re-emergence into the genetics field, dating from 1976. Both of us were honoured to be associated with him in this work. Neither of us can remember the precise details, but the three-way interaction between the UK, the USA and Australia, carried out mainly by the now out-of-date flimsy blue aerogrammes, must have started in 1976, and continued during the time of Kingman's intense involvement in population genetics. This note is a personal account, focusing on this interaction: many others were working in the field at the same time.

One of Kingman's research activities during the period 1961-1976 leads to our first 'background' theme. In 1974 he established (Kingman, 1975) a surprising and beautiful result, found in the context of storage strategies. It is well known that the symmetric K-dimensional Dirichlet distribution

$$\frac{\Gamma(K\alpha)}{\Gamma(\alpha)^K}(x_1 x_2 \cdots x_K)^{\alpha-1} \, dx_1 \, dx_2 \ldots dx_{K-1}, \qquad (2.1)$$

where $x_i \geq 0$, $\sum x_j = 1$, does not have a non-trivial limit as $K \to \infty$, for given fixed α. Despite this, if we let $K \to \infty$ and $\alpha \to 0$ in such a way that the product $K\alpha$ remains fixed at a constant value θ, then the distribution of the *order statistics* $x_{(1)} \geq x_{(2)} \geq x_{(3)} \geq \cdots$ converges to a non-degenerate limit. (The parameter θ will turn out to have an important genetical interpretation, as discussed below.) Kingman called this the Poisson–Dirichlet distribution, but we suggest that its true author be honoured and that it be called the 'Kingman distribution'. We refer to it by this name in this paper. So important has the distribution become in mathematics generally that a book has been written devoted entirely to it (Feng, 2010). This distribution has a rather complex form, and aspects of this form are given below.

The Kingman distribution appears, at first sight, to have nothing to do with population genetics theory. However, as we show below, it turns out, serendipitously, to be central to that theory. To see why this is so, we turn to our second 'background' theme, namely the development of population theory in the 1960s and 1970s.

The nature of the gene was discovered by Watson and Crick in 1953. For our purposes the most important of their results is the fact that a gene is in effect a DNA sequence of, typically, some 5000 bases, each base being one of four types, A, G, C or T. Thus the number of types, or alleles, of a gene consisting of 5000 bases is $4^{5,000}$. Given this number, we may for many practical purposes suppose that there are infinitely many different alleles possible at any gene locus. However, gene sequencing methods took some time to develop, and little genetic information at the fundamental DNA level was available for several decades after Watson and Crick.

The first attempt at assessing the degree of genetic variation from one person to another in a population at a less fundamental level depended on the technique of gel electrophoresis, developed in the 1960s. In loose terms, this method measures the electric charge on a gene, with the charge levels usually thought of as taking integer values only. Genes having different electric charges are of different allelic types, but it can well happen that genes of different allelic types have the same electric charge. Thus there is no one-to-one relation between charge level and allelic type. A simple mutation model assumes that a mutant gene has a charge differing from that of its parent gene by either ± 1. We return to this model in a moment.

In 1974 Kingman travelled to Australia, and while there met Pat Moran (as it happens, the PhD supervisor of both authors of this paper), who was working at that time on this 'charge-state' model. The two of them discussed the properties of a stochastic model involving a population of N individuals, and hence $2N$ genes at any given locus. The population is assumed to evolve by random sampling: any daughter generation of genes is found by sampling, with replacement, from the genes from the parent generation. (This is the well-known 'Wright–Fisher' model of population genetics, introduced into the population genetics literature independently by Wright (1931) and Fisher (1922).) Further, each daughter generation gene is assumed to inherit the same charge as that of its parent with probability $1 - u$, and with probability u is a charge-changing mutant, the change in charge being equally likely to be $+1$ and -1.

At first sight it might seem that, as time progresses, the charge levels on the genes in future generations become dispersed over the entire array of positive and negative integers. But this is not so. Kingman recognized that there is a coherency to the locations of the charges on the genes brought about by common ancestry and the genealogy of the genes in any generation. In Kingman's words (Kingman 1976), amended here to our terminology, "The probability that [two genes in generation t] have a common ancestor gene [in generation s, for $s < t$,] is $1 - (1 - (2N)^{-1})^{t-s}$, which is near unity when $(t - s)$ is large compared to $2N$. Thus the [locations of the charges in any generation] form a coherent group, $\ldots$, and the relative distances between the [charges] remain stochastically bounded". We do not dwell here on the elegant theory that Kingman developed for this model, and note only that in the above quotation we see here the beginnings of the idea of looking backward in time to discuss properties of genetic variation observed in a contemporary generation. This viewpoint is central to Kingman's concept of the coalescent, discussed in detail below.

Parenthetically, the question of the mean number of 'alleles', or occupied charge states, in a population of size N ($2N$ genes) is of some mathematical interest. This depends on the mutation rate u and the population size N. It was originally conjectured by Kimura and Ohta (1978) that this mean remains bounded as $N \to \infty$. However, Kesten (1980a,b) showed that it increases indefinitely as $N \to \infty$, but at an extraordinarily slow rate. More exactly, he found the following astounding result. Define $\gamma_0 = 1$, $\gamma_{k+1} = e^{\gamma_k}$, $k = 1, 2, 3, \ldots$, and $\lambda(2N)$ as the largest k such that $\gamma_k < 2N$. Suppose that $4Nu = 0.2$. Then the random number of 'alleles' in the population divided by $\lambda(2N)$ converges in probability to a constant whose value is approximately 2 as $N \to \infty$. Some idea of the slowness of the divergence of the mean number of alleles can be found by observing that if $2N = 10^{1656520}$, then $\lambda(2N) = 3$.

In a later paper (Kingman 1977a), Kingman extended the theory to the multi-dimensional case, where it is assumed that data are available on a vector of measurements on each gene. Much of the theory for the one-dimensional charge-state model carries through more or less immediately to the multi-dimensional case. As the number of dimensions increases, some of this theory established by Kingman bears on the 'infinitely many alleles' model discussed in the next paragraph, although as Kingman himself noted, the geometrical structure inherent in the model implies that a convergence of his results to those of the infinitely-many-alleles model does not occur, since the latter model has no geometrical structure.

The infinitely-many-alleles model, introduced in the 1960s, forms the second background development that we discuss. This model has two components. The first is a purely demographic, or genealogical, model of the population. There are many such models, and here we consider only the Wright–Fisher model referred to above. (In the contemporary literature many other such models are discussed in the context of the infinitely-many-alleles model, particularly those of Moran (1958) and Cannings (1974), discussed in Section 4.) The second component refers to the mutation assumption, superimposed on this model. In the infinitely-many-alleles model this assumption is that any new mutant gene is of an allelic type never before seen in the population. (This is motivated by the very large number of alleles possible at any gene locus, referred to above.) The model also assumes that the probability that any gene is a mutant is some fixed value u, independent of the allelic type of the parent and of the type of the mutant gene.

From a practical point of view, the model assumes a technology (relevant to the 1960s) which is able to assess whether any two genes are of the same or are of different allelic types (unlike the charge-state model, which does not fully possess this capability), but which is not able to distinguish any further between two genes (as would be possible, for example, if the DNA sequences of the two genes were known). Further, since an entire generation of genes is never observed in practice, attention focuses on the allelic configuration of the genes in a sample of size n, where n is assumed to be small compared to $2N$, the number of genes in the entire population.

Given the nature of the mechanism assumed in this model for distinguishing the allelic types of the n genes in the sample, the data in effect consist of a partition of the integer n described by the vector $(a_1, a_2, \ldots, a_n)$, where a_i is the number of allelic types observed in the sample exactly i times each. It is necessary that $\sum i a_i = n$, and it turns out that under this condition, and to a close approximation, the stationary probability of observing this vector is

$$\frac{n! \theta^{\sum a_i}}{1^{a_1} 2^{a_2} \cdots n^{a_n} a_1! a_2! \cdots a_n! S_n(\theta)},\tag{2.2}$$

where θ is defined as $4Nu$ and $S_n(\theta) = \theta(\theta + 1)(\theta + 2) \cdots (\theta + n - 1)$, (Ewens (1972), Karlin and McGregor (1972)).

The marginal distribution of the number $K = \sum a_i$ of distinct alleles in the sample is found from (2.2) as

$$\text{Prob}(K = k) = |S_n^k| \theta^k / S_n(\theta),\tag{2.3}$$

where S_N^k is a Stirling number of the first kind. It follows from (2.2) and (2.3) that K is a sufficient statistic for θ, so that the conditional distribution of $(a_1, a_2, \ldots, a_n)$ given K is independent of θ.

The relevance of this observation is as follows. As noted above, the extent of genetic variation in a population was, by electrophoresis and other methods, beginning to be understood in the 1960s. As a result of this knowledge, and for reasons not discussed here, Kimura advanced (Kimura 1968) the so-called 'neutral theory', in which it was claimed that much of the genetic variation observed did not have a selective basis. Rather, it was claimed that it was the result of purely random changes in allelic frequency inherent in the random sampling evolutionary model outlined above. This (neutral) theory then becomes the null hypothesis in a statistical testing procedure, with some selective mechanism being the alternative hypothesis. Thus the expression in (2.2) is the null hypothesis allelic-partition distribution of the alleles in a sample of size n. The fact that the conditional distribution of $(a_1, a_2, \ldots, a_n)$ given K is independent of θ implies that an objective testing procedure for the neutral theory can be found free of unknown parameters.

Both authors of this paper worked on aspects of this statistical testing theory during the period 1972–1978, and further reference to this is made below. The random sampling evolutionary scheme described above is no doubt a simplification of real evolutionary processes, so in order for the testing theory to be applicable to more general evolutionary models it is natural to ask: "To what extent does the expression in (2.2) apply for evolutionary models other than that described above?" One of us (GAW) worked on this question in the mid-1970s (Watterson, 1974a, 1974b). This question is also discussed below.

3 Putting it together

One of us (GAW) read Kingman's 1975 paper soon after it appeared and recognized its potential application to population genetics theory. In the 1970s the joint density function (2.1) was well known to arise in that theory when some fixed finite number K of alleles is possible at the gene locus of interest, with symmetric mutation between these alleles. In population genetics theory one considers, as mentioned above, infinitely many possible alleles at any gene locus, so that the relevance of King-man's limiting ($K \to \infty$) procedure to the infinitely many alleles model, that is the relevance of the Kingman distribution, became immediately apparent.

This observation led (Watterson 1976) to a derivation of an explicit form for the joint density function of the first r order statistics $x_{(1)}, x_{(2)}, \ldots, x_{(r)}$ in the Kingman distribution. (There is an obvious printer's error in equation (8) of Watterson's paper.) This joint density function was shown to be of the form

$$f(x_{(1)}, x_{(2)}, \ldots, x_{(r)}) = \theta^r \Gamma(\theta) e^{\gamma \theta} g(y) \{x_{(1)} x_{(2)} \cdots x_{(r)}\}^{-1} x_{(r)}^{\theta-1}, \quad (3.1)$$

where $y = (1 - x_{(1)} - x_{(2)} - \cdots - x_{(r)})/x_{(r)}$, γ is Euler's constant $0.57721\ldots$, and $g(y)$ is best defined through the Laplace transform equation (Watterson and Guess (1977))

$$\int_0^\infty e^{-ty} g(y) dy = \exp\left(\theta \int_0^1 u^{-1}(e^{-tu} - 1) \, du\right). \quad (3.2)$$

The expression (3.1) simplifies to

$$f(x_{(1)}, \ldots, x_{(r)}) = \theta^r \{x_{(1)} \cdots x_{(r)}\}^{-1} (1 - x_{(1)} - \cdots - x_{(r)})^{\theta-1} \quad (3.3)$$

when $x_{(1)} + x_{(2)} + \cdots + x_{(r-1)} + 2x_{(r)} \geq 1$, and in particular,

$$f(x_{(1)}) = \theta(x_{(1)})^{-1}(1 - x_{(1)})^{\theta-1} \tag{3.4}$$

when $\frac{1}{2} \leq x_{(1)} \leq 1$.

Population geneticists are interested in the probability of 'population monomorphism', defined in practice as the probability that the most frequent allele arises in the population with frequency in excess of 0.99. Equation (3.4) implies that this probability is close to $1 - (0.01)^{\theta}$.

Kingman himself had placed some special emphasis on the largest of the order statistics, which in the genetics context is the allele frequency of the most frequent allele. This leads to interesting questions in genetics. For instance, Crow (1973) had asked: "What is the probability that the most frequent allele in a population at any time is also the oldest allele in the population at that time?" A nice application of reversibility arguments for suitable population models allowed Watterson and Guess (1977) to obtain a simple answer to this question. In models where all alleles are equally fit, the probability that any nominated allele will survive longest into the future is (by a simple symmetry argument) its current frequency. For time-reversible processes, this is also the probability that it is the oldest allele in the population. Thus conditional on the current allelic frequencies, the probability that the most frequent allele is also the oldest is simply its frequency $x_{(1)}$. Thus the answer to Crow's question is simply the mean frequency of the most frequent allele. A formula for this mean frequency, as a function of the mutation parameter θ, together with some numerical values, were given in Watterson and Guess (1977), and a partial listing is given in the first row of Table 3.1. (We discuss the entries in the second row of this table in Section 7.)

Table 3.1 *Mean frequency of (a) the most frequent allele, (b) the oldest allele, in a population as a function of θ. The probability that the most frequent allele is the oldest allele is also its mean frequency.*

θ	0.1	0.2	0.5	1.0	2.0	5.0	10.0	20.0
Most frequent	0.936	0.882	0.758	0.624	0.476	0.297	0.195	0.122
Oldest	0.909	0.833	0.667	0.500	0.333	0.167	0.091	0.048

As will be seen from the table, the mean frequency $E(x_{(1)})$ of the most frequent allele decreases as θ increases. Watterson and Guess (1977) provided the bounds $(\frac{1}{2})^{\theta} \leq E(x_{(1)}) \leq 1 - \theta(1 - \theta)\log 2$, which give an idea of the value of $E(x_{(1)})$ for small values of θ, and also showed that $E(x_{(1)})$ decreases asymptotically like $(\log \theta)/\theta$, giving an idea of the value of $E(x_{(1)})$ for large θ.

From the point of view of testing the neutral theory of Kimura, Watterson (1977, 1978) subsequently used properties of these order statistics for testing the null hypothesis that there are no selective forces determining observed allelic frequencies. He considered various alternatives, particularly heterozygote advantage or the presence of some deleterious alleles. For instance, in (Watterson 1977) he investigated the situation when all heterozygotes had a slight selective advantage over all homozygotes. The population truncated homozygosity $\sum_i^r x_i^2$ figures prominently in the allelic distribution corresponding to (3.1) and was thus studied as a test statistic for the null hypothesis of no selective advantage. Similarly, when only a random sample of n genes is taken from the population, the sample homozygosity can be used as a test statistic of neutrality.

Here we make a digression to discuss two of the values in the first row of Table 3.1. It is well known that in the case $\theta = 1$, the allelic partition formula (2.2) describes the probabilistic structure of the lengths of the cycles in a random permutation of the numbers $\{1, 2, \ldots, n\}$. Each cycle corresponds to an allelic type and in the notation a_j thus indicates the number of cycles of length j. Various limiting ($n \to \infty$) properties of random permutations have long been of interest (see for example Finch (2003)). Finch (page 284) gives the limiting mean of the normalized length of the longest cycle as $0.624\ldots$ in such a random permutation, and this agrees with the value listed in Table 3.1 for the case $\theta = 1$. (Finch also in effect gives the standard deviation of this normalized length as $0.1921\ldots$.) Next, (3.4) shows that the limiting probability that the (normalized) length of the longest cycle exceeds $\frac{1}{2}$ is $\log 2$. This is the limiting value of the exact probability for a random permutation of the numbers $\{1, 2, \ldots, n\}$, which from (2.2) is $1 - \frac{1}{2} + \frac{1}{3} - \cdots \pm \frac{1}{n}$.

Finch also considers aspects of a random mapping of $\{1, 2, \ldots, n\}$ to $\{1, 2, \ldots, n\}$. Any such a mapping forms a random number of 'components', each component consisting of a cycle with a number (possibly zero) of branches attached to it. Aldous (1985) provides a full description of these, with diagrams which help in understanding them. Finch takes up the question of finding properties of the normalized size of the largest component of such a random mapping, giving (page 289) a limiting mean of $0.758\ldots$ for this. This agrees with the value in Table 3.1 for the case $\theta = 0.5$. This is no coincidence: Aldous (1985) shows that in a limiting sense (2.2) provides the limiting distribution of the number and (unnormalized) sizes of the components of this mapping, with now a_j indicating the number of components of size j. As a further result, (3.4) shows that the limiting probability that the (normalized) size of the largest component of a random mapping exceeds $\frac{1}{2}$ is $\log(1 + \sqrt{2}) \approx 0.881374$.

Arratia *et al.* (2003) show that (2.2) provides, for various values of θ, the partition structure of a variety of other combinatorial objects for finite n, and presumably the Kingman distribution describes appropriate limiting ($n \to \infty$) results. Thus the genetics-based equation (2.2) and the Kingman distribution provide a unifying theme for these objects.

The allelic partition formula (2.2) was originally derived without reference to the K-allele model (2.1), but was also found (Watterson, 1976) from that model as follows. We start with a population whose allele frequencies are given by the Dirichlet distribution (2.1). If a random sample of n genes is taken from such a population, then given the population's allele frequencies, the sample allele frequencies have a multinomial distribution. Averaging this distribution over the population distribution (2.1), and then introducing the alternative order-statistic sample description $(a_1, a_2, \ldots, a_n)$ as above, the limiting distribution is the partition formula (2.2), found by letting $K \to \infty$ and $\alpha \to 0$ in (2.1) in such a way that the product $K\alpha$ remains fixed at a constant value θ.

4 Robustness

As stated above, the expression (2.2) was first found by assuming a random sampling evolutionary model. As also noted, it can also be arrived at by assuming that a random sample of genes has been taken from an infinite population whose allele frequencies have the Dirichlet distribution (2.1). It applies, however, to further models. Moran (1958) introduced a 'birth-and-death' model in which, at each unit time point, a gene is chosen at random from the population to die. Another gene is chosen at random to reproduce. The new gene either inherits the allelic type of its parent (probability $1 - u$), or is of a new allelic type, not so far seen in the population, with probability u. Trajstman (1974) showed that (2.2) applies as the stationary allelic partition distribution exactly for Moran's model, but with n replaced by the finite population number of genes $2N$ and with θ defined as $2Nu/(1 - u)$. More than this, if a random sample of size n is taken without replacement from the Moran model population, it too has an exact description as in (2.2). This result is a consequence of Kingman's (1978b) study of the consistency of the allelic properties of sub-samples of samples. (In practice, of course, the difference between sampling with, or without, replacement is of little consequence for small samples from large populations.) Kingman (1977a, 1977b) followed up this result by showing that random sampling from various other population models, including significant cases of the Cannings (1974) model, could also be approximated by (2.2). This was important because several consequences of (2.2) could then be applied more generally than was first thought, especially for the purposes of testing of the neutral alleles postulate. He also used the concept of 'non-interference' (see the concluding comments in Section 6) as a further reason for the robustness of (2.2).

5 A convergence result

It was noted in Section 3 that Watterson (1976) was able to arrive at both the Kingman distribution and the allelic partition formula (2.2) from the same starting point (the 'K-allele' model). This makes it clear that there must be a close connection between the two, and in this section we outline Kingman's work (Kingman 1977b) which made this explicit. Kingman imagined a sequence of populations in which the size of population i, ($i = 1, 2, \ldots$) tends to infinity as $i \to \infty$. For any fixed i and any fixed sample size n of genes taken from the population, there will be some probability of the partition $\{a_1, a_2, \ldots, a_n\}$, where a_j has the definition given in Section 2. Kingman then stated that this sequence of populations would have the *Ewens sampling property* if, for each fixed n, this corresponding sequence of probabilities of $\{a_1, a_2, \ldots, a_n\}$ approached that given in (2.2) as $i \to \infty$. In a parallel fashion, for each fixed i there will also be a probability distribution for the order statistics $(p_1, p_2, \ldots)$, where p_j denotes the frequency of the jth most frequent allele in the population. Kingman then stated that this sequence would have the *Poisson–Dirichlet limit* if this sequence of probabilities approached that given by the Poisson–Dirichlet distribution. (We would replace 'Poisson–Dirichlet' in this sentence by 'Kingman'.) He then showed that this sequence of populations has the *Ewens sampling property* if and only if it has the Poisson–Dirichlet (Kingman distribution) limit.

The proof is quite technical and we do not discuss it here. We have noted that the Kingman distribution may be thought of as the distribution of the (ordered) allelic frequencies in an infinitely large population evolving as the random sampling infinitely-many-allele process, so this result provides a beautiful (and useful) relation between population and sample properties of such a population.

6 Partition structures

By 1977 Kingman was in full flight in his investigation of various genetics problems. One line of his work started with the probability distribution (2.2), and his initially innocent-seeming observation that the size n of the sample of genes bears further consideration. The size of a sample is generally taken in Statistics as being comparatively uninteresting, but Kingman (1978b) noted that a sample of n genes could be regarded as having arisen from a sample of $n+1$ genes, one of which was accidently lost, and that this observation induces a consistency property on the probability of any partition of the number n. Specifically, he observed that if we write $P_n(a_1, a_2, \ldots)$ for the probability of the sample partition in a sample of size n, we require

$$P_n(a_1, a_2, \ldots) = \frac{a_1 + 1}{n+1} P_{n+1}(a_1 + 1, a_2, \ldots) +$$
$$\sum_{j=2}^{n+1} \frac{j(a_j + 1)}{n+1} P_{n+1}(a_1, \ldots, a_{j-1} - 1, a_j + 1, \ldots). \quad (6.1)$$

Fortunately, the distribution (2.2) does satisfy this equation. But Kingman went on to ask a deeper question: "What are the most general distributions that satisfy equation (6.1)?" These distributions he called 'partition structures'. He showed that all such distributions that are of interest in genetics could be represented in the form

$$P_n(a_1, a_2, \ldots) = \int P_n(a_1, a_2, \ldots | \mathbf{x}) \, \mu(d\mathbf{x}) \quad (6.2)$$

where μ is some probability measure over the space of infinite sequences $(x_1, x_2, x_3 \ldots)$ satisfying $x_1 \geq x_2 \geq x_3 \cdots$, $\sum_{n=1}^{\infty} x_n = 1$.

An intuitive understanding of this equation is the following. One way to obtain a consistent set of distributions satisfying (6.1) is to imagine a hypothetically infinite population of types, with a proportion x_1 of the most frequent type, a proportion x_2 of the second most frequent type, and so on, forming a vector $\mathbf{x}$. For a fixed value of n, one could then imagine taking a sample of size n from this population, and write $P_n(a_1, a_2, \ldots | \mathbf{x})$ for the (effectively multinomial) probability that the configuration of the sample is $(a_1, a_2, \ldots)$. It is clear that the resulting sampling probabilities will automatically satisfy the consistency property in (6.1). More generally one could imagine the composition of the infinite population itself being random, so that first one chooses its composition $\mathbf{x}$ from μ, and then conditional on $\mathbf{x}$ one takes a sample of size n with probability $P_n(a_1, a_2, \ldots | \mathbf{x})$. The right-hand side in (6.2) is then the probability of obtaining the sample configuration $(a_1, a_2, \ldots)$ averaged over the composition of the population. Kingman's remarkable result was that all partition structures arising in genetics must have the form (6.2), for some μ. Kingman called partition structures that could be expressed as in (6.2) 'representable partition structures' and μ the 'representing measure', and later (Kingman 1978c) found a representation generalizing (6.2) applying for any partition structure.

The similarity between (6.2) and the celebrated de Finetti representation theorem for exchangeable sequences might be noted. This has been explored by Aldous (1985) and Kingman (1978a), but we do not pursue the details of this here.

In the genetics context, the results of Section 4 show that samples from Moran's infinitely many neutral alleles model, as well as the population as a whole, have the partition structure property. So do samples of genes from other genetical models. This makes it natural to ask: "What is the representing measure μ for the allelic partition distribution (2.2)?" And here we come full circle, since he showed that the required representing measure is the Kingman distribution, found by him in (Kingman, 1975) in quite a different context!

The relation between the Kingman distribution and the sampling distribution (2.2) is of course connected to the convergence results discussed in the previous section. From the point of view of the geneticist, the Kingman distribution is then regarded as applying for an infinitely large population, evolving essentially via the random sampling process that led to (2.2). This was made precise by Kingman in (1978b), and it makes it unfortunate that the Kingman distribution does not have a 'nice' mathematical form. However, we see in Section 7 that a very pretty analogue of the Kingman distribution exists when we label alleles not by their frequencies but by their ages in the population. This in turn leads to the capstone of Kingman's work in genetics, namely the coalescent process.

Before discussing these matters we mention another property enjoyed by the distribution (2.2) that Kingman investigated, namely that of non-interference. Suppose that we take a gene at random from the sample of n genes, and find that there are in all r genes of the allelic type of this gene in the sample. These r genes are now removed, leaving $n - r$ genes. The non-interference requirement is that the probability structure of these $n - r$ genes should be the same as that of an original sample of $n - r$ genes, simply replacing n wherever found by $n - r$. Kingman showed that of all partition structures of interest in genetics, the only one also satisfying this non-interference requirement is (2.2). This explains in part the robustness properties of (2.2) to various evolutionary genetic models. However, it also has a natural interpretation in terms of the coalescent process, to be discussed in Section 8.

We remark in conclusion that the partition structure concept has become influential not only in the genetics context, but in Bayesian statistics, mathematics and various areas of science, as the papers of Aldous (2010) and of Gnedin, Haulk and Pitman (2010) in this Festschrift show. That this should be so is easily understood when one considers the natural logic of the ideas leading to it.

7 'Age' properties and the GEM distribution

We have noted above that the Kingman distribution is not user-friendly. This makes it all the more interesting that a *size-biased* distribution closely related to it, namely the GEM distribution, named for Griffiths (1980), Engen (1975) and McCloskey (1965), who established its salient properties, is both simple and elegant, thus justifying the acronym 'GEM'. More important, it has a central interpretation with respect to the ages of the alleles in a population. We now describe this distribution.

We have shown that the ordered allelic frequencies in the population follow the Kingman distribution. Suppose that a gene is taken at random from the population. The probability that this gene will be of an allelic type whose frequency in the population is x is just x. This allelic type was thus sampled by this choice in a size-biased way. It can be shown from properties of the Kingman distribution that the probability density of the frequency of the allele determined by this randomly chosen gene is

$$f(x) = \theta(1 - x)^{\theta - 1}, \quad 0 < x < 1. \tag{7.1}$$

This result was also established by Ewens (1972).

Suppose now that all genes of the allelic type just chosen are removed from the population. A second gene is now drawn at random from the population and its allelic type observed. The frequency of the allelic type of this gene among the genes remaining at this stage is also given by (7.1). All genes of this second allelic type are now also removed from the population. A third gene then drawn at random from the genes remaining, its allelic type observed, and all genes of this (third) allelic type removed from the population. This process is continued indefinitely. At any stage, the distribution of the frequency of the allelic type of any gene just drawn among the genes left when the draw takes place is given by (7.1). This leads to the following representation. Denote by w_j the population frequency of the jth allelic type drawn. Then we can write

$$w_1 = x_1, \ \ldots, \ w_j = (1 - x_1)(1 - x_2) \cdots (1 - x_{j-1})x_j, \quad (j = 2, 3, \ldots),$$
(7.2)

where the x_j are independent random variables, each having the distribution (7.1). The random vector $(w_1, w_2, \ldots)$ then has the GEM distribution.

All the alleles in the population at any time eventually leave the population, through the joint processes of mutation and random drift, and any allele with current population frequency x survives the longest with probability x. That is, since the GEM distribution was found according to a size-biased process, it also arises when alleles are labelled according to the length of their future persistence in the population. Time reversibility arguments then show that the GEM distribution also applies when the alleles in the population are labelled by their age. In other words, the vector $(w_1, w_2, \ldots)$ can be thought of as the vector of allelic frequencies when alleles are ordered with respect to their ages in the population (with allele 1 being the oldest).

The Kingman coalescent, to be discussed in the following section, is concerned among other things with 'age' properties of the alleles in the population. We thus present some of these properties here as an introduction to the coalescent: a more complete list can be found in Ewens (2004). The elegance of many age-ordered formulae derives directly from the simplicity and tractability of the GEM distribution.

Given the focus on retrospective questions, it is natural to ask questions about the oldest allele in the population. The GEM distribution shows that the mean population frequency of the oldest allele in the population is

$$\theta \int_0^1 x(1 - x)^{\theta-1} \, dx = \frac{1}{1 + \theta}.$$
(7.3)

This implies that when θ is very small, this mean frequency is approximately $1 - \theta$. It is interesting to compare this with the mean frequency of the most frequent allele when θ is small, found in effect from the Kingman distribution to be approximately $1 - \theta \log 2$. A more general set of comparisons of these two mean frequencies, for representative values of θ, is given in Table 3.1.

More generally, the mean population frequency of the jth oldest allele in the population is

$$\frac{1}{1+\theta}\left(\frac{\theta}{1+\theta}\right)^{j-1}.$$

For the case $\theta = 1$, Finch (2003) gives the mean frequencies of the second and third most frequent alleles as $0.20958\ldots$ and $0.088316\ldots$ respectively, which may be compared to the mean frequencies of the second and third oldest alleles, namely 0.25 and 0.125. For $\theta = 1/2$ the mean frequency of the second most frequent allele is $0.170910\ldots$, while the mean frequency of the second oldest allele is 0.22222.

Next, the probability that a gene drawn at random from the population is of the type of the oldest allele is the mean frequency of the oldest allele, namely $1/(1+\theta)$, as just shown (see also Table 3.1). More generally the probability that n genes drawn at random from the population are all of the type of the oldest allele in the population is

$$\theta \int_0^1 x^n (1-x)^{\theta-1} \, dx = \frac{n!}{(1+\theta)(2+\theta)\cdots(n+\theta)}. \tag{7.4}$$

The GEM distribution has a number of interesting mathematical properties, of which we mention here only one. It is a so-called 'residual allocation' model (Halmos 1944). Halmos envisaged a king with one kilogram of gold dust, and an infinitely long line of beggars asking for gold. To the first beggar the king gives w_1 kilogram of gold, to the second w_2 kilogram of gold, and so on, as specified in (7.2), where the x_j are independently and identically distributed (i.i.d.) random variables, each having some probability distribution over the interval $(0, 1)$.

Different forms of this distribution lead to different properties of the distribution of the 'residual allocations' w_1, w_2, w_3, $\ldots$. One such property is that the distribution of w_1, w_2, w_3, $\ldots$ be invariant under size-biased sampling. It can be shown that the GEM distribution is the only residual allocation model having this property. This fact had been exploited by Hoppe (1986, 1987) to derive various results of interest in genetics and ecology.

We now turn to sampling results. The probability that n genes drawn at random from the population are all of the same allelic type as the oldest allele in the population is given in (7.4). The probability that n genes drawn at random from the population are all of the same unspecified allelic type is

$$\theta \int_0^1 x^{n-1}(1-x)^{\theta-1}\, dx = \frac{(n-1)!}{(1+\theta)(2+\theta)\cdots(n+\theta-1)},$$

in agreement with (2.2) for the case $a_j = 0$, $j = 1, 2, \ldots, n-1$, $a_n = n$. From this result and that in (7.4), given that n genes drawn at random are all of the same allelic type, the probability that they are all of the allelic type of the oldest allele is $n/(n+\theta)$. The similarity of this expression with that deriving from a Bayesian calculation is of some interest.

Perhaps the most important sample distribution concerns the frequencies of the alleles in the sample when ordered by age. This distribution was found by Donnelly and Tavaré (1986), who showed that the probability that the number of alleles in the sample takes the value k, and that the age-ordered numbers of these alleles in the sample are, in age order, $n_{(1)}, n_{(2)}, \ldots, n_{(k)}$, is

$$\frac{\theta^k (n-1)!}{S_n(\theta)n_{(k)}(n_{(k)} + n_{(k-1)})\cdots(n_{(k)} + n_{(k-1)} + \cdots n_{(2)})}, \qquad (7.5)$$

where $S_j(\theta)$ is defined below (2.2). This formula can be found in several ways, one being as the size-biased version of (2.2).

These are many interesting results connecting the oldest allele in the sample to the oldest allele in the population. For example, Kelly (1976) showed that the probability that the oldest allele in the sample is represented j times in the sample is

$$\frac{\theta}{n}\binom{n}{j}\binom{n+\theta-1}{j}^{-1}, \qquad j = 1, 2, \ldots, n. \qquad (7.6)$$

He also showed that the probability that the oldest allele in the population is observed at all in the sample is $n/(n + \theta)$. The probability that a gene seen j times in the sample is of the oldest allelic type in the population is $j/(n + \theta)$. When $j = n$, so that there is only one allelic type present in the sample, this probability is $n/(n+\theta)$. Donnelly (1986) showed, more generally, that the probability that the oldest allele in the population is observed j times in the sample is

$$\frac{\theta}{n+\theta}\binom{n}{j}\binom{n+\theta-1}{j}^{-1}, \qquad j = 0, 1, 2, \ldots, n. \qquad (7.7)$$

This is of course closely connected to Kelly's result. For the case $j = 0$ the probability (7.7) is $\theta/(n+\theta)$, confirming the complementary probability $n/(n+\theta)$ found above. Conditional on the event that the oldest allele in the population does appear in the sample, a straightforward calculation using (7.7) shows that this conditional probability and that in (7.6) are identical.

It will be expected that various exact results hold for the Moran model, with θ defined as $2Nu/(1-u)$. The first of these is an exact representation of the GEM distribution, analogous to (7.2). This has been provided by Hoppe (1987). Denote by N_1, N_2, ... the numbers of genes of the oldest, second-oldest, ... alleles in the population. Then N_1, N_2, ... can be defined in turn by

$$N_i = 1 + M_i, \quad i = 1, 2, \ldots, \tag{7.8}$$

where M_i has a binomial distribution with index $2N - N_1 - N_2 - \cdots - N_{i-1} - 1$ and parameter x_i, where x_1, x_2, ... are i.i.d. continuous random variables each having the density function (7.1). Eventually $N_1 + N_2 + \cdots + N_k = 2N$ and the process stops, the final index k being identical to the number K_{2N} of alleles in the population.

It follows directly from this representation that the mean of N_1 is

$$1 + (2N-1)\theta \int_0^1 x(1-x)^{\theta-1}\, dx \;=\; \frac{2N+\theta}{1+\theta}.$$

If there is only one allele in the population, so that the population is strictly monomorphic, this allele must be the oldest one in the population. The above representation shows that the probability that the oldest allele arises $2N$ times in the population is

$$\mathrm{Prob}\,(M_1 = 2N-1) = \theta \int_0^1 x^{2N-1}(1-x)^{\theta-1}\, dx,$$

and this reduces to the exact monomorphism probability

$$\frac{2N-1}{(1+\theta)(2+\theta)\cdots(2N-1+\theta)}$$

for the Moran model.

More generally, Kelly (1977) has shown that the probability that the oldest allele in the population is represented by j genes is, exactly,

$$\frac{\theta}{2N}\binom{2N}{j}\binom{2N+\theta-1}{j}^{-1}. \tag{7.9}$$

The case $j = 2N$ considered above is a particular example of (7.9), and the mean number $(2N+\theta)/(1+\theta)$ also follows from (7.9).

We now consider 'age' questions. It is found that the mean time, into the past, that the oldest allele in the population entered the population (by a mutation event) is

$$\text{Mean age of oldest allele } = \sum_{j=1}^{2N} \frac{4N}{j(j+\theta-1)} \text{ generations.} \qquad (7.10)$$

It can be shown (see Watterson and Guess (1977) and Kelly (1977)) that not only the mean age of the oldest allele, but indeed the entire probability distribution of its age, is independent of its current frequency and indeed of the frequency of all alleles in the population.

If an allele is observed in the population with frequency p, its mean age is

$$\sum_{j=1}^{2N} \frac{4N}{j(j+\theta-1)} \left(1 - (1-p)^j\right) \text{ generations.} \qquad (7.11)$$

This is a generalization of the expression in (7.10), since if $p = 1$ only one allele exists in the population, and it must then be the oldest allele.

Our final calculation concerns the mean age of the oldest allele in a sample of n genes. This is

$$4N \sum_{j=1}^{n} \frac{1}{j(j+\theta-1)} \text{ generations.} \qquad (7.12)$$

Except for small values of n, this is close to the mean age of the oldest allele in the population, given in (7.10). In other words, unless n is small, it is likely that the oldest allele in the population is represented in the sample.

We have listed the various results given in this section not only because of their intrinsic interest, but because they form a natural lead-in to Kingman's celebrated coalescent process, to which we now turn.

8 The coalescent

The concept of the coalescent is now discussed at length in many text-books, and entire books (for example Hein, Schierup and Wiuf (2005) and Wakeley (2009)) and book chapters (for example Marjoram and Joyce (2009) and Nordborg (2001)) have been written about it. Here we can do no more than outline the salient aspects of the process.

The aim of the coalescent is to describe the common ancestry of the sample of n genes at various times in the past through the concept of an equivalence class. To do this we introduce the notation τ, indicating a time τ in the past (so that if $\tau_1 > \tau_2$, time τ_1 is further in the past than time τ_2). The sample of n genes is assumed taken at time $\tau = 0$.

Two genes in the sample of n are in the same equivalence class at time τ if they have a common ancestor at this time. Equivalence classes are denoted by parentheses: Thus if $n = 8$ and at time τ genes 1 and 2 have one common ancestor, genes 4 and 5 a second, and genes 6 and 7 a third, and none of the three common ancestors are identical and none is identical to the ancestor of gene 3 or of gene 8 at time τ, the equivalence classes at time τ are

$$\{(1,2),(3),(4,5),(6,7),(8)\}. \tag{8.1}$$

We call any such set of equivalence classes an equivalence relation, and denote any such equivalence relation by a Greek letter. As two particular cases, at time $\tau = 0$ the equivalence relation is $\phi_1 = \{(1),(2),(3),(4),(5),(6),(7),(8)\}$, and at the time of the most recent common ancestor of all eight genes, the equivalence relation is $\phi_n = \{(1,2,3,4,5,6,7,8)\}$. The Kingman coalescent process is a description of the details of the ancestry of the n genes moving from ϕ_1 to ϕ_n. For example, given the equivalence relation in (8.1), one possibility for the equivalence relation following a coalescence is $\{(1,2),(3),(4,5),(6,7,8)\}$. Such an amalgamation is called a coalescence, and the process of successive such amalgamations is called the coalescence process.

Coalescences are assumed to take place according to a Poisson process, but with a rate depending on the number of equivalence classes present. Suppose that there are j equivalence classes at time τ. It is assumed that no coalescence takes places between time τ and time $\tau + \delta\tau$ with probability $1 - \frac{1}{2}j(j-1)\delta\tau$. (Here and throughout we ignore terms of order $(\delta\tau)^2$.) The probability that the process moves from one nominated equivalence class (at time τ) to some nominated equivalence class which can be derived from it is $\delta\tau$. In other words, a coalescence takes place in this time interval with probability $\frac{1}{2}j(j-1)\delta\tau$, and all of the $j(j-1)/2$ amalgamations possible at time τ are equally likely to occur.

In order for this process to describe the 'random sampling' evolutionary model described above, it is necessary to scale time so that unit time corresponds to $2N$ generations. With this scaling, the time T_j between the formation of an equivalence relation with j equivalence classes to one with $j - 1$ equivalence classes has an exponential distribution with mean $2/j(j-1)$.

The (random) time $T_{\text{MRCAS}} = T_n + T_{n-1} + T_{n-2} + \cdots + T_2$ until all genes in the sample first had just one common ancestor has mean

$$E(T_{\text{MRCAS}}) = 2 \sum_{j=2}^{n} \frac{1}{j(j-1)} = 2 \left(1 - \frac{1}{n}\right). \tag{8.2}$$

(The suffix 'MRCAS' stands for 'most recent common ancestor of the sample.) This is, of course close to 2 coalescent time units, or 4N generations, when n is large. Tavaré (2004) has found the (complicated) distribution of T_{MRCAS}. Kingman (1982a,b,c) showed that for large populations, many population models (including the 'random sampling' model) are well approximated in their sampling attributes by the coalescent process. The larger the population the more accurate is this approximation.

We now introduce mutation into the coalescent. Suppose that the probability that any particular ancestral gene mutates in the time interval $(\tau + \delta\tau, \tau)$ is $\frac{\theta}{2}\delta\tau$. All mutants are assumed to be of new allelic types (the infinitely many alleles assumption). If at time τ in the coalescent there are j equivalence classes, the probability that either a mutation or a coalescent event had occurred in $(\tau + \delta\tau, \tau)$ is

$$j\frac{\theta}{2}\delta\tau + \frac{j(j-1)}{2}\delta\tau = \frac{1}{2}j(j + \theta - 1)\delta\tau. \tag{8.3}$$

We call such an occurrence a defining event, and given that a defining event did occur, the probability that it was a mutation is $\theta/(j + \theta - 1)$ and that it is a coalescence is $(j - 1)/(j + \theta - 1)$.

The probability that k different allelic types are seen in the sample is then the probability that k of these defining events were mutations. The above reasoning shows that this probability must be proportional to $\theta^k / S_n(\theta)$, where $S_n(\theta)$ is defined below (2.2), the constant of proportionality being independent of θ. This argument leads to (2.3).

Using these results and combinatorial arguments counting all possible coalescent paths from a partition $(a_1, a_2, \ldots, a_n)$ back to the original common ancestor, Kingman (1982a) was able to derive the more detailed sample partition probability distribution (2.2), and deriving this distribution from coalescent arguments is perhaps the most pleasing way of arriving at it. For further comments along these lines, see (Kingman (2000)).

The description of the coalescent given above follows the original derivation given by Kingman (1982a). The coalescent is perhaps more naturally understood as a random binary tree. These have now been investigated in great detail: see for example Aldous and Pitman (1999).

Many genetic results can be obtained quite simply by using the coalescent ideas. For example, Watterson and Donnelly (1992) used Kingman's coalescent to discuss the question "Do Eve's Alleles Live On?" To answer this question we assume the infinitely-many-neutral-alleles model for the population and consider a random sample of n genes taken at time 'now'. Looking back in time, the ancestral lines of those genes coalesce to the MRCAS, which may be called the sample's 'Eve'. Of course if Eve's allelic type survives into the sample it would be the oldest, but it may not have survived because of intervening mutation. If we denote by X_n the number of representative genes of the oldest allele, and by Y_n the number of genes having Eve's allele, then Kelly's result (7.6) gives the distribution of X_n. We denote that distribution here by $p_n(j)$, $j = 0, 1, 2, \ldots, n$, and the distribution of Y_n by $q_n(j)$, $j = 0, 1, 2, \ldots, n$. Unlike the simple explicit expression for $p_n(j)$, the corresponding expression for $q_n(j)$ is very complicated: see (2.14) and (2.15) in Watterson and Donnelly (1992), derived using some of Kingman's (1982a) results. Using the relative probabilities of a mutation or a coalescence at a defining event gives rise to a recurrence equation for $q_n(j)$, $j = 0, 1, 2, \ldots, n$ as

$$[n(n-1) + j\theta]q_n(j)$$
$$= n(j-1)q_{n-1}(j-1) + n(n-j-1)q_{n-1}(j) + (j+1)\theta q_n(j+1) \tag{8.4}$$

for $j = 0, 1, 2, \ldots, n$, (provided that we interpret $q_n(j)$ as zero outside this range), and for $n = 2, 3, \ldots$. The boundary conditions $q_1(j) = 1$ for $j = 1$, $q_1(j) = 0$ for $j > 1$, and

$$q_n(n) = p_n(n) = \prod_{k=2}^{n} \frac{k-1}{k + \theta - 1}$$

apply, the latter because if $X_n = n$ then all sample genes descend from a gene having the oldest allele, and 'she' must be Eve. The recurrence (8.4) is a special case of one found by Griffiths (1989) in his equation (3.7).

The expected number of genes of Eve's allelic type was given by Griffiths (1986), (see also Beder (1988)), as

$$\mathrm{E}(Y_n) = \sum_{j=0}^{n} j q_n(j) = n \prod_{j=2}^{n} \frac{j(j-1)}{j(j-1) + \theta}. \tag{8.5}$$

Watterson and Donnelly (1992) gave some numerical examples, some asymptotic results, and some bounds for the distribution $q_n(j)$, $j = 0$,

1, 2, ..., n. One result of interest is that $q_n(0)$, the probability of Eve's allele being extinct in the sample, increases with n, to $q_\infty(0)$ say. One reason for this is that a larger sample may well have its 'Eve' further back in the past than a smaller sample. We might interpret $q_\infty(0)$ as being the probability that an infinitely large population has lost its 'Eve's' allele. Note that the bounds

$$\frac{\theta^2}{(2+\theta)(1+\theta)} < q_\infty(0) \leq \frac{\theta e^\theta - \theta}{\theta e^\theta + 1}, \tag{8.6}$$

for $0 < \theta < \infty$, indicate that for all θ in this range, $q_\infty(0)$ is neither 0 nor 1. Thus, in contrast to the situation in branching processes, there are no sub-critical or super-critical phenomena here.

9 Other matters

There are many other topics that we could mention in addition to those described above. On the mathematical side, the Kingman distribution has a close connection to prime factorization of large integers. On the genetical side, we have not mentioned the 'infinitely many sites' model, now frequently used by geneticists, in which the DNA structure of the gene plays a central role. It is a tribute to Kingman that his work opened up more topics than can be discussed here.

Acknowledgements Our main acknowledgement is to John Kingman himself. The power and beauty of his work was, and still is, an inspiration to us both. His generosity, often ascribing to us ideas of his own, was unbounded. For both of us, working with him was an experience never to be forgotten. More generally the field of population genetics owes him an immense and, fortunately, well-recognized debt. We also thank an anonymous referee for suggestions which substantially improved this paper.

References

Aldous, D. J. 1985. Exchangeability and related topics. Pages 1–198 of: *Ecole d'Été de probabilités de Saint-Flour XIII. Lecture Notes in Math.*, vol. 1117. Berlin: Springer-Verlag.

Aldous, D. J. 2010. More uses of exchangeability: representations of complex random structures. In: Bingham, N. H., and Goldie, C. M. (eds),

Probability and Mathematical Genetics: Papers in Honour of Sir John Kingman. London Math. Soc. Lecture Note Ser. Cambridge: Cambridge Univ. Press.

Aldous, D. J., and Pitman, J. 1999. A family of random trees with random edge lengths. *Random Structures & Algorithms*, **15**, 176–195.

Arratia, R., Barbour, A. D., and Tavaré, S. 2003. *Logarithmic Combinatorial Structures: A Probabilistic Approach.* European Mathematical Society Monographs in Mathematics. Zurich: EMS Publishing House.

Atkinson, F. V, Watterson, G. A., and Moran, P. A. P. 1960. A matrix inequality. *Quart. J. Math. Oxford Ser. (2)*, **12**, 137–140.

Beder, B. 1988. Allelic frequencies given the sample's common ancestral type. *Theor. Population Biology*, **33**, 126–137.

Cannings, C. 1974. The latent roots of certain Markov chains arising in genetics: a new approach. 1. Haploid models. *Adv. in Appl. Probab.*, **6**, 260–290.

Crow, J. F. 1973. The dilemma of nearly neutral mutations: how important are they for evolution and human welfare? *J. Heredity*, **63**, 306–316.

Donnelly, P. J. 1986. Partition structures, Pólya urns, the Ewens sampling formula, and the ages of alleles. *Theor. Population Biology*, **30**, 271–288.

Donnelly, P. J., and Tavaré, S. 1986. The ages of alleles and a coalescent. *Adv. in Appl. Probab.*, **18**, 1–19.

Engen, S. 1975. A note on the geometric series as a species frequency model. *Biometrika*, **62**, 694–699.

Ewens, W. J. 1972. The sampling theory of selectively neutral alleles. *Theor. Population Biology*, **3**, 87–112.

Ewens, W. J. 2004. *Mathematical Population Genetics.* New York: Springer-Verlag.

Feng, S. 2010. *The Poisson–Dirichlet Distribution and Related Topics.* New York: Springer-Verlag.

Finch, S. R. 2003. *Mathematical Constants.* Cambridge: Cambridge Univ. Press.

Fisher, R. A. 1922. On the dominance ratio. *Proc. Roy. Soc. Edinburgh*, **42**, 321–341.

Gnedin, A. V., Haulk, C., and Pitman, J. 2010. Characterizations of exchangeable random partitions by deletion properties. In: Bingham, N. H., and Goldie, C. M. (eds), *Probability and Mathematical Genetics: Papers in Honour of Sir John Kingman.* London Math. Soc. Lecture Note Ser. Cambridge: Cambridge Univ. Press.

Griffiths, R. C. 1980. Unpublished notes.

Griffiths, R. C. 1986. Family trees and DNA sequences. Pages 225–227 of: Francis, I. S., Manly, B. F. J., and Lam, F. C. (eds), *Proceedings of the Pacific Statistical Congress.* Amsterdam: Elsevier Science Publishers.

Griffiths, R. C. 1989. Genealogical-tree probabilities in the infinitely-many-sites model. *J. Math. Biol.*, **27**, 667–680.

Halmos, P. R. 1944. Random alms. *Ann. Math. Statist.*, **15**, 182–189.

Hein, J., Schierup, M. H., and Wiuf, C. 2005. *Gene Genealogies, Variation and Evolution.* Oxford: Oxford Univ. Press.

Hoppe, F. 1986. Size-biased sampling of Poisson–Dirichlet samples with an application to partition structures in population genetics. *J. Appl. Probab.*, **23**, 1008–1012.

Hoppe, F. 1987. The sampling theory of neutral alleles and an urn model in population genetics. *J. Math. Biol.*, **25**, 123–159.

Karlin, S., and McGregor, J. L. 1972. Addendum to a paper of W. Ewens. *Theor. Population Biology*, **3**, 113–116.

Kelly, F. P. 1976. On stochastic population models in genetics. *J. Appl. Probab.*, **13**, 127–131.

Kelly, F. P. 1977. Exact results for the Moran neutral allele model. *J. Appl. Probab.*, **14**, 197–201.

Kesten, H. 1980a. The number of alleles in electrophoretic experiments. *Theor. Population Biology*, **18**, 290–294.

Kesten, H. 1980b. The number of distinguishable alleles according to the Ohta–Kimura model of neutral evolution. *J. Math. Biol.*, **10**, 167–187.

Kimura, M. 1968. Evolutionary rate at the molecular level. *Nature*, **217**, 624–626.

Kimura, M., and Ohta, T. 1978. Stepwise mutation model and distribution of allelic frequencies in a finite population. *Proc. Natl. Acad. Sci. USA*, **75**, 2868–72.

Kingman, J. F. C. 1961a. On an inequality in partial averages. *Quart. J. Math. Oxford Ser. (2)*, **12**, 78–80.

Kingman, J. F. C. 1961b. A mathematical problem in population genetics. *Proc. Cambridge Philos. Soc.*, **57**, 574–582.

Kingman, J. F. C. 1975. Random discrete distributions. *J. Roy. Statist. Soc. Ser. B*, **37**, 1–22.

Kingman, J. F. C. 1976. Coherent random walks arising in some genetical models. *Proc. R. Soc. Lond. Ser. A.*, **351**, 19–31.

Kingman, J. F. C. (1977a). A note on multi-dimensional models of neutral mutation. *Theor. Population Biology*, **11**, 285–290.

Kingman, J. F. C. (1977b). The population structure associated with the Ewens sampling formula. *Theor. Population Biology*, **11**, 274–283.

Kingman, J. F. C. (1978a). Uses of exchangeability. *Ann. Probab.*, **6**, 183–197.

Kingman, J. F. C. (1978b). Random partitions in population genetics. *Proc. R. Soc. Lond. Ser. A*, **361**, 1–20.

Kingman, J. F. C. (1978c). The representation of partition structures. *J. Lond. Math. Soc.*, **18**, 374–380.

Kingman, J. F. C. (1982a). The coalescent. *Stochastic Process Appl.*, **13**, 235–248.

Kingman, J. F. C. (1982b). Exchangeability and the evolution of large populations. Pages 97–112 of: Koch, G., and Spizzichino, F. (eds), *Exchangeability in Probability and Statistics*. Amsterdam: North-Holland.

Kingman, J. F. C. (1982c). On the genealogy of large populations. *J. Appl. Probab.*, **19A**, 27–43.

Kingman, J. F. C. 2000. Origins of the coalescent: 1974-1982. *Genetics* **156**, 1461–1463.

Mandel, S. P. H., and Hughes, I. M. 1958. Change in mean viability at a multi-allelic locus in a population under random mating. *Nature*, **182**, 63–64.

Marjoram, P., and Joyce, P. 2009. Practical implications of coalescent theory. In: Lenwood, S., and Ramakrishnan, N. (eds), *Problem Solving Handbook in Computational Biology and Bioinformatics*. New York: Springer-Verlag.

McCloskey, J. W. 1965. *A Model for the Distribution of Individuals by Species in an Environment*. Unpublished PhD. thesis, Michigan State University.

Mulholland, H. P., and Smith, C. A. B. 1959. An inequality arising in genetical theory. *Amer. Math. Monthly*, **66**, 673–683.

Moran, P. A. P. 1958. Random processes in genetics. *Proc. Cambridge Philos. Soc.*, **54**, 60–71.

Nordborg, M. 2001. Coalescent theory. In: Balding, D. J, Bishop, M. J., and Cannings, C. (eds), *Handbook of Statistical Genetics*. Chichester: John Wiley & Sons.

Scheuer, P. A. G., and Mandel, S. P. H. 1959. An inequality in population genetics. *Heredity*, **31**, 519–524.

Tavaré, S. 2004. Ancestral inference in population genetics. Pages 1–188 of: Cantoni, O., Tavaré, S., and Zeitouni, O. (eds), *École d'Été de Probabilités de Saint-Flour XXXI-2001*. Berlin: Springer-Verlag.

Trajstman, A. C. 1974. On a conjecture of G. A. Watterson. *Adv. in Appl. Probab.*, **6**, 489–503.

Wakeley, J. 2009. *Coalescent Theory*. Greenwood Village, Colorado: Roberts and Company.

Watterson, G. A. 1974a. Models for the logarithmic species abundance distributions. *Theor. Population Biology*, **6**, 217–250.

Watterson, G. A. 1974b. The sampling theory of selectively neutral alleles. *Adv. in Appl. Probab.*, **6**, 463–488.

Watterson, G. A. 1976. The stationary distribution of the infinitely-many neutral alleles diffusion model. *J. Appl. Probab.*, **13**, 639–651.

Watterson, G. A. 1977. Heterosis or neutrality? *Genetics*, **85**, 789–814.

Watterson, G. A. 1978. An analysis of multi-allelic data, *Genetics*, **88**, 171–179.

Watterson, G. A., and Donnelly, P. J. 1992. Do Eve's alleles live on? *Genet. Res. Camb.*, **60**, 221–234.

Watterson, G. A., and Guess, H. A. 1977. Is the most frequent allele the oldest? *Theor. Population Biology*, **11**, 141–160.

Wright, S. 1931. Evolution in Mendelian populations. *Genetics*, **16**, 97–159.

11

Characterizations of exchangeable partitions and random discrete distributions by deletion properties

Alexander Gnedin[a], Chris Haulk[b] and Jim Pitman[c]

Abstract

We prove a long-standing conjecture which characterizes the Ewens–Pitman two-parameter family of exchangeable random partitions, plus a short list of limit and exceptional cases, by the following property: for each $n = 2, 3, \ldots$, if one of n individuals is chosen uniformly at random, independently of the random partition π_n of these individuals into various types, and all individuals of the same type as the chosen individual are deleted, then for each $r > 0$, given that r individuals remain, these individuals are partitioned according to π'_r for some sequence of random partitions (π'_r) which does not depend on n. An analogous result characterizes the associated Poisson–Dirichlet family of random discrete distributions by an independence property related to random deletion of a frequency chosen by a size-biased pick. We also survey the regenerative properties of members of the two-parameter family, and settle a question regarding the explicit arrangement of intervals with lengths given by the terms of the Poisson–Dirichlet random sequence into the interval partition induced by the range of a homogeneous neutral-to-the right process.

AMS subject classification (MSC2010) 60C05, 60G09, 05A18

[a] Department of Mathematics, Utrecht University, PO Box 80010, 3508 TA Utrecht, The Netherlands; A.V.Gnedin@uu.nl

[b] Department of Statistics, Evans Hall, University of California at Berkeley, Berkeley, CA 94720–3860, USA; haulk@stat.berkeley.edu

[c] Department of Statistics, Evans Hall, University of California at Berkeley, Berkeley, CA 94720–3860, USA; pitman@stat.berkeley.edu. Research supported in part by N.S.F. Award 0806118.

1 Introduction

Kingman [14] introduced the concept of a *partition structure*, that is a family of probability distributions for random partitions π_n of a positive integer n, with a sampling consistency property as n varies. Kingman's work was motivated by applications in population genetics, where the partition of n may be the allelic partition generated by randomly sampling a set of n individuals from a population of size $N \gg n$, considered in a large-N limit which implies sampling consistency. Subsequent authors have established the importance of Kingman's theory of partition structures, and representations of these structures in terms of exchangeable random partitions and random discrete distributions [19], in a number of other settings, which include the theory of species sampling [20], random trees and associated random processes of fragmentation and coalescence [21, 9, 1, 2], Bayesian statistics and machine learning [26, 27]. Kingman [13] showed that the Ewens sampling formula from population genetics defines a particular partition structure (π_n), which he characterized by the following property, together with the regularity condition that $\mathbb{P}(\pi_n = \lambda) > 0$ for every partition λ of n:

for each $n = 2, 3, \ldots$, if an individual is chosen uniformly at random independently of a random partitioning of these individuals into various types according to π_n, and all individuals of the same type as the chosen individual are deleted, then conditionally given that the number of remaining individuals is $r > 0$, these individuals are partitioned according to a copy of π_r.

We establish here a conjecture of Pitman [18] that if this property is weakened by replacing π_r by π'_r for some sequence of random partitions (π'_r), and a suitable regularity condition is imposed, then (π_n) belongs to the two-parameter family of partition structures introduced in [18]. Theorem 2.3 below provides a more careful statement. We also present a corollary of this result, to characterize the two-parameter family of Poisson–Dirichlet distributions by an independence property of a single size-biased pick, thus improving upon [19].

Kingman's characterization of the Ewens family of partition structures by deletion of a type has been extended in another direction by allowing other deletion algorithms but continuing to require that the distribution of the partition structure be preserved. The resulting theory of *regenerative partition structures* [7] is connected to the theory of regenerative sets, including Kingman's regenerative phenomenon [12], on a multiplicative scale. In the last section of the paper we review such deletion properties of the two-parameter family of partition structures, and offer a new proof of a result of Pitman and Winkel [24] regarding the explicit arrangement of intervals with lengths given by the terms of the Poisson–Dirichlet random sequence into the interval partition induced by a multiplicatively regenerative set.

2 Partition structures

This section briefly reviews Kingman's theory of partition structures, which provides the general context of this article. To establish some terminology and notation for use throughout the paper, recall that a *composition* λ of a positive integer n is a sequence of positive integers $\lambda = (\lambda_1, \ldots, \lambda_k)$, with $\sum_{i=1}^{k} \lambda_i = n$. Both $k = k_\lambda$ and $n = n_\lambda$ may be regarded as functions of λ. Each term λ_i is called a *part* of λ. A *partition* λ *of* n is a multiset of positive integers whose sum is n, commonly identified with the composition of n obtained by putting its positive integer parts in decreasing order, or with the infinite sequence of non-negative integers obtained by appending an infinite string of zeros to this composition of n. So

$$\lambda = (\lambda_1, \lambda_2, \ldots) \quad \text{with} \quad \lambda_1 \geq \lambda_2 \geq \cdots \geq 0$$

represents a partition of $n = n_\lambda$ into $k = k_\lambda$ parts, where

$$n_\lambda := \sum_i \lambda_i \text{ and } k_\lambda := \max\{i : \lambda_i > 0\}.$$

Informally, a partition λ describes an unordered collection of n_λ balls of k_λ different colors, with λ_i balls of the ith most frequent color. A *random partition of* n is a random variable π_n with values in the finite set of all partitions λ of n. Kingman [14] defined a *partition structure* to be a sequence of random partitions $(\pi_n)_{n \in \mathbb{N}}$ which is *sampling-consistent* in the following sense:

> if a ball is picked uniformly at random and deleted from n balls randomly colored according to π_n, then the random coloring of the remaining $n - 1$ balls is distributed according to π_{n-1}.

As shown by Kingman [15], the theory of partition structures and associated partition-valued processes is best developed in terms of random partitions of the set of positive integers. Our treatment here follows [18].

If we regard a random partition π_n of a positive integer n as a random coloring of n unordered balls, an associated random partition Π_n of the set $[n] := \{1, \ldots, n\}$ may be obtained by placement of the colored balls in a row. We will assume for the rest of this introduction that this placement is made by a random permutation which given π_n is uniformly distributed over all $n!$ possible orderings of n distinct balls.

Formally, a partition of $[n]$ is a collection of disjoint non-empty blocks $\{B_1, \ldots, B_k\}$ with $\cup_{i=1}^k B_i = n$ for some $1 \leq k \leq n$, where each $B_i \subseteq [n]$ represents the set of places occupied by balls of some particular color. We adopt the convention that the blocks B_i are listed *in order of appearance*, meaning that B_i is the set of places in the row occupied by balls of the ith color to appear. So $1 \in B_1$, and if $k \geq 2$ the least element of B_2 is the least element of $[n] \setminus B_1$, if $k \geq 3$ the least element of B_3 is the least element of $[n] \setminus (B_1 \cup B_2)$, and so on. This enumeration of blocks identifies each partition of $[n]$ with an *ordered partition* $(B_1, \ldots, B_k)$, subject to these constraints. The sizes of parts $(|B_1|, \ldots, |B_k|)$ of this partition form a composition of n. The notation $\Pi_n = (B_1, \ldots, B_k)$ is used to signify that $\Pi_n = \{B_1, \ldots, B_k\}$ for some particular sequence of blocks $(B_1, \ldots, B_k)$ listed in order of appearance. If Π_n is derived from π_n by uniform random placement of balls in a row, then Π_n is *exchangeable*, meaning that its distribution is invariant under every deterministic rearrangement of places by a permutation of $[n]$. Put another way, for each partition $(B_1, \ldots, B_k)$ of $[n]$, with blocks in order of appearance,

$$\mathbb{P}(\Pi_n = (B_1, \ldots, B_k)) = p(|B_1|, \ldots, |B_k|) \tag{2.1}$$

for a function $p = p(\lambda)$ of compositions λ of n which is a symmetric function of its k arguments for each $1 \leq k \leq n$. Then p is called the *exchangeable partition probability function* (EPPF) associated with Π_n, or with π_n, the partition of n defined by the unordered sizes of blocks of Π_n.

As observed by Kingman [15], (π_n) is sampling-consistent if and only if the sequence of partitions (Π_n) can be constructed to be consistent in the sense that for $m < n$ the restriction of Π_n to $[m]$ is Π_m. This amounts to a simple recursion formula satisfied by p, recalled later as (3.1). The sequence $\Pi = (\Pi_n)$ can then be interpreted as a random partition of the set $\mathbb{N}$ of all positive integers, whose restriction to $[n]$ is Π_n for every n. Such Π consists of a sequence of blocks $\mathcal{B}_1, \mathcal{B}_2, \ldots$, which may be identified as random disjoint subsets of $\mathbb{N}$, with $\cup_{i=1}^\infty \mathcal{B}_i = \mathbb{N}$, where the nonempty blocks are arranged by increase of their minimal elements, and if the number of nonempty blocks is some $K < \infty$, then by convention $\mathcal{B}_i = \varnothing$ for $i > K$. Similarly, Π_n consists of a sequence of blocks $\mathcal{B}_{ni} := \mathcal{B}_i \cap [n]$, where $\bigcup_i \mathcal{B}_{ni} = [n]$, and the nonempty blocks are consistently arranged by increase of their minimal elements, for all n.

These considerations are summarized by the following proposition:

Proposition 2.1 (Kingman [15])   *The most general partition structure, defined by a sampling-consistent collection of distributions for partitions π_n of integers n, is associated with a unique probability distribution of an exchangeable partition of positive integers $\Pi = (\Pi_n)$, as determined by an EPPF p according to* (2.1).

We now recall a form of *Kingman's paintbox construction* of such an exchangeable random partition Π of positive integers. Regard the unit interval $[0, 1)$ as a continuous spectrum of distinct colors, and suppose given a sequence of random variables $(P_1^\downarrow, P_2^\downarrow, \ldots)$ called *ranked frequencies*, subject to the constraints

$$1 \geq P_1^\downarrow \geq P_2^\downarrow \geq \ldots \geq 0, \quad P_* := 1 - \sum_{j=1}^{\infty} P_j^\downarrow \geq 0. \tag{2.2}$$

The color spectrum is partitioned into a sequence of intervals $[l_i, r_i)$ of lengths $P_i^\downarrow$, and in case $P_* > 0$ a further interval $[1 - P_*, 1)$ of length P_*. Each point u of $[0, 1)$ is assigned the color $c(u) = l_i$ if $u \in [l_i, r_i)$ for some $i = 1, 2, \ldots$, and $c(u) = u$ if $u \in [1 - P_*, 1)$. This coloring of points of $[0, 1)$, called *Kingman's paintbox* associated with $(P_1^\downarrow, P_2^\downarrow, \ldots)$, is sampled by an infinite sequence of independent uniform$[0, 1]$ variables U_i, to assign a color $c(U_i)$ to the ith ball in a row of balls indexed by $i = 1, 2, \ldots$. The associated *color partition* of $\mathbb{N}$ is generated by the random equivalence relation $\sim$ defined by $i \sim j$ if and only if $c(U_i) = c(U_j)$, meaning that either U_i and U_j fall in the same compartment of the paintbox, or that $i = j$ and U_i falls in $[1 - P_*, 1)$.

Theorem 2.2 (Kingman's paintbox representation of exchangeable partitions [14])   *Each exchangeable partition Π of $\mathbb{N}$ generates a sequence of ranked frequencies $(P_1^\downarrow, P_2^\downarrow, \ldots)$ such that the conditional distribution of Π given these frequencies is that of the color partition of $\mathbb{N}$ derived from $(P_1^\downarrow, P_2^\downarrow, \ldots)$ by Kingman's paintbox construction. The exchangeable partition probability function p associated with Π determines the distribution of $(P_1^\downarrow, P_2^\downarrow, \ldots)$, and vice versa.*

The distributions of ranked frequencies $(P_1^\downarrow, P_2^\downarrow, \ldots,)$ associated with naturally arising partition structures (π_n) are quite difficult to deal with analytically. See for instance [25]. Still, $(P_1^{\downarrow}, P_2^{\downarrow}, \ldots)$ can be constructed as the decreasing rearrangement of the frequencies P_i of blocks $\mathcal{B}_i$ of Π defined as the almost-sure limits

$$P_i = \lim_{n \to \infty} n^{-1} |\mathcal{B}_{ni}|, \tag{2.3}$$

where $i = 1, 2, \ldots$ indexes the blocks in order of appearance, while

$$P_* = 1 - \sum_{i=1}^{\infty} P_i = 1 - \sum_{i=1}^{\infty} P_i^{\downarrow}$$

is the asymptotic frequency of the union of singleton blocks

$$\mathcal{B}_* := \cup_{\{i : |\mathcal{B}_i| = 1\}} \mathcal{B}_i,$$

so that (2.3) holds also for $i = *$. The frequencies are called *proper* if $P_* = 0$ a.s.; then almost surely every nonempty block $\mathcal{B}_i$ of Π has a strictly positive frequency, hence $|\mathcal{B}_i| = \infty$, while every block $\mathcal{B}_i$ with $0 < |\mathcal{B}_i| < \infty$ is a singleton block.

The ranked frequencies $P_1^{\downarrow}$, $P_2^{\downarrow}$, $\ldots$ appear in the sequence (P_j) in the order in which intervals of these lengths are discovered by a process of uniform random sampling, as in Kingman's paintbox construction. If $P_* > 0$ then in addition to the strictly positive terms of $P_1^{\downarrow}$, $P_2^{\downarrow}$, $\ldots$ the sequence (P_i) also contains infinitely many zeros which correspond to singletons in Π. The conditional distribution of (P_j) given $(P_j^{\downarrow})$ can also be described in terms of iteration of a single size-biased pick, defined as follows. For a sequence of non-negative random variables (X_i) with $\sum_i X_i \leq 1$ and a random index $J \in \{1, 2, \ldots, \infty\}$, call X_J a *size-biased pick* from (X_i) if X_J has value X_j if $J = j < \infty$ and $X_J = 0$ if $J = \infty$, with

$$\mathbb{P}(J = j \mid (X_i, i \in \mathbb{N})) = X_j \quad (0 < j < \infty) \tag{2.4}$$

(see [5] for this and another definition of size-biased pick in the case of improper frequencies). The *sequence derived from* (X_i) *by deletion of* X_J *and renormalization* refers to the sequence (Y_i) obtained from (X_i) by first deleting the Jth term X_J, then closing up the gap if $J \neq \infty$, and finally normalizing each term by $1 - X_J$. Here by convention, $(Y_i) = (X_i)$ if $X_J = 0$, and (Y_i) is the the zero sequence if $X_J = 1$. Then P_1 is a size-biased pick from $(P_j^{\downarrow})$, P_2 is a size-biased pick from the sequence derived from $(P_j^{\downarrow})$ by deletion of P_1 and renormalization, and so on. For this reason, (P_i) is said to be a *size-biased permutation* of $(P_i^{\downarrow})$.

The two-parameter family. It was shown in [18] that for each pair of real parameters (α, θ) with

$$0 \le \alpha < 1, \ \theta > -\alpha \tag{2.5}$$

the formula

$$p_{\alpha,\theta}(\lambda) := \frac{\prod_{i=1}^{k-1}(\theta + i\alpha)}{(\theta + 1)_{n-1}} \prod_{j=1}^{k}(1 - \alpha)_{\lambda_j - 1}, \tag{2.6}$$

where $k = k_\lambda$, $n = n_\lambda$, and

$$(x)_n := x(x + 1) \dots (x + n - 1) = \frac{\Gamma(x + n)}{\Gamma(x)}$$

is a rising factorial, defines the EPPF of an exchangeable random partition of positive integers whose block frequencies (P_i) in order of appearance admit the *stick-breaking representation*

$$P_i = W_i \prod_{j=1}^{i-1}(1 - W_j) \tag{2.7}$$

for random variables W_j such that

$$W_1, W_2, \dots \text{ are mutually independent} \tag{2.8}$$

with

$$W_k \overset{d}{=} \beta_{1-\alpha,\theta+k\alpha}, \tag{2.9}$$

where $\overset{d}{=}$ indicates equality in distribution and $\beta_{a,b}$ for $a, b > 0$ denotes a random variable with the beta(a, b) density

$$\mathbb{P}(\beta_{a,b} \in du) = \frac{\Gamma(a + b)}{\Gamma(a)\Gamma(b)} u^{a-1}(1 - u)^{b-1} du \qquad (0 < u < 1), \tag{2.10}$$

which is also characterized by the moments

$$\mathbb{E}[\beta_{a,b}^i(1 - \beta_{a,b})^j] = \frac{(a)_i(b)_j}{(a + b)_{i+j}} \qquad (i, j = 0, 1, 2, \dots). \tag{2.11}$$

Formula (2.6) also defines an EPPF for (α, θ) in the range

$$\alpha < 0, \ \theta = -M\alpha \text{ for some } M \in \mathbb{N}, \tag{2.12}$$

in which case the stick-breaking representation (2.7) with factors as in (2.9) makes sense for $1 \le k \le M$, with the last factor $W_M = 1$. The frequencies $(P_1, \dots, P_M)$ in this case are a size-biased random permutation of $(Q_1, \dots, Q_M)$ with the symmetric Dirichlet distribution with

M parameters equal to $\nu := -\alpha > 0$. It is well known that the Q_i can be constructed as $Q_i = \gamma_\nu^{(i)}/\Sigma$, $1 \le i \le M$, where $\Sigma = \sum_{i=1}^M \gamma_\nu^{(i)}$ and the $\gamma_\nu^{(i)}$ are independent and identically distributed copies of a gamma variable γ_ν with density

$$\mathbb{P}(\gamma_\nu \in dx) = \Gamma(\nu)^{-1} x^{\nu-1} e^{-x} \, dx \qquad (x > 0). \tag{2.13}$$

As shown by Kingman [13], the $(0, \theta)$ EPPF (2.6) for $\alpha = 0$, $\theta > 0$ arises in the limit of random sampling from such symmetric Dirichlet frequencies as $\nu = -\alpha \downarrow 0$ and $M \uparrow \infty$ with $\nu M = \theta$ held fixed. In this case, the distribution of the partition π_n is that determined by the Ewens sampling formula with parameter θ, the residual fractions W_i in the stick-breaking representation are identically distributed like $\beta_{1,\theta}$, and the ranked frequencies $P_i^\downarrow$ can be obtained by normalization of the jumps of a gamma process with stationary independent increments $(\gamma_\nu, 0 \le \nu \le \theta)$. Perman, Pitman and Yor [17] gave extensions of this description to the case $0 < \alpha < 1$ when the distribution of ranked frequencies can be derived from the jumps of a stable subordinator of index α. See also [25, 22, 23] for further discussion and applications to the description of ranked lengths of excursion intervals of Brownian motion and Bessel processes.

In the limit case when $\nu = -\alpha \to \infty$ and $\theta = M\nu \to \infty$, for a fixed positive integer M, the EPPF (2.6) converges to

$$p_M(\lambda) := \frac{M(M-1)\cdots(M-k+1)}{M^n}, \tag{2.14}$$

corresponding to sampling from M equal frequencies

$$P_1 = P_2 = \cdots = P_M = 1/M$$

as in the classical coupon-collector's problem with some fixed number M of equally frequent types of coupon. We refer to the collection of partition structures defined by (2.6) for the parameter ranges (2.5) and (2.12), as well as the limit cases (2.14), as the *extended two-parameter family*.

The partition $\mathbf{0}$ of $\mathbb{N}$ into singletons and the partition $\mathbf{1}$ of $\mathbb{N}$ into a single block both belong to the closure of the two-parameter family. As noticed by Kerov [11], a mixture of these two trivial partitions with mixing proportions t and $1-t$ also belongs to the closure, as is seen from (2.6) by letting $\alpha \to 1$ and $\theta \to -1$ in such a way that $(1-\alpha)/(\theta+1) \to t$ and $(\theta + \alpha)/(\theta + 1) \to 1 - t$.

Characterizations by deletion properties. The main focus of this paper is the following result, which was first conjectured by Pitman [18]. For convenience in presenting this result, we impose the following mild *regularity condition* on the EPPF p associated with a partition structure (π_n):

$$p(2,2,1) > 0 \quad \text{and} \quad \lim_{n\to\infty} p(n) = 0. \tag{2.15}$$

Equivalently, in terms of the frequencies P_i in order of appearance,

$$\mathbb{P}(0 < P_1 < P_1 + P_2 < 1) > 0 \text{ and } \mathbb{P}(P_1 = 1) = 0, \tag{2.16}$$

or again, in terms of the ranked frequencies $P_i^{\downarrow}$,

$$\mathbb{P}(0 < P_2^{\downarrow}, \ P_1^{\downarrow} + P_2^{\downarrow} < 1) > 0 \text{ and } \mathbb{P}(P_1^{\downarrow} = 1) = 0. \tag{2.17}$$

Note that this regularity condition does not rule out the case of improper frequencies. See Section 5 for discussion of how the following results can be modified to accommodate partition structures not satisfying the regularity condition.

Theorem 2.3 *Among all partition structures (π_n) with EPPF p subject to (2.15), the extended two-parameter family is characterized by the following property:*

> *if one of n balls is chosen uniformly at random, independently of a random coloring of these balls according to π_n, then given the number of other balls of the same color as the chosen ball is $m - 1$, for some $1 \le m < n$, the coloring of the remaining $n - m$ balls is distributed according to π'_{n-m} for some sequence of partitions $(\pi'_1, \pi'_2, \ldots)$ which does not depend on n.*

Moreover, if (π_n) has the (α, θ) EPPF (2.6), then (π'_n) has the $(\alpha, \theta + \alpha)$ EPPF (2.6), whereas if (π_n) has the EPPF (2.14) for some M, then the EPPF of (π'_n) has the same form except with M decremented by 1.

Note that it is not assumed as part of the property that (π'_n) is a partition structure. Rather, this is implied by the conclusion. Our formulation of Theorem 2.3 was inspired by Kingman [13], who assumed also that $\pi'_n \overset{d}{=} \pi_n$ for all n. The conclusion then holds with $\alpha = 0$, in which case the distribution of π_n is that determined by the Ewens sampling formula from population genetics.

In Section 4 we offer a proof of Theorem 2.3 by purely combinatorial methods. Some preliminary results which we develop in Section 3 allow

Theorem 2.3 to be reformulated in terms of frequencies as in the following Corollary:

Corollary 2.4   *Let the asymptotic frequencies (P_i) of an exchangeable random partition of positive integers Π be represented in the stick-breaking form (2.7) for some sequence of random variables $W_1 < 1$, W_2, ... The condition*

$$W_1 \text{ is independent of } (W_2, W_3, \ldots) \qquad (2.18)$$

obtains if and only if

$$\text{the } W_i \text{ are mutually independent.} \qquad (2.19)$$

If in addition to (2.18) the regularity condition (2.15) holds, then Π is governed by the extended two-parameter family, either with $W_i \stackrel{d}{=} \beta_{1-\alpha,\theta+i\alpha}$, or with $W_i = 1/(M - i + 1)$ for $1 \leq i \leq M$, as in the limit case (2.14), for some $M = 3, 4, \ldots$.

The characterization of the two-parameter family using (2.19) rather than the weaker condition (2.18) was provided by Pitman [19]. As we show in Section 4, it is possible to derive (2.19) directly from (2.18), without passing via Theorem 2.3.

The law of frequencies (P_i) defined by the stick-breaking scheme (2.7) for independent factors W_i with $W_i \stackrel{d}{=} \beta_{1-\alpha,\theta+i\alpha}$ is known as the the two-parameter Griffiths–Engen–McCloskey distribution, denoted GEM(α, θ). The property of the independence of residual proportions W_i, also known as *complete neutrality*, has also been studied extensively in connection with finite-dimensional Dirichlet distributions [3].

The above results can also be expressed in terms of ranked frequencies. Recall that the distribution of ranked frequencies $(P_k^{\downarrow})$ of an (α, θ)-partition is known as the *two-parameter Poisson–Dirichlet distribution* PD(α, θ). According the the previous discussion, a random sequence $(P_k^{\downarrow})$ with PD(α, θ) distribution is obtained by ranking a sequence (P_i) with GEM(α, θ) distribution. The PD(α, θ) distribution was systematically studied in [25], and has found numerous further applications to random trees and associated processes of fragmentation and coagulation [21, 9, 2].

Corollary 2.5   *Let $(P_k^{\downarrow})$ be a decreasing sequence of ranked frequencies subject to (2.2) and the regularity condition (2.17). For $P_J^{\downarrow}$, a size-biased pick from $(P_k^{\downarrow})$, let $(Q_k^{\downarrow})$ be derived from $(P_k^{\downarrow})$ by deletion of $P_J^{\downarrow}$ and renormalization. The random variable $P_J^{\downarrow}$ is independent of the sequence*

$(Q_k^{\downarrow})$ *if and only if either the distribution of* $(P_k^{\downarrow})$ *is* $\mathrm{PD}(\alpha, \theta)$ *for some* (α, θ), *or* $P_k^{\downarrow} = 1/M$ *for all* $1 \leq k \leq M$, *for some* $M \geq 3$. *In the former case, the distribution of* $(Q_k^{\downarrow})$ *is* $\mathrm{PD}(\alpha, \theta + \alpha)$, *whereas in the latter case, the deletion and renormalization simply decrements* M *by one.*

The 'if' part of this Corollary is Proposition 34 of Pitman–Yor [25], while the 'only if' part follows easily from Corollary 2.4, using Kingman's paintbox representation.

3 Partially exchangeable partitions

We start by recalling from [18] some basic properties of *partially exchangeable partitions of positive integers*, which are consistent sequences $\Pi = (\Pi_n)$, where Π_n is a partition of $[n]$ whose probability distribution is of the form (2.1) for some function $p = p(\lambda)$ of compositions λ of positive integers. The consistency of Π_n as n varies amounts to the *addition rule*

$$p(\lambda) = \sum_{j=1}^{k+1} p(\lambda^{(j)}), \tag{3.1}$$

where $k = k_\lambda$ is the number of parts of λ, and $\lambda^{(j)}$ is the composition of $n_\lambda + 1$ derived from λ by incrementing λ_j to $\lambda_j + 1$, and leaving all other components of λ fixed. In particular, for $j = k_\lambda + 1$ this means appending a 1 to λ. There is also the normalization condition $p(1) = 1$. To illustrate (3.1) for $\lambda = (3, 1, 2)$:

$$p(3, 1, 2) = p(4, 1, 2) + p(3, 2, 2) + p(3, 1, 3) + p(3, 1, 2, 1).$$

The following proposition recalls the analog of Kingman's representation for partially exchangeable partitions:

Proposition 3.1 (Corollary 7 from [18])   *Every partially exchangeable partition of positive integers* Π *is such that for each* $k \geq 1$, *the* k*th block* $\mathcal{B}_k$ *has an almost sure limit frequency* P_k. *The partition probability function* p *can then be presented as*

$$p(\lambda) = \mathbb{E}\left[\prod_{i=1}^{k} P_i^{\lambda_i - 1} \prod_{j=1}^{k-1} R_j \right], \tag{3.2}$$

where $k = k_\lambda$ *and* $R_j := (1 - P_1 - \cdots - P_j)$. *Alternatively, in terms of* *the residual fractions W_k in the stick-breaking representation (2.7):*

$$p(\lambda) = \mathbb{E}\left[\prod_{i=1}^{k} W_i^{\lambda_i - 1}\overline{W}_i^{\Lambda_{i+1}}\right],\qquad(3.3)$$

where $\overline{W}_i := 1 - W_i$, $\Lambda_j := \sum_{i\geq j}\lambda_i$. This formula sets up a correspondence between the probability distribution of Π, encoded by the partition probability function p, and an arbitrary joint distribution of a sequence of random variables $(W_1, W_2, \ldots)$ with $0 \leq W_i \leq 1$ for all i.

In terms of randomly coloring a row of n_λ balls, the product whose expectation appears in (3.3) is the conditional probability given W_1, W_2, $\ldots$ of the event that the first λ_1 balls are colored one color, the next λ_2 balls another color, and so on. So (3.3) reflects the fact that conditionally given W_1, W_2, $\ldots$ the process of random coloring of integers occurs according to the following *residual allocation scheme* [18, Construction 16]:

> Ball 1 is painted a first color, and so is each subsequent ball according to a sequence of independent trials with probability W_1 of painting with color 1. The set of balls so painted defines the first block $\mathcal{B}_1$ of Π. Conditionally given $\mathcal{B}_1$, the first unpainted ball is painted a second color, and so is each subsequent unpainted ball according to a sequence of independent trials with probability W_2 of painting with color 2. The balls colored 2 define $\mathcal{B}_2$, and so on. Given an arbitrary sequence of random variables (W_k) with $0 \leq W_k \leq 1$, this coloring scheme shows how to construct a partially exchangeable partition of $\mathbb{N}$ whose asymptotic block frequencies are given by the stick-breaking scheme (2.7).

Note that the residual allocation scheme terminates at the first k, if any, such that $W_k = 1$, by painting all remaining balls color k. The values of W_i for i larger than such a k have no effect on the construction of Π, so cannot be recovered from its almost-sure limit frequencies. To ensure that a unique joint distribution of $(W_1, W_2, \ldots)$ is associated with each p, the convention may be adopted that the sequence (W_i) terminates at the first k if any such that $W_k = 1$. This convention will be adopted in the following discussion.

For W_i which are independent, formula (3.3) factorizes as

$$p(\lambda) = \prod_{i=1}^{k} \mathbb{E}(W_i^{\lambda_i - 1}\overline{W}_i^{\Lambda_{i+1}}).\qquad(3.4)$$

In particular, for independent W_i with the beta distributions (2.9), this formula is readily evaluated using (2.11) to obtain (2.6). Inspection of (2.6) shows that this function of compositions λ is a symmetric function of its parts. Hence the associated random partition Π is exchangeable.

There is an alternate sequential construction of the two-parameter family of partitions which has become known as the 'Chinese Restaurant Process' (see [23], Chapter 3). Instead of coloring rows of balls, imagine customers entering a restaurant with an unlimited number of tables. Initially customer 1 sits at table 1. At stage n, if there are k occupied tables, the ith of them occupied by λ_i customers for $1 \leq i \leq k$, customer $n+1$ sits at one of the previously occupied tables with probability $(\lambda_i - \alpha)/(n+\theta)$, and occupies a new table $k+1$ with probability $(\theta+k\alpha)/(n+\theta)$. It is then readily checked that for each partition of $[n]$ into blocks B_i with $|B_i| = \lambda_i$, after n customers labeled by $[n]$ have entered the restaurant, the probability that those customers labeled by B_i sat at table i for each $1 \leq i \leq k_\lambda$ is given by the product formula (2.6). Moreover, the stick-breaking description of the limit frequencies P_i is readily derived from the Pólya-urn-scheme description of exchangeable trials which, given a beta(a, b)-distributed variable S, are independent with success probability S.

Continuing the consideration of a partially exchangeable partition Π of positive integers, we record the following Lemma.

Lemma 3.2 *Let Π be a partially exchangeable random partition of $\mathbb{N}$ with partition probability function p, and with blocks $\mathcal{B}_1$, $\mathcal{B}_2$, ... and residual frequencies W_1, W_2, ... such that $W_1 < 1$ almost surely. Let Π' denote the partition of $\mathbb{N}$ derived from Π by deletion of the block $\mathcal{B}_1$ containing 1 and re-labeling of $\mathbb{N} - \mathcal{B}_1$ by the increasing bijection with $\mathbb{N}$. Then the following hold:*

(i) *The partition Π' is partially exchangeable, with partition probability function*

$$p'(\lambda_2, \ldots, \lambda_k) = \sum_{\lambda_1=1}^{\infty} \binom{\lambda_1 + \ldots + \lambda_k - 2}{\lambda_1 - 1} p(\lambda_1, \lambda_2, \ldots, \lambda_k) \quad (3.5)$$

and residual frequencies W_2, W_3, ...

(ii) *If Π is exchangeable, then so is Π'.*

(iii) *For $1 \leq m \leq n$*

$$q(n : m) := \mathbb{P}(\mathcal{B}_1 \cap [n] = [m]) = \mathbb{E}(W_1^{m-1}\overline{W}_1^{n-m}), \quad (3.6)$$

and there is the addition rule

$$q(n : m) = q(n + 1 : m + 1) + q(n + 1 : m). \tag{3.7}$$

(iv) *Let* $T_n := \inf\{m : |[n + m] \setminus \mathcal{B}_1| = n\}$ *which is the number of balls of the first color preceding the nth ball not of the first color. Then*

$$\mathbb{P}(T_n = m) = \binom{m + n - 2}{m - 1} q(n + m : m), \tag{3.8}$$

and consequently

$$\sum_{m=1}^{\infty} \binom{m + n - 2}{m - 1} q(n + m : m) = 1. \tag{3.9}$$

Proof Formula (3.6) is read from the general construction of $\mathcal{B}_1$ given W_1 by assigning each $i \geq 2$ to $\mathcal{B}_1$ independently with the same probability W_1. The formulas (3.5) and (3.8) are then seen to be marginalizations of the following expression for the joint distribution of T_n and Π'_n, the restriction of Π' to $[n]$:

$$\begin{aligned}
&\mathbb{P}(T_n = m, \Pi'_n = (C_1, \ldots, C_{k-1})) \\
&= \binom{m + n - 2}{m - 1} q(n + m : m) p(m, |C_1|, \ldots, |C_{k-1}|)
\end{aligned} \tag{3.10}$$

for every partition $(C_1, \ldots, C_{k-1})$ of $[n]$. To check (3.10), observe that the event in question occurs if and only if $\Pi_{n+m} = (B_1, \ldots, B_k)$ for some blocks B_i with $|B_1| = m$ and $|B_i| = |C_{i-1}|$ for $2 \leq i \leq k$. Once B_1 is chosen, each B_i for $2 \leq i \leq k$ is the image of C_{i-1} via the increasing bijection from $[n]$ to $[n + m] \setminus B_1$. For prescribed C_{i-1}, $2 \leq i \leq k$, the choice of $B_1 \subset [n+m]$ is arbitrary subject to the constraint that $1 \in B_1$ and $n + m \notin B_1$. The number of choices is the binomial coefficient in (3.10), so the conclusion is evident. □

The connection between Theorem 2.3 and Corollary 2.4 is established by the following Lemma:

Lemma 3.3 *Let* Π *be a partially exchangeable partition of* $\mathbb{N}$ *with residual frequencies* W_i *such that* $\mathbb{P}(W_1 < 1) = 1$, *with the convention that the sequence terminates at the first k (if any) such that* $W_k = 1$, *so the joint distribution of* (W_i) *is determined uniquely by the partition probability function p of* Π, *and vice versa, according to formula (3.3). For* $\mathcal{B}_1$ *the first block of* Π *with frequency* W_1, *let* Π' *be derived from* Π *by deleting block* $\mathcal{B}_1$ *and relabeling the remaining elements as in Lemma 3.2. The following four conditions on* Π *are equivalent:*

(i) W_1 *is independent of* $(W_2, W_3, \ldots)$.

(ii) *The partition probability function* p *of* Π *admits a factorization of the following form, for all compositions* λ *of positive integers with* $k \geq 2$ *parts:*

$$p(\lambda) = q(n_\lambda : \lambda_1)p'(\lambda_2, \ldots, \lambda_k) \tag{3.11}$$

for some non-negative functions $q(n : m)$ *and* $p'(\lambda_2, \ldots, \lambda_k)$.

(iii) *For each* $1 \leq m < n$, *the conditional distribution of* Π'_{n-m} *given* $|\mathcal{B}_1 \cap [n]| = m$ *depends only on* $n - m$.

(iv) *The random set* $\mathcal{B}_1$ *is independent of the random partition* Π' *of* $\mathbb{N}$.

Finally, if these conditions hold, then (ii) *holds in particular for* $q(n : m)$ *as in* (3.6) *and* $p'(\lambda_2, \ldots, \lambda_k)$ *the partition probability function of* Π'.

Proof That (i) implies (ii) is immediate by combination of the moment formula (3.3), (3.5) and (3.6). Conversely, if (ii) holds for some $q(n : m)$ and $p'(\lambda_2, \ldots, \lambda_k)$, Lemma 3.2 implies easily that (ii) holds for q and p' as in that Lemma. So (ii) gives a formula of the form

$$\mathbb{E}[f(W_1)g(W_2, W_3, \ldots)] = \mathbb{E}[f(W_1)]\mathbb{E}[g(W_2, W_3, \ldots)], \tag{3.12}$$

where g ranges over a collection of bounded measurable functions whose expectations determine the law of W_2, W_3, $\ldots$, and for the g associated with λ_2, $\ldots$, λ_k, the function $f(w)$ ranges over the polynomials $w^{m-1}(1-w)^n$ where $m = \lambda_1 \in \mathbb{N}$ and $n = n_\lambda - \lambda_1 = \sum_{j=2}^{k} \lambda_j$. But linear combinations of these polynomials can be used to uniformly approximate any bounded continuous function of w on $[0,1]$ which vanishes in a neighborhood of 1. It follows that (3.12) holds for all such f, for each g, hence the full independence condition (i). Lastly, the equivalence of (ii), (iii) and (iv) is easily verified. $\square$

4 Exchangeable partitions

For a block B of a random partition Π_n of $[n]$ with $|B| = m$, let $\Pi_n \setminus B$ denote the partition of $[n-m]$ obtained by first deleting the block B of Π_n, then mapping the restriction of Π_n to $[n] \setminus B$ to a partition of $[n-m]$ via the increasing bijection between $[n] \setminus B$ and $[n-m]$. In terms of a coloring of n balls in a row, this means deleting all m balls of some color, then closing up the gaps between remaining balls, to obtain a coloring of $n - m$ balls in a row. Theorem 2.3 can be formulated a little more sharply as follows:

Theorem 4.1 *Among all exchangeable partitions* (Π_n) *of positive integers with EPPF p subject to* (2.15), *the extended two-parameter family is characterized by the following property:*

if $\mathcal{B}_{n1}$ denotes the random block of Π_n containing 1, then for each $1 \leq m < n$, conditionally given $\mathcal{B}_{n1}$ with $|\mathcal{B}_{n1}| = m$, the partition $\Pi_n \setminus \mathcal{B}_{n1}$ has the same distribution as Π'_{n-m} for some sequence of partitions $\Pi'_1, \Pi'_2, \ldots$ which does not depend on n.

Moreover, if (Π_n) is an (α, θ) partition, then we can take for (Π'_n) the exchangeable $(\alpha, \theta + \alpha)$ partition of $\mathbb{N}$.

For an arbitrary partition Π_n of $[n]$ with blocks listed in the order of appearance, define J_n as the index of the block containing an element chosen from $[n]$ uniformly at random, independently of Π_n. We call the block $\mathcal{B}_{nJ_n}$ a *size-biased pick* from the sequence of blocks. Note that this definition agrees with (2.4) in the sense that the number $|\mathcal{B}_{nJ_n}|/n$ is a size-biased pick from the numerical sequence $(|\mathcal{B}_{nj}|/n, \; j = 1, 2, \ldots)$, because given a sequence of blocks of partition Π_n the value $J_n = j$ is taken with probability $|\mathcal{B}_{nj}|/n$. Assuming Π_n exchangeable, the size of the block $|\mathcal{B}_{n1}|$ has the same distribution as $|\mathcal{B}_{nJ_n}|$ conditionally given the ranked sequence of block-sizes, and the reduced partitions $\Pi_n \setminus \mathcal{B}_{n1}$ and $\Pi_n \setminus \mathcal{B}_{nJ_n}$ also have the same distributions. The equivalence of Theorem 2.3 and Theorem 4.1 is evident from these considerations.

We turn to the proof of Theorem 4.1. The condition considered in Theorem 4.1 is just that considered in Lemma 3.3(iii), so we can work with the equivalent factorization condition (3.11). We now invoke the symmetry of the EPPF for an exchangeable Π. Suppose that an EPPF p admits the factorization (3.11), and re-write the identity (3.11) in the form

$$p(m, \lambda) = \frac{q(|\lambda| + m : m)}{q(|\lambda| + 1 : 1)} p(1, \lambda).$$

For this expression we must have non-zero denominator, but this is assured by $\mathbb{P}(0 < W_1 < 1) > 0$, which is implied by the regularity condition (2.15). Instead of part m in $p(m, \lambda)$, we have now 1 in $p(1, \lambda)$. But p is symmetric, hence we can iterate, eventually reducing each part to 1.

Let $\lambda = (\lambda_1, \ldots, \lambda_k)$ be a generic composition, and denote $\Lambda_j = \lambda_j +$

$\cdots + \lambda_k$ the tail sums; thus $\Lambda_1 = |\lambda|$. Iteration yields

$$p(\lambda) = \frac{q(\Lambda_1 : \lambda_1)}{q(1 + \Lambda_2 : 1)} \frac{q(1 + \Lambda_2 : \lambda_2)}{q(2 + \Lambda_3 : 1)} \cdots$$
$$\frac{q(k - 2 + \Lambda_{k-1} : \lambda_{k-1})}{q(k - 1 + \Lambda_k : 1)} \frac{q(k - 1 + \Lambda_k : \lambda_k)}{q(k : 1)} p(1^k),$$

(4.1)

where $p(1^k)$ is the probability of the singleton partition of $[k]$. This leads to the following lemma, which is a simplification of [19, Lemma 12]:

Lemma 4.2 *Suppose that an EPPF p satisfies the factorization condition (3.11) and the regularity condition (2.15). Then*

(i) *either*

$$q(n : m) = \frac{(a)_{m-1}(b)_{n-m}}{(a + b)_{n-1}}$$

for some $a, b > 0$, corresponding to W_1 with beta(a, b) distribution,

(ii) *or*

$$q(n : m) = c^{m-1}(1 - c)^{n-m}$$

for some $0 < c < 1$, corresponding to $W_1 = c$, in which case necessarily $c = 1/M$ for some $M \geq 3$.

Proof By symmetry and the assumption that $p(2, 2, 1) > 0$, it is easily seen from Kingman's paintbox representation that for each $m = 1, 2, \ldots$ there is some composition μ of m such that

$$p(3, 2, \mu) = p(2, 3, \mu) > 0,$$

where for instance $(3, 2, \mu)$ means the composition of $5 + m$ obtained by concatenation of $(3, 2)$ and μ. Indeed, it is clear that one can take either $\mu = 1^m$ or μ to be a single part of size m, according to whether the probability of at least three non-zero frequencies is zero or greater than zero. Applying (4.1) for suitable $k \geq 3$ with $p(1^k) > 0$, and cancelling some common factors of the form $q(n', m')$, which are all strictly positive because $p(2, 2, 1) > 0$ implies $\mathbb{P}(0 < W_1 < 1) > 0$, we see that for every $m = 1, 2, \ldots$

$$\frac{q(m + 5 : 3)q(m + 3 : 2)}{q(m + 3 : 1)} = \frac{q(m + 5 : 2)q(m + 4 : 3)}{q(m + 4 : 1)}.$$

(4.2)

We have by the addition rule (3.7)

$$q(m + 1 : 2) = q(m : 1) - q(m + 1 : 1),$$
$$q(m + 2 : 3) = q(m : 1) - 2q(m + 1 : 1) + q(m + 2 : 1),$$

and introducing variables $x_m = q(m : 1)$, $n = m + 2$,

$$\frac{(x_{n+1} - 2x_{n+2} + x_{n+3})(x_n - x_{n+1})}{x_{n+1}}$$
$$= \frac{(x_{n+2} - x_{n+3})(x_n - 2x_{n+1} + x_{n+2})}{x_{n+2}}.$$

The recursion is homogeneous; to pass to inhomogeneous variables divide both sides of the equality by x_n, then set $y_n := x_{n+1}/x_n$ and rewrite as

$$(1 - 2y_{n+1} + y_{n+2}y_{n+1})(1 - y_n) = (1 - y_{n+2})(1 - 2y_n + y_n y_{n+1}),$$

which simplifies as

$$-2y_{n+1} + y_{n+1}y_{n+2} + y_n y_{n+1} = -y_n - y_{n+2} + 2y_n y_{n+2}.$$

Finally, use substitution

$$y_n = 1 - \frac{1}{z_n}$$

to arrive at

$$\frac{z_n - 2z_{n+1} + z_{n+2}}{z_n z_{n+1} z_{n+2}} = 0.$$

From this, z_n is a linear function of n, which must be nondecreasing to agree with $0 < y_n < 1$.

If z_n is not constant, then going back to x_ns we obtain

$$q(n : 1) = c_0 \frac{(b)_{n-1}}{(a + b)_{n-1}}, \quad n \geq 3,$$

for some a, b, c_0, where the factor c_0 appears since the relation (4.2) is homogeneous. It is seen from the moments representation

$$q(n : 1) = \int_{[0,1]} (1 - x)^{n-1} \mathbb{P}(P_1 \in dx), \quad n \geq 3,$$

that when a, b are fixed, the factor c_0 is determined from the normalization by choosing a value of $\mathbb{P}(P_1 = 1)$. The condition $p(n) \to 0$ means that $\mathbb{P}(P_1 = 1) = 0$, in which case $c_0 = 1$ and the distribution of P_1 is beta(a, b) with some positive a, b.

If $(z_n, \; n \geq 3)$ is a constant sequence, then $q(n : 1)$ is a geometric progression, and a similar argument shows that the case (ii) prevails. That $c = 1/M$ for some $M \geq 3$ is quite obvious: the only way that a size-biased choice of a frequency can be constant is if there are M equal frequencies for some $M \geq 1$. The regularity assumption (2.15) rules out the cases $M = 1, 2$. □

Proof of Theorem 4.1   In the case (i) of Lemma 4.2, substituting in (4.1) yields

$$\frac{p(\lambda)}{p(1^k)} = \frac{(a)_{\lambda_1-1}(b)_{\Lambda_2}}{(a+b)_{\Lambda_1-1}} \frac{(a+b)_{\Lambda_2}}{(b)_{\Lambda_2}} \frac{(a)_{\lambda_2-1}(b)_{\Lambda_3+1}}{(a+b)_{\Lambda_2}} \frac{(a+b)_{\Lambda_3+1}}{(b)_{\Lambda_3+1}} \cdots$$
$$\frac{(a)_{\lambda_k-1}(b)_{k-1}}{(a+b)_{\Lambda_k+k-2}} \frac{(a+b)_{k-1}}{(b)_{k-1}},$$

provided $p(1^k) > 0$. After cancellation this becomes

$$\frac{p(\lambda)}{p(1^k)} = \frac{(a+b)_{k-1}}{(a+b)_{n-1}} \prod_{j=1}^{k} (a)_{\lambda_j-1},$$

where $n = \Lambda_1 = \lambda_1 + \ldots + \lambda_k = |\lambda|$. Specializing,

$$\frac{p(2,1^{k-1})}{p(1^k)} = \frac{a}{a+b+k-1},$$

and using the addition rule (3.1),

$$p(1^k) = kp(2,1^{k-1}) + p(1^{k+1});$$

we obtain the recursion

$$\frac{p(1^{k+1})}{p(1^k)} = \frac{a+b+k(1-a)-1}{a+b+k-1}, \quad p(1) = 1.$$

Now (2.6) follows readily by re-parametrization $\theta = a+b-1$, $\alpha = 1-a$.

The case (ii) of Lemma 4.2 is even simpler, as it is immediate that $W_1 = 1/M$ implies that the partition is generated as if by coupon collecting with M equally frequent coupons. $\square$

Proof of Corollary 2.4   As observed earlier, Corollary 2.4 characterizing the extended two-parameter family by the condition that

$$W_1 \text{ and } (W_2, W_3, \ldots) \text{ are independent} \tag{4.3}$$

can be read from Theorem 4.1 and Lemma 3.3. We find it interesting nonetheless to provide another proof of Corollary 2.4 based on analysis of the limit frequencies rather than the EPPF. This was in fact the first argument we found, without which we might not have persisted with the algebraic approach of the previous section.

Suppose then that $W_1, W_2, W_3, \ldots$ is the sequence of residual fractions associated with an EPPF p, and that (4.3) holds. The symmetry condition $p(r+1, s+1) = p(s+1, r+1)$ and the moment formula (3.3) give

$$\mathbb{E}(W_1^r \overline{W}_1^{s+1})\mathbb{E}(W_2^s) = \mathbb{E}(W_1^s \overline{W}_1^{r+1})\mathbb{E}(W_2^r), \tag{4.4}$$

for non-negative integers r and s. Setting $r = 0$, this expresses moments of W_2 in terms of the moments of W_1. So the distribution of W_1 determines that of W_2. Assume now the regularity condition (2.15). According to Lemma 4.2 we are reduced either to the case with M equal frequencies with sum 1, or to the case where W_1 has a beta distribution, and hence so does W_2, by consideration of (4.4). There is nothing more to discuss in the first case, so we assume for the rest of this section that

> each of W_1 and W_2 has a non-degenerate beta distribution, with possibly different parameters. (4.5)

Recall that

$$P_1 = W_1 \text{ and } P_2 = (1 - W_1)W_2.$$

As observed in [19],

> the conditional distribution of $(P_3, P_4, \ldots)$ given P_1 and P_2 depends symmetrically on P_1 and P_2.

This can be seen from Kingman's paintbox representation, which implies that conditionally given $P_1^\downarrow, P_2^\downarrow, \ldots$, as well as P_1 and P_2, the sequence $(P_3, P_4, \ldots)$ is derived by a process of random sampling from the frequencies $(P_i^\downarrow)$ with the terms P_1 and P_2 deleted. No matter what $(P_i^\downarrow)$ this process depends symmetrically on P_1 and P_2, so the same is true without the extra conditioning on $(P_i^\downarrow)$.

Since $P_1 + P_2$ is a symmetric function of P_1 and P_2, and $(W_3, W_4, \ldots)$ is a measurable function of $P_1 + P_2$ and $(P_3, P_4, \ldots)$,

> the conditional distribution of $W_3, W_4, \ldots$ given (P_1, P_2) depends symmetrically on P_1 and P_2.

The condition that W_1 is independent of $(W_2, W_3, W_4, \ldots)$ implies easily that

W_1 is conditionally independent of $(W_3, W_4, \ldots)$ given W_2.

Otherwise put:

P_1 is conditionally independent of $(W_3, W_4, \ldots)$ given $P_2/(1 - P_1)$,

hence by the symmetry discussed above

P_2 is conditionally independent of $(W_3, W_4, \ldots)$ given $P_1/(1 - P_2)$.

Let $X := P_2/(1 - P_1)$, $Y := P_1/(1 - P_2)$ and $Z := (W_3, W_4, \ldots)$. Then we have both

$$X \text{ is conditionally independent of } Z \text{ given } Y, \qquad (4.6)$$

and

$$Y \text{ is conditionally independent of } Z \text{ given } X, \qquad (4.7)$$

from which it follows under suitable regularity conditions (see Lemma 4.3 below) that

$$(X, Y) \text{ is independent of } Z, \qquad (4.8)$$

meaning in the present context that

$$W_1, W_2 \text{ and } (W_3, W_4, \ldots) \text{ are independent.} \qquad (4.9)$$

Lauritzen [16, Proposition 3.1] shows that (4.6) and (4.7) imply (4.8) under the assumption that (X, Y, Z) has a positive and continuous joint density relative to a product measure. From (4.5) and strict positivity of the beta densities on $(0, 1)$, we see that (X, Y) has a strictly positive and continuous density relative to Lebesgue measure on $(0, 1)^2$. We are not in a position to assume that (X, Y, Z) has a density relative to a product measure. However, the passage from (4.6) and (4.7) to (4.8) is justified by Lemma 4.3 below without need for a trivariate density. So we deduce that (4.9) holds. By Lemma 3.2, $(W_2, W_3, \ldots)$ is the sequence of residual fractions of an exchangeable partition Π', and W_2 has a beta density. So either $W_3 = 1$ and we are in the case (2.12) with $M = 3$, or W_3 has a beta density, and the previous argument applies to show that

$$W_1, W_2, W_3 \text{ and } (W_4, W_5, \ldots) \text{ are independent.}$$

Continue by induction to conclude the independence of $W_1, W_2, \ldots, W_k$ for all k such that $p(1^k) > 0$. □

Lemma 4.3 *Let X, Y and Z denote random variables with values in arbitrary measurable spaces, all defined on a common probability space, such that (4.6) and (4.7) hold. If the joint distribution of the pair (X, Y) has a strictly positive probability density relative to some product probability measure, then (4.8) holds.*

Proof Let $p(X, Y)$ be a version of $\mathbb{P}(Z \in B \mid X, Y)$ for B a measurable set in the range of Z. By standard measure theory (e.g. Kallenberg [10, 6.8]) the first conditional independence assumption gives $\mathbb{P}(Z \in B \mid X, Y) = \mathbb{P}(Z \in B \mid X)$ a.s. so that

$p(X, Y) = g(X)$ a.s. for some measurable function g.

Similarly from the second conditional independence assumption,

$p(X, Y) = h(Y)$ a.s. for some measurable function h,

and we wish to conclude that

$p(X, Y) = c$ a.s. for some constant c.

To complete the argument it suffices to draw this conclusion from the above two assumptions about a jointly measurable function p, with (X, Y) the identity map on the product space of pairs $\mathcal{X} \times \mathcal{Y}$, and the two almost-sure equalities holding with respect to some probability measure P on this space, with P having a strictly positive density relative to a product probability measure $\mu \otimes \nu$. Fix $u \in (0, 1)$; from the previous assumptions it follows that

$$\{p(X, Y) > u\} = \{X \in A_u\} = \{Y \in C_u\} \qquad \text{a.s.} \qquad (4.10)$$

for some measurable sets A_u, C_u, whence

$$\{p(X, Y) > u\} = \{X \in A_u\} \cap \{Y \in C_u\} \qquad \text{a.s.,} \qquad (4.11)$$

where the almost-sure equalities hold both with respect to the joint distribution P of (X, Y), and with respect to a product probability measure $\mu \otimes \nu$ governing (X, Y). But under $\mu \otimes \nu$ the random variables X and Y are independent. So if $q := (\mu \otimes \nu)(p(X, Y) > u)$, then (4.10) and (4.11) imply that $q = q^2$, so $q = 0$ or $q = 1$. Thus $p(X, Y)$ is constant a.s. with respect to $\mu \otimes \nu$, hence also constant with respect to P. $\qquad \square$

5 The deletion property without the regularity condition

Observe that the property required in Theorem 2.3 is void if π_n happens to be the one-block partition (n). This readily implies that mixing with the trivial one-block partition $\mathbf{1}$ does not destroy the property. Therefore the $\mathbf{1}$-component may be excluded from the consideration, meaning that it is enough to focus on the case

$$P_1 < 1 \text{ a.s., or equivalently } P_1^{\downarrow} < 1 \text{ a.s., or equivalently} \atop \lim_{n \to \infty} p(n) = 0. \qquad (5.1)$$

Suppose then that this condition holds, but that the first condition in (2.15) does not hold, so that $p(2, 2, 1) = 0$. Then

$$\mathbb{P}(P_2^\downarrow = 1 - P_1^\downarrow > 0) + \mathbb{P}(P_1^\downarrow < 1, \ P_2^\downarrow = 0) = 1.$$

If both terms have positive probability then $\mathbb{P}(W_2 = 1 \,|\, W_1 = 0) = 0$ but $\mathbb{P}(W_2 = 1 \,|\, W_1 > 0) > 0$, so the independence of W_1 and W_2 fails. Thus the independence forces either $\mathbb{P}(P_2^\downarrow = 1 - P_1^\downarrow > 0) = 1$ or $\mathbb{P}(P_1^\downarrow < 1, \ P_2^\downarrow = 0) = 1$. The two cases are readily treated:

(i) If $\mathbb{P}(P_2^\downarrow = 1 - P_1^\downarrow > 0) = 1$ then $W_2 = 1$ a.s. and the independence trivially holds. This is the case when Π has two blocks almost surely.

(ii) If $\mathbb{P}(P_1^\downarrow < 1, \ P_2^\downarrow = 0) = 1$ and $\mathbb{P}(P_1^\downarrow > 0) > 0$ then $\mathbb{P}(W_2 > 0 \,|\, W_1 > 0) = 0$ but $\mathbb{P}(W_2 > 0 \,|\, W_1 = 0) > 0$, hence W_1 and W_2 are not independent. Therefore $\mathbb{P}(P_1^\downarrow < 1, \ P_2^\downarrow = 0) = 1$ and the independence imply $P_1^\downarrow = 0$ a.s., meaning that $\Pi = \mathbf{0}$.

We conclude that the most general exchangeable partition Π which has the property in Theorem 4.1 is a two-component mixture, in which the first component is either a partition from the extended two-parameter family, or a two-block partition as in (i) above, or $\mathbf{0}$, and the second component is the trivial partition $\mathbf{1}$.

6 Regeneration and τ-deletion

In this section we partly survey and partly extend the results from [8, 7] concerning characterizations of (α, θ) partitions by regeneration properties. As in Kingman's study of the regenerative processes [12], subordinators (increasing Lévy processes) appear naturally in our framework of *multiplicative* regenerative phenomena. Following [7], we call a partition structure (π_n) *regenerative* if

for each n it is possible to delete a randomly chosen part of π_n in such a way that for each $0 < m < n$, given the deleted part is of size m, the remaining parts form a partition of $n - m$ with the same distribution as π_{n-m}.

In terms of an exchangeable partition $\Pi = (\Pi_n)$ of $\mathbb{N}$, the associated partition structure (π_n) is regenerative if and only if for each n it is possible to select a random block $\mathcal{B}_{nJ_n}$ of Π_n in such a way that for each $0 < m < n$, conditionally given that $|\mathcal{B}_{nJ_n}| = m$ the partition $\Pi_n \setminus \mathcal{B}_{nJ_n}$ of $[n - m]$ is distributed according to the unconditional distribution of Π_{n-m}:

$$(\Pi_n \setminus \mathcal{B}_{nJ_n} \text{ given } |\mathcal{B}_{nJ_n}| = m) \stackrel{d}{=} \Pi_{n-m}, \qquad (6.1)$$

where $\Pi_n \setminus \mathcal{B}_{nJ_n}$ is defined as in the discussion preceding Theorem 4.1. Moreover, there is no loss of generality in supposing further that the conditional distribution of J_n given Π_n is of the form

$$\mathbb{P}(J_n = j \mid \Pi_n = \{B_1, \ldots, B_k\}) = d(|B_1|, \ldots, |B_k|; j) \qquad (6.2)$$

for some *symmetric deletion kernel* d, meaning a non-negative function of a composition λ of n and $1 \le j \le k_\lambda$ such that

$$d(\lambda_1, \lambda_2, \ldots, \lambda_k; j) = d(\lambda_{\sigma(1)}, \lambda_{\sigma(2)}, \ldots, \lambda_{\sigma(k)}; 1) \qquad (6.3)$$

for every permutation σ of $[k]$ with $\sigma(1) = j$. To determine a symmetric deletion kernel, it suffices to specify $d(\lambda; 1)$, which is the conditional probability, given blocks of sizes $\lambda_1, \lambda_2, \ldots, \lambda_k$, of picking the first of these blocks. This is a non-negative symmetric function of $(\lambda_2, \ldots, \lambda_k)$, subject to the further constraint that its extension to arguments $j \ne 1$ via (6.3) satisfies

$$\sum_{j=1}^{k_\lambda} d(\lambda; j) = 1$$

for every composition λ of n. The regeneration condition can now be reformulated in terms of the EPPF p of Π in a manner similar to (3.11):

Lemma 6.1 *An exchangeable random partition Π with EPPF p is regenerative if and only if there exists a symmetric deletion kernel d such that*

$$p(\lambda)d(\lambda; 1) = q(n, \lambda_1)\binom{n}{\lambda_1}^{-1} p(\lambda_2, \ldots, \lambda_k) \qquad (6.4)$$

for every composition λ of n into at least two parts and some non-negative function q. Then

$$q(n, m) = \mathbb{P}(|\mathcal{B}_{n,J_n}| = m) \qquad (m \in [n]) \qquad (6.5)$$

for J_n as in (6.2).

Proof Formula (6.4) offers two different ways of computing the probability of the event that $\Pi_n = \{B_1, \ldots, B_k\}$ and $J_n = 1$ for an arbitrary partition $\{B_1, \ldots, B_k\}$ of $[n]$ with $|B_i| = \lambda_i$ for $i \in [k]$: on the left side, by definition of the symmetric deletion kernel, and on the right side by conditioning on the event $\mathcal{B}_{n, J_n} = B_1$ and appealing to the regeneration property and exchangeability.   □

Consider now the question of whether an (α, θ) partition with EPPF $p = p_{\alpha, \theta}$ as in (2.6) is regenerative with respect to some deletion kernel. By the previous lemma and cancellation of common factors, the question is whether there exists a symmetric deletion kernel $d(\lambda; j)$ such that the value of

$$q(n, \lambda_1) = d(\lambda; 1) \binom{n}{\lambda_1} \frac{(1 - \alpha)_{\lambda_1 - 1} (\theta + (k - 1)\alpha)}{(\theta + n - \lambda_1)_{\lambda_1}} \qquad (6.6)$$

is the same for all compositions λ of n with k parts and a prescribed value of λ_1. But it is easily checked that the formula

$$d_{\alpha, \theta}(\lambda; j) = \frac{\theta \lambda_j + \alpha(n - \lambda_j)}{n(\theta + \alpha(k - 1))} \qquad (6.7)$$

provides just such a symmetric deletion kernel. Note that the kernel depends on (α, θ) only through the ratio $\tau := \alpha/(\alpha + \theta)$, and that the kernel is non-negative for all compositions λ only if both α and θ are non-negative.

To provide a more general context for this and later discussions, let $(x_1, \ldots, x_k)$ be a fixed sequence of positive numbers with sum $s = \sum_{j=1}^{k} x_j$. For a fixed parameter $\tau \in [0, 1]$, define a random variable T with values in $[k]$ by

$$\mathbb{P}(T = j \mid (x_1, \ldots, x_k)) = \frac{(1 - \tau)x_j + \tau(s - x_j)}{s(1 - \tau + \tau(k - 1))}. \qquad (6.8)$$

The random variable x_T is called a τ-*biased pick* from $x_1, \ldots, x_k$. The law of x_T does not depend on the order of the sequence $(x_1, \ldots, x_k)$, and there is also a scaling invariance: $s^{-1} x_T$ is a τ-biased pick from $(s^{-1} x_1, \ldots, s^{-1} x_k)$. Note that a 0-biased pick is a size-biased pick from $(x_1, \ldots, x_k)$, choosing any particular element with probability proportional to its size. A 1/2-biased pick is a uniform random choice from the list, as (6.8) then equals $1/k$ for all j. And a 1-biased pick may be called a co-size biased pick, as it chooses j with probability proportional to its co-size $s - x_j$.

These definitions are now applied to the sequence of block sizes x_j of the restriction to $[n]$ of an exchangeable partition Π of $\mathbb{N}$. We denote by T_n a random variable whose conditional distribution given Π_n with k blocks and $|\mathcal{B}_{nj}| = x_j$ for $j \in [k]$ is defined by (6.8), and denote by $\mathcal{B}_{nT_n}$ the τ-biased pick from the sequence of blocks of Π_n. We call Π τ-*regenerative* if Π_n is regenerative with respect to deletion of the τ-biased pick $\mathcal{B}_{nT_n}$.

Theorem 6.2 ([8, 7])   *For each* $\tau \in [0, 1]$, *apart from the constant partitions* **0** *and* **1**, *the only exchangeable partitions of* $\mathbb{N}$ *that are* τ-*regenerative are the members of the two-parameter family with parameters in the range*

$$\{(\alpha, \theta) \in [0, 1] \times [0, \infty] : \alpha/(\alpha + \theta) = \tau\}.$$

Explicitly, the distribution of the τ-*biased pick for such* (α, θ) *partitions of* $[n]$ *is*

$$\mathbb{P}(|\mathcal{B}_{nT_n}| = m) = \binom{n}{m} \frac{(1 - \alpha)_{m-1}}{(\theta + n - m)_m} \frac{(n - m)\alpha + m\theta}{n}, \qquad m \in [n].$$
(6.9)

Proof   The preceding discussion around (6.6) and (6.7) shows that members of the two-parameter family with parameters in the indicated range are τ-regenerative, and gives the formula (6.9) for the decrement matrix. See [7] for the proof that these are the only non-degenerate exchangeable partitions of $\mathbb{N}$ that are τ-regenerative.   □

In particular, each (α, α) partition is $1/2$-regenerative, meaning regenerative with respect to deletion of a block chosen uniformly at random. The constant partitions **0** and **1** are obviously τ regenerative for every $\tau \in [0, 1]$. This is consistent with the characterization above because the $(1, \theta)$ partition is the **0** partition for every $\theta \geq 0$, and because the partition **1** can be reached as a limit of (α, θ) partitions as $\alpha, \theta \downarrow 0$ with $\alpha(\alpha + \theta)^{-1}$ held fixed.

Multiplicative regeneration.   By Corollary 2.4, if (P_i) is the sequence of limit frequencies for a $(0, \theta)$ partition for some $\theta > 0$ and if the first limit frequency P_1 is deleted and the other frequencies renormalized to sum to 1, then the resulting sequence (Q_j) is independent of P_1 and has the same distribution as (P_j). Because P_1 is a size-biased pick from the sequence (P_i), this regenerative property of the frequencies (P_i) can be seen as an analogue of the 0-regeneration property of the $(0, \theta)$ partitions.

If (P_i) is instead the sequence of limit frequencies of an (α, θ) partition Π for parameters satisfying $0 < \alpha < 1$, $\alpha/(\alpha + \theta) = \tau$, a question arises:

does the regenerative property of Π_n with respect to a τ-biased pick have an analogue in terms of a τ-biased pick from the frequencies (P_i)? This cannot be answered straightforwardly as in the $\tau = 0$ case, because when $\tau > 0$ the formula (6.8) defines a proper probability distribution only for series (x_j) with some finite number k of positive terms. For instance, in the case $\tau = 1/2$ there is no such analogue of (6.8) as 'uniform random choice' from infinitely many terms.

Still, the Ewens case provides a clue if we turn to a *bulk deletion*. Let P_J be a size-biased pick from the frequencies (P_j), as defined by (2.4), and let (Q_j) be a sequence obtained from (P_j) by deleting all $P_1, \ldots, P_J$ and renormalizing. Then (Q_j) is independent of $P_1, \ldots, P_J$, and $(Q_j) \stackrel{d}{=} (P_j)$. The latter assertion follows from the i.i.d. property of the residual fractions and by noting that (2.4) is identical with

$$\mathbb{P}(J = j \mid (W_i, i \in \mathbb{N})) = W_j \prod_{i=1}^{j-1}(1 - W_i).$$

A similar bulk deletion property holds for partitions in the Ewens family, in the form

$$(\Pi_n \setminus (\mathcal{B}_{n1} \cup \cdots \cup \mathcal{B}_{nJ_n}) \text{ given } |\mathcal{B}_1 \cup \cdots \cup \mathcal{B}_{nJ_n}| = m) \stackrel{d}{=} \Pi_{n-m}$$

for all $1 \leq m \leq n$, where $\mathcal{B}_{nJ_n}$ is a size-biased pick from the blocks.

To make the Ansatz of bulk deletion work for $\tau \neq 0$ it is necessary to arrange the frequencies in a more complex manner. To start with, we modify the paintbox construction. Let $\mathcal{U} \subset [0, 1]$ be a random open set canonically represented as the union of its disjoint open component intervals. We suppose that the Lebesgue measure of $\mathcal{U}$, equal to the sum of lengths of the components, is 1 almost surely. We associate with $\mathcal{U}$ an exchangeable partition Π exactly as in Kingman's representation in Theorem 2.2. For each component interval $G \subset \mathcal{U}$ there is an index $i_G := \min\{n : U_n \in G\}$ that is the minimal index of a sequence (U_i) of i.i.d. uniform[0,1] points hitting the interval, and for all j, P_j is the length of the jth component interval when the intervals are listed in order of increasing minimal indices. So (P_j) is a size-biased permutation of the lengths of interval components of $\mathcal{U}$.

Let $\triangleleft$ be the linear order on $\mathbb{N}$ induced by the *interval order* of the components of $\mathcal{U}$, so $j \triangleleft k$ iff the interval of length P_j, which is the home interval of the jth block $\mathcal{B}_j$ to appear in the process of uniform random sampling of intervals, lies to the left of the interval of length P_k associated with block $\mathcal{B}_k$. A convergence argument shows that $\mathcal{U}$ is uniquely determined by (P_j) and $\lhd$. In loose terms, $\mathcal{U}$ is an arrangement of a sequence of tiles of sizes P_j in the order on indices j prescribed by $\lhd$, and this arrangement is constructible by sequentially placing the tile j in the position prescribed by the order $\lhd$ restricted to $[j]$.

For $x \in [0, 1)$ let $(a_x, b_x) \subset \mathcal{U}$ be the component interval containing x. Define $\mathcal{V}_x$ as the open set obtained by deleting the bulk of component intervals to the left of b_x, then linearly rescaling the remaining set $\mathcal{U} \cap [b_x, 1]$ to $[0, 1]$. We say that $\mathcal{U}$ is *multiplicatively regenerative* if for each $x \in [0, 1)$, $\mathcal{V}_x$ is independent of $\mathcal{U} \cap [0, b_x]$ and $\mathcal{V}_x \overset{d}{=} \mathcal{U}$.

An ordered version of the paintbox correspondence yields:

Theorem 6.3 ([8, 7]) *An exchangeable partition Π is regenerative if and only if it has a paintbox representation in terms of some multiplicatively regenerative set $\mathcal{U}$. The deletion operation is then defined by classifying n independent uniform points from $[0, 1]$ according to the intervals of $\mathcal{U}$ into which they fall, and deleting the block of points in the leftmost occupied interval.*

A property of the frequencies (P_j) of an exchangeable regenerative partition Π of $\mathbb{N}$ now emerges: there exists a strict total order $\lhd$ on $\mathbb{N}$, which is a random order, which has some joint distribution with (P_j) such that arranging the intervals of sizes (P_j) in order $\lhd$ yields a multiplicatively regenerative set $\mathcal{U}$. Equivalently, there exists a multiplicatively regenerative set $\mathcal{U}$ that induces a partition with frequencies (P_j) and an associated order $\lhd$. This set $\mathcal{U}$ is then necessarily unique in distribution as a random element of the space of open subsets of $[0, 1]$ equipped with the Hausdorff metric [8] on the complementary closed subsets. A subtle point here is that the joint distribution of (P_j) and $\lhd$ is not unique, and neither is the joint distribution of (P_j) and $\mathcal{U}$, unless further conditions are imposed. For instance, one way to generate $\lhd$ is to suppose that the (P_j) are generated by a process of uniform random sampling from $\mathcal{U}$. But for a $(0, \theta)$ partition, we know that another way is to construct $\mathcal{U}$ from (P_j) by simply placing the intervals in deterministic order $P_1, P_2, \ldots$ from left to right. In the construction by uniform random sampling from $\mathcal{U}$ the interval of length P_1 discovered by the first sample point need not be the leftmost, and need not lie to the left of the second discovered interval P_2.

In [8] we showed that the multiplicative regeneration of $\mathcal{U}$ follows from an apparently weaker property: if (a_U, b_U) is the component interval of $\mathcal{U}$ containing a uniform$[0,1]$ sample U independent of $\mathcal{U}$, and if $\mathcal{V}$ is defined as the open set obtained by deleting the component intervals to the left of b_U and linearly rescaling the remaining set $\mathcal{U} \cap [b_U, 1]$ to [0,1], then given $b_U < 1$, $\mathcal{V}$ is independent of b_U (hence, as we proved, independent of $\mathcal{U} \cap [0, b_U]$ too!) and has distribution equal to the unconditional distribution of $\mathcal{U}$. This independence is the desired analogue for more general regenerative partitions of the bulk-deletion property of Ewens partitions.

The fundamental representation of multiplicatively regenerative sets involves a random process F_t known in statistics as a neutral-to-the right distribution function.

Theorem 6.4 ([8])   *A random open set $\mathcal{U}$ of Lebesgue measure 1 is multiplicatively regenerative if and only if there exists a drift-free subordinator $S = (S_t, t \geq 0)$ with $S_0 = 0$ such that $\mathcal{U}$ is the complement to the closed range of the process $F_t = 1 - \exp(-S_t)$, $t \geq 0$. The Lévy measure of S is determined uniquely up to a positive factor.*

According to Theorems 6.3 and 6.4, regenerative partition structures with proper frequencies are parametrized by a measure $\tilde{\nu}(\mathrm{d}u)$ on $(0, 1]$ with finite first moment, which is the image via the transformation from s to $1 - \exp(-s)$ of the Lévy measure $\nu(\mathrm{d}s)$ on $(0, \infty]$ associated with the subordinator S. The Laplace exponent Φ of the subordinator, defined by the Lévy–Khintchine formula

$$\mathbb{E}[\exp(-aS_t)] = \exp[-t\Phi(a)], \qquad a \geq 0,$$

determines the Lévy measure $\nu(\mathrm{d}s)$ on $(0, \infty]$ and its image $\tilde{\nu}(\mathrm{d}u)$ on $(0, 1]$ via the formulae

$$\Phi(a) = \int_{(0,\infty]} (1 - e^{-ax})\nu(\mathrm{d}x) = \int_{]0,1]} (1 - (1 - x)^a)\tilde{\nu}(\mathrm{d}x).$$

As shown in [8], the decrement matrix q of the regenerative partition structure, as in (6.5), is then

$$q(n, m) = \frac{\Phi(n, m)}{\Phi(n)}, \qquad 1 \leq m \leq n, \ n = 1, 2, \ldots,$$

where

$$\Phi(n, m) = \binom{n}{m} \int_{]0,1]} x^m (1 - x)^{n-m} \tilde{\nu}(\mathrm{d}x).$$

Uniqueness of the parametrization is achieved by a normalization condition, such as $\Phi(1) = 1$.

In [8] the subordinator $S^{\alpha,\theta}$ which produces $\mathcal{U}$ as in Theorem 6.4 for the (α, θ) partition was identified by the following formula for the right tail of its Lévy measure:

$$\nu(x, \infty] = (1 - e^{-x})^{-\alpha} e^{-x\theta}, \quad x > 0. \tag{6.10}$$

The subordinator $S^{(0,\theta)}$ is a compound Poisson process whose jumps are exponentially distributed with rate θ. For $\theta = 0$ the Lévy measure has a unit mass at ∞, so the subordinator $S^{(\alpha,0)}$ is killed at unit rate. The $S^{(\alpha,\alpha)}$ subordinator belongs to the class of Lamperti-stable processes recently studied in [4]. For positive parameters the subordinator $S^{(\alpha,\theta)}$ can be constructed from the $(0, \theta)$ and $(\alpha, 0)$ cases, as follows. First split $\mathbb{R}_+$ by the range of $S^{(0,\theta)}$, that is at points $E_1 < E_2 < \cdots$ of a Poisson process with rate θ. Then run an independent copy of $S^{(\alpha,0)}$ up to the moment the process crosses E_1 at some random time, say t_1. The level-overshooting value is neglected and the process is stopped. At the same time t_1 a new independent copy of $S^{(\alpha,0)}$ is started at value E_1 and run until crossing E_2 at some random time t_2, and so on.

In terms of $F_t = 1 - \exp(-S_t)$, the range of the process in the $(0, \theta)$ case is a stick-breaking set $\{1 - \prod_{i=1}^{j-1}(1 - V_i), i = 0, 1, \ldots\}$ with i.i.d. beta$(1, \theta)$ factors V_i. In the case $(\alpha, 0)$ the range of (F_t) is the intersection of $[0, 1]$ with the α-stable set (the range of α-stable subordinator). In other cases $\mathcal{U}$ is constructible as a cross-breed of the cases $(\theta, 0)$ and $(0, \alpha)$: first $[0, 1]$ is partitioned in subintervals by the beta$(1, \theta)$ stick-breaking, then each subinterval (a, b) of this partition is further split by independent copy of the multiplicatively regenerative $(\alpha, 0)$ set, shifted to start at a and truncated at b.

Constructing the order. Following [7, 24], we shall describe an arrangement which allows us to pass from (α, θ) frequencies (P_j) to the multiplicatively regenerative set associated with the subordinator $S^{(\alpha,\theta)}$. The connection between size-biased permutation and τ-deletion (Lemma 6.6) is new.

A linear order $\lhd$ on $\mathbb{N}$ is conveniently described by a sequence of the initial ranks $(\rho_j) \in [1] \times [2] \times \cdots$, with $\rho_j = i$ if and only if j is ranked ith smallest in the order $\lhd$ among the integers $1, \ldots, j$. For instance, the initial ranks $1, 2, 1, 3, \ldots$ appear when $3 \lhd 1 \lhd 4 \lhd 2$.

For $\xi \in [0, \infty]$ define a random order $\lhd_\xi$ on $\mathbb{N}$ by assuming that the initial ranks ρ_k, $k \in \mathbb{N}$, are independent, with distribution

$$\mathbb{P}(\rho_k = j) = \frac{1}{k + \xi - 1} 1(0 < j < k) + \frac{\xi}{k + \xi - 1} 1(j = k), \qquad k > 1.$$

The edge cases $\xi = 0$, ∞ are defined by continuity. The order $\lhd_1$ is a 'uniformly random order', in the sense that restricting to $[n]$ we have all $n!$ orders equally likely, for every n. The order $\lhd_\infty$ coincides with the standard order $<$ almost surely. For every permutation $i_1, \ldots, i_n$ of $[n]$ we have

$$\mathbb{P}(i_1 \lhd_\xi \cdots \lhd_\xi i_n) = \frac{\xi^r}{\xi(\xi+1)\ldots(\xi+n-1)},$$

where r is the number of upper records in the permutation. See [6] for this and more general permutations with tilted record statistics.

Theorem 6.5 ([24, Corollary 7])   *For $0 \le \alpha < 1$, $\theta \ge 0$ the arrangement of $GEM(\alpha, \theta)$ frequencies (P_j) represented as open intervals in an independent random order $\lhd_{\theta/\alpha}$ is a multiplicatively regenerative open set $\mathcal{U} \subset [0,1]$, where $\mathcal{U}$ is representable as the complement of the closed range of the process $F_t = 1 - \exp(-S_t)$, $t \ge 0$, for S the subordinator with Lévy measure (6.10).*

This result was presented without proof as [24, Corollary 7], in a context where the regenerative ordering of frequencies was motivated by an application to a tree-growth process. Here we offer a proof which exposes the combinatorial structure of the composition of size-biased permutation and a $\lhd_{\theta/\alpha}$ ordering of frequencies.

For a sequence of positive reals $(x_1, \ldots, x_k)$, define the τ-*biased permutation* of this sequence, denoted $\mathrm{perm}_\tau(x_1, \ldots, x_k)$, by iterating a single τ-biased pick, as follows. A number x_T is chosen from $x_1, \ldots, x_k$ without replacement, with T distributed on $[k]$ according to (6.8), and x_T is placed in position 1. Then the next number is chosen from the $k-1$ remaining numbers using again the rule of τ-biased pick, and placed in position 2, etc.

The instance perm_0 is the size-biased permutation, which is defined more widely for finite or infinite summable sequences $(x_1, x_2, \ldots)$, and shuffles them in the same way as it shuffles $(s^{-1}x_1, s^{-1}x_2, \ldots)$ where $s = \sum_j x_j$. Denote by $\lhd_\xi(x_1, \ldots, x_k)$ the arrangement of $x_1, \ldots, x_k$ in succession according to the $\lhd_\xi$-order on $[k]$.

Lemma 6.6   *For $\xi = (1 - \tau)/\tau$ there is the compositional formula*

$$\mathrm{perm}_\tau(x_1, \ldots, x_k) \overset{d}{=} \lhd_\xi(\mathrm{perm}_0(x_1, \ldots, x_k)), \tag{6.11}$$

where on the right-hand side $\lhd_\xi$ and perm_0 are independent.

Proof   On each side of this identity, the distribution of the random permutation remains the same if the sequence $x_1, \ldots, x_k$ is permuted. So it suffices to check that each scheme returns the identity permutation with the same probability. If on the right hand side we set

$$\mathrm{perm}_0(x_1, \ldots, x_k) = (x_{\sigma(1)}, \ldots, x_{\sigma(k)})$$

then the right hand scheme generates the identity permutation with probability

$$\frac{\mathbb{E}\xi^R}{\xi(\xi+1)\cdots(\xi+k-1)}, \qquad (6.12)$$

where R is the number of upper records in the sequence of ranks which generated σ^{-1}, which equals the number of upper records in σ. Now $R = \sum_{j=1}^k X_j$ where X_j is the indicator of the event A_j that j is an upper record level for σ, meaning that there is some $1 \le i \le n$ such that

$$\sigma(i') < j \text{ for all } i' < i \text{ and } \sigma(i) = j.$$

Equivalently, A_j is the event that

$$\sigma^{-1}(j) < \sigma^{-1}(\ell) \text{ for each } j < \ell \le k.$$

Or again, assuming for simplicity that the x_i are all distinct, which involves no loss of generality, because the probability in question depends continuously on $(x_1, \ldots, x_k)$, A_j is the event that x_j precedes x_ℓ in the permutation $(x_{\sigma(1)}, \ldots, x_{\sigma(k)})$ for each $j < \ell \le k$. Now it is easily shown that $(x_{\sigma(1)}, \ldots, x_{\sigma(k)})$ with x_1 deleted is a size-biased permutation of $(x_2, \ldots, x_k)$, and that the same is true conditionally given A_1. It follows by induction that the events A_j are mutually independent, with

$$\mathbb{P}(A_j) = x_j/(x_j + \cdots + x_k) \text{ for } 1 \le j \le k.$$

This allows the probability in (6.12) to be evaluated as

$$\prod_{j=1}^k \frac{(\xi x_j + x_{j+1} + \cdots + x_k)}{(x_j + x_{j+1} + \cdots + x_k)(\xi + j - 1)}.$$

This is evidently the probability that $\mathrm{perm}_\tau(x_1, \ldots, x_k)$ generates the identity permutation, and the conclusion follows. $\qquad\square$

The τ-biased arrangement cannot be defined for an infinite positive summable sequence $(x_1, x_2, \ldots)$, since the '$k = \infty$' instance of (6.8) is not a proper distribution for $\tau \ne 0$. But the right-hand side of (6.11) is well-defined as arrangement of $x_1, x_2, \ldots$ in some total order, hence the composition $\lhd_\xi \circ \mathrm{perm}_0$ is the natural extension of the τ-biased arrangement to infinite series.

Proof of Theorem 6.5   We may represent a finite or infinite positive sequence (x_j) whose sum is 1 as an open subset of $[0, 1]$ composed of contiguous intervals of sizes x_j. The space of open subsets of $[0, 1]$ is endowed with the Hausdorff distance on the complementary compact sets. This topology is weaker than the product topology on positive series summable to 1. The limits below are understood as $n \to \infty$.

Let $\widehat{\mathcal{U}}$ denote the multiplicatively regenerative set associated with $GEM(\alpha, \theta)$ frequencies (P_j) by the results of [8] recalled earlier in this section. Let $\Pi = (\Pi_n)$ be the random partition of positive integers derived by random sampling from $\widehat{\mathcal{U}}$, and let $(|\mathcal{B}_{nj}|, j \geq 1)$ denote the normalized sizes of blocks of Π_n, in order of appearance in the random sampling process. According to [7], these block sizes ordered from left to right, according to the associated subintervals of $\widehat{\mathcal{U}}$, may be represented as $\mathrm{perm}_\tau^n(|\mathcal{B}_{nj}|, j \geq 1)$ for a random permutation perm_τ^n which conditionally given $(|\mathcal{B}_{nj}|, j \geq 1)$ is a τ-biased permutation of these block sizes. Combined with the basic convergence result of [8], this gives

$$\mathrm{perm}_\tau^n(|\mathcal{B}_{nj}|/n, j \geq 1) \to \widehat{\mathcal{U}} \text{ a.s. in the Hausdorff topology.} \quad (6.13)$$

On the other hand, according to Lemma 6.6,

$$\mathrm{perm}_\tau^n(|\mathcal{B}_{nj}|, j \geq 1) \overset{d}{=} \lhd_\xi(\mathrm{perm}_0^n(|\mathcal{B}_{nj}|, j \geq 1)) \overset{d}{=} \lhd_\xi(|\mathcal{B}_{nj}|, j \geq 1). \quad (6.14)$$

Here the random order $\lhd_\xi$ is assumed independent of $(|\mathcal{B}_{nj}|, j \geq 1))$ and of a size-biased permutation perm_0^n of $(|\mathcal{B}_{nj}|, j \geq 1)$. The second equality in distribution holds because the blocks $(|\mathcal{B}_{nj}|, j \geq 1))$ are already in a size-biased random order, by exchangeability of the sampling construction from $\widehat{\mathcal{U}}$. We know by a version of Kingman's correspondence [18] that

$$(|\mathcal{B}_{nj}|/n, j \geq 1) \to (P_j) \text{ a.s. in the product topology.}$$

By looking at the M first terms of (P_j) for M such that these terms sum to at least $1 - \epsilon$ with probability at least $1 - \epsilon$, then sending $\epsilon \to 0$ and $M \to \infty$, we deduce that

$$\lhd_\xi(|\mathcal{B}_{nj}|/n, j \geq 1) \to \mathcal{U} := \lhd_\xi(P_j) \text{ a.s. in the Hausdorff topology,} \quad (6.15)$$

where both sides of (6.15) are identified as discussed earlier with random open subsets of $[0, 1]$. Combining (6.13), (6.14), (6.15), we conclude that $\mathcal{U} \overset{d}{=} \widehat{\mathcal{U}}$, as claimed by the theorem. $\square$

In three special cases, already identified in the previous work [7], the arrangement of $PD(\alpha, \theta)$ (or $GEM(\alpha, \theta)$) frequencies in a multiplicatively regenerative set has a simpler description: in the $(0, \theta)$ case the frequencies are placed in the size-biased order; in the (α, α) case the frequencies are 'uniformly randomly shuffled'; and in the $(\alpha, 0)$ case a size-biased pick is placed contiguously to 1, while the other frequencies are 'uniformly randomly shuffled'. The latter is an infinite analogue of the co-size biased arrangement $perm_1$.

We refer to [9, 24] for further recent developments related to ordered (α, θ) partitions and their regenerative properties.

References

[1] Bertoin, J. 2006. *Random Fragmentation and Coagulation Processes.* Cambridge Stud. Adv. Math., vol. 102. Cambridge: Cambridge Univ. Press.

[2] Bertoin, J. 2008. Two-parameter Poisson-Dirichlet measures and reversible exchangeable fragmentation-coalescence processes. *Combin. Probab. Comput.*, **17**(3), 329–337.

[3] Bobecka, K., and Wesołowski, J. 2007. The Dirichlet distribution and process through neutralities. *J. Theoret. Probab.*, **20**(2), 295–308.

[4] Caballero, M. E., Pardo, J. C., and Pérez, J. L. 2008. *On the Lamperti Stable Processes.* Preprint. http://arxiv.org/abs/0802.0851.

[5] Gnedin, A. V. 1998. On convergence and extensions of size-biased permutations. *J. Appl. Probab.*, **35**(3), 642–650.

[6] Gnedin, A. V. 2007. Coherent random permutations with record statistics. Pages 147–158 of: *2007 Conference on Analysis of Algorithms, AofA 07.* Discrete Math. Theor. Comput. Sci. Proc., vol. AH. Nancy: Assoc. Discrete Math. Theor. Comput. Sci.

[7] Gnedin, A. V., and Pitman, J. 2004. Regenerative partition structures. *Electron. J. Combin.*, **11**(2) **R12**, 1–21.

[8] Gnedin, A. V., and Pitman, J. 2005. Regenerative composition structures. *Ann. Probab.*, **33**(2), 445–479.

[9] Haas, B., Pitman, J., and Winkel, M. 2009. Spinal partitions and invariance under re-rooting of continuum random trees. *Ann. Probab.*, **37**(4), 1381–1411.

[10] Kallenberg, O. 2002. *Foundations of Modern Probability.* 2nd edn. Probab. Appl. (N. Y.). New York: Springer-Verlag.

[11] Kerov, S. 2005. Coherent random allocations, and the Ewens-Pitman formula. *Zap. Nauchn. Sem. S.-Peterburg. Otdel. Mat. Inst. Steklov. (POMI)*, **325**, 127–145, 246.

[12] Kingman, J. F. C. 1972. *Regenerative Phenomena.* New York: John Wiley & Sons.

[13] Kingman, J. F. C. 1978a. Random partitions in population genetics. *Proc. Roy. Soc. London Ser. A*, **361**, 1–20.

[14] Kingman, J. F. C. 1978b. The representation of partition structures. *J. London Math. Soc. (2)*, **18**(2), 374–380.

[15] Kingman, J. F. C. 1982. The coalescent. *Stochastic Process. Appl.*, **13**(3), 235–248.

[16] Lauritzen, S. L. 1996. *Graphical Models*. Oxford Statist. Sci. Ser., vol. 17. New York: Oxford Univ. Press.

[17] Perman, M., Pitman, J., and Yor, M. 1992. Size-biased sampling of Poisson point processes and excursions. *Probab. Theory Related Fields*, **92**, 21–39.

[18] Pitman, J. 1995. Exchangeable and partially exchangeable random partitions. *Probab. Theory Related Fields*, **102**, 145–158.

[19] Pitman, J. 1996a. Random discrete distributions invariant under size-biased permutation. *Adv. in Appl. Probab.*, **28**, 525–539.

[20] Pitman, J. 1996b. Some developments of the Blackwell-MacQueen urn scheme. Pages 245–267 of: Ferguson, T. S., Shapley, L. S., and MacQueen, J. N. (eds), *Statistics, Probability and Game Theory; Papers in Honor of David Blackwell*. IMS Lecture Notes Monogr. Ser., vol. 30. Hayward, CA: Institute of Mathematical Statistics.

[21] Pitman, J. 1999. Coalescents with multiple collisions. *Ann. Probab.*, **27**, 1870–1902.

[22] Pitman, J. 2003. Poisson-Kingman partitions. Pages 1–34 of: Goldstein, D. R. (ed), *Science and Statistics: A Festschrift for Terry Speed*. IMS Lecture Notes Monogr. Ser., vol. 40. Beachwood, OH: Institute of Mathematical Statistics.

[23] Pitman, J. 2006. *Combinatorial Stochastic Processes*. Lecture Notes in Math., vol. 1875. Berlin: Springer-Verlag.

[24] Pitman, J., and Winkel, M. 2009. Regenerative tree growth: binary self-similar continuum random trees and Poisson-Dirichlet compositions. *Ann. Probab.*, **37**(5), 1999–2041.

[25] Pitman, J., and Yor, M. 1997. The two-parameter Poisson-Dirichlet distribution derived from a stable subordinator. *Ann. Probab.*, **25**, 855–900.

[26] Teh, Y. W. 2006. A hierarchical Bayesian language model based on Pitman-Yor processes. Pages 985–992 of: *ACL-44: Proceedings of the 21st International Conference on Computational Linguistics and the 44th annual meeting of the Association for Computational Linguistics*. Morristown, NJ: Association for Computational Linguistics.

[27] Wood, F., Archambeau, C., Gasthaus, J., James, L., and Teh, Y. W. 2009. A stochastic memoizer for sequence data. Pages 1129–1136 of: *ICML '09: Proceedings of the 26th Annual International Conference on Machine Learning*. New York: ACM.

12

Applying coupon-collecting theory to computer-aided assessments

C. M. Goldie[a], R. Cornish[b] and C. L. Robinson[c]

Abstract

Computer-based tests with randomly generated questions allow a large number of different tests to be generated. Given a fixed number of alternatives for each question, the number of tests that need to be generated before all possible questions have appeared is surprisingly low.

AMS subject classification (MSC2010) 60G70, 60K99

1 Introduction

The use of computer-based tests in which questions are randomly generated in some way provides a means whereby a large number of different tests can be generated; many universities currently use such tests as part of the student assessment process. In this paper we present findings that illustrate that, although the number of different possible tests is high and grows very rapidly as the number of alternatives for each question increases, the average number of tests that need to be generated before all possible questions have appeared at least once is surprisingly low. We presented preliminary findings along these lines in Cornish et al. (2006).

A computer-based test consists of q questions, each (independently) selected at random from a separate bank of a alternatives. Let N_q be the

[a] Mathematics Department, Mantell Building, University of Sussex, Brighton BN1 9RF; C.M.Goldie@sussex.ac.uk
[b] Department of Social Medicine, University of Bristol, Canynge Hall, 39 Whatley Road, Bristol BS8 2PS; R.Cornish@bristol.ac.uk
[c] Mathematics Education Centre, Schofield Building, Loughborough University, Loughborough LE11 3TU; C.L.Robinson@lboro.ac.uk

number of tests one needs to generate in order to see all the aq questions in the q question banks at least once. We are interested in how, for fixed a, the random variable N_q grows with the number of questions q in the test. Typically, a might be 10—i.e. each question might have a bank of 10 alternatives—but we shall allow any value of a, and give numerical results for $a = 20$ and $a = 5$ as well as for $a = 10$.

2 Coupon collecting

In the case $q = 1$, i.e. a one-question test, we re-notate N_q as Y, and observe that we have an equivalent to the classic coupon-collector problem: your favourite cereal has a coupon in each packet, and there are a alternative types of coupon. Y is the number of packets you have to buy in order to get at least one coupon of each of the a types. The coupon-collector problem has been much studied; see e.g. (Grimmett and Stirzaker, 2001, p. 55).

We can write Y as

$$Y = Y_1 + Y_2 + \cdots + Y_a$$

where each Y_i is the number of cereal packets you must buy in order to acquire a new type of coupon, when you already have $i - 1$ types in your collection. Thus $Y_1 = 1$, Y_2 is the number of further packets you find you need to gain a second type, and so on. The random variables $Y_1, \ldots, Y_a$ are mutually independent. For the distribution of Y_k, clearly

$$P(Y_k = y) = \frac{a - k + 1}{a} \left(\frac{k - 1}{a} \right)^{y-1} \qquad (y = 1, 2, \ldots).$$

We say that $X \sim \text{Geom}(p)$, or X has a geometric distribution with parameter p, if $P(X = x) = p(1 - p)^{x-1}$ for $x = 1, 2, \ldots$. Thus $Y_k \sim \text{Geom}((a - k + 1)/a)$. As the $\text{Geom}(p)$ distribution has expectation $1/p$ it follows that

$$EY = \sum_{k=1}^{a} EY_k = \sum_{k=1}^{a} \frac{a}{a - k + 1} = a \sum_{k=1}^{a} \frac{1}{k}.$$

For different values of a we therefore have the following.

a	5	10	15	20
EY	11·42	29·29	49·77	71·96

In other words, if there are 10 coupons to collect then an average of 29 packets of cereal would have to be bought in order to obtain all 10 of these coupons. In the context of computer-based tests, if a test had one question selected at random from a bank of 10 alternatives, an average of 29 tests would need to be generated in order to see all the questions at least once.

To apply the theory to tests with more than one question we will also need an explicit expression for $P(Y > y)$. To revert to the language of coupons in cereal packets, let us number the coupon types $1, 2, \ldots, a$, and let A_i be the event that type i does not occur in the first y cereal packets bought. The event that $Y > y$ is then the union of the events $A_1, A_2, \ldots, A_a$. So by the inclusion-exclusion formula,

$$P(Y > y) = P\left(\bigcup_{i=1}^{a} A_i\right)$$

$$= \sum_{i=1}^{a} P(A_i) - \sum_{i<j} P(A_i \cap A_j) + \sum_{i<j<k} P(A_i \cap A_j \cap A_k) - \cdots$$
$$+ (-1)^{a+1} P(A_1 \cap \cdots \cap A_a).$$

Obviously $P(A_i) = (1 - 1/a)^y$ for each i. For distinct i and j, $A_i \cap A_j$ is the event that a particular two of the a coupon types do not occur in the first y purchases, so has probability $(1 - 2/a)^y$. Similarly $A_i \cap A_j \cap A_k$, for distinct i, j and k, has probability $(1 - 3/a)^y$, and so on. We conclude that

$$P(Y > y) = \sum_{k=1}^{a} (-1)^{k+1} \binom{a}{k} \left(1 - \frac{k}{a}\right)^y \tag{2.1}$$

(when $y > 0$ the final term of the sum is zero). Let F be the distribution function for Y; thus the above is equivalent to

$$F(y) := P(Y \leq y) = \sum_{k=0}^{a} (-1)^k \binom{a}{k} \left(1 - \frac{k}{a}\right)^y \quad (y = a, a+1, a+2, \ldots).$$
$$\tag{2.2}$$

This is a classical formula for the probability that all cells are occupied when y balls are distributed at random among a cells; cf. (Feller, 1968, (11.11)). The right-hand side of (2.2) has value 0 when $y = 0, 1, \ldots, a - 1$.

3 How many tests?

We return to the initial question. We have a test containing q questions, each selected at random from a bank of a alternatives. N_q is defined to be the number of tests that need to be generated in order to see all possible aq questions at least once.

For question j of the test, let Y_j be the number of tests needed to see all the a alternatives in its question bank. The random variables Y_1, Y_2, ..., Y_q are mutually independent, each distributed as the Y of the previous section, and N_q is their maximum:

$$N_q = \max\{Y_1, Y_2, \ldots, Y_q\}.$$

We thus have

$$
\begin{aligned}
EN_q &= \sum_{n=0}^{\infty} P(N_q > n) \\
&= \sum_{n=0}^{\infty} \left(1 - P(N_q \leq n)\right) \\
&= \sum_{n=0}^{\infty} \left(1 - \prod_{j=1}^{q} P(Y_j \leq n)\right) \\
&= \sum_{n=0}^{\infty} \left(1 - \prod_{j=1}^{q} (1 - P(Y_j > n))\right) \\
&= \sum_{n=0}^{\infty} \left(1 - (1 - P(Y > n))^q\right).
\end{aligned}
\tag{3.1}
$$

This can be reduced to a finite sum as follows.

$$
\begin{aligned}
EN_q &= \sum_{n=0}^{\infty} \sum_{m=1}^{q} (-1)^{m+1} \binom{q}{m} (P(Y > n))^m \\
&= \sum_{n=0}^{\infty} \sum_{m=1}^{q} (-1)^{m+1} \binom{q}{m} \sum_{j_1=1}^{a} (-1)^{j_1+1} \binom{a}{j_1} \left(1 - \frac{j_1}{a}\right)^n \cdots \\
&\qquad \cdots \sum_{j_m=1}^{a} (-1)^{j_m+1} \binom{a}{j_m} \left(1 - \frac{j_m}{a}\right)^n \\
&= -\sum_{m=1}^{q} \binom{q}{m} \sum_{j_1=1}^{a} \cdots \sum_{j_m=1}^{a} (-1)^{j_1+\cdots+j_m}
\end{aligned}
$$

$$\binom{a}{j_1} \cdots \binom{a}{j_m} \sum_{n=0}^{\infty} \left(\prod_{i=1}^{m} \left(1 - \frac{j_i}{a} \right) \right)^n$$

$$= - \sum_{m=1}^{q} \binom{q}{m} \sum_{j_1=1}^{a} \cdots \sum_{j_m=1}^{a} \frac{(-1)^{j_1+\cdots+j_m} \binom{a}{j_1} \cdots \binom{a}{j_m}}{1 - \prod_{i=1}^{m}(1 - j_i/a)}.$$

This, though, is not well suited to computation, and we have used (3.1) for the numerical results below.

Note The way in which CMG got involved in writing this paper was through chancing on a query posted by RC on Allstat, a UK-based electronic mailing list, asking how to calculate the expected number of tests a student would need to access in order to see the complete bank of questions. CMG immediately recognised the query as a form of coupon-collecting problem, but not quite in standard form. What he should have done then was to think and calculate, following Littlewood's famous advice (Littlewood, 1986, p. 93)

> "It is of course good policy, and I have often practised it, to begin without going too much into the existing literature".

What he actually did was to seek previous work using *Google*. With customary speed and accuracy, *Google* produced a list with Adler and Ross (2001) in position 6. Knowing that Sheldon Ross is unbeatable at combinatorial probability problems, CMG looked up this paper—and was thoroughly led astray. The paper does indeed treat our problem and is an excellent paper, but it is much more general than we needed and sets up a structure that obscures the relatively simple nature of what we needed for this problem. It was better to work the above out from first principles.

4 Asymptotics

We employ Extreme-Value Theory (EVT) to investigate the random variable N_q as the number of questions q becomes large, the number a of alternatives per question staying fixed. It turns out we are in a case identified by C. W. Anderson in 1970, where a limit fails to exist but there are close bounds above and below. Thus despite the absence of a limit we gain asymptotic results of some precision.

The relevant extreme-value distribution will be the *Gumbel distribution*, with (cumulative) distribution function $\Lambda(x) = \exp(-e^{-x})$ for all $x \in \mathbb{R}$; write Z for a random variable with the Gumbel distribution.

Throughout this section $a \geq 2$ is an integer, and we set $\alpha := \log(a/(a-1)) > 0$. Proofs of the results in this section are in §5.

A first goal of EVT for the random variables N_q would be to find a norming sequence $a_q > 0$ and a centring sequence b_q such that $(N_q - b_q)/a_q$ has a limit distribution as $q \to \infty$.

Theorem 4.1 *There do not exist sequences $a_q > 0$ and b_q such that $(N_q - b_q)/a_q$ has a non-degenerate limit distribution as $q \to \infty$. However, with $b_q := \alpha^{-1} \log(aq)$ we have for all $x \in \mathbb{R}$ that*

$$\Lambda(\alpha(x - 1)) = \liminf_{q \to \infty} P(N_q - b_q \leq x)$$

$$\leq \limsup_{q \to \infty} P(N_q - b_q \leq x) = \Lambda(\alpha x). \quad (4.1)$$

Thus $N_q - b_q$, in distribution, is asymptotically between $\alpha^{-1}Z$ and $1 + \alpha^{-1}Z$, with Z Gumbel, and these distributional bounds are sharp.

To describe the local behaviour, let $\lfloor x \rfloor$ denote the integer part of x, $\{x\} := x - \lfloor x \rfloor$ the fractional part, and let $\lceil x \rceil := \lfloor x \rfloor + 1$. Then for each integer n,

$$P(N_q - \lceil b_q \rceil = n) - \Lambda\big(\alpha(n + 1 - \{b_q\})\big) + \Lambda\big(\alpha(n - \{b_q\})\big) \to 0$$

$$as \; q \to \infty. \quad (4.2)$$

We remark that the Gumbel distribution has mean $\gamma \simeq 0.5772$, the Euler–Mascheroni constant, and variance $\pi^2/6$. Its distribution tails decay exponentially or better: $\lim_{x \to \infty} e^x(1 - \Lambda(x)) = 1$ and $\lim_{x \to -\infty} e^{-x}\Lambda(x) = 0$. We use these facts below. We first extend the above stochastic boundedness of the sequence $(N_q - b_q)$ to L^p-boundedness for all p. For the rest of the paper we set $b_q := \alpha^{-1} \log(aq)$ and $R_q := N_q - b_q$.

Theorem 4.2 *For each $p \geq 1$, $\sup_{q \in \mathbb{N}} E(|R_q|^p) < \infty$.*

Theorem 4.2 implies that the distributional asymptotics of Theorem 4.1 will extend to give asymptotic bounds on moments. Moment convergence in EVT is treated in (Resnick, 1987, §2.1), and we use some of the ideas from the proofs there in proving the results below.

Theorem 4.3

$$\frac{\gamma + \log a}{\alpha} \leq \liminf_{q \to \infty} \left(EN_q - \frac{\log q}{\alpha} \right)$$

$$\leq \limsup_{q \to \infty} \left(EN_q - \frac{\log q}{\alpha} \right) \leq \frac{\gamma + \log a}{\alpha} + 1.$$

By similar methods one may obtain bounds on higher moments. We content ourselves with those on the second moment, leading to good bounds on $\operatorname{var} N_q$, the variance of N_q.

Lemma 4.4

$$E\big((1 + \alpha^{-1}Z)^2 \mathbf{1}_{1+\alpha^{-1}Z \leq 0} + (\alpha^{-1}Z)^2 \mathbf{1}_{Z>0}\big) \leq \liminf_{q \to \infty} E(R_q^2)$$

$$\leq \limsup_{q \to \infty} E(R_q^2) \leq E\big((\alpha^{-1}Z)^2 \mathbf{1}_{Z \leq 0} + (1 + \alpha^{-1}Z)^2 \mathbf{1}_{1+\alpha^{-1}Z > 0}\big). \quad (4.3)$$

Theorem 4.5

$$\limsup_{q \to \infty} \left| \operatorname{var} N_q - \frac{\pi^2}{6\alpha^2} \right| \leq \theta(\alpha) + 1 - e^{-1} + \frac{2\big(\gamma + E_1(1)\big)}{\alpha},$$

where $\theta(\alpha) = E\big((1 + \alpha^{-1}Z)^2 \mathbf{1}_{0<1+\alpha^{-1}Z \leq 1}\big)$ *satisfies* $0 < \theta(\alpha) < 1$, *and* $E_1(1) = \int_1^\infty t^{-1} e^{-t}\, dt \simeq 0\cdot2194$.

Here, $E_1(1)$ is a value of the exponential integral (cf. (Abramowitz and Stegun, 1965, §5.1)) $E_n(x) = \int_1^\infty t^{-n} e^{-xt}\, dt$.

5 Proofs for §4

Proof of Theorem 4.1 In (2.1) the $k = 1$ term dominates for large y, so

$$P(Y > y) = a(1 - 1/a)^y(1 + o(1)) \tag{5.1}$$

as $y \to \infty$ through integer values. As noted in (Anderson, 1970, §1), the fact that the integer-valued random variable Y has

$$\frac{P(Y > y)}{P(Y > y + 1)} \to \frac{a}{a - 1} > 1 \quad \text{as } y \to \infty$$

prevents it from belonging to the 'domain of attraction' for maxima of any extreme-value distribution, and so no non-trivial limit distribution for $(N_q - b_q)/a_q$, for any choices of a_q and b_q, can exist.

For the rest of the proof, $b_q := \alpha^{-1} \log(aq)$. Via the definition of α,

(5.1) gives that $F(y) = 1 - ae^{-\alpha y}(1 + o(1))$ as $y \to \infty$ through integer values. So for each fixed $x \in \mathbb{R}$,

$$P(N_q - b_q \le x) = F^q(\lfloor x + b_q \rfloor) = \left(1 - ae^{-\alpha \lfloor x + b_q \rfloor}(1 + o(1))\right)^q \quad (5.2)$$

as $q \to \infty$. Then

$$P(N_q - b_q \le x) \le \left(1 - ae^{-\alpha(x + b_q)}(1 + o(1))\right)^q$$
$$= \left(1 - \frac{e^{-\alpha x}(1 + o(1))}{q}\right)^q \to \Lambda(\alpha x) \quad \text{as } q \to \infty.$$

With $x \in \mathbb{R}$ still fixed we define the sequence $\big(q(k)\big)_{k=1}^{\infty}$ to be those q for which the interval $(x + b_{q-1}, x + b_q]$ contains one or more integers, i.e. for which $x + b_{q-1} < \lfloor x + b_q \rfloor$. Since $b_q \to \infty$ this is an infinite sequence, and since $b_{q+1} - b_q \to 0$ we have $x + b_{q(k)} - \lfloor x + b_{q(k)} \rfloor \to 0$ as $k \to \infty$, whence with (5.2) we conclude that $P(N_{q(k)} - b_{q(k)} \le x) \to \Lambda(\alpha x)$ as $k \to \infty$. Thus $\limsup_{q \to \infty} P(N_q - b_q \le x) = \Lambda(\alpha x)$.

For the limit inferior,

$$P(N_q - b_q \le x) \ge \left(1 - ae^{-\alpha(x - 1 + b_q)}(1 + o(1))\right)^q$$
$$= \left(1 - \frac{e^{-\alpha(x-1)}(1 + o(1))}{q}\right)^q$$
$$\to \Lambda(\alpha(x - 1)) \quad \text{as } q \to \infty.$$

With the same sequence $\big(q(k)\big)$ as above, note that $x + b_{q(k)-1} - \lfloor x + b_{q(k)-1} \rfloor \to 1$ as $k \to \infty$, so

$$P(N_{q(k)-1} - b_{q(k)-1} \le x) = \left(1 - ae^{-\alpha \lfloor x + b_{q(k)-1} \rfloor}(1 + o(1))\right)^{q(k)-1}$$
$$= \left(1 - ae^{-\alpha(x + b_{q(k)} - 1 - 1)}(1 + o(1))\right)^{q(k)-1}$$

by (5.2). The right-hand side converges to $\Lambda(\alpha(x - 1))$. Thus $\liminf_{q \to \infty} P(N_q - b_q \le x) = \Lambda(\alpha(x - 1))$. This establishes (4.1).

The extension to local behaviour is due to Anderson (1980). To gain the conclusion as we formulate it, (4.2), we may argue directly: fix an integer n and start from

$$P(N_q - \lceil b_q \rceil \le n) = F^q(n + \lceil b_q \rceil) = \left(1 - ae^{-\alpha(n + \lceil b_q \rceil)}(1 + o(1))\right)^q$$

as $q \to \infty$. Now

$$ae^{-\alpha(n + \lceil b_q \rceil)} = \frac{e^{-\alpha(n + \lceil b_q \rceil - b_q)}}{q} = \frac{e^{-\alpha(n + 1 - \{b_q\})}}{q},$$

and as the convergence in $(1 - c/q)^q \to e^{-c}$ is locally uniform in c we deduce that

$$P(N_q - \lceil b_q \rceil \le n) - \Lambda\big(\alpha(n + 1 - \{b_q\})\big) \to 0 \quad \text{as } q \to \infty.$$

Subtract from this the corresponding formula with n replaced by $n - 1$, and (4.2) follows. $\qquad\square$

For the next result we need a uniform bound on expressions of the form $1 - (1 - u/n)^n$:

Lemma 5.1 *For any $u_0 > 0$ there exists a positive integer $n_1 = n_1(u_0)$ such that for $n \ge n_1$ and $0 \le u \le u_0$,*

$$1 - \left(1 - \frac{u}{n}\right)^n \le 2u.$$

Proof There exists $t_0 > 0$ (its value is about $0 \cdot 7968$) such that $\log(1 - t_0) = -2t_0$, so $\log(1 - t) \ge -2t$ for $0 \le t \le t_0$. Take $n_1 \ge u_0/t_0$, then $1 - (1 - u/n)^n \le 1 - e^{-2u}$ for $n \ge n_1$ and $0 \le u \le u_0$, and as $1 - e^{-2u} \le 2u$ the result follows. $\qquad\square$

Proof of Theorem 4.2 We write $\mathbf{1}_T := 1$ if statement T is true, $\mathbf{1}_T := 0$ if T is false. Fix $n \in \mathbb{N}$. The distribution of N_q is such that $E(R_q^{2n}) < \infty$ for all q. We prove that $\sup_{q \in \mathbb{N}} E(R_q^{2n}) < \infty$. Now

$$E(R_q^{2n}) = \int_{(-\infty, 0]} x^{2n} \, dP(R_q \le x) - \int_{(0, \infty)} x^{2n} \, dP(R_q > x),$$

and so, on integrating by parts,

$$E(R_q^{2n}) = -2n \int_{-\infty}^{0} x^{2n-1} P(R_q \le x) \, dx + 2n \int_{0}^{\infty} x^{2n-1} P(R_q > x) \, dx$$

$$=: A + B,$$

say.

In (2.1) the right-hand side is asymptotic to its first term, $ae^{-\alpha y}$. There exists y_0 such that for real $y \ge y_0$ (not just integer y), $P(Y > y) \le 2ae^{-\alpha(y-1)}$. So for $x \ge 0$ and $q \ge a^{-1}e^{\alpha y_0}$,

$$P(Y > x + b_q) \le 2ae^{-\alpha(x + b_q - 1)} = \frac{2}{q} e^{\alpha - \alpha x},$$

and hence

$$P(R_q > x) = 1 - \big(1 - P(Y > x + b_q)\big)^q \le 1 - \left(1 - \frac{2}{q} e^{\alpha - \alpha x}\right)^q.$$

Now apply Lemma 5.1. It follows that there exists q_1 such that for $q \geq q_1$ and $x \geq 0$,

$$P(R_q > x) \leq 4e^{\alpha - \alpha x}.$$

Therefore, for $q \geq q_1$,

$$B = 2n \int_0^\infty x^{2n-1} P(R_q > x) \, dx \leq 8n \int_0^\infty x^{2n-1} e^{\alpha - \alpha x} \, dx < \infty.$$

It remains to bound A. Returning again to (2.1), observe that we may find y_1 so that $P(Y > y) \geq \frac{1}{2} a e^{-\alpha y}$ for all real $y \geq y_1$. Therefore for $x \geq y_1 - b_q$ we have

$$P(Y > x + b_q) \geq \frac{1}{2} a e^{-\alpha(x + b_q)} = \frac{1}{2q} e^{-\alpha x},$$

and so

$$\begin{aligned}
P(R_q \leq x) &= \left(1 - P(Y > x + b_q)\right)^q \\
&\leq \exp\left(-q P(Y > x + b_q)\right) \\
&\leq \exp\left(-\frac{1}{2} e^{-\alpha x}\right) \quad \text{for } x \geq y_1 - b_q.
\end{aligned} \tag{5.3}$$

In $A = -2n \int_{-\infty}^0 x^{2n-1} P(R_q \leq x) \, dx$, the lower endpoint of the interval of integration may be taken to be $-b_q$, as the integrand vanishes below this point, and we then choose further to split the integral to obtain

$$\begin{aligned}
A &= -2n \int_{y_1 - b_q}^0 x^{2n-1} P(R_q \leq x) \, dx - 2n \int_{-b_q}^{y_1 - b_q} x^{2n-1} P(R_q \leq x) \, dx \\
&=: A_1 + A_2,
\end{aligned}$$

say. If we take q so large that $b_q > y_1$, (5.3) gives

$$\begin{aligned}
A_1 &\leq - \int_{y_1 - b_q}^0 x^{2n-1} \exp\left(-\frac{1}{2} e^{-\alpha x}\right) dx \\
&< -2n \int_{-\infty}^0 x^{2n-1} \exp\left(-\frac{1}{2} e^{-\alpha x}\right) dx < \infty.
\end{aligned}$$

Finally,

$$
\begin{aligned}
A_2 &= -2n \int_{b_q}^{y_1 - b_q} x^{2n-1} F^q (x + b_q) \, dx \\
&= -2n \int_0^{y_1} (u - b_q)^{2n-1} F^q (u) \, du \\
&\leq 2n y_1 b_q^{2n-1} F^q (y_1) \\
&= 2n y_1 \left(\frac{\log(aq)}{\alpha} \right)^{2n-1} F^q (y_1).
\end{aligned}
$$

This tends to 0 as $q \to \infty$, because $0 < F(y_1) < 1$.

We have shown that $\limsup_{q \to \infty} E(R_q^{2n}) < \infty$, so $\sup_{q \in \mathbb{N}} E(R_q^{2n}) < \infty$ as claimed, and the result follows. $\qquad\square$

Before proving Theorem 4.3 we note that (4.1) says that for each $x \in \mathbb{R}$,

$$
\Lambda(\alpha(x-1)) = \liminf_{q \to \infty} P(R_q \leq x) \leq \limsup_{q \to \infty} P(R_q \leq x) = \Lambda(\alpha x), \quad (5.4)
$$

and that what we have to prove is

$$
E(\alpha^{-1} Z) \leq \liminf_{q \to \infty} E R_q \leq \limsup_{q \to \infty} E R_q \leq E(1 + \alpha^{-1} Z). \quad (5.5)
$$

We use (5.4) mostly in the form

$$
\begin{aligned}
P(\alpha^{-1} Z > x) &= \liminf_{q \to \infty} P(R_q > x) \\
&\leq \limsup_{q \to \infty} P(R_q > x) = P(1 + \alpha^{-1} Z > x), \quad (5.6)
\end{aligned}
$$

obtained by subtracting each component from 1. We make much use of Fatou's Lemma, that for non-negative f_n,

$$
\liminf_{n \to \infty} \int f_n \geq \int \liminf_{n \to \infty} f_n,
$$

and also of its extended form: that if $f_n \leq f$ and f is integrable then

$$
\limsup_{n \to \infty} \int f_n \leq \int \limsup_{n \to \infty} f_n.
$$

The latter may be deduced from the former by considering $f - f_n$.

Proof of Theorem 4.3 We use the fact that for a random variable X with finite mean, and any constant c,

$$
E(X \mathbf{1}_{X > -c}) = -c P(X > -c) + \int_{-c}^{\infty} P(X > x) \, dx, \quad (5.7)
$$

as may be proved by integrating $\int_{(-c,\infty)} x\, dP(X \leq x)$ by parts. We thus have, for $c > 0$,

$$
ER_q \leq E(R_q \mathbf{1}_{R_q > -c})
$$
$$
= -cP(R_q > -c) + \int_{-c}^{c} P(R_q > x)\, dx + \int_{c}^{\infty} P(R_q > x)\, dx
$$
$$
=: A + B + C,
$$

say. First, by the left-hand equality in (5.6), $\limsup_{q \to \infty} A = -cP(\alpha^{-1}Z > -c)$. Second, from the right-hand equality in (5.6), and the extended Fatou Lemma (take the dominating integrable function to be 1),

$$
\limsup_{q \to \infty} B \leq \int_{-c}^{c} P(1 + \alpha^{-1}Z > x)\, dx \leq \int_{-c}^{\infty} P(1 + \alpha^{-1}Z > x)\, dx.
$$

Combining the bounds on A and B yields

$$
\limsup_{q \to \infty}(A + B) \leq -cP(1 + \alpha^{-1}Z > -c) + \int_{-c}^{\infty} P(1 + \alpha^{-1}Z > x)\, dx
$$
$$
+ c\big(P(1 + \alpha^{-1}Z > -c) - P(\alpha^{-1}Z > -c)\big)
$$
$$
= E\big((1 + \alpha^{-1}Z)\mathbf{1}_{1+\alpha^{-1}Z>-c}\big)
$$
$$
+ cP(-c - 1 < \alpha^{-1}Z \leq -c)
$$
$$
< E\big((1 + \alpha^{-1}Z)\mathbf{1}_{1+\alpha^{-1}Z>-c}\big) + cP(\alpha^{-1}Z \leq -c).
$$

For the third upper bound, on C, we note (with an eye to the next proof as well) that by Theorem 4.2, $K := \sup_{q \in \mathbb{N}} E(|R_q|^3) < \infty$. Then for $x > 0$, $P(R_q > x) \leq K/x^3$, hence $C \leq K/(2c^2)$. On combining this bound with that on $A + B$ we gain an upper bound on $\limsup_{q \to \infty} ER_q$ that converges to $E(1 + \alpha^{-1}Z)$ as $c \to \infty$, concluding the proof of the upper bound in (5.5).

For the lower bound we again use (5.7), this time to write

$$
ER_q = E(R_q \mathbf{1}_{R_q \leq -c}) + E(R_q \mathbf{1}_{R_q > -c})
$$
$$
= E(R_q \mathbf{1}_{R_q \leq -c}) - cP(R_q > -c) + \int_{-c}^{\infty} P(R_q > x)\, dx
$$
$$
=: \tilde{A} + \tilde{B} + \tilde{C},
$$

say. First, Fatou's Lemma and then the left-hand equality in (5.6) give

$$
\liminf_{q \to \infty} \tilde{C} \geq \int_{-c}^{\infty} \liminf_{q \to \infty} P(R_q > x)\, dx = \int_{-c}^{\infty} P(\alpha^{-1}Z > x)\, dx.
$$

Second,

$$\liminf_{q\to\infty} \tilde{B} = -c \limsup_{q\to\infty} P(R_q > -c) = -cP(1 + \alpha^{-1}Z > -c),$$

this time by the right-hand equality in (5.6). Combining, we find that

$$\liminf_{q\to\infty}(\tilde{B} + \tilde{C}) \geq -cP(\alpha^{-1}Z > -c) + \int_{-c}^{\infty} P(\alpha^{-1}Z > x)\,dx$$

$$- c\big(P(1 + \alpha^{-1}Z > -c) - P(\alpha^{-1}Z > -c)\big)$$

$$= E(\alpha^{-1}Z\mathbf{1}_{\alpha^{-1}Z>-c}) - cP(-c - 1 < \alpha^{-1}Z \leq -c)$$

$$\geq E(\alpha^{-1}Z) - cP(\alpha^{-1}Z \leq -c).$$

Finally, to put a lower bound on $\tilde{A}$ we may again use the 'Markov inequality' method used above for C, obtaining $\tilde{A} \geq -K/(2c^2)$. Combining this with the above, we gain a lower bound on $\liminf_{q\to\infty} ER_q$ that converges to $E(\alpha^{-1}Z)$ as $c \to \infty$. We thus obtain the lower bound in (5.5). $\qquad\square$

Proof of Lemma 4.4 We use variants of the decompositions in the previous proof. First, the upper bound. With $c > 0$ fixed,

$$E(R_q^2) = E(R_q^2\mathbf{1}_{R_q>-c}) + E(R_q^2\mathbf{1}_{R_q\leq-c})$$

$$= c^2 P(R_q > -c) + 2\int_{-c}^{\infty} xP(R_q > x)\,dx + E(R_q^2\mathbf{1}_{R_q\leq-c})$$

$$= c^2 P(R_q > -c) + 2\int_{-c}^{0} xP(R_q > x)\,dx + 2\int_{0}^{c} xP(R_q > x)\,dx$$

$$+ 2\int_{c}^{\infty} xP(R_q > x)\,dx + E(R_q^2\mathbf{1}_{R_q\leq-c})$$

$$=: A + B_1 + B_2 + C + D,$$

say. By the right-hand equality in (5.6), $\limsup_{q\to\infty} A = c^2 P(1+\alpha^{-1}Z > -c)$. By the left-hand equality and Fatou's Lemma, followed by an integration by parts,

$$\limsup_{q\to\infty} B_1 \leq 2\int_{-c}^{0} xP(\alpha^{-1}Z > x)\,dx$$

$$= -c^2 P(\alpha^{-1}Z > -c) + E\big((\alpha^{-1}Z)^2\mathbf{1}_{-c<\alpha^{-1}Z\leq0}\big).$$

Combining,

$$\limsup_{q\to\infty}(A + B_1) \le c^2 P(-c - 1 < \alpha^{-1}Z \le -c)$$

$$+ E\big((\alpha^{-1}Z)^2 \mathbf{1}_{-c<\alpha^{-1}Z\le 0}\big)$$

$$\le c^2 P(\alpha^{-1}Z \le -c) + E\big((\alpha^{-1}Z)^2 \mathbf{1}_{Z\le 0}\big). \qquad (5.8)$$

Next, by the right-hand equality in (5.6), and the extended Fatou Lemma,

$$\limsup_{q\to\infty} B_2 \le 2 \int_0^c x P(1 + \alpha^{-1}Z > x)\,dx$$

$$\le 2 \int_0^\infty x P(1 + \alpha^{-1}Z > x)\,dx$$

$$= E\big((1 + \alpha^{-1}Z)^2 \mathbf{1}_{1+\alpha^{-1}Z>0}\big).$$

On combining this with (5.8) and letting $c \to \infty$ we conclude that

$$\lim_{c\to\infty} \limsup_{q\to\infty}(A + B_1 + B_2)$$

$$\le E\big((\alpha^{-1}Z)^2 \mathbf{1}_{Z\le 0} + (1 + \alpha^{-1}Z)^2 \mathbf{1}_{1+\alpha^{-1}Z>0}\big).$$

The upper bound in (4.3) will follow if we can show that $\lim_{c\to\infty} \limsup_{q\to\infty} C = 0$, and likewise for D. For C this follows by inserting into its defining formula the bound $P(R_q > x) \le K/x^3$ developed in the proofs above, while for D it follows from Theorem 4.2 via the uniform integrability of the family $(R_q^2)_{q\in\mathbb{N}}$. The upper bound in (4.3) is proved.

For the lower bound we fix $c > 0$ and write

$$E(R_q^2) \ge E(R_q^2 \mathbf{1}_{R_q>-c})$$

$$= c^2 P(R_q > -c) + 2 \int_{-c}^\infty x P(R_q > x)\,dx$$

$$\ge c^2 P(R_q > -c) + 2 \int_{-c}^0 x P(R_q > x)\,dx + 2 \int_0^c x P(R_q > x)\,dx.$$

In this right-hand side, use the left-hand equality in (5.6) on the first term, use the right-hand equality and the extended Fatou Lemma on the second term, and use the left-hand equality and Fatou's Lemma on the third term, to give

$$\liminf_{q\to\infty} E(R_q^2) \geq c^2 P(\alpha^{-1}Z > -c) + 2\int_{-c}^{0} xP(1 + \alpha^{-1}Z > x)\,dx$$

$$+ 2\int_{0}^{c} xP(\alpha^{-1}Z > x)\,dx.$$

By two integrations by parts this becomes

$$\liminf_{q\to\infty} E(R_q^2) \geq c^2 P(\alpha^{-1}Z > -c) - c^2 P(1 + \alpha^{-1}Z > -c)$$

$$+ E\big((1 + \alpha^{-1}Z)^2 \mathbf{1}_{-c<1+\alpha^{-1}Z\leq 0}\big)$$
$$+ c^2 P(\alpha^{-1}Z > c) + E\big((\alpha^{-1}Z)^2 \mathbf{1}_{0<\alpha^{-1}Z\leq c}\big)$$
$$= -c^2 P(-c - 1 < \alpha^{-1}Z \leq -c)$$
$$+ E\big((1 + \alpha^{-1}Z)^2 \mathbf{1}_{-c<1+\alpha^{-1}Z\leq 0}\big)$$
$$+ c^2 P(\alpha^{-1}Z > c) + E\big((\alpha^{-1}Z)^2 \mathbf{1}_{0<\alpha^{-1}Z\leq c}\big)$$
$$\geq -c^2 P(\alpha^{-1}Z \leq -c) + E\big((1 + \alpha^{-1}Z)^2 \mathbf{1}_{-c<1+\alpha^{-1}Z\leq 0}\big)$$
$$+ E\big((\alpha^{-1}Z)^2 \mathbf{1}_{0<\alpha^{-1}Z\leq c}\big).$$

On letting $c \to \infty$ we obtain the lower bound in (4.3). $\qquad\square$

Proof of Theorem 4.5 By Lemma 4.4,

$$\limsup_{q\to\infty} E(R_q^2)$$

$$\leq E\big((\alpha^{-1}Z)^2 \mathbf{1}_{Z\leq 0}\big) + E\big((1 + \alpha^{-1}Z)^2 \mathbf{1}_{0<1+\alpha^{-1}Z\leq 1}\big)$$
$$+ E\big((1 + \alpha^{-1}Z)^2 \mathbf{1}_{Z>0}\big)$$
$$= E\big((\alpha^{-1}Z)^2\big) + \theta(\alpha) + P(Z > 0) + \frac{2}{\alpha}E(Z\mathbf{1}_{Z>0}). \qquad (5.9)$$

Now $R_q = N_q - b_q$, so $\operatorname{var} N_q = \operatorname{var} R_q = E(R_q^2) - (ER_q)^2$. From (5.5) we have $\liminf_{q\to\infty} ER_q \geq E(\alpha^{-1}Z) = \gamma/\alpha > 0$, so $\liminf_{q\to\infty}(ER_q)^2 \geq \big(E(\alpha^{-1}Z)\big)^2$. With (5.9) this gives

$$\limsup_{q\to\infty} \operatorname{var} N_q \leq \operatorname{var}(\alpha^{-1}Z) + \theta(\alpha) + P(Z > 0) + \frac{2}{\alpha}E(Z\mathbf{1}_{Z>0}).$$

We have $\operatorname{var} Z = \pi^2/6$, while $P(Z > 0) = 1 - e^{-1}$. Also $E(Z\mathbf{1}_{Z>0}) = \gamma - E(Z\mathbf{1}_{Z\leq 0})$, and

$$-E(Z\mathbf{1}_{Z\leq 0}) = \int_{-\infty}^{0} (-z)e^{-z}\exp(-e^{-z})\,dz$$

$$= \int_{1}^{\infty} (\log t)e^{-t}\,dt = \int_{1}^{\infty} \frac{e^{-t}}{t}\,dt = E_1(1).$$

The bound

$$\limsup_{q\to\infty} \operatorname{var} N_q \le \frac{\pi^2}{6\alpha^2} + \theta(\alpha) + 1 - e^{-1} + \frac{2(\gamma + E_1(1))}{\alpha}$$

follows.

For the lower bound, the lower bound in Lemma 4.4 may be written

$$\liminf_{q\to\infty} E(R_q^2) \ge E\big((1+\alpha^{-1}Z)^2 \mathbf{1}_{Z\le 0}$$

$$- E\big((1+\alpha^{-1}Z)^2 \mathbf{1}_{0<1+\alpha^{-1}Z\le 1}\big) + E\big((\alpha^{-1}Z)^2 \mathbf{1}_{Z>0}\big)$$
$$= E\big((\alpha^{-1}Z)^2\big) - \theta(\alpha) + P(Z\le 0) + 2E(\alpha^{-1}Z\mathbf{1}_{Z\le 0}).$$

From (5.5) we have $0 < \limsup_{q\to\infty} ER_q \le E(1+\alpha^{-1}Z)$, so $\limsup_{q\to\infty}$ $(ER_q)^2 \le 1 + 2E(\alpha^{-1}Z) + \big(E(\alpha^{-1}Z)\big)^2$, which with the above gives

$$\liminf_{q\to\infty} \operatorname{var} R_q$$

$$\ge \operatorname{var}(\alpha^{-1}Z) - \theta(\alpha) - 1 + P(Z\le 0) - 2\alpha^{-1}\big(EZ - E(Z\mathbf{1}_{Z\le 0})\big).$$

Thus, since $\operatorname{var} N_q = \operatorname{var} R_q$,

$$\liminf_{q\to\infty} \operatorname{var} N_q \ge \frac{\pi^2}{6\alpha^2} - \theta(\alpha) - 1 + e^{-1} - \frac{2(\gamma + E_1(1))}{\alpha},$$

which is the required lower bound on $\liminf_{q\to\infty} \operatorname{var} N_q$ and completes the proof. $\qquad\square$

6 Numerical results

Matlab and *Pascal* were used to evaluate EN_q for different values of a and q. Fig. 6.1 shows values for EN_q for different values of a for tests with up to 20 questions. For example, for a test with 10 alternatives for each question EN_q ranges from 29 when there is one question in the test to only 56 when there are 20 questions. Contrast this with the total number of possible tests, which increases from 10 to 10^{20} in this range.

These results led the authors to extend the investigation to consider tests containing up to 200 questions. Fig. 6.2 demonstrates that, as the number of questions in a test is increased, the average number of tests required in order for all possible questions to have appeared increases quite slowly. In a 200-question test with 10 alternatives for each question, there are 10^{200} different possible tests and a total bank of 2000 questions; however, on average all questions will have appeared at least once by the

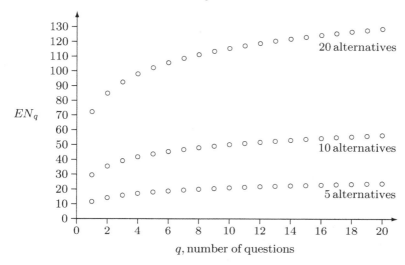

Figure 6.1 EN_q, the expected number of tests that need to be generated in order for all questions to have appeared at least once, for tests with up to 20 questions and 5, 10, and 20 alternatives for each question.

time only 78 tests have been generated. Table 6.1 summarises the results from Fig. 6.2, giving EN_q for different values of a and q.

Table 6.1 *Values of EN_q for various values of a and q.*

Number of alternatives for each question(a)	Number of questions in test (q)						
	1	5	10	20	50	100	200
5	11·4	17·8	20·8	23·8	27·9	31·0	34·1
10	29·3	43·5	49·9	56·4	65·0	71·6	78·1
20	72·0	102·0	115·3	128·7	146·5	160·0	173·5

7 Discussion

The asymptotics concern the behaviour of the random variable N_q, defined in §3, as the number of questions, q, grows. There is also dependence on a, the number of alternative answers per question in the multiple choice, but we regard a as fixed; it is any integer at least 2, and

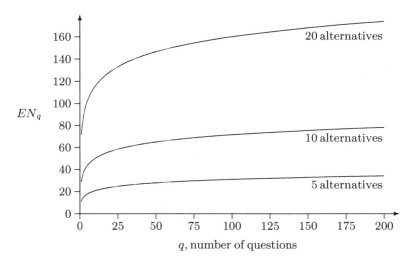

Figure 6.2 EN_q, the average number of tests that need to be generated in order for all questions to have appeared at least once, for tests with up to 200 questions and 5, 10, and 20 alternatives for each question.

we set

$$\alpha := \log\left(\frac{a}{a-1}\right),$$

so $\alpha > 0$. Theorem 4.1 first says that N_q cannot be centred and normed so that its distribution properly converges (one could get convergence to 0, of course, just by heavy norming). However it then says that by centring (translation) alone, N_q comes very close to looking like the random variable Z/α, where Z has the Gumbel distribution. The difference is a 'wobble' of between 0 and 1 in the limit; persistence of discreteness is responsible for this.

Theorem 4.3 establishes that the expected value of N_q behaves accordingly, growing like $\alpha^{-1} \log q$. More exactly, after centring by $b_q := \alpha^{-1} \log(aq)$ it differs from $\alpha^{-1} EZ$ by a number between 0 and 1 in the limit. Table 7.1 gives values of $b_q + \alpha^{-1} EZ$ for different values of a and q. For $q \geq 20$ the actual values of EN_q in the previous table exceed these by 0·5–0·6, exactly as Theorem 4.3 predicts.

What about the variance of N_q as q grows? Theorem 4.5 says that it does not tend to infinity, but is trapped as $q \to \infty$ between bounds that do not depend on q. The precision is pleasing, given that N_q does not

Table 7.1 *Values of* $b_q + \alpha^{-1}EZ$ *for various values of a and q.*

Number of alternatives for each question (a)	Number of questions in test (q)						
	1	5	10	20	50	100	200
5	9·8	17·0	20·1	23·2	27·3	30·4	33·5
10	27·3	42·6	49·2	55·8	64·5	71·0	77·6
20	68·7	101·0	114·5	128·1	145·9	159·4	173·0

converge, in any sense. The asymptotic bounds on the variance of N_q, var N_q, are $\frac{\pi^2}{6\alpha^2} \pm \Delta$ where Δ is a strange jumble of constants:

$$\Delta = \theta(\alpha) + 1 - e^{-1} + \frac{2(\gamma + E_1(1))}{\alpha}$$

(the bounds are not claimed to be sharp).

Table 7.2 *Asymptotic bounds on the standard deviation of N_q.*

a	2	3	4	5	10	20
Min s.d.	0·641	2·323	3·697	5·024	11·507	24·362
Max s.d.	2·537	3·823	5·107	6·390	12·804	25·630

The amount of variability can be better appreciated through the standard deviation. The asymptotic bounds on the standard deviation of N_q are

$$\sqrt{\frac{\pi^2}{6\alpha^2} \pm \Delta},$$

and some values for these are in Table 7.2. The lower bound is non-trivial, i.e. positive, in each case.

Acknowledgements We are grateful to Dave Pidcock, a colleague in the Mathematics Education Centre at Loughborough University, for raising the query in the first place. As a member of staff using computer-based tests to assess students, he was concerned about this issue from a practical viewpoint. That led RC to post a query on Allstat. CMG was not the only person to respond to the query, and we also acknowledge the others who responded, particularly Simon Bond.

References

Abramowitz, M., and Stegun, I. A. (ed.) 1965. *Handbook of Mathematical Functions.* New York: Dover.

Adler, I., and Ross, S. M. 2001. The coupon subset collection problem. *J. Appl. Probab.*, **38**, 737–746.

Anderson, C. W. 1970. Extreme value theory for a class of discrete distributions with applications to some stochastic processes. *J. Appl. Probab.*, **7**, 99–113.

Anderson, C. W. 1980. Local limit theory for the maxima of discrete random variables. *Math. Proc. Cambridge Philos. Soc.* **88**, 161–165.

Cornish, R., Goldie, C. M., and Robinson, C. L. 2006. Computer-assisted assessment: how many questions are enough? *Computer-Aided Assessment in Mathematics*, 9pp.; `http://mathstore.ac.uk/articles/maths-caa-series/feb2006`.

Feller, W. 1968. *An Introduction to Probability and its Applications*, vol. 1, 3rd edn. New York: John Wiley & Sons.

Grimmett, G. R., and Stirzaker, D. R. 2001. *Probability and Random Processes*, 3rd edn. Oxford: Oxford Univ. Press.

Littlewood, J. E. (ed. Bollobás, B.). 1986. *Littlewood's Miscellany.* Cambridge: Cambridge Univ. Press.

Resnick, S. I. 1987. *Extreme Values, Regular Variation, and Point Processes.* New York: Springer-Verlag.

13

Colouring and breaking sticks: random distributions and heterogeneous clustering

Peter J. Green[a]

Abstract

We begin by reviewing some probabilistic results about the Dirichlet Process and its close relatives, focussing on their implications for statistical modelling and analysis. We then introduce a class of simple mixture models in which clusters are of different 'colours', with statistical characteristics that are constant within colours, but different between colours. Thus cluster identities are exchangeable only within colours. The basic form of our model is a variant on the familiar Dirichlet process, and we find that much of the standard modelling and computational machinery associated with the Dirichlet process may be readily adapted to our generalisation. The methodology is illustrated with an application to the partially-parametric clustering of gene expression profiles.

Keywords Bayesian nonparametrics, gene expression profiles, hierarchical models, loss functions, MCMC samplers, optimal clustering, partition models, Pólya urn, stick breaking

AMS subject classification (MSC2010) 60G09, 62F15, 62G99, 62H30, 62M99

1 Introduction

The purpose of this note is four-fold: to remind some Bayesian nonparametricians gently that closer study of some probabilistic literature might

---

[a] School of Mathematics, University of Bristol, Bristol BS8 1TW;
P.J.Green@bristol.ac.uk

be rewarded, to encourage probabilists to think that there are statistical modelling problems worth of their attention, to point out to all another important connection between the work of John Kingman and modern statistical methodology (the role of the coalescent in population genetics approaches to statistical genomics being the most important example; see papers by Donnelly, Ewens and Griffiths in this volume), and finally to introduce a modest generalisation of the Dirichlet process.

The most satisfying basis for statistical clustering of items of data is a probabilistic model, which usually takes the form of a mixture model, broadly interpreted. In most cases, the statistical characteristics of each cluster or mixture component are the same, so that cluster identities are *a priori* exchangeable. In Section 5 we will introduce a class of simple mixture models in which clusters are of different categories, or colours as we shall call them, with statistical characteristics that are constant within colours, but different between colours. Thus cluster identities are exchangeable only within colours.

2 Mixture models and the Dirichlet process

Many statistical models have the following character. Data $\{Y_i\}$ are available on n units that we shall call *items*, indexed $i = 1, 2, \ldots, n$. There may be item-specific covariates, and other information, and the distribution of each Y_i is determined by an unknown parameter $\phi_i \in \Omega$, where we will take Ω here to be a subset of a Euclidean space. Apart from the covariates, the items are considered to be exchangeable, so we assume the $\{Y_i\}$ are conditionally independent given $\{\phi_i\}$, and model the $\{\phi_i\}$ as exchangeable random variables. Omitting covariates for simplicity, we write $Y_i|\phi_i \sim f(\cdot|\phi_i)$.

It is natural to take $\{\phi_i\}$ to be independent and identically distributed random variables, with common distribution G, where G itself is unknown, and treated as random. We might be led to this assumption whether we are thinking of a de Finetti-style representation theorem (de Finetti (1931, 1937); see also Kingman (1978), Kallenberg (2005)), or by following hierarchical modelling principles (Gelman et al., 1995; Green et al., 2003), Thus, unconditionally, $Y_i|G \sim \int f(\cdot|\phi)G(d\phi)$, independently given G.

This kind of formulation enables us to borrow strength across the units in inference about unknown parameters, with the aim of controlling the degrees of freedom, capturing the idea that while the $\{\phi_i\}$ may be different from item to item, we nevertheless understand that, through exchangeability, knowing the value of one of them would tell us something about the others.

There are still several options. One is to follow a standard parametric formulation, and to assume a specific parametric form for G, with parameters, or rather 'hyperparameters', in turn given a hyperprior distribution. However, many would argue that in most practical contexts, we would have little information to build such a model for G, which represents variation in the population of possible items of the parameter ϕ that determines the distribution of the data Y.

Thus we would be led to consider more flexible models, and one of several approaches might occur to us:

- a nonparametric approach, modelling uncertainty about G without making parametric assumptions;
- a mixture model representation for G;
- a partition model, where the $\{\phi_i\}$ are grouped together, in a way determined *a posteriori* by the data.

One of the things we will find, below, is that taking natural choices in each of these approaches can lead to closely related formulations in the end, so long as both modelling and inference depend solely on the $\{\phi_i\}$. These connections, not novel but not entirely well-known either, shed some light on the nature and implications of the different modelling approaches.

2.1 Ferguson definition of the Dirichlet process

Much Bayesian nonparametric distributional modelling (Walker et al., 1999) begins with the Dirichlet process (Ferguson, 1973). Building on earlier work by Dubins, Freedman and Fabius, Ferguson intended this model to provide a nonparametric prior model for G with a large support, yet one remaining capable of tractable prior-to-posterior analysis.

Given a probability distribution G_0 on an arbitrary measure space Ω, and a positive real θ, we say the random distribution G on Ω follows a Dirichlet process,

$$G \sim DP(\theta, G_0),$$

if for all partitions $\Omega = \bigcup_{j=1}^{m} B_j$ ($B_j \cap B_k = \emptyset$ if $j \neq k$), and for all m,

$$(G(B_1), \ldots, G(B_m)) \sim \text{Dirichlet}(\theta G_0(B_1), \ldots, \theta G_0(B_m)),$$

where Dirichlet$(\alpha_1, \alpha_2, \ldots, \alpha_m)$ denotes the distribution on the m-dimensional simplex with density at $(x_1, x_2, \ldots, x_m)$ proportional to $\prod_{j=1}^{m} x_j^{\alpha_j - 1}$.

The base measure G_0 gives the *expectation* of G:

$$E(G(B)) = G_0(B).$$

Even if G_0 is continuous, G is a.s. discrete (Kingman, 1967; Ferguson, 1973; Blackwell, 1973; Kingman, 1975), so i.i.d. draws $\{\phi_i, i = 1, 2, \ldots, n\}$ from G exhibit ties. The parameter θ measures (inverse) *concentration*: given i.i.d. draws $\{\phi_i, i = 1, 2, \ldots, n\}$ from G,

- as $\theta \to 0$, all ϕ_i are equal, a single draw from G_0;
- as $\theta \to \infty$, the ϕ_i are drawn i.i.d. from G_0.

2.2 The stick-breaking construction

A draw G from a Dirichlet process is a discrete distribution on Ω, so an alternative way to define the Dirichlet process would be via a construction of such a random distribution, through specification of the joint distribution of the locations of the atoms, and their probabilities. Such a construction was given by Ferguson (1973): in this, the locations are i.i.d. draws from G_0, with probabilities forming a decreasing sequence constructed from increments of a gamma process.

This is not the explicit construction that is most commonly used today, which is that known in the Bayesian nonparametric community as Sethuraman's stick-breaking model (Sethuraman and Tiwari, 1982; Sethuraman, 1994). This leads to this algorithm for generating the $\{\phi_i\}$:

1. draw $\phi_j^\star \sim G_0$, i.i.d., $j = 1, 2, \ldots$;
2. draw $V_j \sim \text{Beta}(1, \theta)$, i.i.d., $j = 1, 2, \ldots$;
3. define G to be the discrete distribution putting probability $(1 - V_1)(1 - V_2) \ldots (1 - V_{j-1})V_j$ on $\phi_j^\star$;
4. draw ϕ_i i.i.d. from G, $i = 1, 2, \ldots, n$.

This construction can be found considerably earlier in the probability literature, especially in connection with models for species sampling. The earliest reference seems to be in McCloskey (1965); for more readily accessible sources, see Patil and Taillie (1977) and Donnelly and Joyce (1989), where it is described in the context of size-biased sampling and the GEM (Generalised Engen–McCloskey) distributions. See also Section 3.3 below.

2.3 Limits of finite mixtures

A more direct, classical approach to modelling the distribution of Y in a flexible way would be to use a finite mixture model. Suppose that Y_i are i.i.d. with density $\sum_j w_j f_0(\cdot|\phi_j^\star)$ for a prescribed parametric density family $f_0(\cdot|\phi)$, and consider a Bayesian formulation with priors on the component weights $\{w_j\}$ and the component-specific parameters $\{\phi_j^\star\}$. The simplest formulation (e.g. Richardson and Green (1997)) uses a Dirichlet prior on the weights, and takes the $\{\phi_j^\star\}$ to be i.i.d. *a priori*, but with arbitrary distribution, so in algorithmic form:

1. draw $(w_1, w_2, \ldots, w_k) \sim \text{Dirichlet}(\delta, \ldots, \delta)$;
2. draw $c_i \in \{1, 2, \ldots, k\}$ with $P\{c_i = j\} = w_j$, i.i.d., $i = 1, \ldots, n$;
3. draw $\phi_j^\star \sim G_0$, i.i.d., $j = 1, \ldots, k$;
4. set $\phi_i = \phi_{c_i}^\star$.

It is well known that if we take the limit $k \to \infty$, $\delta \to 0$ such that $k\delta \to \theta$, then the joint distribution of the $\{\phi_i\}$ is the same as that obtained via the Dirichlet process formulation in the previous subsections (see for example Green and Richardson (2001)). This result is actually a corollary of a much stronger statement due to Kingman (1975), about the convergence of discrete probability measures. For more recent results in this direction see Muliere and Secchi (2003) and Ishwaran and Zarepour (2002).

We are still using the formulation $Y_i|G \sim \int f(\cdot|\phi)G(d\phi)$, independently given G, but note that G is invisible in this view; it has implicitly been integrated out.

2.4 Partition distribution

Suppose that, as above, G is drawn from $DP(\theta, G_0)$, and then $\{\phi_i : i = 1, 2, \ldots, n\}$ drawn i.i.d. from G. We can exploit the conjugacy of the Dirichlet with respect to multinomial sampling to integrate out G. For a fixed partition $\{B_j\}_{j=1}^m$ of Ω, and integers $c_i \in \{1, 2, \ldots, m\}$, we can write

$$P\{\phi_i \in B_{c_i}, i = 1, 2, \ldots, n\} = \frac{\Gamma(\theta)}{\Gamma(\theta + n)} \prod_{j=1}^m \frac{\Gamma(\theta G_0(B_j) + n_j)}{\Gamma(\theta G_0(B_j))},$$

where $n_j = \#\{i : c_i = j\}$. The jth factor in the product above is 1 if $n_j = 0$, and otherwise $\theta G_0(B_j)(\theta G_0(B_j)+1)(\theta G_0(B_j)+2)\ldots(\theta G_0(B_j)+n_j-1)$, so we find that if the partition becomes increasingly refined, and

G_0 is non-atomic, then the joint distribution of the $\{\phi_i\}$ can equivalently be described by

1. partitioning $\{1, 2, \ldots, n\} = \bigcup_{j=1}^{d} C_j$ at random, so that

$$p(C_1, C_2, \ldots, C_d) = \frac{\Gamma(\theta)}{\Gamma(\theta + n)} \theta^d \prod_{j=1}^{d} (n_j - 1)! \qquad (2.1)$$

   where $n_j = \#C_j$;

2. drawing $\phi_j^\star \sim G_0$, i.i.d., $j = 1, \ldots, d$, and then

3. setting $\phi_i = \phi_j^\star$ if $i \in C_j$.

Note that the partition model (2.1) shows extreme preference for unequal cluster sizes. If we let $a_r = \#\{j : n_j = r\}$, then the joint distribution of $(a_1, a_2, \ldots)$ is

$$\frac{n!}{n_1! n_2! \cdots n_d!} \times \frac{1}{\prod_r a_r!} \times p(C_1, C_2, \ldots, C_d). \qquad (2.2)$$

This is equation (A3) of Ewens (1972), derived in a context where n_j is the number of genes in a sample of the jth allelic type, in sampling from a selectively neutral population process. The first factor in (2.2) is the multinomial coefficient accounting for the number of ways the n items can be allocated to clusters of the required sizes, and the second factor accounts for the different sets of $\{n_1, n_2 \ldots, n_d\}$ leading to the same $(a_1, a_2, \ldots)$. Multiplying all this together, a little manipulation leads to the familiar Ewens sampling formula:

$$p(a_1, a_2, \ldots) = \frac{n! \Gamma(\theta)}{\Gamma(\theta + n)} \prod_r \frac{\theta^{a_r}}{r^{a_r} a_r!}. \qquad (2.3)$$

See also Kingman (1993), page 97.

This representation of the partition structure implied by the Dirichlet process was derived by Antoniak (1974), in the form (2.3). He noted that a consequence of this representation is that the joint distribution of the $\{\phi_i\}$ given d is independent for θ; thus given observed $\{\phi_i\}$, d is sufficient for θ. A similar observation was also made by Ewens (1972) in the genetics context of his work.

Note that as in the previous section, G has been integrated out, and so is invisible in this view of the Dirichlet process model.

2.5 Reprise

Whichever of the points of view is taken, items are clustered, according to a tractable distribution parametrised by $\theta > 0$, and for each cluster the cluster-specific parameter ϕ is an independent draw from G_0. Much statistical methodology built on the Dirichlet-process model uses only this joint distribution of the $\{\phi_i\}$, and so should hardly be called 'non-parametric'. Of course, even though G itself is invisible in two of the derivations above, the Dirichlet-process model does support inference about G, but this is seldom exploited in applications.

2.6 Multiple notations for partitions

In what follows, we will need to make use of different notations for the random partition induced by the Dirichlet-process model, or its relatives. We will variously use

- c is a partition of $\{1, 2, \ldots, n\}$;
- clusters of partition are $C_1, C_2, \ldots, C_d$ (d is the *degree* of the partition): $\bigcup_{j=1}^{d} C_j = \{1, 2, \ldots, n\}$, $C_j \cap C_{j'} = \emptyset$ if $j \neq j'$;
- c is the allocation vector: $c_i = j$ if and only if $i \in C_j$.

Note that the first of these makes no use of the (arbitrary) labelling of the clusters used in the second and third. We have to take care with multiplicities, and the distinction between (labelled) allocations and (unlabelled) partitions.

3 Applications and generalisations

3.1 Some applications of the Dirichlet process in Bayesian nonparametrics

Lack of space precludes a thorough discussion of the huge statistical methodology literature exploiting the Dirichlet process in Bayesian nonparametric procedures, so we will only review a few highlights.

Lo (1984) proposed density estimation procedures devised by mixing a user-defined kernel function $K(y, u)$ with respect to a Dirichlet process; thus i.i.d. data $\{Y_i\}$ are assumed distributed as $\int K(\cdot, u) G(du)$ with G drawn from a Dirichlet process. This is now known as the Dirichlet process mixture model (a better terminology than the formerly-used 'mixture of Dirichlet processes'). The formulation is identical to that we started with in Section 2, but for the implicit assumption that y and u lie in the same space, and that the kernel $K(\cdot, u)$ is a unimodal density located near u.

In the 1990s there was a notable flourishing of applied Bayesian non-parametrics, stimulated by interest in the Dirichlet process, and the rapid increase in computational power available to researchers, allowing almost routine use of the Pólya urn sampler approach (see Section 4) to posterior computation. For example, Escobar (1994) re-visited the Normal Means problem, West et al. (1994) discussed regression and density estimation, and Escobar and West (1995) further developed Bayesian density estimation. Müller et al. (1996) ingeniously exploited multivariate density estimation using Dirichlet process mixtures to perform Bayesian curve fitting of one margin on the others.

3.2 Example: clustered linear models for gene expression profiles

Let us consider a substantial and more specific application in some detail, to motivate the Dirichlet process (DP) set-up as a natural elaboration of a standard parametric Bayesian hierarchical model approach.

A remarkable aspect of modern microbiology has been the dramatic development of novel high-throughput assays, capable of delivering very high dimensional quantitative data on the genetic characteristics of organisms from biological samples. One such technology is the measurement of gene expression using Affymetrix gene chips. In Lau and Green (2007), we work with possibly replicated gene expression measures. The data are $\{Y_{isr}\}$, indexed by

- genes $i = 1, 2, \ldots, n$,
- conditions $s = 1, 2, \ldots, S$, and
- replicates $r = 1, 2, \ldots, R_s$.

Typically R_s is very small, S is much smaller than n, and the 'conditions' represent different subjects, different treatments, or different experimental settings.

We suppose there is a k-dimensional ($k \leq S$) covariate vector x_s describing each condition, and model parametric dependence of Y on x; the focus of interest is on the pattern of variation in these gene-specific parameters across the assayed genes.

Although other variants are easily envisaged, we suppose here that

$$Y_{isr} \sim N(x_s'\beta_i, \tau_i^{-1}), \quad \text{independently.}$$

Here $\phi_i = (\beta_i, \tau_i) \in \mathcal{R}^{k+1}$ is a gene-specific parameter vector characterising the dependence of gene expression on the condition-specific covariates. A priori, the genes can be considered exchangeable, and a standard hierarchical formulation would model the $\{\phi_i\}$ as i.i.d. draws from a parametric prior distribution G, say, whose (hyper)parameters have unknown values. This set-up allows borrowing of strength across genes in the interest of stability and efficiency of inference.

The natural nonparametric counterpart to this would be to suppose instead that G, the distribution describing variation of ϕ across the population of genes, does not have prescribed parametric form, but is modelled as a random distribution from a 'nonparametric' prior such as the Dirichlet process, specifically

$$G \sim DP(\theta, G_0).$$

A consequence of this assumption, as we have seen, is that G is atomic, so that the genes will be clustered together into groups sharing a common value of ϕ. A posteriori we obtain a probabilistic clustering of the gene expression profiles.

Lau and Green (2007) take a standard normal–inverse Gamma model, so that $\phi = (\beta, \tau) \sim G_0$ means

$$\tau \sim \Gamma(a_0, b_0) \quad \text{and} \quad \beta | \tau \sim N_k(m_0, (\tau t_0)^{-1} I).$$

This is a conjugate set-up, so that (β, τ) can be integrated out *in each cluster*. This leads easily to explicit within-cluster parameter posteriors:

$$\tau_j^\star | Y \sim \Gamma(a_j, b_j),$$
$$\beta_j^\star | \tau_j^\star, Y \sim N_k(m_j, (\tau_j^\star t_j)^{-1}),$$

where

$$a_j = a_0 + 1/2 \#\{isr : c_i = j\},$$
$$b_j = b_0 + 1/2 (Y_{C_j} - X_{C_j} m_0)'(X_{C_j} t_0^{-1} X_{C_j}')^{-1}(Y_{C_j} - X_{C_j} m_0),$$
$$m_j = (X_{C_j}' X_{C_j} + t_0 I)^{-1}(X_{C_j}' Y_{C_j} + t_0 m_0),$$
$$t_j = X_{C_j}' X_{C_j} + t_0 I.$$

The marginal likelihoods $p(Y_{C_j})$ are multivariate t distributions.

We continue this example later, in Sections 5.4 and 5.5.

3.3 Generalisations of the Dirichlet process, and related models

Viewed as a nonparametric model or as a basis for probabilistic clustering, the Dirichlet process is simple but inflexible—a single real parameter θ controls both variation and concentration, for example. And although the space Ω where the base measure G_0 lies and in which ϕ lives can be rather general, it is essentially a model for 'univariate' variation and unable to handle in a flexible way, for example, time-series data.

Driven both by such considerations of statistical modelling (Walker et al., 1999), or curious pursuit of more general mathematical results, the Dirichlet process has proved a fertile starting point for numerous generalisations, and we touch on just a few of these here.

The Poisson–Dirichlet distribution and its two-parameter generalisation. Kingman (1975) observed and exploited the fact that the limiting behaviour of random discrete distributions could become nontrivial and accessible through permutation of the components to be in ranked (decreasing) order. The limit law is the Poisson–Dirichlet distribution, implicitly defined and later described (Kingman, 1993, page 98) as 'rather less than user-friendly'.

Donnelly and Joyce (1989) elucidated the role of both ranking and size-biased sampling in establishing limit laws for random distributions; see also Holst (2001) and Arratia et al. (2003, page 107). The two-parameter generalisation of the Poisson–Dirichlet model was discovered by Pitman and co-workers, see for example Pitman and Yor (1997). This has been a rich topic for probabilistic study to the present day; see chapters by Gnedin, Haulk and Pitman, and by Aldous in this volume. The simplest view to take of the two-parameter Poisson–Dirichlet model $PD(\alpha, \theta)$ is to go back to stick-breaking (Section 2.2) and replace the $\mathrm{Beta}(1, \theta)$ distribution for the variables V_j there by $\mathrm{Beta}(1 - \alpha, \theta + j\alpha)$.

Ishwaran and James (2001) have considered Bayesian statistical applications of stick-breaking priors defined in this way, and implementation of Gibbs sampling for computing posterior distributions.

Dirichlet process relations in structured dependent models. Motivated by the need to build statistical models for structured data of various kinds, there has been a huge effort in generalising Dirichlet process models for such situations—indeed, there is now an 'xDP' for nearly every letter of the alphabet.

This has become a rich and sometimes confusing area; perhaps the most important current models are Dependent Dirichlet processes (MacEachern, 1999; MacEachern et al., 2001), Order-based dependent Dirichlet processes (Griffin and Steel, 2006), Hierarchical Dirichlet processes (Teh et al., 2006), and Kernel stick breaking processes (Dunson and Park, 2007). Many of the models are based on stick-breaking representations, but in which the atoms and/or the weights for the representations of different components of the process are made dependent on each other, or on covariates. The new book by Hjort et al. (2010) provides an excellent introduction and review of these developments.

Pólya trees. Ferguson's definition of the Dirichlet process focussed on the (random) probabilities to be assigned to arbitrary partitions (Section 2.1). As we have seen, the resulting distributions G are almost surely discrete. An effective way to modify this process to control continuity properties is to limit the partitions to which elementary probabilities are assigned, and in the case of Pólya tree processes this is achieved by imposed a fixed binary partition of Ω, and assigning probabilities to successive branches in the tree through independent Beta distributions. The parameters of these distributions can be set to obtain various degrees of smoothness of the resulting G. This approach, essentially beginning with Ferguson himself, has been pursued by Lavine (1992, 1994); see also Walker et al. (1999).

4 Pólya urn schemes and MCMC samplers

There is a huge literature on Markov chain Monte Carlo methods for posterior sampling in Dirichlet mixture models (MacEachern, 1994; Escobar and West, 1995; Müller et al., 1996; MacEachern and Müller, 1998; Neal, 2000; Green and Richardson, 2001). Although these models have 'variable dimension', the posteriors can be sampled without necessarily using reversible jump methods (Green, 1995).

Cases where G_0 is not conjugate to the data model $f(\cdot|\phi)$ demand keeping $\{\phi_i\}$ in the state vector, to be handled through various augmentation or reversible jump schemes. In the conjugate case, however, it is obviously appealing to integrate ϕ out, and target the Markov chain on the posterior solely of the partition, generating ϕ values subsequently as needed. To discuss this, we first go back to probability theory.

4.1 The Pólya urn representation of the Dirichlet process

The Pólya urn is a simple and well-known discrete probability model for a reinforcement process: coloured balls are drawn sequentially from an urn; after each is drawn it is replaced, together with a new ball of the same colour. This idea can be seen in a generalised form, in a recursive definition of the joint distribution of the $\{\phi_i\}$.

Suppose that for each $n = 0, 1, 2, \ldots,$

$$\phi_{n+1}|\phi_1, \phi_2, \ldots, \phi_n \sim \frac{1}{n+\theta} \sum_{i=1}^{n} \delta_{\phi_i} + \frac{\theta}{n+\theta} G_0, \qquad (4.1)$$

where $\theta > 0$, G_0 is an arbitrary probability distribution, and δ_ϕ is a point probability mass at ϕ. Blackwell and MacQueen (1973) termed such a sequence a Pólya sequence; they showed that the conditional distribution on the right hand side of (4.1) converges to a random probability distribution G distributed as $DP(\theta, G_0)$, and that, given G, ϕ_1, ϕ_2, ... are i.i.d. distributed as G. See also Antoniak (1974) and Pitman (1995).

Thus we have yet another approach to defining the Dirichlet process, at least in so far as specifying the joint distribution of the $\{\phi_i\}$ is concerned. This representation has a particular role, of central importance in computing inferences in DP models. This arises directly from (4.1) and the exchangeability of the $\{\phi_i\}$, for it follows that

$$\phi_i|\phi_{-i} \sim \frac{1}{n-1+\theta} \sum_{j\neq i} \delta_{\phi_j} + \frac{\theta}{n-1+\theta} G_0, \qquad (4.2)$$

where ϕ_{-i} means $\{\phi_j : j = 1, 2, \ldots, n, j \neq i\}$. In this form, the statement has an immediate role as the *full conditional* distribution for each component of $(\phi_i)_{i=1}^{n}$, and hence defines a Gibbs sampler update in a Markov chain Monte Carlo method aimed at this target distribution. By conjugacy this remains true, with obvious changes set out in the next section, for posterior sampling as well.

The Pólya urn representation of the Dirichlet process has been the point of departure for yet another class of probability models, namely species sampling models (Pitman, 1995, 1996), that are beginning to find a use in statistical methodology (Ishwaran and James, 2003).

4.2 The Gibbs sampler for posterior sampling of allocation variables

We will consider posterior sampling in the conjugate case in a more general setting, specialising back to the Dirichlet process mixture case later. The set-up we will assume is based on a partition model: it consists of a prior distribution $p(\mathbf{c}|\theta)$ on partitions $\mathbf{c}$ of $\{1, 2, \ldots, n\}$ with hyperparameter θ, together with a conjugate model within each cluster. The prior on the cluster-specific parameter ϕ_j has hyperparameter ψ, and is conjugate to the likelihood, so that for any subset $C \subseteq \{1, 2, \ldots, n\}$, $p(Y_C|\psi)$ is known explicitly, where Y_C is the subvector of $(Y_i)_{i=1}^n$ with indices in C. We have

$$p(Y_C|\psi) = \int \prod_{i \in C} p(Y_i|\phi)p(\phi|\psi)\,d\psi.$$

We first consider only re-allocating a single item at a time (the single-variable Gibbs sampler for c_i). Then repeatedly we withdraw an item, say i, from the model, and reallocate it to a cluster according to the full conditional for c_i, which is proportional to $p(\mathbf{c}|Y, \theta, \psi)$. It is easy to see that we have two choices:

- allocate Y_i to a new cluster $C_\star$, with probability

$$\propto p(\mathbf{c}^{i \to \star}|\theta) \times p(Y_i|\psi),$$

where $\mathbf{c}^{i \to \star}$ denotes the current partition $\mathbf{c}$ with i moved to $C_\star$, or
- allocate Y_i to cluster C_j^{-i}, with probability

$$\propto p(\mathbf{c}^{i \to j}|\theta) \times p(Y_{C_j^{-i} \cup \{i\}}|\psi)/p(Y_{C_j^{-i}}|\psi),$$

where $\mathbf{c}^{i \to j}$ denotes the partition $\mathbf{c}$, with i moved to cluster C_j.

The ratio of marginal likelihoods $p(Y|\psi)$ in the second expression can be interpreted as the posterior predictive distribution of Y_i given those observations already allocated to the cluster, i.e. $p(Y_i|Y_{C_j^{-i}}, \psi)$ (a multivariate t for the Normal–inverse gamma set-up from Section 3.2).

For Dirichlet mixtures we have, from (2.1),

$$p(\mathbf{c}|\theta) = \frac{\Gamma(\theta)}{\Gamma(\theta + n)}\theta^d \prod_{j=1}^{d}(n_j - 1)!,$$

where $n_j = \#C_j$ and $\mathbf{c} = (C_1, C_2, \ldots, C_d)$, so the re-allocation probabilities are explicit and simple in form.

But the same sampler can be used for many other partition models, and the idea is not limited to moving one item at a time.

4.3 When the Pólya urn sampler applies

All we require of the model for the Pólya urn sampler to be available for posterior simulation are that

1. a partition c of $\{1, 2, \ldots, n\}$ is drawn from a prior distribution with parameter θ;
2. conditionally on c, parameters $(\phi_1, \phi_2, \ldots, \phi_d)$ are drawn independently from a distribution G_0 (possibly with a hyperparameter ψ);
3. conditional on c and on $\phi = (\phi_1, \phi_2, \ldots, \phi_d)$, $\{y_1, y_2, \ldots, y_n\}$ are drawn independently, from not necessarily identical distributions $p(y_i|c, \phi) = f_i(y_i|\phi_j)$ for $i \in C_j$, for which G_0 is conjugate.

If these all hold, then the Pólya urn sampler can be used; we see from Section 4.2 that it will involve computing only marginal likelihoods, and ratios of the partition prior, up to a multiplicative constant. The first factor depends only on G_0 and the likelihood, the second only on the partition model.

Examples. $p(c^{i \to \star}|\theta)$ and $p(c^{i \to j}|\theta)$ are proportional simply to

- θ and $\#C_j^{-i}$ for the DP mixture model,
- $(k - d(c^{-i}))\delta$ and $\#C_j^{-i} + \delta$ for the Dirichlet–multinomial finite mixture model,
- $\theta + \alpha d(c^{-i})$ and $\#C_j^{-i} - \alpha$ for the Kingman–Pitman–Yor two-parameter Poisson–Dirichlet process (Section 3.3).

It is curious that the ease of using the Pólya urn sampler has often been cited as motivation to use Dirichlet process mixture models, when the class of models for which it is equally readily used is so wide.

4.4 Simultaneous re-allocation

There is no need to restrict to updating only one c_i at a time: the idea extends to simultaneously re-allocating any subset of items *currently in the same cluster*.

The notation can be rather cumbersome, but again the subset forms a new cluster, or moves to an existing cluster, with relative probabilities that are each products of two terms:

- the relative (new) partition prior probabilities, and
- the predictive density of the moved set of item data, given those already in the receiving cluster.

A more sophisticated variant on this scheme has been proposed by Nobile and Fearnside (2007), and studied in the case of finite mixture models.

5 A coloured Dirichlet process

For the remainder of this note, we focus on the use of these models for clustering, rather than density estimation or other kinds of inference. There needs to be a small caveat—mixture models are commonly used either for clustering, or for fitting non-standard distributions; in a problem demanding *both*, we cannot expect to be able meaningfully to identify clusters with the components of the mixture, since multiple components may be needed to fit the non-standard distributional shape within each cluster. Clustered Dirichlet process methodology in which there is clustering at two levels that can be used for such a purpose is under development by Dan Merl and Mike West at Duke (personal communication).

Here we will not pursue this complication, and simply consider a mixture model used for clustering in the obvious way.

In many domains of application, practical considerations suggest that the clusters in the data do not have equal standing; the most common such situation is where there is believed to be a 'background' cluster, and one or several 'foreground' clusters, but more generally, we can imagine there being several classes of cluster, and our prior beliefs are represented by the idea that cluster labels are exchangeable within these classes, but not overall. It would be common, also, to have different beliefs about cluster-specific parameters within each of these classes.

In this section, we present a variant on standard mixture/cluster models of the kinds we have already discussed, aimed at modelling this situation of partial exchangeability of cluster labels. We stress that it will remain true that, *a priori*, item labels are exchangeable, and that we have no prior information that particular items are drawn to particular classes of cluster; the analysis is to be based purely on the data $\{Y_i\}$.

We will describe the class of a cluster henceforth as its 'colour'. To define a variant on the DP in which not all clusters are exchangeable:

1. for each 'colour' $k = 1, 2, \ldots$, draw G_k from a Dirichlet process $\text{DP}(\theta_k, G_{0k})$, independently for each k;
2. draw weights (w_k) from the Dirichlet distribution $\text{Dir}(\gamma_1, \gamma_2, \ldots)$, independently of the G_k;
3. define G on $\{k\} \times \Omega$ by $G(k, B) = w_k G_k(B)$;
4. draw colour–parameter pairs (k_i, ϕ_i) i.i.d. from G, $i = 1, 2, \ldots, n$.

This process, denoted $\text{CDP}(\{(\gamma_k, \theta_k, G_{0k})\})$, is a Dirichlet mixture of Dirichlet processes (with different base measures), $\sum_k w_k \text{DP}(\theta_k, G_{0k})$, with the added feature that the the colour of each cluster is identified (and indirectly observed), while labelling of clusters within colours is arbitrary.

It can be defined by a 'stick-breaking-and-colouring' construction:

1. colour segments of the stick using the $\text{Dirichlet}(\{\gamma_k\})$-distributed weights;
2. break each coloured segment using an infinite sequence of independent $\text{Beta}(1, \theta_k)$ variables V_{jk};
3. draw $\phi_{jk}^\star \sim G_{0k}$, i.i.d., $j = 1, 2, \ldots$; $k = 1, 2, \ldots$;
4. define G_k to be the discrete distribution putting probability $(1 - V_{1k})(1 - V_{2k}) \cdots (1 - V_{j-1,k})V_{jk}$ on $\phi_{jk}^\star$.

Note that in contrast to other elaborations to more structured data of the Dirichlet process model, in which the focus is on nonparametric analysis and more sharing of information would be desirable, here, where the focus is on clustering, we are content to leave the atoms and weights within each colour completely uncoupled *a priori*.

5.1 Coloured partition distribution

The coloured Dirichlet process (CDP) generates the following partition model: partition $\{1, 2, \ldots, n\} = \bigcup_k \bigcup_{j=1}^{d_k} C_{kj}$ at random, where C_{kj} is the jth cluster of colour k, so that

$$p(C_{11}, C_{12}, \ldots, C_{1d_1}; C_{21}, \ldots, C_{2d_2}; C_{31}, \ldots) =$$

$$\frac{\Gamma(\sum_k \gamma_k)}{\Gamma(n + \sum_k \gamma_k)} \prod_k \left(\frac{\Gamma(\theta_k)\Gamma(n_k + \gamma_k)}{\Gamma(n_k + \theta_k)\Gamma(\gamma_k)} \theta_k^{d_k} \prod_{j=1}^{d_k} (n_{kj} - 1)! \right),$$

where $n_{kj} = \#C_{kj}$, $n_k = \sum_j n_{kj}$.

It is curious to note that this expression simplifies when $\theta_k \equiv \gamma_k$, although such a choice seems to have no particular significance in the probabilistic construction of the model. Only when it is also true that the θ_k are independent of k (and the colours are ignored) does the model degenerate to an ordinary Dirichlet process.

The clustering remains exchangeable over items. To complete the construction of the model, analogously to Section 2.4, for $i \in C_{kj}$, we set $k_i = k$ and $\phi_i = \phi_j^\star$, where $\phi_j^\star$ are drawn i.i.d. from G_{0k}.

5.2 Pólya urn sampler for the CDP

The explicit availability of the (coloured) partition distribution immediately allows generalisation of the Pólya-urn Gibbs sampler to the CDP.

In reallocating item i, let n_{kj}^{-i} denote the number *among the remaining items* currently allocated to C_{kj}, and define n_k^{-i} accordingly. Then reallocate i to

- a new cluster of colour k, with probability $\propto \theta_k \times (\gamma_k + n_k^{-i})/(\theta_k + n_k^{-i}) \times p(Y_i|\psi)$, for $k = 1, 2, \ldots$;
- the existing cluster C_{kj}, with probability $\propto n_{kj}^{-i} \times (\gamma_k + n_k^{-i})/(\theta_k + n_k^{-i}) \times p(Y_i|Y_{C_{kj}^{-i}}, \psi)$, for $j = 1, 2, \ldots, n_k^{-i}$; $k = 1, 2, \ldots$.

Again, the expressions simplify when $\theta_k \equiv \gamma_k$.

5.3 A Dirichlet process mixture with a background cluster

In many applications of probabilistic clustering, including the gene expression example from Section 3.2, it is natural to suppose a 'background' cluster that is not *a priori* exchangeable with the others. One way to think about this is to adapt the 'limit of finite mixtures' view from Section 2.3:

1. draw $(w_0, w_1, w_2, \ldots, w_k) \sim \text{Dirichlet}(\gamma, \delta, \ldots, \delta)$;
2. draw $c_i \in \{0, 1, \ldots, k\}$ with $P\{c_i = j\} = w_j$, i.i.d., $i = 1, \ldots, n$;
3. draw $\phi_0^\star \sim H_0$, $\phi_j^\star \sim G_0$, i.i.d., $j = 1, \ldots, k$;
4. set $\phi_i = \phi_{c_i}^\star$.

Now let $k \to \infty$, $\delta \to 0$ such that $k\delta \to \theta$, but leave γ fixed. The cluster labelled 0 represents the 'background'.

The background cluster model is a special case of the CDP, specifically $\text{CDP}(\{(\gamma, 0, H_0), (\theta, \theta, G_0)\})$. The two colours correspond to the background and regular clusters. The limiting-case $\text{DP}(0, H_0)$ is a point mass, randomly drawn from H_0. We can go a little further in a regression set-ting, and allow different regression models for each colour.

The Pólya urn sampler for prior or posterior simulation is readily adapted. When re-allocating item i, there are three kinds of choice: a new cluster C_*, the 'top table' C_0, or a regular cluster $C_j, j \neq 0$: the corresponding prior probabilities $p(c^{i \to *}|\theta)$, $p(c^{i \to 0}|\theta)$ and $p(c^{i \to j}|\theta)$ are proportional to θ, $(\gamma + \#C_0^{-i})$ and $\#C_j^{-i}$ for the background cluster CDP model.

5.4 Using the CDP in a clustered regression model

As a practical illustration of the use of the CDP background cluster model, we discuss a regression set-up that expresses a vector of meas-urements $\mathbf{y}_i = (y_{i1}, \ldots, y_{iS})$ for $i = 1, \ldots, n$, where S is the number of samples, as a linear combination of known covariates, $(\mathbf{z}_1 \cdots \mathbf{z}_S)$ with dimension K' and $(\mathbf{x}_1 \cdots \mathbf{x}_S)$ with dimension K. These two collections of covariates, and the corresponding regression coefficients $\boldsymbol{\delta}_j$ and $\boldsymbol{\beta}_j$, are distinguished since we wish to hold one set of regression coefficients fixed in the background cluster. We assume

$$\mathbf{y}_i = \begin{bmatrix} y_{i1} \\ \vdots \\ y_{iS} \end{bmatrix} = \sum_{k'=1}^{K'} \delta_{jk'} \begin{bmatrix} z_{1k'} \\ \vdots \\ z_{Sk'} \end{bmatrix} + \sum_{k=1}^{K} \beta_{jk} \begin{bmatrix} x_{1k} \\ \vdots \\ x_{Sk} \end{bmatrix} + \begin{bmatrix} \epsilon_{j1} \\ \vdots \\ \epsilon_{jS} \end{bmatrix}$$
$$= [\mathbf{z}_1 \cdots \mathbf{z}_S]'\boldsymbol{\delta}_j + [\mathbf{x}_1 \cdots \mathbf{x}_S]'\boldsymbol{\beta}_j + \boldsymbol{\epsilon}_j \qquad (5.1)$$

where $\boldsymbol{\epsilon}_j \sim N(\mathbf{0}_{S \times 1}, \tau_j^{-1} \mathbf{I}_{S \times S})$, $\mathbf{0}_{S \times 1}$ is the S–dimension zero vector and $\mathbf{I}_{S \times S}$ is the order–S identity matrix. Here, $\boldsymbol{\delta}_j$, $\boldsymbol{\beta}_j$ and τ_j are cluster-specific parameters. The profile of measurements for individual i is $\mathbf{y}_i = [y_{i1} \cdots y_{iS}]'$ for $i = 1, \ldots, n$. Given the covariates $\mathbf{z}_s = [z_{s1} \cdots z_{sK'}]'$, $\mathbf{x}_s = [x_{s1} \cdots x_{sK}]'$, and the cluster j, the parameters/latent variables are $\boldsymbol{\delta}_j = [\delta_{j1} \cdots \delta_{jK'}]'$, $\boldsymbol{\beta}_j = [\beta_{j1} \cdots \beta_{jK}]'$ and τ_j. The kernel is now represented as $k(\mathbf{y}_i|\boldsymbol{\delta}_j, \boldsymbol{\beta}_j, \tau_j)$, which is a multivariate Normal density, $N([\mathbf{z}_1 \cdots \mathbf{z}_S]'\boldsymbol{\delta}_j + [\mathbf{x}_1 \cdots \mathbf{x}_S]'\boldsymbol{\beta}_j, \tau_j^{-1}\mathbf{I}_{S \times S})$. In particular, we take differ-ent probability measures, the parameters of heterogeneous DP, for the background and regular clusters,

$$\mathbf{u}_0 = (\boldsymbol{\delta}_0, \boldsymbol{\beta}_0, \tau_0) \sim H_0(d\boldsymbol{\delta}_0, d\boldsymbol{\beta}_0, d\tau_0)$$
$$= \delta_{\boldsymbol{\delta}_0}(d\boldsymbol{\delta}_0) \times \text{Normal–Gamma}(d\boldsymbol{\beta}_0, d\tau_0^{-1});$$

$$\mathbf{u}_j = (\boldsymbol{\delta}_j, \boldsymbol{\beta}_j, \tau_j) \sim G_0(d\boldsymbol{\delta}_j, d\boldsymbol{\beta}_j, d\tau_j)$$
$$= \text{Normal–Gamma}(d(\boldsymbol{\delta}_j', \boldsymbol{\beta}_j')', d\tau_j^{-1})$$
$$\text{for } j = 1, \ldots, n(\mathbf{p}) - 1.$$

Here H_0 is a probability measure that includes a point mass at $\boldsymbol{\delta}_0$ and a Normal–Gamma density for $\boldsymbol{\beta}_0$ and τ_0^{-1}. On the other hand, we take G_0 to be a probability measure that is a Normal–Gamma density for $(\boldsymbol{\delta}_j', \boldsymbol{\beta}_j')'$ and τ_j^{-1}. Thus the regression parameters corresponding to the z covariates are held fixed at $\boldsymbol{\delta}_0$ in the background cluster, but not in the others.

We will first discuss the marginal distribution for the regular clusters. Given τ_j, $(\boldsymbol{\delta}_j', \boldsymbol{\beta}_j')'$ follows the $(K' + K)$-dimensional multivariate Normal with mean $\widetilde{\mathbf{m}}$ and variance $(\tau_j \widetilde{\mathbf{t}})^{-1}$ and τ_j follows the univariate Gamma with shape $\widetilde{a}$ and scale $\widetilde{b}$. We denote the joint distribution $G_0(d(\boldsymbol{\delta}_j', \boldsymbol{\beta}_j')', d\tau_j)$ as a joint Gamma and Normal distribution, Normal–Gamma$(\widetilde{a}, \widetilde{b}, \widetilde{\mathbf{m}}, \widetilde{\mathbf{t}})$, and further we take

$$\widetilde{\mathbf{m}} = \begin{bmatrix} \widetilde{\mathbf{m}}_\delta \\ \widetilde{\mathbf{m}}_\beta \end{bmatrix} \text{ and } \widetilde{\mathbf{t}} = \begin{bmatrix} \widetilde{\mathbf{t}}_\delta & 0 \\ 0 & \widetilde{\mathbf{t}}_\beta \end{bmatrix}. \tag{5.2}$$

Based on this set-up, we have

$$m_{G_0}(\mathbf{y}_{C_j}) =$$

$$t_{2\widetilde{a}}(\mathbf{Y}_{C_j} | \mathbf{Z}_{C_j} \widetilde{\mathbf{m}}_\delta + \mathbf{X}_{C_j} \widetilde{\mathbf{m}}_\beta, \frac{\widetilde{b}}{\widetilde{a}}(\mathbf{Z}_{C_j} \widetilde{\mathbf{t}}_\delta^{-1} \mathbf{Z}_{C_j}' + \mathbf{X}_{C_j} \widetilde{\mathbf{t}}_\beta^{-1} \mathbf{X}_{C_j}' + \mathbf{I}_{e_j S \times e_j S})), \tag{5.3}$$

where $\mathbf{Y}_{C_j} = [\mathbf{y}_{i_1}' \cdots \mathbf{y}_{i_{e_j}}']'$, $\mathbf{X}_{C_j} = [[\mathbf{x}_1 \cdots \mathbf{x}_S] \cdots [\mathbf{x}_1 \cdots \mathbf{x}_S]]'$ and $\mathbf{Z}_{C_j} = [[\mathbf{z}_1 \cdots \mathbf{z}_S] \cdots [\mathbf{z}_1 \cdots \mathbf{z}_S]]'$ for $C_j = \{i_1, \ldots, i_{e_j}\}$. Note that $\mathbf{Y}_{C_j}$ is a $e_j S$ vector, $\mathbf{Z}_{C_j}$ is a $e_j S \times K'$ matrix and $\mathbf{X}_{C_j}$ is a $e_j S \times K$ matrix. Moreover, $m_{G_0}(\mathbf{y}_{C_j})$ is a multivariate t density with mean $\mathbf{Z}_{C_j} \widetilde{\mathbf{m}}_\delta + \mathbf{X}_{C_j} \widetilde{\mathbf{m}}_\beta$, scale

$$\frac{\widetilde{b}}{\widetilde{a}}(\mathbf{Z}_{C_j} \widetilde{\mathbf{t}}_\delta^{-1} \mathbf{Z}_{C_j}' + \mathbf{X}_{C_j} \widetilde{\mathbf{t}}_\beta^{-1} \mathbf{X}_{C_j}' + \mathbf{I}_{e_j S \times e_j S}))$$

and degree of freedom $2\widetilde{a}$.

For the background cluster, we take H_0 to be a joint Gamma and Normal distribution, Normal–Gamma$(\overline{a}, \overline{b}, \overline{\mathbf{m}}_\beta, \overline{\mathbf{t}}_\beta)$. The precision τ_0 follows the univariate Gamma with shape $\overline{a}$ and scale $\overline{b}$. Given τ_0, $\boldsymbol{\beta}_0$ follows the K-dimension multivariate Normal with mean $\overline{\mathbf{m}}_\beta$ and variance $(\tau_0 \overline{\mathbf{t}}_\beta)^{-1}$ and τ_0 follows the univariate Gamma with shape $\overline{a}$ and scale $\overline{b}$. The marginal distribution becomes

$$m_{H_0}(\mathbf{y}_{C_0}) = t_{2\overline{a}}(\mathbf{Y}_{C_j}|\mathbf{Z}_{C_j}\boldsymbol{\delta}_0 + \mathbf{X}_{C_j}\overline{\mathbf{m}}_\beta, \frac{\overline{b}}{\overline{a}}(\mathbf{X}_{C_j}\overline{\mathbf{t}}_\beta^{-1}\mathbf{X}'_{C_j} + \mathbf{I}_{e_jS \times e_jS})).$$
(5.4)

So $m_{H_0}(\mathbf{y}_{C_0})$ is a multivariate t density with mean $\mathbf{Z}_{C_j}\boldsymbol{\delta}_0 + \mathbf{X}_{C_j}\overline{\mathbf{m}}_\beta$, scale

$$\frac{\overline{b}}{\overline{a}}(\mathbf{X}_{C_j}\overline{\mathbf{t}}_\beta^{-1}\mathbf{X}'_{C_j} + \mathbf{I}_{e_jS \times e_jS})$$

and degree of freedom $2\overline{a}$.

In some applications, the xs and βs are not needed and so can be omitted, and we consider the following model,

$$\mathbf{y}_i = \begin{bmatrix} y_{i1} \\ \vdots \\ y_{iS} \end{bmatrix} = \sum_{k'=1}^{K'} \delta_{jk'} \begin{bmatrix} z_{1k'} \\ \vdots \\ z_{Sk'} \end{bmatrix} + \begin{bmatrix} \epsilon_{j1} \\ \vdots \\ \epsilon_{jS} \end{bmatrix} = [\mathbf{z}_1 \cdots \mathbf{z}_S]'\boldsymbol{\delta}_j + \boldsymbol{\epsilon}_j;$$
(5.5)

here we assume that $K = 0$ or $[\mathbf{x}_1 \cdots \mathbf{x}_S]' = \mathbf{0}_{S \times K}$ where $\mathbf{0}_{S \times K}$ is the $S \times K$ matrix with all zero entries of the model (5.1). We can derive the marginal distributions analogous to (5.3) and (5.4),

$$m_{G_0}(\mathbf{y}_{C_j}) = t_{2\widetilde{a}}(\mathbf{Y}_{C_j}|\mathbf{Z}_{C_j}\widetilde{\mathbf{m}}_\delta, \frac{\widetilde{b}}{\widetilde{a}}(\mathbf{Z}_{C_j}\widetilde{\mathbf{t}}_\delta^{-1}\mathbf{Z}'_{C_j} + \mathbf{I}_{e_jS \times e_jS})), \quad (5.6)$$

$$m_{H_0}(\mathbf{y}_{C_0}) = t_{2\overline{a}}(\mathbf{Y}_{C_j}|\mathbf{Z}_{C_j}\boldsymbol{\delta}_0, \frac{\overline{b}}{\overline{a}}\mathbf{I}_{e_jS \times e_jS}). \quad (5.7)$$

Here $t_\nu(\mathbf{x}|\boldsymbol{\mu}, \boldsymbol{\Sigma})$ is a multivariate t density in d dimensions with mean $\boldsymbol{\mu}$ and scale $\boldsymbol{\Sigma}$ with degrees of freedom ν;

$$t_\nu(\mathbf{x}|\boldsymbol{\mu}, \boldsymbol{\Sigma}) = \frac{\Gamma((\nu + d)/2)}{\Gamma((\nu)/2)} \frac{|\boldsymbol{\Sigma}|^{-1/2}}{(\nu\pi)^{d/2}}(1 + \frac{1}{\nu}(\mathbf{x} - \boldsymbol{\mu})'\boldsymbol{\Sigma}^{-1}(\mathbf{x} - \boldsymbol{\mu}))^{-(\nu+d)/2}.$$
(5.8)

5.5 Time-course gene expression data

We consider the application of this methodology to data from a gene expression time-course experiment. Wen et al. (1998) studied the central nervous system development of the rat; see also Yeung et al. (2001). The mRNA expression levels of 112 genes were recorded over the period of development of the central nervous system. In the dataset, there are 9 records for each gene over 9 time points; they are from embryonic days 11, 13, 15, 18, 21, postnatal days 0, 7, 14, and the 'adult' stage (postnatal day 90).

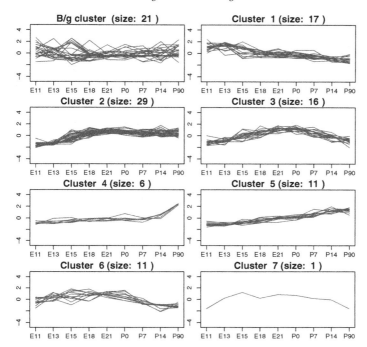

Figure 5.1 Profile plot of our partition estimate for the Rat data set of Wen et al. (1998).

In their analysis, Wen et al. (1998) obtained 5 clusters/waves (totally 6 clusters), taken to characterize distinct phases of development. The data set is available at `http://faculty.washington.edu/kayee/cluster/GEMraw.txt`. We take $S = 9$ and $K' = 5$. The design matrix of covariates is taken to be

$$[\mathbf{z}_1 \cdots \mathbf{z}_S]' = \begin{bmatrix} 1 & 1 & 1 & 1 & 1 & 0 & 0 & 0 & 0 \\ 11 & 13 & 15 & 18 & 21 & 0 & 0 & 0 & 0 \\ 0 & 0 & 0 & 0 & 0 & 1 & 1 & 1 & 0 \\ 0 & 0 & 0 & 0 & 0 & 0 & 7 & 14 & 0 \\ 0 & 0 & 0 & 0 & 0 & 0 & 0 & 0 & 1 \end{bmatrix}',$$

representing piecewise linear dependence on time, within three separate phases (embryonic, postnatal and adult).

In our analysis of these data, we take $\theta = 1$, $\gamma = 5$, $\widetilde{a} = \overline{a} = 0.01$, $\widetilde{b} = \overline{b} = 0.01$, $\widetilde{\mathbf{m}}_\delta = \overline{\mathbf{m}}_\delta = [0 \cdots 0]'$, $\widetilde{\mathbf{t}}_\delta = \overline{\mathbf{t}}_\delta = 0.01\mathbf{I}$, $\widetilde{\mathbf{m}}_\beta = [0 \cdots 0]'$, $\widetilde{\mathbf{t}}_\beta = 0.01\mathbf{I}$ and $\boldsymbol{\delta}_0 = [0 \cdots 0]'$. The Pólya urn sampler was implemented, and run for 20000 sweeps starting from the partition consisting of all singleton clusters, 10000 being discarded as burn-in. We then use the last 10000 partitions sampled as in Lau and Green (2007), to estimate the optimal Bayesian partition on a decision-theoretic basis, using a pairwise coincidence loss function that equally weights false 'positives' and 'negatives'.

We present some views of the resulting posterior analysis of this data set.

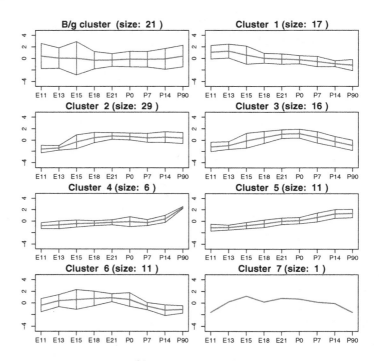

Figure 5.2 Mean and 95% CI of genes across clusters of our partition estimate.

Figure 5.1 shows the profiles in the inferred clusters plotted, and Figure 5.2 the mean and the 95% CI of the clusters. Figure 5.3 cross-tabulates the clusters with the biological functions attributed to the relevant genes by Wen et al. (1998).

Acknowledgements I am grateful to Sonia Petrone and Simon Tavaré for some pointers to the literature, John Lau for the analysis of the gene

| | General gene class | | | | Neurotransmitter receptors | | | | | |
| | | | | | Ligand class | | | | Sequence class | |
	% peptide signaling	% neurotr. receptors	% neuroglial markers	% diverse	% ACh	% GABA	% Glu	% 5HT	% ion channel	% G protein coupled
1	**41% (7)**	6% (1)	24% (4)	29% (5)	**100% (1)**				**100% (1)**	
2	3% (1)	**62% (18)**	31% (9)	3% (1)	17% (3)	**39% (7)**	33% (6)	11% (2)	**61% (11)**	39% (7)
3	6% (1)	**63% (10)**	19% (3)	13% (2)	20% (2)	20% (2)	**40% (4)**	20% (2)	50% (5)	50% (5)
4		33% (2)	17% (1)	**50% (3)**			**100% (2)**		50% (1)	50% (1)
5	18% (2)	27% (3)	**45% (5)**	9% (1)	33% (1)	**67% (2)**			**67% (2)**	33% (1)
6	27% (3)	18% (2)	18% (2)	**36% (4)**	50% (1)		50% (1)		**100% (2)**	
7		**100% (1)**			**100% (1)**				**100% (1)**	
B/g	**62% (13)**	10% (2)	5% (1)	24% (5)	**100% (2)**				**100% (2)**	

Figure 5.3 Biological functions of our Bayesian partition estimate for the genes in the data set of Wen et al. (1998), showing correspondence between inferred clusters and the functional categories of the genes. All genes are classified into 4 general gene classes. Additionally, the Neurotransmitter genes have been further categorised by ligand class and functional sequence class. Boldface type represents the dominant class in the cluster, in each categorisation.

expression data, and John Kingman for his undergraduate lectures in Measure and Probability.

References

Antoniak, C. E. 1974. Mixtures of Dirichlet processes with applications to Bayesian nonparametric problems. *Ann. Statist.*, **2**, 1152–1174.

Arratia, R., Barbour, A. D., and Tavaré, S. 2003. *Logarithmic Combinatorial Structures: A Probabilistic Approach.* EMS Monogr. Math. Zurich: European Math. Soc. Publishing House.

Blackwell, D. 1973. Discreteness of Ferguson selections. *Ann. Statist.*, **1**, 356–358.

Blackwell, D., and MacQueen, J. B. 1973. Ferguson distributions via Pólya urn schemes. *Ann. Statist.*, **1**, 353–355.

Donnelly, P. J., and Joyce, P. 1989. Continuity and weak convergence of ranked and size-biased permutations on the infinite simplex. *Stochastic Process. Appl.*, **31**, 89–103.

Dunson, D. B., and Park, J-H. 2007. Kernel stick breaking processes. *Biometrika*, **95**, 307–323.

Escobar, M. D. 1994. Estimating normal means with a Dirichlet process prior. *J. Amer. Statist. Assoc.*, **89**, 268–277.

Escobar, M. D., and West, M. 1995. Bayesian density estimation and inference using mixtures. *J. Amer. Statist. Assoc.*, **90**, 577–588.

Ewens, W. J. 1972. The sampling theory of selectively neutral alleles. *Theor. Population Biology*, **3**, 87–112.

Ferguson, T. S. 1973. A Bayesian analysis of some nonparametric problems. *Ann. Statist.*, **1**, 209–230.

Finetti, B. de. 1931. Funzione caratteristica di un fenomeno aleatorio. *Atti della R. Academia Nazionale dei Lincei, ser. 6*, **4**, 251–299. Memorie, Classe di Scienze Fisiche, Mathematiche e Naturali.

Finetti, B. de. 1937. La prévision: ses lois logiques, ses sources subjectives. *Ann. Inst. H. Poincaré*, **7**, 1–68.

Gelman, A., Carlin, J. B., Stern, H. S., and Rubin, D. B. 1995. *Bayesian Data Analysis*. London: Chapman and Hall.

Green, P. J. 1995. Reversible jump Markov chain Monte Carlo computation and Bayesian model determination. *Biometrika*, **82**, 711–732.

Green, P. J., and Richardson, S. 2001. Modelling heterogeneity with and without the Dirichlet process. *Scand. J. Statist.*, **28**, 355–375.

Green, P. J., Hjort, N. L., and Richardson, S. (eds). 2003. *Highly Structured Stochastic Systems*. Oxford: Oxford Univ. Press.

Griffin, J. E., and Steel, M. F. J. 2006. Order-based dependent Dirichlet processes. *J. Amer. Statist. Assoc.*, **101**, 179–194.

Hjort, N. L., Holmes, C., Müller, P., and Walker, S. G. (eds). 2010. *Bayesian Nonparametrics*. Camb. Ser. Stat. Probab. Math., vol. 28. Cambridge: Cambridge Univ. Press.

Holst, L. 2001. *The Poisson–Dirichlet Distribution and its Relatives Revisited.* Tech. rept. Department of Mathematics, Royal Institute of Technology, Stockholm.

Ishwaran, H., and James, L. F. 2001. Gibbs sampling methods for stick-breaking priors. *J. Amer. Statist. Assoc.*, **96**, 161–173.

Ishwaran, H., and James, L. F. 2003. Generalized weighted Chinese Restaurant processes for species sampling mixture models. *Statist. Sinica*, **13**, 1211–1235.

Ishwaran, H., and Zarepour, M. 2002. Dirichlet prior sieves in finite Normal mixtures. *Statist. Sinica*, **12**, 941–963.

Kallenberg, O. 2005. *Probabilistic Symmetries and Invariance Principles*. New York: Springer-Verlag.

Kingman, J. F. C. 1967. Completely random measures. *Pacific J. Math.*, **21**, 59–78.

Kingman, J. F. C. 1975. Random discrete distributions (with discussion and response). *J. Roy. Statist. Soc. Ser. B*, **37**, 1–22.

Kingman, J. F. C. 1978. Uses of exchangeability. *Ann. Probab.*, **6**, 183–197.

Kingman, J. F. C. 1993. *Poisson Processes*. Oxford: Oxford Univ. Press.

Lau, J. W., and Green, P. J. 2007. Bayesian model-based clustering procedures. *J. Comput. Graph. Statist.*, **16**, 526–558.

Lavine, M. 1992. Some aspects of Pólya tree distributions for statistical modelling. *Ann. Statist.*, **20**, 1222–1235.

Lavine, M. 1994. More aspects of Pólya tree distributions for statistical modelling. *Ann. Statist.*, **22**, 1161–1176.

Lo, A. Y. 1984. On a class of Bayesian nonparametric estimates, I: Density estimates. *Ann. Statist.*, **12**, 351–357.

MacEachern, S. N. 1994. Estimating normal means with a conjugate style Dirichlet process prior. *Commun. Statist. Simulation and Computation*, **23**, 727–741.

MacEachern, S. N. 1999. Dependent nonparametric processes. In: *Proceedings of the Section on Bayesian Statistical Science*. American Statistical Association.

MacEachern, S. N., and Müller, P. 1998. Estimating mixture of Dirichlet process models. *J. Comput. Graph. Statist.*, **7**, 223–238.

MacEachern, S. N., Kottas, A., and Gelfand, A. 2001. *Spatial Nonparametric Bayesian Models*. Tech. rept. 01–10. Institute of Statistics and Decision Sciences, Duke University.

McCloskey, J. W. 1965. *A Model for the Distribution of Species in an Environment*. Ph.D. thesis, Michigan State University.

Müller, P., Erkanli, A., and West, M. 1996. Bayesian curve fitting using multivariate normal mixtures. *Biometrika*, **83**, 67–79.

Muliere, P., and Secchi, P. 2003. Weak convergence of a Dirichlet-multinomial process. *Georgian Math. J.*, **10**, 319–324.

Neal, R. M. 2000. Markov chain sampling methods for Dirichlet process mixture models. *J. Comput. Graph. Statist.*, **9**, 249–265.

Nobile, A., and Fearnside, A. T. 2007. Bayesian finite mixtures with an unknown number of components: the allocation sampler. *Statist. Comput.*, **17**, 147–162.

Patil, C. P., and Taillie, C. 1977. Diversity as a concept and its implications for random communities. *Bull. Int. Statist. Inst.*, **47**, 497–515.

Pitman, J. 1995. Exchangeable and partially exchangeable random partitions. *Probab. Theory Related Fields*, **102**, 145–158.

Pitman, J. 1996. Some developments of the Blackwell-MacQueen urn scheme. Pages 245–267 of: Ferguson, T. S., Shapley, L. S., and MacQueen, J. B. (eds), *Statistics, Probability and Game Theory; Papers in Honor of David Blackwell*. Hayward, CA: Institute of Mathematical Statistics.

Pitman, J., and Yor, M. 1997. The two-parameter Poisson–Dirichlet distribution derived from a stable subordinator. *Ann. Probab.*, **25**, 855–900.

Richardson, S., and Green, P. J. 1997. On Bayesian analysis of mixtures with an unknown number of components (with discussion and response). *J. Roy. Statist. Soc. Ser. B*, **59**, 731–792.

Sethuraman, J. 1994. A constructive definition of Dirichlet priors. *Statist. Sinica*, **4**, 639–650.

Sethuraman, J., and Tiwari, R. C. 1982. Convergence of Dirichlet measures and the interpretation of their parameters. Pages 305–315 of: Gupta, S. S., and Berger, J. O. (eds), *Statistical Decision Theory and Related Topics III*, vol. 2. New York: Academic Press.

Teh, Y. W., Jordan, M. I., Beal, M. J., and Blei, D. M. 2006. Hierarchical Dirichlet processes. *J. Amer. Statist. Assoc.*, **101**, 1566–1581.

Walker, S. G., Damien, P., Laud, P. W., and Smith, A. F. M. 1999. Bayesian nonparametric inference for random distributions and related functions (with discussion). *J. Roy. Statist. Soc. Ser. B*, **61**, 485–527.

Wen, X., Fuhrman, S., Michaels, G. S., Carr, D. B., Smith, S., Barker, J. L., and Somogyi, R. 1998. Large-scale temporal gene expression mapping of central nervous system development. *Proc. Natl. Acad. Sci. USA*, **95**, 334–339.

West, M., Müller, P., and Escobar, M. D. 1994. Hierarchical priors and mixture models, with application in regression and density estimation. In: Freeman, P. R., and Smith, A. F. M. (eds), *Aspects of Uncertainty: A Tribute to D. V. Lindley*. Chichester: Wiley.

Yeung, K. Y., Haynor, D. R., and Ruzzo, W. L. 2001. Validating clustering for gene expression data. *Bioinformatics*, 309–318.

14

The associated random walk and martingales in random walks with stationary increments

D. R. Grey[a]

Abstract

We extend the notion of the associated random walk and the Wald martingale in random walks where the increments are independent and identically distributed to the more general case of stationary ergodic increments. Examples are given where the increments are Markovian or Gaussian, and an application in queueing is considered.

AMS subject classification (MSC2010) 60G50, 60G42, 60G10, 60K25

1 Introduction and definition

Let X_1, X_2, ... be independent identically distributed (i.i.d.) random variables with distribution function (d.f.) F and positive mean. If $S_n = X_1 + X_2 + \cdots + X_n$ for each $n = 0, 1, 2, \ldots$ then the process $\{S_n\}$ is called the random walk with increments X_1, X_2, ... Because the increments have positive mean, by the strong law of large numbers the random walk will in the long run drift upwards to infinity.

There may exist $\theta \neq 0$ such that

$$\hat{F}(\theta) := E(e^{-\theta X_1}) = \int_{-\infty}^{\infty} e^{-\theta x} \, dF(x) = 1,$$

in which case θ is unique, and necessarily positive because of the upward drift. In this case, if we define a new increment distribution by

$$dF^*(x) := e^{-\theta x} \, dF(x),$$

[a] Department of Probability and Statistics, Hicks Building, University of Sheffield, Sheffield S3 7RH; D.Grey@sheffield.ac.uk

then we obtain the *associated random walk*, which has downward drift. Because of the definition, probabilities in one random walk may easily be expressed in terms of those in the other. For instance, renewal theory in the associated random walk yields the Cramér estimate of the probability of ruin in the original random walk, the parameter θ determining the rate of exponential decay (Feller (1971), XI.7, XII.6; see also Asmussen (2000)). Note also that since $\hat{F}^*(-\theta) = \int_{-\infty}^{\infty} e^{\theta x} \, dF^*(x) = 1$ the association is a duality relationship in the sense that if we perform the analogous transformation on the associated random walk with θ replaced by $-\theta$ then we obtain the original one.

For the same value of θ, it may easily be shown that $V_n := e^{-\theta S_n}$ defines a martingale, known as the *Wald martingale*. This is also useful in the investigation of hitting probabilities.

The question arises to what extent the concepts of the associated random walk and the Wald martingale may be generalized to the case where the increments are no longer necessarily i.i.d. but merely stationary and ergodic. In such cases, because of the ergodic theorem, the random walk still drifts upwards to infinity and so we might still be interested in, for example, the probability of ruin (hitting a low level). The following is a suggested way forward. It should be noted that, once one goes beyond the ergodic theorem, the general stationary ergodic process behaves quite differently from the independent-increments case. For example, the convergence rate in the ergodic theorem may be arbitrarily slow, however many moments may be finite. For background, see for example Eberlein & Taqqu (1986).

The work below is motivated by that of Lu (1991) on branching processes in random environments (Smith & Wilkinson (1969), Athreya & Karlin (1971)) and by a convergence result given a straightforward proof by the author (Grey (2001)). We obtain generalizations of the associated random walk and the Wald martingale, under certain assumptions. Three applications are considered in Section 2, to the Markov and Gaussian cases, and to a queueing problem. Our work is also relevant to random walks in random environments; see for example Révész (1990), Part III.

To proceed, we need to make the following assumptions. In Section 2 we shall show that these assumptions are satisfied in some important cases of interest.

Assumption 1 There exists $\theta > 0$ such that

$$q := \lim_{n \to \infty} E(e^{-\theta S_n})$$

exists and is positive and finite.

This assumption is trivially satisfied in the i.i.d. case, with θ as identified earlier and $q = 1$. It is important to note that since $S_n \to \infty$ with probability one, there can be at most one value of θ satisfying the assumption. This is because, if such θ exists, for any positive constant K

$$E(e^{-\theta S_n}; S_n \le -K) \to q$$

as the remaining contribution to the expectation tends to zero, by the bounded convergence theorem. From here it is easy to see that if $0 < \phi < \theta$,

$$\limsup_{n \to \infty} E(e^{-\phi S_n}; S_n \le -K) \le e^{-(\theta - \phi)K} q$$

whence by similar reasoning

$$\limsup_{n \to \infty} E(e^{-\phi S_n}) \le e^{-(\theta - \phi)K} q$$

and so, since K is arbitrary, $E(e^{-\phi S_n}) \to 0$. Similarly if $\phi > \theta$ then $E(e^{-\phi S_n}) \to \infty$.

For our next assumption we extend our sequence of increments to a doubly infinite one $\ldots, X_{-1}, X_0, X_1, \ldots$, as is always possible with a stationary sequence (Breiman (1968), Proposition 6.5). We also define the more general partial sum

$$S_{m,n} := \sum_{r=m}^{n} X_r.$$

Let $\mathcal{F}_{m,n}$ denote the σ-field generated by $\{X_r; r = m, \ldots, n\}$.

Assumption 2 For all $k = 1, 2, \ldots$ and for all $B \in \mathcal{F}_{-k,k}$,

$$q(B) := \lim_{m,n \to \infty} E(e^{-\theta S_{-m,n}}; B)$$

exists, where θ is as defined in Assumption 1.

Again we refer immediately to the i.i.d. case, where the assumption is easily seen to be satisfied, the limiting operation being essentially trivial, and we may write explicitly

$$q(B) = E(e^{-\theta S_{-k,k}}; B).$$

If we now fix k and define for m, $n \geq k$

$$P^*_{m,n}(B) := \frac{E(e^{-\theta S_{-m,n}}; B)}{E(e^{-\theta S_{-m,n}})}$$

for $B \in \mathcal{F}_{-k,k}$, then clearly $P^*_{m,n}$ is a probability measure on $\mathcal{F}_{-k,k}$. Moreover if Assumptions 1 and 2 hold then

$$P^*_{m,n}(B) \to \frac{q(B)}{q} \quad \text{as} \quad m, n \to \infty$$

for all $B \in \mathcal{F}_{-k,k}$. We make use of the result, given a straightforward proof in Grey (2001), rather simpler than for the more general case of signed measures considered by Halmos (1950, p. 170), and a special case of the Vitali–Hahn–Saks theorem (Dunford & Schwartz (1958), III.7.2–4), that if $\{P_n\}$ is a sequence of probability measures on a space $(\Omega, \mathcal{F})$ such that the limit $P(A) := \lim_{n \to \infty} P_n(A)$ exists for all $A \in \mathcal{F}$, then P is a probability measure on $(\Omega, \mathcal{F})$. It follows that

$$P^*(B) := \frac{q(B)}{q}$$

defines a probability measure on $\mathcal{F}_{-k,k}$. Since this definition obviously does not depend upon k, we have consistency between different values of k and therefore a probability measure defined on $\bigcup_{k=1}^{\infty} \mathcal{F}_{-k,k}$. The Carathéodory extension theorem (Durrett (1996), Appendix A.2) now ensures that P^* can be extended to a probability measure defined on the whole σ-field $\mathcal{F} := \mathcal{F}_{-\infty,\infty}$.

It is the probability measure P^* which we use to define the distribution of the increments of the *associated random walk*. Note that, because the original process is stationary and because of the double-ended limiting process involved in the definition of P^*, the associated process of increments is also stationary; it would not have been possible to achieve this with a single-ended sequence. Whether the associated process is necessarily ergodic and whether duality occurs is left open here; some further remarks are given in Section 3. Note that ergodicity and duality do occur in the two special cases studied in detail in Section 2. Note also that in the i.i.d. case the associated random walk as defined here coincides with the one which we have already met, since, for example, in the discrete case

$$P^*(X_1 = x_1, \ldots, X_k = x_k) = e^{-\theta \sum_{i=1}^{k} x_i} P(X_1 = x_1, \ldots, X_k = x_k)$$

$$= \prod_{i=1}^{k} (e^{-\theta x_i} P(X_i = x_i))$$

as expected.

To construct our martingale, we need to replace Assumption 2 by the following one-sided equivalent, which again is to be read in conjunction with Assumption 1. Write $\mathcal{F}_k := \mathcal{F}_{1,k}$.

Assumption 2* For all $k = 1, 2, \ldots$ and for all $B \in \mathcal{F}_k$,

$$r(B) := \lim_{n \to \infty} E(e^{-\theta S_n}; B)$$

exists, where θ is as defined in Assumption 1.

If this assumption holds, then by the aforementioned convergence theorem, for each k, r is a measure on $\mathcal{F}_k$ with total mass q. Also it is absolutely continuous with respect to P, since

$$P(B) = 0 \implies \int_B e^{-\theta S_n} \, dP = 0 \implies r(B) = 0.$$

So r restricted to $\mathcal{F}_k$ has a Radon–Nikodým derivative V_k with respect to P:

$$\int_B V_k \, dP = r(B) \quad \text{for all} \quad B \in \mathcal{F}_k,$$

where V_k is $\mathcal{F}_k$-measurable.

But if $B \in \mathcal{F}_k$, then $B \in \mathcal{F}_{k+1}$, and therefore

$$\int_B V_k \, dP = r(B) = \int_B V_{k+1} \, dP \quad \text{for all} \quad B \in \mathcal{F}_k.$$

So, by definition of conditional expectation, $V_k = E(V_{k+1}|\mathcal{F}_k)$ almost surely, which shows that $\{V_k\}$ is a martingale with respect to $\{\mathcal{F}_k\}$.

Note In many cases it will be true that $V_k = \lim_{n \to \infty} E(e^{-\theta S_n}|\mathcal{F}_k)$ almost surely, but it seems difficult to try to use this equation as a definition of V_k in general.

In the case of i.i.d. increments it is easy to see that $V_k = e^{-\theta S_k}$ almost surely, and so our definition generalizes that of the Wald martingale.

2 Three examples

In this section we demonstrate the existence of the associated random walk in two important cases of interest: stationary Markov chain increments and stationary Gaussian increments. In both cases, as indeed in the simpler i.i.d. case, certain regularity conditions will be required. The corresponding martingale is mentioned more briefly in each case. We also consider an application in queueing theory.

2.1 Stationary Markov chain increments

Here we suppose that the increments $\{X_n\}$ perform a stationary irreducible (and therefore ergodic) aperiodic Markov chain with countable state space S. We shall use labels such as i and j to represent the actual sizes of the increments, so that they need not be integer-valued or nonnegative; however this will not prevent us from also using them to denote positions in vectors and matrices, since this non-standard notation will not lead to confusion. The associated Markov chain constructed here has been considered in a rather more general context by, for example, Arjas & Speed (1973).

Let the transition matrix of the Markov chain be $\boldsymbol{P} = (p_{ij})$ and let its equilibrium distribution be given by the column vector $\boldsymbol{\pi} = (\pi_i)$. Note that if $\mathbf{1}$ denotes the vector consisting entirely of ones, then $\boldsymbol{P}$ has Perron–Frobenius eigenvalue 1 with $\boldsymbol{\pi}^T$ and $\mathbf{1}$ as corresponding left and right eigenvectors respectively; also $\boldsymbol{\pi}^T \mathbf{1} = 1$ and $\boldsymbol{P}^n \to \mathbf{1}\boldsymbol{\pi}^T$ as $n \to \infty$.

The regularity conditions which we impose are as follows. For some $\theta > 0$ the matrix $\boldsymbol{Q}$ with elements $(p_{ij}e^{-\theta j})$ has Perron–Frobenius eigenvalue 1 and corresponding left and right eigenvectors $\boldsymbol{v}^T$ (with components (v_i)) and $\boldsymbol{c}$ (with components (c_i)) respectively; also $\boldsymbol{v}^T \boldsymbol{c} = 1$ and $\boldsymbol{Q}^n \to \boldsymbol{c}\boldsymbol{v}^T$ as $n \to \infty$. This requirement is not especially stringent when the state space S is finite; the Perron–Frobenius eigenvalue $\lambda(\theta)$ of $\boldsymbol{Q}$ for general θ behaves rather like the Laplace transform $\hat{F}(\theta)$ in the i.i.d. case (Lu (1991)).

We firstly show that Assumption 1 holds, with θ as defined above.

$$E(e^{-\theta S_n}) = \sum_{i_0 \in S} \sum_{i_1 \in S} \cdots \sum_{i_n \in S} \pi_{i_0} p_{i_0 i_1} \cdots p_{i_{n-1} i_n} e^{-\theta(i_1 + \cdots + i_n)}$$

$$= \boldsymbol{\pi}^T \boldsymbol{Q}^n \mathbf{1} \;\to\; \boldsymbol{\pi}^T \boldsymbol{c}\boldsymbol{v}^T \mathbf{1} \quad \text{as} \quad n \to \infty.$$

To check Assumption 2, we shall evaluate

$$E(e^{-\theta S_{-m,n}}; X_{-k} = i_{-k}, \ldots, X_k = i_k)$$

for given $k < m$, n and $i_{-k}, \ldots, i_k \in S$. By the Markov property, this may be written as the product of the three factors $e^{-\theta(i_{-k} + \cdots + i_k)}$ $P(X_{-k} = i_{-k}, \ldots, X_k = i_k)$ together with $E(e^{-\theta S_{k+1,n}} | X_k = i_k)$ and $E(e^{-\theta S_{-m,-k-1}} | X_{-k} = i_{-k})$. The first factor may be written

$$e^{-\theta(i_{-k} + \cdots + i_k)} \pi_{i_{-k}} p_{i_{-k} i_{-k+1}} \cdots p_{i_{k-1} i_k}.$$

The second factor may be written

$$\sum_{i_{k+1}\in S}\cdots\sum_{i_n\in S}e^{-\theta(i_{k+1}+\cdots+i_n)}p_{i_k i_{k+1}}\cdots p_{i_{n-1}i_n}.$$

Because the reverse Markov chain has transition probabilities $\pi_j p_{ji}/\pi_i$, the third factor may be written

$$\sum_{i_{-k-1}\in S}\cdots\sum_{i_{-m}\in S}e^{-\theta(i_{-k-1}+\cdots+i_{-m})}\frac{\pi_{i_{-k-1}}}{\pi_{i_{-k}}}p_{i_{-k-1}i_{-k}}\cdots\frac{\pi_{i_{-m}}}{\pi_{i_{-m+1}}}p_{i_{-m}i_{-m+1}}.$$

The second factor is seen to be the i_k component of the vector $Q^{n-k}\mathbf{1}$ and so converges to $c_{i_k}v^T\mathbf{1}$ as $n\to\infty$.

Writing $\mu_i := \pi_i e^{-\theta i}$ for each $i\in S$ and letting μ be the corresponding vector, after cancellation and rearrangement the third factor is seen to be $\mu_{i_{-k}}^{-1}$ times the i_{-k} component of the vector $\mu^T Q^{m-k}$ and so converges to $\mu_{i_{-k}}^{-1}v_{i_{-k}}\mu^T c$ as $m\to\infty$. Note that

$$\mu^T c = \sum_{j\in S}\pi_j e^{-\theta j}c_j$$

$$= \sum_{j\in S}\sum_{i\in S}\pi_i p_{ij}e^{-\theta j}c_j = \sum_{i\in S}\pi_i\sum_{j\in S}p_{ij}e^{-\theta j}c_j = \sum_{i\in S}\pi_i c_i = \pi^T c.$$

Using this fact and putting all the preceding results together, after cancellation we see that $P^*(X_{-k} = i_{-k},\ldots,X_k = i_k)$ exists and is equal to

$$e^{-\theta(i_{-k+1}+\cdots+i_k)}p_{i_{-k}i_{-k+1}}\cdots p_{i_{k-1}i_k}c_{i_k}v_{i_{-k}}.$$

Writing $p_{ij}^* := p_{ij}e^{-\theta j}c_j/c_i$ and $\pi_i^* := c_i v_i$ for each $i, j\in S$, the above may also be written

$$P^*(X_{-k} = i_{-k},\ldots,X_k = i_k) = \pi_{i_{-k}}^* p_{i_{-k}i_{-k+1}}^*\cdots p_{i_{k-1}i_k}^*.$$

It is a routine matter to check that the numbers (p_{ij}^*) form the transition probabilities of a Markov chain and that (π_i^*) is an equilibrium distribution for it. We have thus established that the associated random walk exists and its increments perform a stationary Markov chain, which is also obviously irreducible and aperiodic like the original. Duality is left as an exercise.

The martingale may be constructed similarly. Letting $B = \{X_1 = i_1,\ldots,X_k = i_k\}$ it may be calculated that

$$E(e^{-\theta S_n}; B)\to P(B)e^{-\theta(i_1+\cdots+i_k)}c_{i_k}v^T\mathbf{1}\qquad \text{as}\qquad n\to\infty$$

and so Assumption 2* holds; moreover, since B is an atom of the σ-field

$\mathcal{F}_k$ it follows that $V_k = e^{-\theta S_k} c_{X_k} v^T 1$. We may take $v^T 1 = 1$ since v and c have so far only been scaled relative to each other. This gives the martingale $V_k = c_{X_k} e^{-\theta S_k}$ which has also been used by Lu (1991).

2.2 Stationary Gaussian increments

Now let $\{X_n\}$ be a stationary Gaussian process in which each X_n has normal distribution with mean $\mu > 0$ and variance $\sigma^2 > 0$. For each $r = 1, 2, \ldots$ let ρ_r be the correlation coefficient between X_n and X_{n+r} for any n. These parameters completely determine the behaviour of the process, since the joint distribution of any finite collection of the X_n is multivariate normal.

The regularity condition we need here is that

$$\sum_{r=1}^{\infty} r|\rho_r| < \infty.$$

This is an asymptotic independence property more than sufficient for ergodicity, and one which is easily satisfied by commonly studied processes such as autoregressive and moving average processes.

Under this condition, let $R := \sum_{r=1}^{\infty} \rho_r$ and let $S := \sum_{r=1}^{\infty} r\rho_r$; these will both be finite. Below we shall see that $R \geq -\frac{1}{2}$ necessarily, and that we need to exclude the extreme case $R = -\frac{1}{2}$.

We firstly find θ such that Assumption 1 is satisfied. Since $S_n = X_1 + \cdots + X_n$ has a normal distribution with mean $n\mu$ and variance $\sigma^2[n + 2\sum_{r=1}^{n-1}(n-r)\rho_r]$, by the standard formula for the Laplace transform of the normal distribution we have that

$$E(e^{-\theta S_n}) = \exp\left\{-n\mu\theta + \tfrac{1}{2}\sigma^2[n + 2\sum_{r=1}^{n-1}(n-r)\rho_r]\theta^2\right\}.$$

Under our regularity condition

$$\sum_{r=1}^{n-1}(n-r)\rho_r = nR - S + o(1) \quad \text{as} \quad n \to \infty$$

and so in particular

$$\text{var } S_n = \sigma^2(n[1 + 2R] - 2S) + o(1) \quad \text{as} \quad n \to \infty,$$

whence $1 + 2R \geq 0$. It is possible to construct examples with $1 + 2R = 0$ (such as $X_n := \mu + Z_n - Z_{n-1}$ where $\{Z_n\}$ are i.i.d. $N(0, \tfrac{1}{2}\sigma^2)$ random variables) but if we exclude this rather extreme case then we see that convergence of $E(e^{-\theta S_n})$ to a positive limit will occur if and only if

$$-\mu\theta + \tfrac{1}{2}\sigma^2[1 + 2R]\theta^2 = 0.$$

This yields

$$\theta = \frac{2\mu}{\sigma^2[1 + 2R]}$$

and then it is easy to compute that for this value of θ

$$E(e^{-\theta S_n}) \to \exp\left\{-\frac{4\mu^2 S}{\sigma^2[1 + 2R]^2}\right\} \quad \text{as} \quad n \to \infty.$$

We turn to Assumption 2. Fix k, and take θ as just identified. For $m, n > k$ it is evidently relevant to look at the distribution of $Y :=$ $S_{-m,-k-1} + S_{k+1,n}$ conditional on $X_{-k} = x_{-k}, \ldots, X_k = x_k$, or $\boldsymbol{X} = \boldsymbol{x}$ say. If the unconditional distribution of Y is denoted by $N(\nu, \tau^2)$, the vector of covariances of Y with the components of $\boldsymbol{X}$ is denoted by $\boldsymbol{v}$, and the mean and covariance matrix of $\boldsymbol{X}$ are denoted by $\boldsymbol{\mu}$ and $\boldsymbol{\Sigma}$ respectively, then by multivariate normal theory (Mardia, Kent & Bibby (1979), Theorem 3.2.4), the conditional distribution is $N(\nu + \boldsymbol{v}^T\boldsymbol{\Sigma}^{-1}(\boldsymbol{x} - \boldsymbol{\mu}), \tau^2 - \boldsymbol{v}^T\boldsymbol{\Sigma}^{-1}\boldsymbol{v})$. (For ease of notation, we suppress the dependence of Y, ν, τ^2 and $\boldsymbol{v}$ on m and n.)

A typical component of $\boldsymbol{v}$ is of the form

$$\sigma^2\left(\sum_{r=i+k+1}^{i+m} \rho_r + \sum_{r=k+1-i}^{n-i} \rho_r\right)$$

for some i, and so $\boldsymbol{v}$ converges to a finite limit as $m, n \to \infty$. Also $\nu = (m + n - 2k)\mu$. Then, since

$$\text{var } S_{-m,n} = \text{var } (\mathbf{1}^T\boldsymbol{X} + Y) = \mathbf{1}^T\boldsymbol{\Sigma}\mathbf{1} + 2\mathbf{1}^T\boldsymbol{v} + \tau^2,$$

we can use the estimate of $\text{var } S_n$ obtained in checking Assumption 1 to deduce that

$$\tau^2 - \sigma^2[1 + 2R](m + n + 1)$$

converges to a finite limit as $m, n \to \infty$. Putting these results together we see that

$$E(e^{-\theta Y}|\boldsymbol{X} = \boldsymbol{x}) = \exp\left\{-[\nu + \boldsymbol{v}^T\boldsymbol{\Sigma}^{-1}(\boldsymbol{x} - \boldsymbol{\mu})]\theta + \tfrac{1}{2}[\tau^2 - \boldsymbol{v}^T\boldsymbol{\Sigma}^{-1}\boldsymbol{v}]\theta^2\right\}$$

converges to a positive limit as $m, n \to \infty$, since because of the value of θ the difference between the large terms in ν and τ^2 converges to a finite limit, and the other terms also converge to finite limits. We may denote the limit in the above by $\exp(\boldsymbol{\alpha}^T \boldsymbol{x} + \beta)$ for some constants $\boldsymbol{\alpha}$ and β. It is then easy to see that for any $B \in \mathcal{F}_{-k,k}$,

$$E(e^{-\theta S_{-m,n}}; B) \to E(\exp\{-\theta \mathbf{1}^T \boldsymbol{X} + \boldsymbol{\alpha}^T \boldsymbol{X} + \beta\}; B) \quad \text{as} \quad m, n \to \infty,$$

and we have established Assumption 2. It is not hard in this case to see that the associated random walk has the same covariance structure as the original, but downward drift $-\mu$.

Calculations similar to the above may be used to check that Assumption 2^* holds in this case also, and that the associated martingale is of the form $V_k = \exp(\boldsymbol{\gamma}^T \boldsymbol{X} + \delta)$ for some $\boldsymbol{\gamma}$, δ, where now $\boldsymbol{X} = (X_1, \ldots, X_k)^T$.

2.3 A queueing application

Suppose we have a $G/GI/1$ queue in which the inter-arrival times $\{T_n\}$ form a stationary ergodic sequence and the independent service times $\{U_n\}$ are i.i.d. with $ET_n > EU_n > 0$. Then the waiting times $\{W_n\}$ satisfy

$$W_{n+1} = (W_n + U_n - T_n)^+$$

and it follows by a standard argument, dating back to Lindley (1952) in the case of i.i.d. inter-arrival times and exploited, among others, by Kingman (1964), that W_n has an equilibrium distribution which is the same as the distribution of minus the all-time minimum of an unrestricted random walk started at zero in state zero, with increments $X_n := T_{-n} - U_{-n}$. The tail of this distribution is therefore intimately related to the probability of ruin in this random walk, and, in particular, the parameter θ, if it exists, has an important part to play.

The simplest example is the $M/M/1$ queue where the T_n are independent exponential with parameter λ and the U_n are exponential with parameter μ, where $\mu > \lambda > 0$. In this case the X_n are i.i.d. and $E(e^{-\theta X_n}) = \lambda\mu/\{(\lambda + \theta)(\mu - \theta)\}$, which is easily seen to be equal to one when $\theta = \mu - \lambda$. So the key parameter θ depends upon both the arrival rate and the service rate in a simple and obvious way.

As another example, suppose that the U_n have some arbitrary distribution with Laplace transform $\phi(\theta) = E(e^{-\theta U_n})$, and that there is a regular appointments system such that customer n arrives at clock time $\lambda^{-1}n + \epsilon_n$, where $\{\epsilon_n\}$ is a sequence of i.i.d. errors with Laplace transform $\psi(\theta) = E(e^{-\theta \epsilon_n})$. In this case it is possible to compute

$$E(e^{-\theta S_n}) = [\phi(-\theta)]^n \exp(-\lambda^{-1}n\theta)\psi(\theta)\psi(-\theta)$$

and so the key parameter θ satisfies the equation

$$\phi(-\theta)\exp(-\lambda^{-1}\theta) = 1$$

which does not involve the distribution of the ϵ_n. By considering the special case $\phi(\theta) = \mu/(\mu + \theta)$, so that the U_n are exponentially distributed, and the mean inter-arrival time and mean service time are λ^{-1} and μ^{-1} respectively as in the previous $M/M/1$ example, it is possible to compare an appointments system with random (Poisson) arrivals. For the value $\theta = \mu - \lambda$ found in the case of random arrivals, we may compute $\phi(-\theta)\exp(-\lambda^{-1}\theta) = (\mu/\lambda)\exp(1 - (\mu/\lambda)) < 1$, and so the actual value of the key parameter θ for the appointments system must be larger than for random arrivals. This suggests a thinner tail for the equilibrium waiting time distribution, and a more efficient system.

3 Some remarks on duality and asymptotic independence

If we wish for duality to occur, then, replacing θ by $-\theta$ and denoting expectation with respect to P^* by E^*, we require for $B \in \mathcal{F}_{-k,k}$ that

$$\frac{E^*(e^{\theta S_{-m,n}}; B)}{E^*(e^{\theta S_{-m,n}})} \to P(B) \quad \text{as} \quad m, n \to \infty.$$

Now, denoting $E(e^{-\theta S_{-r,s}})$ by $q_{r,s}$ and noting that $P^*_{r,s}$ has Radon–Nikodým derivative $q_{r,s}^{-1}e^{-\theta S_{-r,s}}$ with respect to P, we have that

$$E^*(e^{\theta S_{-m,n}}; B) = \int_B e^{\theta S_{-m,n}}\, dP^*$$

$$= \lim_{r,s \to \infty} \int_B e^{\theta S_{-m,n}}\, dP^*_{r,s}$$

$$= \lim_{r,s \to \infty} \int_B e^{\theta S_{-m,n}} q_{r,s}^{-1} e^{-\theta S_{-r,s}}\, dP$$

$$= q^{-1} \lim_{r,s \to \infty} \int_B e^{-\theta S_{-r,-m-1}} e^{-\theta S_{n+1,s}}\, dP.$$

For the required convergence to occur, it seems therefore that for large m and n there must be approximate independence between $S_{-r,-m-1}$, $S_{n+1,s}$ and the event B, so that we can say that the above integral is approximately $P(B)E(e^{-\theta S_{-r,-m-1}})E(e^{-\theta S_{n+1,s}})$ which converges to

$P(B)q^2$ as r, $s \to \infty$. Hence under these circumstances

$$\frac{E^*(e^{\theta S_{-m,n}}; B)}{E^*(e^{\theta S_{-m,n}})} \sim \frac{q^{-1}P(B)q^2}{q^{-1}q^2} \to P(B) \quad \text{as} \quad m, n \to \infty.$$

The asymptotic independence is a kind of mixing condition which is already stronger than ergodicity, suggesting that the latter is not the most appropriate property to be considering in this context. See Bradley (2005) on mixing conditions.

Acknowledgements The author would like to thank Nick Bingham and an anonymous referee for helpful remarks on the presentation of this paper.

References

Arjas, E., and Speed, T. P. 1973. An extension of Cramér's estimate for the absorption probability of a random walk. *Proc. Cambridge Philos. Soc.*, **73**, 355–359.

Asmussen, S. 2000. *Ruin Probabilities*. River Edge, NJ: World Scientific.

Athreya, K. B., and Karlin, S. 1971. On branching processes with random environments, I, II. *Ann. Math. Statist.*, **42**, 1499–1520, 1843–1858.

Bradley, R. C. 2005. *Introduction to Strong Mixing Conditions*, vols 1–3. Bloomington, IN: Custom Publishing, Indiana University.

Breiman, L. 1968. *Probability*. Reading, MA: Addison–Wesley.

Dunford, N., and Schwartz, J. T. 1958. *Linear Operators, Part I: General Theory*. New York: Interscience. [Reprinted 1988, Wiley Classics Library. Wiley–Interscience, John Wiley & Sons.]

Durrett, R. 1996. *Probability: Theory and Examples*, 2nd edn. Belmont, CA: Duxbury Press.

Eberlein, E., and Taqqu, M. (eds). 1986. *Dependence in Probability and Statistics*. Boston, MA: Birkhäuser.

Feller, W. 1971. *An Introduction to Probability Theory and its Applications*, vol. II, 2nd edn. New York: John Wiley & Sons.

Grey, D. R. 2001. A note on convergence of probability measures. *J. Appl. Probab.*, **38**, 1055–1058.

Halmos, P. R. 1950. *Measure Theory*. Princeton, NJ: Van Nostrand.

Kingman, J. F. C. 1964. A martingale inequality in the theory of queues. *Proc. Cambridge Philos. Soc.*, **60**, 359–361.

Lindley, D. V. 1952. The theory of queues with a single server. *Proc. Cambridge Philos. Soc.*, **48**, 277–289.

Lu, Z. 1991. *Survival of Reproducing Populations in Random Environments*. PhD thesis, University of Sheffield.

Mardia, K. V., Kent, J. T., and Bibby, J. M. 1979. *Multivariate Analysis.* New York: Academic Press.

Révész, P. 1990. *Random Walk in Random and Non-Random Environments.* Hackensack, NJ: World Scientific.

Smith, W. L., and Wilkinson, W. E. 1969. On branching processes in random environments. *Ann. Math. Statist.*, **40**, 814–827.

15

Diffusion processes and coalescent trees

Robert C. Griffiths[a] and Dario Spanó[b]

Abstract

We dedicate this paper to Sir John Kingman on his 70th Birthday.

In modern mathematical population genetics the ancestral history of a population of genes back in time is described by John Kingman's coalescent tree. Classical and modern approaches model gene frequencies by diffusion processes. This paper, which is partly a review, discusses how coalescent processes are dual to diffusion processes in an analytic and probabilistic sense.

Bochner (1954) and Gasper (1972) were interested in characterizations of processes with Beta stationary distributions and Jacobi polynomial eigenfunctions. We discuss the connection with Wright–Fisher diffusions and the characterization of these processes. Subordinated Wright–Fisher diffusions are of this type. An Inverse Gaussian subordinator is interesting and important in subordinated Wright–Fisher diffusions and is related to the Jacobi Poisson Kernel in orthogonal polynomial theory. A related time-subordinated forest of non-mutant edges in the Kingman coalescent is novel.

AMS subject classification (MSC2010) 92D25, 60J70, 92D15

[a] Department of Statistics, University of Oxford, 1 South Parks Rd, Oxford OX1 3TG; griff@stats.ox.ac.uk
[b] Department of Statistics, University of Warwick, Coventry CV4 7AL; D.Spano@warwick.ac.uk

1 Introduction

The Wright–Fisher diffusion process $\{X(t), t \geq 0\}$ models the relative frequency of type a genes in a population with two types of genes a and A. Genes are subject to random drift and mutation over time. The generator of the process is

$$\mathcal{L} = \frac{1}{2}x(1-x)\frac{\partial^2}{\partial x^2} + \frac{1}{2}\left(-\alpha x + \beta(1-x)\right)\frac{\partial}{\partial x}, \qquad (1.1)$$

where the mutation rate $A \to a$ is $\frac{1}{2}\alpha$ and the rate $a \to A$ is $\frac{1}{2}\beta$. If α and β are zero then zero and one are absorbing states where either A or a becomes fixed in the population. If α, $\beta > 0$ then $\{X(t), t \geq 0\}$ is a reversible process with a Beta stationary density

$$f_{\alpha,\beta}(y) = B(\alpha,\beta)^{-1}y^{\alpha-1}(1-y)^{\beta-1}, \ 0 < y < 1. \qquad (1.2)$$

The transition density has an eigenfunction expansion

$$f(x,y;t) = f_{\alpha,\beta}(y)\left\{1 + \sum_{n=1}^{\infty} \rho_n^\theta(t)\widetilde{P}_n^{(\alpha,\beta)}(x)\widetilde{P}_n^{(\alpha,\beta)}(y)\right\}, \qquad (1.3)$$

where $\theta = \alpha + \beta$,

$$\rho_n^\theta(t) = \exp\left\{-\frac{1}{2}n(n+\theta-1)t\right\}, \qquad (1.4)$$

and $\{\widetilde{P}_n^{(\alpha,\beta)}(y), n \in \mathbb{Z}_+\}$ are orthonormal Jacobi polynomials on the Beta (α,β) distribution, scaled so that

$$\mathbb{E}\left[\widetilde{P}_m^{(\alpha,\beta)}(Y)\widetilde{P}_n^{(\alpha,\beta)}(Y)\right] = \delta_{mn}, \ m,n \in \mathbb{Z}_+$$

under the stationary distribution (1.2). The Wright–Fisher diffusion is also known as the Jacobi diffusion because of the eigenfunction expansion (1.3). The classical Jacobi polynomials, orthogonal on

$$(1-x)^\alpha(1+x)^\beta, \ -1 < x < 1,$$

can be expressed as

$$P_n^{(\alpha,\beta)}(x) = \frac{(\alpha+1)_{(n)}}{n!}\,{}_2F_1(-n, n+\alpha+\beta+1; \alpha+1; (1-x)/2), \quad (1.5)$$

where ${}_2F_1$ is a hypergeometric function. The relationship between the two sets of polynomials is that

$$\widetilde{P}_n^{(\alpha,\beta)}(x) = c_n P_n^{(\beta-1,\alpha-1)}(2x-1),$$

where

$$c_n = \sqrt{\frac{(2n + \alpha + \beta - 1)(\alpha + \beta)_{(n-1)}n!}{\alpha_{(n)}\beta_{(n)}}}.$$

Define

$$\bar{\mathcal{L}} = \frac{1}{2}\frac{\partial^2}{\partial x^2}x(1 - x) - \frac{\partial}{\partial x}\frac{1}{2}(-\alpha x + \beta(1 - x)), \qquad (1.6)$$

the forward generator of the process. The Jacobi polynomials are eigenfunctions satisfying, for $n \in \mathbb{Z}_+$,

$$\mathcal{L}\widetilde{P}_n^{(\alpha,\beta)}(x) = -\frac{1}{2}n(n + \theta - 1)\widetilde{P}_n^{(\alpha,\beta)}(x);$$

$$\bar{\mathcal{L}}f_{\alpha,\beta}(x)\widetilde{P}_n^{(\alpha,\beta)}(x) = -\frac{1}{2}n(n + \theta - 1)f_{\alpha,\beta}(x)\widetilde{P}_n^{(\alpha,\beta)}(x). \qquad (1.7)$$

The well known fact that the Jacobi polynomials $\{\widetilde{P}_n^{(\alpha,\beta)}(x)\}$ satisfy (1.7) implies that they are eigenfunctions with corresponding eigenvalues $\{\rho_n^\theta(t)\}$.

In modern mathematical population genetics the ancestral history of a population back in time is described by John Kingman's elegant coalescent process [19]. The connection between the coalescent and Fleming–Viot diffusion processes is made explicit by Donnelly and Kurtz in [7], [8] by their look-down process. An approach by Ethier and Griffiths [10] uses duality to show that a 'non-mutant lines of descent' process which considers a forest of trees back in time to their first mutations is dual to the Fleming–Viot infinitely-many-alleles diffusion process. The two-allele process $\{X(t), t \geq 0\}$ is recovered from the Fleming–Viot process by a 2-colouring of alleles in the infinitely-many-alleles model. If there is no mutation then the dual process is the same as the Kingman coalescent process with an entrance boundary at infinity. The dual process approach leads to a transition density expansion in terms of the transition functions of the process which counts the number of non-mutant lineages back in time. It is interesting to make a connection between the eigenfunction expansion (1.3) and dual process expansion of the transition densities of $\{X(t), t \geq 0\}$. Bochner [6] and Gasper [13] find characterizations of processes which have Beta stationary distributions and Jacobi polynomial eigenfunctions. Subordinated Jacobi processes $\{X(Z(t)), t \geq 0\}$, where $\{Z(t), t \geq 0\}$ is a Lévy process, fit into this class, because subordination does not change the eigenvectors or the stationary distribution of the process. The subordinated processes are jump diffusions. A particular class of importance is when $\{Z(t), t \geq 0\}$

is an Inverse Gaussian process. Griffiths [18] obtains characterizations of processes with stationary distributions in the Meixner class, as well as for Jacobi processes. The current paper is partly a review paper describing connections between Jacobi diffusions, eigenfunction expansions of transition functions, coalescent trees, and Bochner characterizations. Novel results describe the subordinated non-mutant lines-of-descent process when the subordination is with an Inverse Gaussian process.

2 A coalescent dual process

A second form of the transition density (1.3) derived in Ethier and Griffiths [10] is

$$f(x, y; t) = \sum_{k=0}^{\infty} q_k^\theta(t) \sum_{l=0}^{k} \mathcal{B}(l; k, x) f_{\alpha+l, \beta+k-l}(y), \quad (2.1)$$

where

$$\mathcal{B}(l; k, x) = \binom{l}{k} x^k (1 - x)^{l-k}, \ k = 0, 1, \ldots, l$$

is the Binomial distribution and $\{q_k^\theta(t)\}$ are the transition functions of a death process with an entrance boundary of infinity, and death rates $k(k + \theta - 1)/2$, $k \geq 1$. The death process represents the number of non-mutant ancestral lineages back in time in the coalescent process with mutation. The number of lineages decreases from k to $k - 1$ from coalescence at rate $\binom{k}{2}$ or mutation at rate $k\theta/2$. If there is no mutation, $\{q_k^0(t), t \geq 0\}$ are transition functions of the number of edges in a Kingman coalescent tree. There is an explicit expression for the transition functions beginning with the entrance boundary of infinity [16, 21, 17] of

$$q_k^\theta(t) = \sum_{j=k}^{\infty} \rho_j^\theta(t)(-1)^{j-k} \frac{(2j + \theta - 1)(k + \theta)_{(j-1)}}{k!(j - k)!}, \quad (2.2)$$

recalling that $\rho_n^\theta(t)$ is defined by (1.4). A complex-variable representation of (2.2) is found in [17]. Let $\{X_t, t \geq 0\}$ be standard Brownian motion so X_t is $N(0, t)$. Denote $Z_t = e^{iX_t}$ and $\omega_t = e^{-\frac{1}{2}\theta t}$, then

$$q_k^\theta(t) = e^{\frac{1}{8}t} \frac{\Gamma(2k + \theta)}{\Gamma(k + \theta)k!} \mathbb{E}\left[\frac{(\omega_t Z_t)^k (1 - \omega_t Z_t)}{\sqrt{Z_t}(1 + \omega_t Z_t)^{2k+\theta}} \right], \quad (2.3)$$

for $k = 0, 1, \ldots$. The transition functions for the process beginning at n, rather than infinity, are

$$q_{nk}^\theta(t) = \sum_{j=k}^{n} \rho_j^\theta(t)(-1)^{j-k} \frac{(2j + \theta - 1)(k + \theta)_{(j-1)} n_{[j]}}{k!(j-k)!(n+\theta)_{(j)}}, \quad (2.4)$$

for $k = 0, 1, \ldots, n$. An analogous complex-variable representation to (2.3) is

$$q_{nk}^\theta(t) = \frac{\Gamma(n+\theta)\Gamma(2k+\theta)}{\Gamma(k+\theta)\Gamma(n+k+\theta)} \binom{n}{k} e^{\frac{1}{8}(\theta-1)^2 t} \mathbb{E}\big[Z_t^{k+(\theta-1)/2}(1 - Z_t)$$
$$\times\, {}_2F_1(-n + k + 1, \theta + 2k; n + k + \theta; Z_t)\big] \quad (2.5)$$

for $k = 0, 1, \ldots, n$. The expansion (2.1) is derived from a two-dimensional dual death process $\{L^\theta(t) \in \mathbb{Z}_+^2, t \geq 0\}$ which looks back in time in the diffusion process $\{X(t), t \geq 0\}$. A derivation in this paper is from [9], which follows more general analytic derivations in [10] for a Fleming–Viot model and [3] for a diffusion model with selection. Etheridge and Griffiths [9] give a very clear probabilistic derivation in a Moran model with selection that provides an understanding of earlier derivations. A sketch of a derivation of (2.1) from [9] is the following. Let $x_1 = x$, $x_2 = 1 - x$ and define for $k \in \mathbb{Z}_+^2$

$$g_k(x) = \frac{\theta_{(|k|)}}{\alpha_{(k_1)}\beta_{(k_2)}} x_1^{k_1} x_2^{k_2},$$

then

$$\mathcal{L}g_k(x) = \frac{1}{2}(|k| + \theta - 1)\big[k_1 g_{k-e_1}(x) + k_2 g_{k-e_2}(x) - |k| g_k(x)\big]. \quad (2.6)$$

Here and elsewhere we use the notation $|y| = \sum_{j=1}^{d} y_j$ for a d-dimensional vector y. In this particular case $|k| = k_1 + k_2$. To obtain a dual process the generator is regarded as acting on $k = (k_1, k_2)$, rather than x. The dual process is a two-dimensional death process $\{L^\theta(t), t \geq 0\}$, the rates of which are read off from the coefficients of the functions g on the right-hand side of (2.6);

$$k \to k - e_i \quad \text{at rate} \quad \frac{1}{2}\frac{k_i}{|k|} \cdot |k|(|k| + \theta - 1). \quad (2.7)$$

The total size, $|L^\theta(t)|$, is a 1-dimensional death process in which

$$|k| \to |k| - 1 \quad \text{at rate} \quad \frac{1}{2}|k|(|k| + \theta - 1)$$

with transition functions denoted by $\{q^\theta_{ml}(t), t \geq 0\}$. There is hypergeometric sampling of types which do not die, so

$$P(L(t) = l \mid L(0) = m) = q^\theta_{ml}(t) = q^\theta_{|m||l|}(t) \frac{\binom{m_1}{l_1}\binom{m_2}{l_2}}{\binom{|m|}{|l|}}, \tag{2.8}$$

where $q^\theta_{|m||l|}(t)$ is defined in (2.4). The dual equation obtained by regarding $\mathcal{L}$ as acting on x or k in (2.6) is

$$\mathbb{E}_{X(0)}\Big[g_{L(0)}\big(X(t)\big)\Big] = \mathbb{E}_{L(0)}\Big[g_{L(t)}\big(X(0)\big)\Big], \tag{2.9}$$

where expectation on the left is with respect to the distribution of $X(t)$, and on the right with respect to the distribution of $L(t)$. Partitioning the expectation on the right of (2.9) by values taken by $L(t)$,

$$\mathbb{E}_x\left[\binom{|m|}{m_1} X_1(t)^{m_1} X_2(t)^{m_2}\right] \tag{2.10}$$

$$= \binom{|m|}{m_1} \frac{\alpha_{(m_1)}\beta_{(m_2)}}{\theta_{(m_1+m_2)}} \sum_{l \leq m} x_1^{l_1} x_2^{l_2} \frac{\theta_{(|l|)}}{\alpha_{(l_1)}\beta_{(l_2)}} q^\theta_{|m||l|}(t) \binom{|l|}{l_1} \frac{m_{1[l_1]} m_{2[l_2]}}{|m|_{[|l|]}}.$$

The transition distribution of $X(t)$ now has an expansion derived from an inversion formula applied to (2.10). Letting $m_1, m_2 \to \infty$ with $m_1/|m| \to y_1$, $m_2/|m| \to y_2$ gives

$$f(x, y; t) = \sum_{l \in \mathbb{Z}_+^2} q^\theta_{|l|}(t) \binom{|l|}{l_1} x_1^{l_1} x_2^{l_2} B(\alpha + l_1, \beta + l_2)^{-1} y_1^{l_1+\alpha-1} y_2^{l_2+\beta-1},$$

which is identical to (2.1).

The two-allele Wright–Fisher diffusion is a special case of a much more general Fleming–Viot measure-valued diffusion process which has $\mathcal{P}(S)$, the probability measures on S, a compact metric space, as a state space. The mutation operator in the process is

$$(Af)(x) = \frac{\theta}{2} \int_S \big(f(\xi) - f(x)\big)\nu_0(d\xi),$$

where $\nu_0 \in \mathcal{P}(S)$ and $f : S \to \mathbb{R}$. The stationary measure is a Poisson–Dirichlet (Ferguson–Dirichlet) random measure

$$\mu = \sum_{i=1}^\infty x_i \delta_{\xi_i},$$

where $\{x_i\}$ is a Poisson–Dirichlet point process, $\mathcal{PD}(\theta)$, independent of $\{\xi_j\}$ which are *i.i.d.* $\nu_0 \in \mathcal{P}(S)$. A description of the $\mathcal{PD}(\theta)$ distribution is contained in Kingman [20].

Denote the stationary distribution of the random measure as

$$\Pi_{\theta,\nu_0}(\cdot) = \mathbb{P}(\mu \in \cdot).$$

Ethier and Griffiths [10] derive a transition function expansion for $P(t, \mu, d\nu)$ with given initial $\mu \in \mathcal{P}(S)$ of

$$\begin{aligned}
\mathbb{P}(t, \mu, .) = {} & q_0^\theta(t)\Pi_{\theta,\nu_0}(\cdot) \\
& + \sum_{n=1}^{\infty} q_n^\theta(t) \int_{S^n} \mu^n(dx_1 \times \cdots \times dx_n) \\
& \times \Pi_{n+\theta,(n+\theta)^{-1}\{n\eta_n(x_1,\ldots,x_n)+\theta\nu_0\}}(\cdot), \quad (2.11)
\end{aligned}$$

where $\eta_n(x_1, \ldots, x_n)$ is the empirical measure of points $x_1, \ldots, x_n \in S$:

$$\eta_n(x_1, \ldots, x_n) = n^{-1}(\delta_{x_1} + \cdots + \delta_{x_n}).$$

There is the famous Kingman coalescent process tree [19] behind the pretty representation (2.11). The coalescent tree has an entrance boundary at infinity and a coalescence rate of $\binom{k}{2}$ while there are k edges in the tree. Mutations occur according to a Poisson process of rate $\theta/2$ along the edges of the coalescent tree. $\{q_n^\theta(t)\}$ is the distribution of the number of non-mutant edges in the tree at time t back. The number of non-mutant edges is the same as the number of edges in a forest where coalescence occurs to non-mutant edges and trees are rooted back in time when mutations occur on an edge. If the time origin is at time t back and there are n non-mutant edges at the origin then the leaves of the infinite-leaf tree represent the population at t forward in time divided into relative frequencies of families of types which are either the n non-mutant types chosen at random from time zero, or mutant types chosen from ν_0 in $(0,t)$. The frequencies of non-mutant families, scaled to have a total frequency unity, have a Dirichlet distribution with unit index parameters, and the new mutation families, scaled to have total frequency unity, are distributed according to a Poisson–Dirichlet random measure with rate θ and type measure ν_0. The total frequency of new mutations has a Beta $(\theta, n-1)$ distribution. An extended description of the tree process is in Griffiths [17].

A d-dimensional reversible diffusion process model for gene frequencies which arises as a limit from the Wright–Fisher model has a backward generator

$$\mathcal{L} = \frac{1}{2}\sum_{i=1}^{d}\sum_{j=1}^{d} x_i(\delta_{ij} - x_j)\frac{\partial^2}{\partial x_i \partial x_j} + \frac{1}{2}\sum_{i=1}^{d}(\epsilon_i - \theta x_i)\frac{\partial}{\partial x_i}, \quad (2.12)$$

where $\theta = |\epsilon|$. In this model mutation is parent-independent from type $i \to j$ at rate $\frac{1}{2}\epsilon_j$, $i, j = 1, \ldots, d$. Assuming that $\epsilon > 0$, the stationary density is the Dirichlet density

$$\frac{\Gamma(\theta)}{\Gamma(\epsilon_1)\cdots\Gamma(\epsilon_d)}x_1^{\epsilon_1-1}\cdots x_d^{\epsilon_d-1}, \tag{2.13}$$

for $x_1, \ldots, x_d > 0$ and $\sum_1^d x_i = 1$. Griffiths [15] shows that the transition density in the model has eigenvalues

$$\rho_{|n|}(t) = e^{-\frac{1}{2}|n|(|n|+\theta-1)t}$$

repeated

$$\binom{|n| + d - 2}{|n|}$$

times corresponding to eigenvectors $\{Q_n^\circ(x), n \in \mathbb{Z}_+^{d-1}\}$ which are multitype orthonormal polynomials of total degree $|n|$ in x. As eigenfunctions the polynomials satisfy

$$\mathcal{L}Q_n^\circ(x) = -\frac{1}{2}|n|(|n| + \theta - 1)Q_n^\circ(x). \tag{2.14}$$

The eigenvalues $\{\rho_k(t), k \in \mathbb{Z}_+\}$ do not depend on the dimension d. The transition density with $X(0) = x$, $X(t) = y$ has the form

$$f(x, y, t) = \mathcal{D}(y, \epsilon)\left\{1 + \sum_{|n|=1}^{\infty} \rho_{|n|}(t)Q_{|n|}(x, y)\right\}. \tag{2.15}$$

The kernel polynomials on the Dirichlet $\{Q_{|n|}(x, y)\}$ appearing in (2.15) are defined as

$$Q_{|n|}(x, y) = \sum_{\{n:|n| \text{ fixed}\}} Q_n^\circ(x)Q_n^\circ(y) \tag{2.16}$$

for *any* complete orthonormal polynomial set $\{Q_n^\circ(x)\}$ on the Dirichlet distribution (2.13). If $d = 2$,

$$Q_{|n|}(x, y) = \widetilde{P}_{|n|}^{(\epsilon_1,\epsilon_2)}(x)\widetilde{P}_{|n|}^{(\epsilon_1,\epsilon_2)}(y)$$

where $\{\widetilde{P}_{|n|}^{(\epsilon_1,\epsilon_2)}(x)\}$ are orthonormal Jacobi polynomials on the Beta distribution on $[0, 1]$. In general n is just a convenient index system for the polynomials since the number of polynomials of total degree $|n|$ is always the same as the number of solutions of $n_1 + \cdots + n_{d-1} = |n|$,

$$\binom{|n| + d - 2}{|n|}.$$

$Q_{|n|}(x, y)$ is invariant under the choice of which orthonormal polynomial set is used. The individual polynomials $Q_n^\circ(x)$ are uniquely determined by their leading coefficients of degree $|n|$ and $Q_{|n|}(x, y)$. A specific form is

$$Q_{|n|}(x, y) = (\theta + 2|n| - 1) \sum_{m=0}^{|n|} (-1)^{|n|-m} \frac{(\theta + m)_{(|n|-1)}}{m!(|n| - m)!} \xi_m, \qquad (2.17)$$

where

$$\xi_m = \sum_{|l|=m} \binom{m}{l} \frac{\theta_{(m)}}{\prod_1^d \epsilon_{i(l_i)}} \prod_1^d (x_i y_i)^{l_i}. \qquad (2.18)$$

An inverse relationship is

$$\xi_m = 1 + \sum_{|n|=1}^{m} \frac{m_{[|n|]}}{(\theta + m)_{(|n|)}} Q_{|n|}(x, y). \qquad (2.19)$$

The transition distribution (2.15) is still valid if any or all elements of ϵ are zero. The constant term in the expansion then vanishes as the diffusion process is transient and there is not a stationary distribution. For example, if $\epsilon = 0$,

$$f(x, y, t) = \prod_{j=1}^{d} y_j^{-1} \left\{ \sum_{|n|\geq d}^{\infty} \rho_{|n|}(t) Q_{|n|}^0(x, y) \right\}, \qquad (2.20)$$

where

$$Q_{|n|}^0(x, y) = (2|n| - 1) \sum_{m=1}^{n} (-1)^{|n|-m} \frac{(m)_{(|n|-1)}}{m!(|n| - m)!} \xi_m^0, \qquad (2.21)$$

with

$$\xi_m^0 = \sum_{\{l:l>0,|l|=m\}} \binom{m}{l} \frac{(m - 1)!}{\prod_1^d (l_i - 1)!} \prod_1^d (x_i y_i)^{l_i}. \qquad (2.22)$$

The derivation of (2.15) is a very classical approach. The same process can be thought of as arising from an infinite-leaf coalescent tree similar to the description in the Fleming–Viot infinitely-many-alleles process. The coalescent rate while there are k edges in the tree is $\binom{k}{2}$ and mutations occur along edges at rate $\theta/2$. In this model there are d types, $1, 2, \ldots, d$ and the probability of mutation $i \to j$, given a mutation, is ϵ_j/θ. This is equivalent to a d-colouring of alleles in the Fleming–Viot infinitely-many-alleles model. Think backwards from time t back to time 0. Let $y = (y_1, \ldots, y_d)$ be the relative frequencies of types in the infinite number of leaves at the current time t forward and $x = (x_1, \ldots, x_d)$ be the frequencies in the population at time 0. Let l be the number of non-mutant edges at time 0 which have families at time t in the leaves of the tree. Given these l edges let $U = (U_1, \ldots, U_l)$ be their relative family sizes in the leaves, and $V = (V_1, \ldots, V_d)$ be the frequencies of families derived from new mutations on the tree edges in $(0, t)$. The distribution of $U \oplus V = (U_1, \ldots, U_l, V_1, \ldots, V_d)$ is $\mathcal{D}(u \oplus v, (1, \ldots, 1) \oplus \epsilon)$. The type of the l lines, and therefore their families, is chosen at random from the frequencies x. The distribution of the number of non-mutant lines at time 0 from the population at t is $q_l^\theta(t)$. The transition density in the diffusion (2.15) is identical to the mixture distribution arising from the coalescent

$$f(x, y, t) = \sum_{|l|=0}^{\infty} q_{|l|}^\theta(t) \sum_{\{l : |l| \text{ fixed}\}} \mathcal{M}(l, x) \mathcal{D}(y, \epsilon + l), \qquad (2.23)$$

by considering types of non-mutant lines, and adding Dirichlet variables and parameters according to l_i non-mutant families being of type i. $\mathcal{M}(l, x)$ is the multinomial distribution describing the choice of the initial line types from the population at time 0. The expansion when $d = 2$ corresponds to (1.3). The argument is valid if any elements of ϵ are zero, considering a generalized Dirichlet distribution $\mathcal{D}(x, \epsilon)$ where if $\epsilon_i = 0$, then $X_i = 0$ with probability 1.

The algebraic identity of (2.23) and (2.15) is easy to see by expressing $Q_{|n|}(x, y)$ in terms of $\{\xi_m\}$, then collecting coefficients of $\xi_{|l|}$ in (2.15) to obtain (2.23). Setting $\rho_0(t) = 1$ and $Q_0(x, y) = 1$, the transition density is

$$f(x, y, t) = \mathcal{D}(y, \epsilon) \sum_{|n|=0}^{\infty} \rho_{|n|}(t) Q_{|n|}(x, y)$$

$$= \sum_{l \in \mathbb{Z}_+^d} \left[\sum_{|n|=|l|}^{\infty} \rho_{|n|}(t)(\theta + 2|n| - 1)(-1)^{|n|-|l|} \frac{(\theta + |l|)_{(|n|-1)}}{|l|!(|n| - |l|)!} \right]$$

$$\times \mathcal{D}(y, \epsilon) \xi_l(x, y)$$

$$= \sum_{|l|=0}^{\infty} q_{|l|}^\theta(t) \sum_{\{l : |l| \text{ fixed}\}} \mathcal{M}(l, x) \mathcal{D}(y, \epsilon + l). \qquad (2.24)$$

The non-mutant line-of-descent process with transition probabilities $\{q_n^\theta(t)\}$ appears in all the Wright–Fisher diffusion processes mentioned in this section as a fundamental dual process. The process does not depend on the dimension of the diffusion, partly because the d-dimensional process can be recovered from the measure-valued process as a special case by colouring new mutations into d classes with probabilities $(\epsilon_1/\theta, \epsilon_2/\theta, \ldots, \epsilon_d/\theta)$ with $\theta = \sum_{j=1}^{d} \epsilon_j$. It is also interesting to see the derivation of the d-dimensional transition density expansion as a mixture in terms of $\{q_n^\theta(t)\}$ via the orthogonal-function expansion of the transition density in (2.24).

3 Processes with beta stationary distributions and Jacobi polynomial eigenfunctions

In this section we consider 1-dimensional processes which have Beta stationary distributions and Jacobi polynomial eigenfunctions, and their connection with Wright–Fisher diffusion processes. We begin by considering Bochner [6] and Gasper's [13] characterization of bivariate Beta distributions.

A class of bivariate distributions with Beta marginals and Jacobi polynomial eigenfunctions has the form

$$f(x,y) = f_{\alpha\beta}(x)f_{\alpha\beta}(y)\left\{1 + \sum_{n=1}^{\infty} \rho_n \widetilde{P}_n^{(\alpha,\beta)}(x)\widetilde{P}_n^{(\alpha,\beta)}(y)\right\}, \qquad (3.1)$$

where $\{\rho_n, n \in \mathbb{Z}_+\}$ is called a correlation sequence. The transition density (1.3) in the Jacobi diffusion has the form of the conditional density of Y given $X = x$ in (3.1) with $\rho_n \equiv \rho_n^\theta(t)$. Bochner [6] and Gasper [13] worked on characterizations of sequences $\{\rho_n\}$ such that the expansion (3.1) is positive, and thus a probability distribution. It is convenient to normalize the Jacobi polynomials by taking

$$R_n^{(\alpha,\beta)}(x) = \frac{\widetilde{P}_n^{(\alpha,\beta)}(x)}{\widetilde{P}_n^{(\alpha,\beta)}(1)}$$

so that $R_n^{(\alpha,\beta)}(1) = 1$; denote

$$h_n^{-1} = \mathbb{E}\big[R_n^{(\alpha,\beta)}(X)^2\big] = \frac{(2n+\alpha+\beta-1)(\alpha+\beta)_{(n-1)}\beta_{(n)}}{\alpha_{(n)}n!},$$

and write

$$f(x,y) = f_{\alpha\beta}(x)f_{\alpha\beta}(y)\left\{1 + \sum_{n=1}^{\infty} \rho_n h_n R_n^{(\alpha,\beta)}(x)R_n^{(\alpha,\beta)}(y)\right\}. \qquad (3.2)$$

Bochner [6] defined a bounded sequence $\{c_n\}$ to be positive definite with respect to the Jacobi polynomials if

$$\sum_{n\geq 0} a_n h_n R_n^{(\alpha,\beta)}(x) \geq 0, \quad \sum_{n\geq 0} |a_n| h_n < \infty$$

implies that

$$\sum_{n\geq 0} a_n c_n h_n R_n^{(\alpha,\beta)}(x) \geq 0.$$

Then $\{\rho_n\}$ is a correlation sequence *if and only if* it is a positive definite sequence. The *only if* proof follows from

$$\sum_{n\geq 0} a_n \rho_n R_n^{(\alpha,\beta)}(x) = \mathbb{E}\left[\sum_{n\geq 0} a_n h_n R_n^{(\alpha,\beta)}(Y) \,\Big|\, X = x\right] \geq 0,$$

where (X, Y) has the distribution (3.2). The *if* proof follows at least heuristically by noting that

$$\sum_{n\geq 0} h_n R_n^{(\alpha,\beta)}(x) R_n^{(\alpha,\beta)}(y) = \frac{\delta(x-y)}{f_{\alpha,\beta}(x)} \geq 0,$$

where $\delta(\cdot)$ has a unit point mass at zero, so if $\{\rho_n\}$ is a positive definite sequence then

$$\sum_{n\geq 0} \rho_n h_n R_n^{(\alpha,\beta)}(x) R_n^{(\alpha,\beta)}(y) \geq 0$$

and (3.2) is non-negative. A careful proof is given in [14].

Under the conditions that

$$\alpha < \beta \text{ and either } 1/2 \leq \alpha \text{ or } \alpha + \beta \geq 2, \tag{3.3}$$

it is shown in [13] that a sequence ρ_n is positive definite if and only if

$$\rho_n = \mathbb{E}\left[R_n^{(\alpha,\beta)}(Z)\right] \tag{3.4}$$

for some random variable Z in $[0, 1]$. The sufficiency rests on showing that under the conditions (3.3) for x, y, $z \in [0, 1]$,

$$K(x, y, z) = \sum_{n=0}^{\infty} h_n R_n^{(\alpha,\beta)}(x) R_n^{(\alpha,\beta)}(y) R_n^{(\alpha,\beta)}(z) \geq 0. \tag{3.5}$$

The sufficiency of (3.4) is then clear by mixing over a distribution for Z in (3.5) to get positivity. (If the conditions (3.3) do not hold then there exist x, y, z such that $K(x, y, z) < 0$.) The necessity follows by setting $x = 1$ in

$$\rho_n R_n^{(\alpha,\beta)}(x) = \mathbb{E}\left[R_n^{(\alpha,\beta)}(Y) \mid X = x\right],$$

and recalling that $R_n^{(\alpha,\beta)}(1) = 1$, so that Z is distributed as Y conditional on $X = 1$. This implies that extreme correlation sequences in exchangeable bivariate Beta distributions with Jacobi polynomial eigenfunctions are the scaled Jacobi polynomials $\{R_n^{(\alpha,\beta)}(z), z \in [0,1]\}$. Bochner [6] was the original author to consider such problems for the ultraspherical polynomials, essentially orthogonal polynomials on Beta distributions with equal parameters.

A characterization of reversible Markov processes with stationary Beta distribution and Jacobi polynomial eigenfunctions, from [13], under (3.3), is that they have transition functions of the form

$$f(x, y; t) = f_{\alpha\beta}(y)\left\{1 + \sum_{n=1}^{\infty} c_n(t)h_n R_n^{(\alpha,\beta)}(x)R_n^{(\alpha,\beta)}(y)\right\}, \qquad (3.6)$$

with $c_n(t) = \exp\{-d_n t\}$, where

$$d_n = \sigma n(n + \alpha + \beta - 1) + \int_0^{1-} \frac{1 - R_n^{(\alpha,\beta)}(z)}{1 - z}\, \nu(dz), \qquad (3.7)$$

$\sigma \geq 0$, and ν is a finite measure on $[0,1)$. If $\nu(\cdot) \equiv 0$, a null measure, then $f(x, y; t)$ is the transition function of a Jacobi diffusion.

Eigenvalues of a general reversible time-homogeneous Markov process with countable spectrum must satisfy Bochner's consistency conditions:

(i) $\{c_n(t)\}$ is a correlation sequence for each $t \geq 0$,
(ii) $c_n(t)$ is continuous in $t \geq 0$,
(iii) $c_n(0) = c_0(t) = 1$, and
(iv) $c_n(t + s) = c_n(t)c_n(s)$ for $t, s \geq 0$.

If there is a spectrum $\{c_n(t)\}$ with corresponding eigenfunctions $\{\xi_n\}$ then

$$c_n(t + s)\xi_n(X(0)) = \mathbb{E}\left[\xi_n(X(t + s)) \mid X(0)\right]$$
$$= \mathbb{E}\left[\mathbb{E}[\xi_n(X(t + s)) \mid X(s)] \mid X(0)\right]$$
$$= c_n(t)\mathbb{E}\left[\xi_n(X(s)) \mid X(0)\right]$$
$$= c_n(t)c_n(s)\xi_n(X(0)),$$

showing (iv). If a stationary distribution exists and $X(0)$ has this distribution then the eigenfunctions can be scaled to be orthonormal on this distribution and the eigenfunction property is then

$$\mathbb{E}\left[\xi_m(X(t))\xi_n(X(0))\right] = c_n(t)\delta_{mn}.$$

$\{X(t), t \geq 0\}$ is a Markov process such that the transition distribution of $Y = X(t)$ given $X(0) = x$ is

$$f(x, y; t) = f(y) \left\{ 1 + \sum_{n=1}^{\infty} c_n(t) \xi_n(x) \xi_n(y) \right\}, \tag{3.8}$$

where $f(y)$ is the stationary distribution. In our context $\{\xi_n\}$ are the orthonormal Jacobi polynomials. A Jacobi process $\{X(t), t \geq 0\}$ with transition distributions (3.6) can be constructed in the following way, which is analogous to constructing a general Lévy process from a compound Poisson process. Let $\{X_k, k \in \mathbb{Z}_+\}$ be a Markov chain with stationary distribution $f_{\alpha\beta}(y)$ and transition distribution of Y given $X = x$ corresponding to (3.1), with (3.3) holding, and $\{N(t), t \geq 0\}$ be an independent Poisson process of rate λ. Then (X_0, X_k) has a correlation sequence $\{\rho_n^k\}$ and the transition functions of $X(t) = X_{N(t)}$ have the form (3.8), with

$$d_n = \lambda \int_0^1 \left(1 - R_n^{(\alpha, \beta)}(z) \right) \mu(dz), \tag{3.9}$$

where μ is a probability measure on $[0, 1]$. The general form (3.7) is obtained by choosing a pair (λ, μ_λ) such that

$$d_n = \lim_{\lambda \to \infty} \lambda \int_0^1 \left(1 - R_n^{(\alpha, \beta)}(z) \right) \mu_\lambda(dz) = \int_0^1 \frac{1 - R_n^{(\alpha, \beta)}(z)}{1 - z} \nu(dz). \tag{3.10}$$

Equation (3.10) agrees with (3.7) when any atom $\nu(\{1\})$ is taken out of the integral, because

$$\lim_{z \to 1} \frac{1 - R_n^{(\alpha, \beta)}(z)}{1 - z} = cn(n + \theta - 1),$$

where $c \geq 0$ is a constant.

4 Subordinated Jacobi diffusion processes

Let $\{X(t), t \geq 0\}$ be a process with transition functions (3.6), and $\{Z(t), t \geq 0\}$ be a non-negative Lévy process with Laplace transform

$$\mathbb{E}\left[e^{-\lambda Z(t)} \right] = \exp \left\{ -t \int_0^\infty \frac{1 - e^{-\lambda y}}{y} H(dy) \right\}, \tag{4.1}$$

where $\lambda \geq 0$ and H is a finite measure. The subordinated process $\{\widetilde{X}(t) = X(Z(t)), t \geq 0\}$ is a Markov process which belongs to the same class of processes with correlation sequences

$$\tilde{c}_n(t) = \mathbb{E}\big[c_n\big(Z(t)\big)\big] = \exp\left\{-t\int_0^\infty \frac{1 - e^{-d_n y}}{y} H(dy)\right\}, \qquad (4.2)$$

where H is a finite measure. $\tilde{c}_n(t)$ necessarily has a representation as $e^{-\tilde{d}_n t}$, where $\tilde{d}_n$ has the form (3.10) for some measure $\tilde{\nu}$. We describe the easiest case from which the general case can be obtained as a limit. Suppose

$$\lambda = \int_0^\infty \frac{H(dy)}{y} < \infty,$$

and write

$$G(dy) = \frac{H(dy)}{\lambda y},$$

so that G is a probability measure. Let

$$K(dz) = f_{\alpha\beta}(z)\,dz\left\{1 + \sum_{n=1}^\infty h_n R_n^{(\alpha,\beta)}(z)\int_0^\infty e^{-d_n y} G(dy)\right\}.$$

Then K is a probability measure and

$$\lambda\int_0^1\Big(1 - R_n^{(\alpha,\beta)}(z)\Big)K(dz) = \lambda\int_0^\infty\Big(1 - e^{-d_n y}\Big)G(dy).$$

The representation (3.10) is now obtained by setting

$$\tilde{\nu}(dz) = \lambda(1 - z)K(dz).$$

We now consider subordinated Jacobi diffusion processes. The subordinated process is no longer a diffusion process because $\{Z(t), t \geq 0\}$ is a jump process and therefore $\{\widetilde{X}(t), t \geq 0\}$ has discontinuous sample paths. It is possible to construct processes such that (4.2) holds with $d_n = n$ by showing that e^{-tn} is a correlation sequence and thus so is $\mathbb{E}\big[e^{-Z(t)n}\big]$. The construction follows an idea in [6]. The Jacobi–Poisson kernel in orthogonal polynomial theory is

$$1 + \sum_{n=1}^\infty r^n h_n R_n^{(\alpha,\beta)}(x) R_n^{(\alpha,\beta)}(y), \qquad (4.3)$$

which is non-negative for all $\alpha,\ \beta > 0$, $x,\ y \in [0,1]$, and $0 \leq r \leq 1$, for which see [1], p112. The series (4.3) is a classical one evaluated early in research on Jacobi polynomials (see [2]). In terms of the original Jacobi polynomials, (1.5)

$$\sum_{n=0}^{\infty} r^n \phi_n P_n^{(\alpha,\beta)}(x) P_n^{(\alpha,\beta)}(y)$$

$$= \frac{\Gamma(\alpha+\beta+2)(1-r)}{2^{\alpha+\beta+1}\Gamma(\alpha+1)\Gamma(\beta+1)(1+r)^{\alpha+\beta+2}}$$

$$\times \sum_{m,n=0}^{\infty} \frac{((\alpha+\beta+2)/2)_{(m+n)}((\alpha+\beta+3)/2)_{(m+n)}}{(\alpha+1)_{(m)}(\beta+1)_{(m)}m!n!} \left(\frac{a^2}{k^2}\right)^m \left(\frac{b^2}{k^2}\right)^n,$$

$$(4.4)$$

where

$$\phi_n^{-1} = \frac{2^{\alpha+\beta+1}}{2n+\alpha+\beta+1} \frac{\Gamma(n+\alpha+1)\Gamma(n+\beta+1)}{\Gamma(n+1)\Gamma(n+\alpha+\beta+1)},$$

$x = \cos 2\varphi$, $y = \cos 2\theta$, $a = \sin \varphi \sin \theta$, $b = \cos \varphi \cos \theta$, $k = (r^{1/2} + r^{-1/2})/2$. The series (4.4) is positive for $-1 \le x, y \le 1$, $0 \le r < 1$ and $\alpha, \beta > -1$.

A Markov process analogy to the Jacobi–Poisson kernel is when the eigenvalues $c_n(t) = \exp\{-nt\}$. Following [6] let $\widetilde{X}(t) = X(Z(t))$, where $\{Z(t), t \ge 0\}$ is a Lévy process with Laplace transform

$$\mathbb{E}\left[e^{-\lambda Z(t)}\right]$$

$$= \exp\left\{-t\left[\sqrt{2\lambda + (\theta-1)^2/4} - \sqrt{(\theta-1)^2/4}\right]\right\} \qquad (4.5)$$

$$= \exp\left\{-\frac{t}{\sqrt{2\pi}} \int_0^{\infty} \frac{e^{-x(\theta-1)^2/8}}{x^{3/2}} \left(1 - e^{-x\lambda}\right) dx\right\}.$$

$\{Z(t), t \ge 0\}$ is a tilted positive stable process with index $\frac{1}{2}$ such that $Z(t)$ has an Inverse Gaussian density

$$IG\left(\frac{2t}{|\theta-1|}, t^2\right), \ \theta \ne 1;$$

that is,

$$\frac{t}{\sqrt{2\pi z^3}} \exp\left\{-\frac{1}{2z}\left(\frac{|\theta-1|}{2} z - t\right)^2\right\}, \ z > 0. \qquad (4.6)$$

The usual stable density is obtained when $\theta = 1$ and (4.6) is a tilted density in the sense that it is proportional to $\exp\{-z(\theta-1)^2/8\}$ times the stable density. See [12] XIII, §11, Problem 5 for an early derivation. $Z(t)$ is distributed as the first passage time

$$T_t = \inf\left\{u > 0; B(u) + \frac{|\theta-1|}{2}u = t\right\},$$

where $\{B(u), u \geq 0\}$ is standard Brownian motion. The eigenvalues of $\widetilde{X}(t)$ are

$$
\begin{aligned}
\widetilde{c}_n(t) &= \mathbb{E}\Big[\exp\Big\{-\frac{1}{2}n(n + \theta - 1)Z(t)\Big\}\Big] \\
&= \exp\Big\{-t\Big[\sqrt{n(n + \theta - 1) + (\theta - 1)^2/4} - \sqrt{(\theta - 1)^2/4}\Big]\Big\} \\
&= \exp\Big\{-t\Big[n + (\theta - 1)/2 - |\theta - 1|/2\Big]\Big\} \\
&= \begin{cases}
\exp\{-nt\} & \text{if } \theta \geq 1, \\
\exp\{-nt\} \times \exp\{t(1 - \theta)\} & \text{if } \theta < 1.
\end{cases}
\end{aligned}
\tag{4.7}
$$

The process $\{\widetilde{X}(t), t \geq 0\}$ is a jump diffusion process, discontinuous at the jumps of $\{Z(t), t \geq 0\}$. Jump sizes increase as θ decreases. If $\theta < 1$ then for $n \geq 1$

$$
\mathbb{E}\Big[\exp\Big\{-\frac{1}{2}n(n + \theta - 1)Z(t)\Big\}\Big] = \exp\{-nt\} \times \exp\{t(1 - \theta)\},
$$

so subordination does not directly produce eigenvalues e^{-nt}. Let $\widetilde{f}(x, y; t)$ be the transition density of $\widetilde{X}(t)$, then the transition density with eigenvalues $\exp\{-nt\}$, $n \geq 0$ is

$$
e^{-t(1-\theta)}\widetilde{f}(x, y; t) + \big(1 - e^{-t(1-\theta)}\big)f_{\alpha\beta}(y).
$$

The subordinated process with this transition density is $X(\widehat{Z}(t))$, where $\widehat{Z}(t)$ is a similar process to $Z(t)$ but has an extra state infinity. $Z(t)$ is killed by a jump to infinity at a rate $(1-\theta)$. Another possible construction does not kill the process $\widetilde{X}$, but restarts it in a stationary state drawn from the Beta distribution. It is convenient to use the notation that a process $\{Z^\circ(t), t \geq 0\}$ is $\{Z(t), t \geq 0\}$ if $\theta \geq 1$, or $\{\widehat{Z}(t), t \geq 0\}$ if $0 < \theta < 1$, and use the single notation $\{X(Z^\circ(t)), t \geq 0\}$ for the subordinated process. The transition density (3.6), where $c_n(t)$ has the general form

$$
\exp\left\{-t \int_0^\infty \frac{(1 - e^{-ny})}{y} H(dy)\right\},
$$

can then be obtained by a composition of subordinators from the Jacobi diffusion with any $\alpha, \beta > 0$.

There is a question as to which processes with transition densities (3.6) and eigenvalues $c_n(t)$ described by (3.7) are subordinated Jacobi diffusion processes. We briefly consider this question. Substituting

$$
R_n^{(\alpha,\beta)}(y) = {}_2F_1(-n, n + \theta - 1; \beta; 1 - y)
$$

in the eigenvalue expression (3.7),

$$\int_0^{1-} \frac{1 - R_n^{(\alpha,\beta)}(y)}{1 - y} \nu(dy)$$

$$= -c \sum_{k=1}^n \frac{(-n)_{(k)}(n + \theta - 1)_{(k)}}{\beta_{(k)}} \frac{\mu_{k-1}}{k!}$$

$$= -c \sum_{k=1}^n \frac{\prod_{j=0}^{k-1}\left(-n(n + \theta - 1) + j(j + \theta - 1)\right)}{\beta_{(k)}} \frac{\mu_{k-1}}{k!},$$

where $\int_0^{1-}(1-y)^k \nu(dy) = c\mu_k$. The generator corresponding to a process with these eigenvalues is

$$\widehat{\mathcal{L}} = c \sum_{k=1}^\infty \frac{\prod_{j=0}^{k-1}\left(2\mathcal{L} + j(j + \theta - 1)\right)}{\beta_{(k)}} \frac{\mu_{k-1}}{k!},$$

where $\mathcal{L}$ is the Jacobi diffusion process generator (1.1). The structure of the class of stochastic processes with the generator $\widehat{\mathcal{L}}$ needs to be understood better. It includes all subordinated Jacobi diffusion processes, but it seems to be a bigger class. A process with generator $\widehat{\mathcal{L}}$ is a subordinated Jacobi diffusion process if and only if the first derivative of

$$-\sum_{k=1}^\infty \frac{\prod_{j=0}^{k-1}\left(-2\lambda + j(j + \theta - 1)\right)}{\beta_{(k)}} \frac{\mu_{k-1}}{k!} \tag{4.8}$$

is a completely monotone function of λ. Factorizing

$$-2\lambda + j(j + \theta - 1) = (j + r_1(\lambda))(j + r_2(\lambda)),$$

where $r_1(\lambda), r_2(\lambda)$ are

$$(\theta - 1)/2 \pm \sqrt{2\lambda + (\theta - 1)^2/4},$$

(4.8) is equal to

$$-\int_0^{1-}\left[{}_2F_1(r_1(\lambda), r_2(\lambda); \beta; 1 - y) - 1\right](1 - y)^{-1}\nu(dy). \tag{4.9}$$

5 Subordinated coalescent process

Subordinating the Jacobi diffusion process $\{X(t), t \geq 0\}$ leads to subordinating the coalescent dual process, which we investigate in this section. A subordinated process $\{\widetilde{X}(t) = X(Z(t)), t \geq 0\}$ has a similar form for the transition density as (2.1), with $q_l^\theta(t)$ replaced by $\mathbb{E}(q_l^\theta(Z(t)),$ which are transition functions of the subordinated death process $A^\theta(Z(t))$. The subordinated process comes from subordinating the forest of non-mutant lineages in a coalescent tree.

If $\tilde{A}^\theta(t) = A^\theta(Z^\circ(t))$, with $Z^\circ(t)$ defined in the last section, we will show that the probability distribution of $\tilde{A}^\theta(t)$, $\theta > 0$ is

$$\binom{2k+\theta-1}{k}\left(\frac{z}{1+z}\right)^k\left(\frac{1}{1+z}\right)^{k+\theta}(1-z), \qquad (5.1)$$

for $k \in \mathbb{Z}_+$, where $z = e^{-t}$. The distribution (5.1) is the distribution of the number of edges in a time-subordinated forest. Note that if $0 < \theta < 1$ we still invoke a subordinator with a possible jump to infinity at rate $1 - \theta$, so

$$\mathbb{E}\left[q_k^\theta(Z^\circ(t))\right] = e^{-(1-\theta)t}\mathbb{E}\left[q_k^\theta(Z(t))\right] + (1 - e^{-(1-\theta)t})\delta_{k0},$$

because $q_k^\theta(\infty) = \delta_{k0}$. Although θ is greater than zero in (5.1), it is interesting to consider the subordinated Kingman coalescent with no mutation. Then $A^0(t) \geq 1$, and

$$\mathbb{E}\left[q_k^0(Z^\circ(t))\right] = e^{-t}\mathbb{E}\left[q_k^0(Z(t))\right] + (1 - e^{-t})\delta_{k1},$$

because a jump to infinity is made at rate 1, and $q_k^\circ(\infty) = \delta_{k1}$. The distribution of $\tilde{A}^0(t)$ is then, for $k \geq 1$,

$$\binom{2k-1}{k}\left(\frac{z}{1+z}\right)^k\left(\frac{1}{1+z}\right)^k(1-z) + \delta_{k1}(1-z). \qquad (5.2)$$

The proof of (5.1) ($\theta > 0$) and (5.2) (with $\theta = 0$) follows directly from the expansion (2.2).

$$\begin{aligned}
\mathbb{E}\left[q_k^\theta(Z^\circ(t))\right] &= \sum_{j=k}^\infty z^j(-1)^{j-k}\frac{(2j+\theta-1)(k+\theta)_{(j-1)}}{k!(j-k)!} \\
&= \frac{\Gamma(2k+\theta)}{k!\Gamma(k+\theta)}z^k \\
&\quad \times \left\{1 + \sum_{j=1}^\infty(-1)^j(2j+2k+\theta-1)\frac{(2k+\theta)_{(j-1)}}{j!}z^j\right\} \\
&= \frac{\Gamma(2k+\theta)}{k!\Gamma(k+\theta)}z^k(1-z)(1+z)^{-(2k+\theta)} \\
&= \binom{2k+\theta-1}{k}\left(\frac{z}{1+z}\right)^k\left(\frac{1}{1+z}\right)^{k+\theta}(1-z). \qquad (5.3)
\end{aligned}$$

Effectively, in the expansion (2.2) of $q_k^\theta(t)$, terms $\rho_j(t) = \exp\{-\frac{1}{2}j(j+$

$\theta - 1)t\}$ are replaced by $z^j = \exp\{-jt\}$. The third line of (5.3) follows from the identity, with $|z| < 1$ and $\alpha = 2k + \theta$, that

$$(1 - z)(1 + z)^{-\alpha} = 1 + \sum_{j=1}^{\infty} (-1)^j (2j + \alpha - 1) \frac{\alpha_{(j-1)}}{j!} z^j,$$

proved by equating coefficients of z^j on both sides. Of course, for any $|z| < 1$, since (5.1) is a probability distribution,

$$\sum_{k=0}^{\infty} \binom{2k + \theta - 1}{k} \left(\frac{z}{1+z}\right)^k \left(\frac{1}{1+z}\right)^{k+\theta} (1 - z) = 1. \qquad (5.4)$$

The probability generating function of (5.1) is

$$G_{\widetilde{A}^\theta(t)}(s) = \left(\frac{1 - 4pqs}{1 - 4pq}\right)^{-\frac{1}{2}} \left(\frac{1 - \sqrt{1 - 4pqs}}{2ps}\right)^{\theta - 1}, \quad \theta > 0, \qquad (5.5)$$

where $p = e^{-t}/(1 + e^{-t})$ and $q = 1/(1 + e^{-t})$. The calculation needed to show (5.5) comes from the identity

$$\sum_{k=0}^{\infty} \binom{2k + \theta - 1}{k} w^k = 2^{\theta - 1} \frac{(1 + \sqrt{1 - 4w})^{-(\theta-1)}}{\sqrt{1 - 4w}}, \qquad (5.6)$$

which is found by substituting

$$w = \frac{z}{(1 + z)^2} \quad \text{or } z = \frac{1 - \sqrt{1 - 4w}}{1 + \sqrt{1 - 4w}}$$

in (5.4), then setting

$$w = \frac{sz}{(1 + z)^2}$$

in (5.6). The calculations used in obtaining the distribution and probability generating function are the same as those used in obtaining the formula (2.3) in Griffiths [17]. There is a connection with a simple random walk on $\mathbb{Z}$ with transitions $j \to j+1$ with probability p and $j \to j-1$ with probability $q = 1 - p$, when $q \geq p$. Let the number of steps to hit $-\theta$, starting from 0, be ξ. Then ξ has a probability generating function of

$$H(s) = \left(\frac{1 - \sqrt{1 - 4pqs^2}}{2ps}\right)^\theta,$$

and $\frac{1}{2}(\xi + \theta)$ has a probability generating function

$$K(s) = \left(\frac{1 - \sqrt{1 - 4pqs}}{2p}\right)^\theta.$$

$\widetilde{A}^\theta(t) + \theta$ has the same distribution as the size-biased distribution of $\frac{1}{2}(\xi + \theta)$, with probability generating function

$$G_{\widetilde{A}^\theta(t)}(s) = \frac{sK'(s)}{K'(1)},$$

identical to (5.5). In the random walk interpretation θ is assumed to be an integer; however $H(s)$ is infinitely divisible, so we use the same description for all $\theta > 0$. Another interpretation is that $K(s)$ is the probability generating function of the total number of progeny in a Galton–Watson branching process with geometric offspring distribution qp^k, $k \in \mathbb{Z}_+$, and extinction probability 1, beginning with θ individuals. See [11] Sections X.13 and XII.5 for details of the random walk and branching process descriptions. An analogous calculation to (5.3) which is included in Theorem 2.1 of [17] is that

$$\mathbb{P}\Big(\widetilde{A}^\theta(s+t) = j \,\Big|\, \widetilde{A}^\theta(s) = i\Big)$$
$$= \binom{i}{j} \frac{\Gamma(i+\theta)\Gamma(2j+\theta)}{\Gamma(j+\theta)\Gamma(i+j+\theta)} z^j (1-z)$$
$$\times {}_2F_1(-i+j+1, 2j+\theta; i+j+\theta; z), \qquad (5.7)$$

where $z = e^{-t}$. The jump rate from $i \to j$ found from (5.7) is

$$\binom{i}{j} \frac{\Gamma(i+\theta)\Gamma(2j+\theta)}{\Gamma(j+\theta)\Gamma(i+j+\theta)} {}_2F_1(-i+j+1, 2j+\theta; i+j+\theta; 1),$$
$$= \binom{i}{j} B(j+\theta, i-j)^{-1} \int_0^1 x^{2j+\theta-1}(1-x)^{2(i-j)-2}\, dx$$
$$= \begin{cases} \binom{i}{j} \frac{\Gamma(2i-2j-1)\Gamma(2j+\theta)\Gamma(i+\theta)}{\Gamma(i-j)\Gamma(j+\theta)\Gamma(2i+\theta-1)} & \text{if } j = i-1, i-2, \ldots, \\ \frac{\Gamma(2j+\theta)}{\Gamma(j+\theta)j!}\left(\frac{1}{2}\right)^{2j+\theta} & \text{if } i = \infty. \end{cases} \qquad (5.8)$$

Bertoin [4], [5] studies the genealogical structure of trees in an infinitely-many-alleles branching process model. In a limit from a large initial population size with rare mutations the genealogy is described by a continuous-state branching process in discrete time with an Inverse Gaussian reproduction law. We expect that there is a fascinating connection with the process $\{\widetilde{A}^\theta(t), t \geq 0\}$. A potential class of transition functions of Markov processes $\{\widehat{q}_k^\theta(t), t \geq 0\}$ which are more general than subordinated processes and related to Bochner's characterization comes from replacing by $\rho_n^\theta(t)$ by $c_n(t)$ described by (3.7); however it is not clear that all such potential transition functions are positive, apart from those derived by subordination.

References

[1] Andrews, G. E., Askey, R., and Roy, R. 1999. *Special Functions*. Encyclopedia Math. Appl., vol. 71. Cambridge: Cambridge Univ. Press.

[2] Bailey, W. N. 1938. The generating function of Jacobi polynomials. *J. Lond. Math. Soc.*, **13**, 8–11.

[3] Barbour, A. D., Ethier, S. N., and Griffiths, R. C. 2000. A transition function expansion for a diffusion model with selection. *Ann. Appl. Probab.*, **10**, 123–162.

[4] Bertoin, J. 2009. The structure of the allelic partition of the total population for Galton–Watson processes with neutral mutations, *Ann. Probab.*, **37**, 1052–1523.

[5] Bertoin, J. 2010. A limit theorem for trees of alleles in branching processes with rare mutations, *Stochastic Process. Appl.*, to appear.

[6] Bochner, S. 1954. Positive zonal functions on spheres. *Proc. Natl. Acad. Sci. USA*, **40**, 1141–1147.

[7] Donnelly, P. J., and Kurtz, T. G. 1996. A countable representation of the Fleming–Viot measure-valued diffusion. *Ann. Appl. Probab.*, **24**, 698–742.

[8] Donnelly, P. J., and Kurtz, T. G. 1999. Particle representations for measure-valued population models. *Ann. Probab.*, **24**, 166–205.

[9] Etheridge, A. M., and Griffiths, R. C. 2009. A coalescent dual process in a Moran model with genic selection, *Theor. Popul. Biol.*, **75**, 320–330.

[10] Ethier, S. N., and Griffiths, R. C. 1993. The transition function of a Fleming–Viot process. *Ann. Probab.*, **21**, 1571–1590.

[11] Feller, W., 1968. *An Introduction to Probability Theory and its Applications*, vol. I, 3rd edn. New York: John Wiley & Sons.

[12] Feller, W., 1971. *An Introduction to Probability Theory and its Applications*, vol. II, 2nd edn. New York: John Wiley & Sons.

[13] Gasper, G. 1972. Banach algebras for Jacobi series and positivity of a kernel. *Ann. of Math. (2)*, **95**, 261–280.

[14] Griffiths, R. C. 1970. Positive definite sequences and canonical correlation coefficients. *Austral. J. Statist.*, **12**, 162–165.

[15] Griffiths, R. C. 1979. A transition density expansion for a multi-allele diffusion model. *Adv. in Appl. Probab.*, **11**, 310–325.

[16] Griffiths, R. C. 1980. Lines of descent in the diffusion approximation of neutral Wright–Fisher models. *Theor. Popul. Biol.*, **17**, 37–50.

[17] Griffiths, R. C. 2006. Coalescent lineage distributions. *Adv. in Appl. Probab.*, **38**, 405–429.

[18] Griffiths, R. C. [Griffiths, B.] 2009. Stochastic processes with orthogonal polynomial eigenfunctions. *J. Comput. Appl. Math.*, **23**, 739–744.

[19] Kingman, J. F. C. 1982. The coalescent. *Stochastic Process. Appl.*, **13**, 235–248.

[20] Kingman, J. F. C. 1993. *Poisson Processes*. Oxford: Oxford Univ. Press.

[21] Tavaré, S. 1984. Line-of-descent and genealogical processes, and their application in population genetics models. *Theor. Popul. Biol.*, **26**, 119–164.

16

Three problems for the clairvoyant demon

Geoffrey Grimmett[a]

Abstract

A number of tricky problems in probability are discussed, having in common one or more infinite sequences of coin tosses, and a representation as a problem in dependent percolation. Three of these problems are of 'Winkler' type, that is, they are challenges for a clairvoyant demon.

Keywords clairvoyant demon, dependent percolation, multiscale analysis, percolation, percolation of words, random walk

AMS subject classification (MSC2010) 60K35, 82B20, 60E15

1 Introduction

Probability theory has emerged in recent decades as a crossroads where many sub-disciplines of mathematical science meet and interact. Of the many examples within mathematics, we mention (not in order): analysis, partial differential equations, mathematical physics, measure theory, discrete mathematics, theoretical computer science, and number theory. The International Mathematical Union and the Abel Memorial Fund have recently accorded acclaim to probabilists. This process of recognition by others has been too slow, and would have been slower without the efforts of distinguished mathematicians including John Kingman.

JFCK's work looks towards both theory and applications. To single

[a] Statistical Laboratory, Centre for Mathematical Sciences, Cambridge University, Wilberforce Road, Cambridge CB3 0WB; g.r.grimmett@statslab.cam.ac.uk, http://www.statslab.cam.ac.uk/~grg/

out just two of his theorems: the subadditive ergodic theorem [22, 23] is a piece of mathematical perfection which has also proved rather useful in practice; his 'coalescent' [24, 25] is a beautiful piece of probability, now a keystone of mathematical genetics. John is also an inspiring and devoted lecturer, who continued to lecture to undergraduates even as the Bristol Vice-Chancellor, and the Director of the Isaac Newton Institute in Cambridge. Indeed, the current author learned his measure and probability from partial attendance at John's course in Oxford in 1970/71.

To misquote Frank Spitzer [35, Sect. 8], we turn to a very down-to-earth problem: consider an infinite sequence of light bulbs. The basic commodity of probability is an infinite sequence of coin tosses. Such a sequence has been studied for so long, and yet there remain 'simple to state' problems that appear very hard. We present some of these problems here. Sections 3–5 are devoted to three famous problems for the so-called clairvoyant demon, a presumably non-human being to whom is revealed the (infinite) realization of the sequence, and who is permitted to plan accordingly for the future.

Each of these problems may be phrased as a geometrical problem of percolation type. The difference with classical percolation [13] lies in the *dependence* of the site variables. Percolation is reviewed briefly in Section 2. This article ends with two short sections on related problems, namely: other forms of dependent percolation, and the question of 'percolation of words'.

2 Site percolation

We set the scene by reminding the reader of the classical 'site percolation model' of Broadbent and Hammersley [9]. Consider a countably infinite, connected graph $G = (V, E)$. To each 'site' $v \in V$ we assign a Bernoulli random variable $\omega(v)$ with density p. That is, $\omega = \{\omega(v) : v \in V\}$ is a family of independent, identically distributed random variables taking the values 0 and 1 with respective probabilities $1 - p$ and p. A vertex v is called *open* if $\omega(v) = 1$, and *closed* otherwise.

Let 0 be a given vertex, called the *origin*, and let $\theta(p)$ be the probability that the origin lies in an infinite open self-avoiding path of G. It is clear that θ is non-decreasing in p, and $\theta(0) = 0$, $\theta(1) = 1$. The *critical*

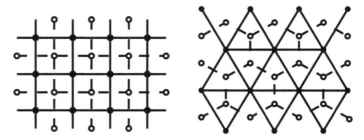

Figure 2.1 The square lattice $\mathbb{Z}^2$ and the triangular lattice $\mathbb{T}$, with their dual lattices.

probability is given as

$$p_c = p_c(G) := \sup\{p : \theta(p) = 0\}.$$

It is a standard exercise to show that the value of p_c does not depend on the choice of origin, but only on the graph G.

One may instead associate the random variables with the *edges* of the graph, rather than the *vertices*, in which case the process is termed 'bond percolation'. Percolation is recognised as a fundamental model for a random medium. It is important in probability and statistical physics, and it continues to be the source of beautiful and apparently hard mathematical problems, of which the most outstanding is to prove that $\theta(p_c) = 0$ for the three-dimensional lattice $\mathbb{Z}^3$. Of the several recent accounts of the percolation model, we mention [13, 14].

Most attention has been paid to the case when G is a crystalline lattice in two or more dimensions. The current article is entirely concerned with aspects of *two-dimensional* percolation, particularly on the square and triangular lattices illustrated in Figure 2.1. Site percolation on the triangular lattice has featured prominently in the news in recent years, owing to the work of Smirnov, Lawler–Schramm–Werner, and others on the relationship of this model (with $p = p_c = \frac{1}{2}$) to the process of random curves in $\mathbb{R}^2$ termed *Schramm–Löwner evolutions* (SLE), and particularly the process denoted SLE_6. See [36].

When G is a directed graph, one may ask about the existence of an infinite open *directed* path from the origin, in which case the process is referred to as *directed* (or *oriented*) *percolation*.

Variants of the percolation model are discussed in the following sections, with the emphasis on models with site/bond variables that are *dependent*.

3 Clairvoyant scheduling

Let $G = (V, E)$ be a finite connected graph. A symmetric random walk on G is a Markov chain $X = (X_k : k = 0, 1, 2, \dots)$ on the state space V, with transition matrix

$$\mathbb{P}(X_{k+1} = w \mid X_k = v) = \begin{cases} \dfrac{1}{\Delta_v} & \text{if } v \sim w, \\ 0 & \text{if } v \not\sim w, \end{cases}$$

where Δ_v is the degree of vertex v, and $\sim$ denotes the adjacency relation of G. Random walks on general graphs have attracted much interest in recent years; see [14, Chap. 1] for example.

Let X and Y be independent random walks on G with distinct starting sites x_0, y_0, respectively. We think of X (respectively, Y) as describing the trajectory of a particle labelled X (respectively, Y) around G. A clairvoyant demon is set the task of keeping the walks apart from one another for all time. To this end, (s)he is permitted to schedule the walks in such a way that exactly one walker moves at each epoch of time. Thus, the walks may be delayed, but they are required to follow their prescribed trajectories.

More precisely, a *schedule* is defined as a sequence $Z = (Z_1, Z_2, \dots)$ in the space $\{X, Y\}^{\mathbb{N}}$, and a given schedule Z is implemented in the following way. From the X and Y trajectories, we construct the rescheduled walks $Z(X)$ and $Z(Y)$, where:

1. If $Z_1 = X$, the X-particle takes one step at time 1, and the Y-particle remains stationary. If $Z_1 = Y$, it is the Y-particle that moves, and the X-particle that remains stationary. Thus,

$$\begin{aligned} \text{if } Z_1 = X \quad &\text{then} \quad Z(X)_1 = X_1, \ Z(Y)_1 = Y_0, \\ \text{if } Z_1 = Y \quad &\text{then} \quad Z(X)_1 = X_0, \ Z(Y)_1 = Y_1. \end{aligned}$$

2. Assume that, after time k, the X-particle has made r moves and the Y-particle $k - r$ moves, so that $Z(X)_k = X_r$ and $Z(Y)_k = Y_{k-r}$.

$$\begin{aligned} \text{If } Z_{k+1} = X \quad &\text{then} \quad Z(X)_{k+1} = X_{r+1}, \ Z(Y)_{k+1} = Y_{k-r}, \\ \text{if } Z_{k+1} = Y \quad &\text{then} \quad Z(X)_{k+1} = X_r, \quad\ Z(Y)_{k+1} = Y_{k-r+1}. \end{aligned}$$

We call the schedule Z *good* if $Z(X)_k \neq Z(Y)_k$ for all $k \geq 1$, and we say that the demon *succeeds* if there exists a good schedule $Z = Z(X, Y)$. (We overlook issues of measurability here.) The probability of success is

$$\theta(G) := \mathbb{P}(\text{there exists a good schedule}),$$

and we ask: for which graphs G is it the case that $\theta(G) > 0$? This question was posed by Peter Winkler (see the discussion in [10, 11]). Note that the answer is independent of the choice of (distinct) starting points x_0, y_0.

The problem takes a simpler form when G is the complete graph on some number, M say, of vertices. In order to simplify it still further, we add a loop to each vertex. Write $V = \{1, 2, \ldots, M\}$, and $\theta(M) := \theta(G)$. A random walk on G is now a sequence of independent, identically distributed points in $\{1, 2, \ldots, M\}$, each with the uniform distribution. It is known[1] that $\theta(M)$ is non-decreasing in M. Also, it is not too hard to show that $\theta(3) = 0$.

Question 3.1 *Is it the case that $\theta(M) > 0$ for sufficiently large M? Perhaps $\theta(4) > 0$?*

This problem has a geometrical formulation of percolation type. Consider the positive quadrant $\mathbb{Z}_+^2 = \{(i, j) : i, j = 0, 1, 2, \ldots\}$ of the square lattice $\mathbb{Z}^2$. A *path* is taken to be an infinite sequence (u_n, v_n), $n \geq 0$, with $(u_0, v_0) = (0, 0)$ such that, for all $n \geq 0$,

$$\text{either} \quad (u_{n+1}, v_{n+1}) = (u_n + 1, v_n) \quad \text{or} \quad (u_{n+1}, v_{n+1}) = (u_n, v_n + 1).$$

With X, Y the random walks as above, we declare the vertex (i, j) to be *open* if $X_i \neq Y_j$. It may be seen that the demon succeeds if and only if there exists a path all of whose vertices are open.

Some discussion of this problem may be found in [11]. The law of the open vertices is 3-wise independent but not 4-wise independent, in the sense of language introduced in Section 6.

The problem becomes significantly easier if paths are allowed to be undirected. For the totally undirected problem, it is proved in [3, 37] that there exists an infinite open path with strictly positive probability if and only if $M \geq 4$.

4 Clairvoyant compatibility

Let $p \in (0, 1)$, and let X_1, X_2, ... and Y_1, Y_2, ... be independent sequences of independent Bernoulli variables with common parameter p. We say that a *collision* occurs at time n if $X_n = Y_n = 1$. The demon is now charged with the removal of collisions, and to this end (s)he is permitted to remove 0s from the sequences.

[1] Ioan Manolescu has pointed out that this holds by a coupling argument.

Let $\mathbb{N} = \{1, 2, \dots\}$ and $\mathcal{W} = \{0, 1\}^{\mathbb{N}}$, the set of singly-infinite sequences of 0s and 1s. Each $w \in \mathcal{W}$ is considered as a *word* in an alphabet of two letters, and we generally write w_n for its nth letter. For $w \in \mathcal{W}$, there exists a sequence $i(w) = (i(w)_1, i(w)_2, \dots)$ of non-negative integers such that $w = 0^{i_1} 10^{i_2} 1 \cdots$, that is, there are exactly $i_j = i(w)_j$ zeros between the $(j-1)$th and jth appearances of 1. For $x, y \in \mathcal{W}$, we write $x \to y$ if $i(x)_j \geq i(y)_j$ for $j \geq 1$. That is, $x \to y$ if and only if y may be obtained from x by the removal of 0s.

Two infinite words v, w are said to be *compatible* if there exist v' and w' such that $v \to v'$, $w \to w'$, and $v'_n w'_n = 0$ for all n. For given realizations X, Y, we say that the demon *succeeds* if X and Y are compatible. Write

$$\psi(p) = \mathbb{P}_p(X \text{ and } Y \text{ are compatible}).$$

Note that, by a coupling argument, ψ is a non-increasing function.

Question 4.1 *For what p is it the case that $\psi(p) > 0$?*

It is easy to see as follows that $\psi(\frac{1}{2}) = 0$. When $p = \frac{1}{2}$, there exists almost surely an integer N such that

$$\sum_{i=1}^{N} X_i > \tfrac{1}{2} N, \quad \sum_{i=1}^{N} Y_i > \tfrac{1}{2} N.$$

With N chosen thus, it is not possible for the demon to prevent a collision in the first N values. By working more carefully, one may obtain that $\psi(\frac{1}{2} - \epsilon) = 0$ for small positive ϵ, see the discussion in [12].

Gács has proved in [12] that $\psi(10^{-400}) > 0$, and he has noted that there is room for improvement.

5 Clairvoyant embedding

The clairvoyant demon's third problem stems from work on long-range percolation of words (see Section 7). Let X_1, X_2, ... and Y_1, Y_2, ... be independent sequences of independent Bernoulli variables with parameter $\frac{1}{2}$. Let $M \in \{2, 3, \dots\}$. The demon's task is to find a monotonic embedding of the X_i within the Y_j in such a way that the gaps between successive terms are no greater than M.

Let v, $w \in \mathcal{W}$. We say that v is *M-embeddable* in w, and we write $v \subseteq_M w$, if there exists an increasing sequence $(m_i : i \geq 1)$ of positive integers such that $v_i = w_{m_i}$ and $1 \leq m_i - m_{i-1} \leq M$ for all $i \geq 1$. (We set $m_0 = 0$.) A similar definition is made for *finite* words v lying in one of the spaces $\mathcal{W}_n = \{0,1\}^n$, $n \geq 1$.

The demon succeeds in the above task if $X \subseteq_M Y$, and we let

$$\rho(M) = \mathbb{P}(X \subseteq_M Y).$$

It is elementary that $\rho(M)$ is non-decreasing in M.

Question 5.1   *Is it the case that $\rho(M) > 0$ for sufficiently large M?*

This question is introduced and discussed in [17], and partial but limited results proved. One approach is to estimate the first two moments of the number $N_n(w)$ of M-embeddings of the finite word $w = w_1 w_2 \cdots w_n \in \mathcal{W}_n$ within the random word Y. It is elementary that $E(N_n(w)) = (M/2)^n$ for any such w, and it may be shown that

$$\frac{E(N_n(X)^2)}{E(N_n(X))^2} \sim A_M c_M^n \qquad \text{as } n \to \infty,$$

where $A_M > 0$ and $c_M > 1$ for $M \geq 2$. The fact that $E(N_n(w)) \equiv 1$ when $M = 2$ is strongly suggestive that $\rho(2) = 0$, and this is part of the next theorem.

Theorem 5.2   [17] *We have that $\rho(2) = 0$. Furthermore, for $M = 2$,*

$$\mathbb{P}(w \subseteq_2 Y) \leq \mathbb{P}(a_n \subseteq_2 Y) \qquad \text{for all } w \in \mathcal{W}_n, \tag{5.1}$$

where $a_n = 0101 \cdots$ is the alternating *word of length n.*

It is immediate that (5.1) implies $\rho(2) = 0$ on noting that, for any infinite periodic word π, $\mathbb{P}(\pi \subseteq_M Y) = 0$ for all $M \geq 2$. One may estimate such probabilities more exactly through solving appropriate difference equations. For example, $v_n(M) = \mathbb{P}(a_n \subseteq_M Y)$ satisfies

$$v_{n+1}(M) = (\alpha + (M-1)\beta)v_n - \beta(M - 2\alpha)v_{n-1}, \qquad n \geq 1, \tag{5.2}$$

with boundary conditions $v_0(M) = 1$, $v_1(M) = \alpha$. Here,

$$\alpha + \beta = 1, \quad \beta = 2^{-M}.$$

The characteristic polynomial associated with (5.2) is a quadratic with one root in each of the disjoint intervals $(0, M\beta)$ and $(\alpha, 1)$. The larger root equals $1 - (1 + o(1))2^{1-2M}$ for large M, so that, in rough terms

$$v_n(M) \approx (1 - 2^{1-2M})^n.$$

Herein lies a health warning for simulators. One knows that, almost surely, $a_n \not\subseteq_M Y$ for large n, but one has to look on scales of order 2^{2M-1} if one is to observe its extinction with reasonable probability.

One may ask about the 'best' and 'worst' words. Inequality (5.1) asserts that an alternating word a_n is the most easily embedded word when $M = 2$. It is not known which word is best when $M > 2$. Were this a periodic word, it would follow that $\rho(M) = 0$. Unsurprisingly, the worst word is a constant word c_n (of which there are of course two). That is, for all $M \geq 2$,

$$\mathbb{P}(w \subseteq_M Y) \geq \mathbb{P}(c_n \subseteq_M Y) \qquad \text{for all } w \in \mathcal{W}_n,$$

where, for definiteness, we set $c_n = 1^n \in \mathcal{W}_n$.

Let $M = 2$, so that the mean number $E(N_n(w))$ of embeddings of any word of length n is exactly 1 (as remarked above). A further argument is required to deduce that $\rho(2) = 0$. Peled [33] has made rigorous the following alternative to that used in the proof of Theorem 5.2. Assume that the word $w \in \mathcal{W}_n$ satisfies $w \subseteq_2 Y$. For some small $c > 0$, one may identify (for most embeddings, with high probability) cn positions at which the embedding may be altered, independently of each other. This gives 2^{cn} possible 'local variations' of the embedding. It may be deduced that the probability of embedding a word $w \in \mathcal{W}_n$ is exponentially small in n, and also $\rho(2) = 0$.

The sequences X, Y have been taken above with parameter $\frac{1}{2}$. Little changes with Question 5.1 in a more general setting. Let the two (respective) parameters be p_X, $p_Y \in (0,1)$. It turns out that the validity of the statement 'for all $M \geq 2$, $\mathbb{P}(X \subseteq_M Y) = 0$' is independent of the values of p_X, p_Y. On the other hand, (5.1) is not generally true. See [17].

A number of easier variations on Question 5.1 spring immediately to mind, of which two are mentioned here.

1. Suppose the gap between the embeddings of X_{i-1} and X_i must be bounded above by some M_i. How slow a growth on the M_i suffices in order that the embedding probability be strictly positive? [An elementary bound follows by the Borel–Cantelli lemma.]
2. Suppose that the demon is allowed to look only boundedly into the future. How much clairvoyance may (s)he be allowed without the embedding probability becoming strictly positive?

Further questions (and variations thereof) have been proposed by others.

1. In a 'penalised embedding' problem, we are permitted mismatches by paying a (multiplicative) penalty b for each. What is the cost of the

'cheapest' penalised embedding of the first n terms, and what can be said as $b \to \infty$? [Erwin Bolthausen]

2. What can be said if we are required to embed only the 1s? That is, a '1' must be matched to a '1', but a '0' may be matched to either '0' or '1'. [Simon Griffiths]

3. The above problems may be described as embedding $\mathbb{Z}$ in $\mathbb{Z}$. In this language, might it be possible to embed $\mathbb{Z}^m$ in $\mathbb{Z}^n$ for some $m, n \geq 2$? [Ron Peled][2]

Question 5.1 may be expressed as a geometrical problem of percolation type. With X and Y as above, we declare the vertex $(i, j) \in \mathbb{N}^2$ to be *open* if $X_i = Y_j$. A *path* is defined as an infinite sequence (u_n, v_n), $n \geq 0$, of vertices such that:

$$(u_0, v_0) = (0, 0), \quad (u_{n+1}, v_{n+1}) = (u_n + 1, v_n + d_n),$$

for some d_n satisfying $1 \leq d_n \leq M$. It is easily seen that $X \subseteq_M Y$ if and only if there exists a path all of whose vertices are open. (We declare $(0, 0)$ to be open.)

With this formulation in mind, the above problem may be represented by the icon at the top left of Figure 5.1. The further icons of that figure represent examples of problems of similar type. Nothing seems to be known about these except that:

1. the argument of Peled [33] may be applied to problem (b) with $M = 2$ to obtain that $\mathbb{P}(w \subseteq_2 Y) = 0$ for all $w \in \mathcal{W}$;

2. problem (e) is easily seen to be trivial.

It is, as one might expect, much easier to embed words in two dimensions than in one, and indeed this may be done along a path of $\mathbb{Z}^2$ that is directed in the north–easterly direction. This statement is made more precise as follows. Let $Y = (Y_{i,j} : i, j = 1, 2, \dots)$ be a two-dimensional array of independent Bernoulli variables with parameter $p \in (0, 1)$, say. A word $v \in \mathcal{W}$ is said to be *M-embeddable* in Y, written $v \subseteq_M Y$, if there exist strictly increasing sequences $(m_i : i \geq 1)$, $(n_i : i \geq 1)$ of positive integers such that $v_i = Y_{m_i, n_i}$ and

$$1 \leq (m_i - m_{i-1}) + (n_i - n_{i-1}) \leq M, \qquad i \geq 1.$$

(We set $m_0 = n_0 = 0$.) The following answers a question posed in [30].

---

[2] *Note added at revision*: See [15] for an affirmative answer to this question, with $n = m + 1$.

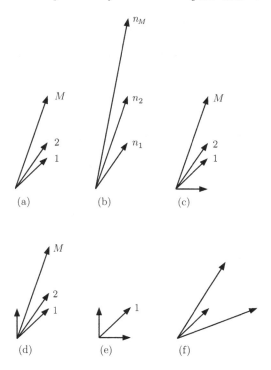

Figure 5.1 Icons describing a variety of embedding problems.

Note added at revision: A related result has been discovered independently in [31].

Theorem 5.3 [14] *Suppose $R \geq 1$ is such that $1 - p^{R^2} - (1-p)^{R^2} > \vec{p}_c$, the critical probability of directed site percolation on $\mathbb{Z}^2$. With strictly positive probability, every infinite word w satisfies $w \subseteq_{5R} Y$.*

The identification of the set of words that are 1-embeddable in the two-dimensional array Y, with positive probability, is much harder. This is a problem of *percolation of words*, and the results to date are summarised in Section 7.

Proof We use a block argument. Let $R \in \{2, 3, \dots\}$. For $(i, j) \in \mathbb{N}^2$, define the block $B_R(i, j) = ((i-1)R, iR] \times ((j-1)R, jR] \subseteq \mathbb{N}^2$. On the graph of blocks, we define the (directed) relation $B_R(i, j) \to B_R(m, n)$ if (m, n) is either $(i+1, j+1)$ or $(i+1, j+2)$. By drawing a picture, one sees that the ensuing directed graph is isomorphic to $\mathbb{N}^2$ directed north–easterly. Note that the L^1-distance between two vertices lying in adjacent blocks is no more than $5R$.

We call a block B_R *good* if it contains at least one 0 and at least one 1. It is trivial that

$$\mathbb{P}_p(B_R \text{ is good}) = 1 - p^{R^2} - (1-p)^{R^2}.$$

If the right side exceeds the critical probability $\vec{p}_c$ of directed site percolation on $\mathbb{Z}^2$, then there is a strictly positive probability of an infinite directed path of good blocks in the block graph, beginning at $B_R(1,1)$. Such a path contains $5R$-embeddings of all words. $\square$

The problem of clairvoyant embedding is connected to a question concerning isometries of random metric spaces discussed in [34]. In broad terms, two metric spaces (S_i, μ_i), $i = 1, 2$, are said to be 'quasi-isometric' (or 'roughly isometric') if their metric structure is the same up to multiplicative and additive constants. That is, there exists a mapping $T : S_1 \rightarrow S_2$ and positive constants M, D, R such that:

$$\frac{1}{M}\mu_1(x,y) - D \leq \mu_2(T(x),T(y)) \leq M\mu_1(x,y) + D, \qquad x,y \in S_1,$$

and, for $x_2 \in S_2$, there exists $x_1 \in S_1$ with $\mu_2(x_2, T(x_1)) \leq R$.

It has been asked whether two Poisson processes on the line, viewed as random sets with metric inherited from $\mathbb{R}$, are quasi-isometric. This question is open at the time of writing. A number of related results are proved in [34], where a history of the problem may be found also. It turns out that the above question is equivalent to the following. Let $X = (\ldots, X_{-1}, X_0, X_1, \ldots)$ be a sequence of independent Bernoulli variables with common parameter p_X. The sequence X generates a random metric space with points $\{i : X_i = 1\}$ and metric inherited from $\mathbb{Z}$. Is it the case that two independent sequences X and Y generate quasi-isometric metric spaces? A possibly important difference between this problem and clairvoyant embedding is that quasi-isometries of metric subspaces of $\mathbb{Z}$ need not be monotone.

6 Dependent percolation

Whereas there is only one type of independence, there are many types of dependence, too many to be summarised here. We mention just three further types of dependent percolation in this section, of which the first (at least) arises in the context of processes in random environments. In each, the dependence has infinite range, and in this sense these problems have something in common with those treated in Sections 3–5.

For our first example, let $X = \{X_i : i \in \mathbb{Z}\}$ be independent, identically distributed random variables taking values in $[0, 1]$. Conditional on X, the vertex (i, j) of $\mathbb{Z}^2$ is declared *open* with probability X_i, and different vertices receive (conditionally) independent states. The ensuing measure possesses a dependence that extends without limit in the vertical direction. Let p_c denote the critical probability of site percolation on $\mathbb{Z}^2$. If the law μ of X_0 places probability both below and above p_c, there exist (almost surely) vertically-unbounded domains that consider themselves subcritical, and others that consider themselves supercritical. Depending on the choice of μ, the process may or may not possess infinite open paths, and necessary and sufficient conditions have proved elusive. The most successful technique for dealing with such problems seems to be the so-called 'multiscale analysis'. This leads to sufficient conditions under which the process is subcritical (respectively, supercritical). See [26, 27].

There is a variety of models of physics and applied probability for which the natural random environment is exactly of the above type. Consider, for example, the contact model in d dimensions with recovery rates δ_x and infection rates λ_e; see [28, 29]. Suppose that the environment is randomised through the assumption that the δ_x (respectively, λ_e) are independent and identically distributed. The graphical representation of this model may be viewed as a 'vertically directed' percolation model on $\mathbb{Z}^d \times [0, \infty)$, in which the intensities of infections and recoveries are dependent in the vertical direction. See [1, 8, 32] for further discussion.

Vertical dependence arises naturally in certain models of statistical physics also, of which we present one example. The 'quantum Ising model' on a graph G may be formulated as a problem in stochastic geometry on a product space of the form $G \times [0, \beta]$, where β is the inverse temperature. A fair bit of work has been done on the quantum model in a random environment, that is, when its parameters vary randomly around different vertices/edges of G. The corresponding stochastic model on $G \times [0, \beta]$ has 'vertical dependence' of infinite range. See [7, 16].

It is easy to adapt the above structure to provide dependencies in both horizontal and vertical directions, although the ensuing problems may be considered (so far) to have greater mathematical than physical interest. For example, consider bond percolation on $\mathbb{Z}^2$, in which the states of horizontal edges are correlated thus, and similarly those of vertical edges. A related three-dimensional system has been studied by Jonasson, Mossel, and Peres [19]. Draw planes in $\mathbb{R}^3$ orthogonal to the

x-axis, such that they intersect the x-axis at points of a Poisson process with given intensity λ. Similarly, draw independent families of planes orthogonal to the y- and z-axes. These three families define a 'stretched' copy of $\mathbb{Z}^3$. An edge of this stretched lattice, of length l, is declared to be open with probability e^{-l}, independently of the states of other edges. It is proved in [19] that, for sufficiently large λ, there exists (a.s.) an infinite open directed percolation cluster that is transient for simple random walk. The method of proof is interesting, proceeding as it does by the method of 'exponential intersection tails' (EIT) of [6]. When combined with an earlier argument of Häggström, this proves the existence of a percolation phase transition for the model.

The method of EIT is invalid in two dimensions, because random walk is recurrent on $\mathbb{Z}^2$. The corresponding percolation question in two dimensions was answered using different means by Hoffman [18].

In our final example, the dependence comes without geometrical information. Let $k \geq 2$, and call a family of random variables k-*wise independent* if any k-subset is independent. Note that the vertex states arising in the clairvoyant scheduling problem of Section 3 are 3-wise independent but not 4-wise independent.

Benjamini, Gurel-Gurevich, and Peled [5] have investigated various properties of k-wise independent Bernoulli families, and in particular the following percolation question. Consider the n-box $B_n = [1, n]^d$ in $\mathbb{Z}^d$ with $d \geq 2$, in which the measure governing the site variables $\{\omega(v) : v \in B_n\}$ has local density p and is k-wise independent. Let L_n be the event that two given opposite faces are connected by an open path in the box. Thus, for large n, the probability of L_n under the product measure $\mathbb{P}_p$ has a sharp threshold around $p = p_c(\mathbb{Z}^d)$. The problem is to find bounds on the smallest value of k such that the probability of L_n is close to its value $\mathbb{P}_p(L_n)$ under product measure.

This question may be formalised as follows. Let $\Pi = \Pi(n, k, p)$ be the set of probability measures on $\{0, 1\}^{B_n}$ that have density p and are k-wise independent. Let

$$\epsilon_n(p, k) = \max_{\mathbb{P} \in \Pi} \mathbb{P}(L_n) - \min_{\mathbb{P} \in \Pi} \mathbb{P}(L_n),$$

and

$$K_n(p) = \min\{k : \epsilon_n(p, k) \leq \delta\},$$

where for definiteness we may take $\delta = 0.01$ as in [5]. Thus, roughly speaking, $K_n(p)$ is a quantification of the amount of independence required in order that, for all $\mathbb{P} \in \Pi$, $\mathbb{P}(L_n)$ differs from $\mathbb{P}_p(L_n)$ by at most δ.

Benjamini, Gurel-Gurevich, and Peled have proved, in an ongoing project [5], that $K_n(p) \leq c \log n$ when $d = 2$ and $p \neq p_c$ (and when $d > 2$ and $p < p_c$), for some constant $c = c(p, d)$. They have in addition a lower bound for $K_n(p)$ that depends on p, d, and n, and goes to ∞ as $n \to \infty$.

7 Percolation of words

Recall the set $\mathcal{W} = \{0, 1\}^{\mathbb{N}}$ of words in the alphabet comprising the two letters 0, 1. Consider the site percolation process of Section 2 on a countably infinite connected graph $G = (V, E)$, and write $\omega = \{\omega(v) : v \in V\}$ for the ensuing configuration. Let $v \in V$ and let $\mathcal{S}_v$ be the set of all self-avoiding walks starting at v. Each $\pi \in \mathcal{S}_v$ is a path v_0, v_1, v_2, ... with $v_0 = v$. With the path π we associate the word $w(\pi) = \omega(v_1)\omega(v_2)\cdots$, and we write $\mathcal{W}_v = \{w(\pi) : \pi \in \mathcal{S}_v\}$ for the set of words 'visible from v'. The central question of site percolation concerns the probability that $\mathcal{W}_v \ni 1^\infty$, where 1^∞ denotes the infinite word $111\cdots$. The so-called AB-percolation problem concerns the existence in $\mathcal{W}_v$ of the infinite alternating word $01010\cdots$; see [2].

More generally, for given p, we ask which words lie in the random set $\mathcal{W}_v$. Partial answers to this question may be found in three papers [4, 20, 21] of Kesten and co-authors Benjamini, Sidoravicius, and Zhang, and their results are summarised here as follows.

For $\mathbb{Z}^d$, with $p = \frac{1}{2}$ and d sufficiently large, we have from [4] that

$$\mathbb{P}_{\frac{1}{2}}(\mathcal{W}_0 = \mathcal{W}) > 0,$$

and indeed there exists (a.s.) some vertex v for which $\mathcal{W}_v = \mathcal{W}$. Partial results are obtained for $\mathbb{Z}^d$ with edge orientations in increasing coordinate directions.

For the triangular lattice $\mathbb{T}$ and $p = \frac{1}{2}$, we have from [20] that

$$\mathbb{P}_{\frac{1}{2}}\left(\bigcup_{v \in V} \mathcal{W}_v \text{ contains almost every word}\right) = 1, \qquad (7.1)$$

where the set of words seen includes all periodic words apart from 0^∞ and 1^∞. The measure on $\mathcal{W}$ can be taken in (7.1) as any non-trivial product measure. This extends the observation that AB-percolation takes place at $p = \frac{1}{2}$, whereas there is no infinite cluster in the usual site percolation model.

Finally, for the 'close-packed' lattice $\mathbb{Z}_{\mathrm{cp}}^2$ obtained from $\mathbb{Z}^2$ by adding both diagonals to each face,

$$\mathbb{P}_p(\mathcal{W}_0 = \mathcal{W}) > 0$$

for $1 - p_{\mathrm{c}} < p < p_{\mathrm{c}}$, with $p_{\mathrm{c}} = p_{\mathrm{c}}(\mathbb{Z}^2)$. Moreover, every word is (a.s.) seen along some self-avoiding path in the lattice. See [21].

Acknowledgements The author acknowledges the contributions of his co-authors Tom Liggett and Thomas Richthammer. He profited from discussions with Alexander Holroyd while at the Department of Mathematics at the University of British Columbia, and with Ron Peled and Vladas Sidoravicius while visiting the Institut Henri Poincaré–Centre Emile Borel, both during 2008. This article was written during a visit to the Section de Mathématiques at the University of Geneva, supported by the Swiss National Science Foundation. The author thanks Ron Peled for his comments on a draft.

References

[1] Andjel, E. 1992. Survival of multidimensional contact process in random environments. *Bull. Braz. Math. Soc.*, **23**, 109–119.

[2] Appel, M. J. B., and Wierman, J. C. 1993. AB percolation on bond-decorated graphs. *J. Appl. Probab.*, **30**, 153–166.

[3] Balister, P. N., Bollobás, B., and Stacey, A. M. 2000. Dependent percolation in two dimensions. *Probab. Theory Related Fields*, **117**, 495–513.

[4] Benjamini, I., and Kesten, H. 1995. Percolation of arbitrary words in $\{0, 1\}^{\mathbb{N}}$. *Ann. Probab.*, **23**, 1024–1060.

[5] Benjamini, I., Gurel-Gurevich, O., and Peled, R. *On k-wise Independent Events and Percolation*. In preparation. http://cims.nyu.edu/~peled/.

[6] Benjamini, I., Pemantle, R., and Peres, Y. 1998. Unpredictable paths and percolation. *Ann. Probab.*, **26**, 1198–1211.

[7] Björnberg, J. E., and Grimmett, G. R. 2009. The phase transition of the quantum Ising model is sharp. *J. Stat. Phys.*, **136**, 231–273.

[8] Bramson, M., Durrett, R. T., and Schonmann, R. H. 1991. The contact processes in a random environment. *Ann. Probab.*, **19**, 960–983.

[9] Broadbent, S. R., and Hammersley, J. M. 1957. Percolation processes I. Crystals and mazes. *Proc. Cambridge Philos. Soc.*, **53**, 629–641.

[10] Coppersmith, D., Tetali, P., and Winkler, P. 1993. Collisions among random walks on a graph. *SIAM J. Discrete Math.*, **6**, 363–374.

[11] Gács, P. 2000. The clairvoyant demon has a hard task. *Combin. Probab. Comput.*, **9**, 421–424.

[12] Gács, P. 2004. Compatible sequences and a slow Winkler percolation. *Combin. Probab. Comput.*, **13**, 815–856.

[13] Grimmett, G. R. 1999. *Percolation.* 2nd edn. Berlin: Springer-Verlag.

[14] Grimmett, G. R. 2008. *Probability on Graphs.* http://www.statslab.cam.ac.uk/~grg/books/pgs.html.

[15] Grimmett, G. R., and Holroyd, A. E. 2010. *Lattice Embeddings in Percolation.* In preparation.

[16] Grimmett, G. R., Osborne, T. J., and Scudo, P. F. 2008. Entanglement in the quantum Ising model. *J. Stat. Phys.*, **131**, 305–339.

[17] Grimmett, G. R., Liggett, T. M., and Richthammer, T. 2010. Percolation of arbitrary words in one dimension. *Random Structures & Algorithms.* http://arxiv.org/abs/0807.1676.

[18] Hoffman, C. 2005. Phase transition in dependent percolation. *Comm. Math. Phys.*, **254**, 1–22.

[19] Jonasson, J., Mossel, E., and Peres, Y. 2000. Percolation in a dependent random environment. *Random Structures & Algorithms*, **16**, 333–343.

[20] Kesten, H., Sidoravicius, V., and Zhang, Y. 1998. Almost all words are seen in critical site percolation on the triangular lattice. *Electron. J. Probab.*, **3**, 1–75. Paper #10.

[21] Kesten, H., Sidoravicius, V., and Zhang, Y. 2001. Percolation of arbitrary words on the close-packed graph of $\mathbb{Z}^2$. *Electron. J. Probab.*, **6**, 1–27. Paper #4.

[22] Kingman, J. F. C. 1968. The ergodic theory of subadditive stochastic processes. *J. Roy. Statist. Soc. Ser. B*, **30**, 499–510.

[23] Kingman, J. F. C. 1973. Subadditive ergodic theory. *Ann. Probab.*, **1**, 883–909.

[24] Kingman, J. F. C. 1982. On the genealogy of large populations. *J. Appl. Probab.*, **19A**, 27–43.

[25] Kingman, J. F. C. 2000. Origins of the coalescent: 1974–1982. *Genetics*, **156**, 1461–1463.

[26] Klein, A. 1994a. Extinction of contact and percolation processes in a random environment. *Ann. Probab.*, **22**, 1227–1251.

[27] Klein, A. 1994b. Multiscale analysis in disordered systems: percolation and contact process in random environment. Pages 139–152 of: Grimmett, G. R. (ed), *Disorder in Physical Systems.* Dordrecht: Kluwer.

[28] Liggett, T. M. 1985. *Interacting Particle Systems.* Berlin: Springer-Verlag.

[29] Liggett, T. M. 1999. *Stochastic Interacting Systems: Contact, Voter and Exclusion Processes.* Berlin: Springer-Verlag.

[30] Lima, B. N. B. de. 2008. A note about the truncation question in percolation of words. *Bull. Braz. Math. Soc.*, **39**, 183–189.

[31] Lima, B. N. B. de, Sanchis, R., and Silva, R. W. C. 2009. *Percolation of Words on $\mathbb{Z}^d$ with Long Range Connections.* http://arxiv.org/abs/0905.4615.

[32] Newman, C. M., and Volchan, S. 1996. Persistent survival of one-dimensional contact processes in random environments. *Ann. Probab.*, **24**, 411–421.

[33] Peled, R. 2008. Personal communication.

[34] Peled, R. 2010. On rough isometries of Poisson processes on the line. *Ann. Appl. Probab.*, **20**(2), 462–494.

[35] Spitzer, F. L. 1976. *Principles of Random Walk*. 2nd edn. New York: Springer-Verlag.

[36] Werner, W. 2009. Lectures on two-dimensional critical percolation. Pages 297–360 of: Sheffield, S., and Spencer, T. (eds), *Statistical Mechanics*. IAS/Park City Mathematics Series. Providence, RI: AMS.

[37] Winkler, P. 2000. Dependent percolation and colliding random walks. *Random Structures & Algorithms*, **16**, 58–84.

17

Homogenization for advection-diffusion in a perforated domain

P. H. Haynes[a], V. H. Hoang[b], J. R. Norris[c] and K. C. Zygalakis[d]

Abstract

The volume of a Wiener sausage constructed from a diffusion process with periodic, mean-zero, divergence-free velocity field, in dimension 3 or more, is shown to have a non-random and positive asymptotic rate of growth. This is used to establish the existence of a homogenized limit for such a diffusion when subject to Dirichlet conditions on the boundaries of a sparse and independent array of obstacles. There is a constant effective long-time loss rate at the obstacles. The dependence of this rate on the form and intensity of the obstacles and on the velocity field is investigated. A Monte Carlo algorithm for the computation of the volume growth rate of the sausage is introduced and some numerical results are presented for the Taylor–Green velocity field.

AMS subject classification (MSC2010) 60G60, 60G65, 35B27, 65C05

[a] Department of Applied Mathematics and Theoretical Physics, Centre for Mathematical Sciences, Wilberforce Road, Cambridge CB3 0WA; phh1@cam.ac.uk

[b] Division of Mathematical Sciences, Nanyang Technological University, SPMS–MAS–04–19, 21 Nanyang Link, Singapore 637371; VHHOANG@ntu.edu.sg

[c] Statistical Laboratory, Centre for Mathematical Sciences, Wilberforce Road, Cambridge, CB3 0WB; j.r.norris@statslab.cam.ac.uk

[d] Mathematical Institute, 24–29 St Giles', Oxford OX1 3LB; zygalakis@maths.ox.ac.uk. Supported by a David Crighton Fellowship and by Award No. KUK–C1–013–04 made by King Abdullah University of Science and Technology (KAUST).

1 Introduction

We consider the problem of the existence and characterization of a homogenized limit for advection-diffusion in a perforated domain. This problem was initially motivated for us as a model for the transport of water vapour in the atmosphere, subject to molecular diffusion and turbulent advection, where the vapour is also lost by condensation on suspended ice crystals. It is of interest to determine the long-time rate of loss and in particular whether this is strongly affected by the advection. In this article we address a simple version of this set-up, where the advection is periodic in space and constant in time and where the ice crystals remain fixed in space.

Let K be a compact subset of $\mathbb{R}^d$ of positive Newtonian capacity. We assume throughout that $d \geqslant 3$. Let $\rho \in (0, \infty)$. We consider eventually the limit $\rho \to 0$. Construct a random perforated domain $D \subseteq \mathbb{R}^d$ by removing all the sets $K + p$, where p runs over the support P of a Poisson random measure μ on $\mathbb{R}^d$ of intensity ρ. Let v be a $\mathbb{Z}^d$-periodic, Lipschitz, mean-zero, divergence-free vector field on $\mathbb{R}^d$. Our aim is to determine the long-time behaviour, over times of order $\sigma^2 = \rho^{-1}$, of advection-diffusion in the domain D corresponding to the operator[1]

$$\mathcal{L} = \tfrac{1}{2}\Delta + v(x).\nabla$$

with Dirichlet boundary conditions. It is well known (see Section 2) that the long-time behaviour of advection-diffusion in the whole space $\mathbb{R}^d$ can be approximated by classical, homogeneous, heat-flow, with a constant diffusivity matrix $\bar{a} = \bar{a}(v)$. The effect of placing Dirichlet boundary conditions on the sets $K + p$ is to induce a loss of heat. The homogenization problem in a perforated domain has been considered already in the case of Brownian motion [5], [7], [12], [15] and Brownian motion with constant drift [3]. The novelty here is to explore the possible interaction between inhomogeneity in the drift and in the domain. We will show that as $\rho \to 0$ there exists an effective constant loss rate $\bar{\lambda}(v, K)$ in the time-scale σ^2. We will also identify the limiting values of $r^{2-d}\bar{\lambda}(v, rK)$ as $r \to 0$ and $r \to \infty$ and we will compute numerically this function of r for one choice of v and K.

Fix a function $f \in L^2(\mathbb{R}^d)$. Write $u = u(t, x)$ for the solution to the Cauchy problem for $\mathcal{L}$ in $[0, \infty) \times D$ with initial data f, and with

---

[1] All results to follow extend to the case of the operator $\frac{1}{2}\operatorname{div} a\nabla + v(x).\nabla$, where a is a constant positive-definite symmetric matrix, by a straightforward scaling transformation. We simplify the presentation by taking $a = I$. Results for the case $a = \varepsilon^2 I$ are stated in Section 7 for easy reference.

Dirichlet conditions on the boundary of D. Thus, for suitably regular K and f, u is continuous on $[0, \infty) \times D$ and on $(0, \infty) \times \bar{D}$, and is $C^{1,2}$ on $(0, \infty) \times D$; we have $u(0, x) = f(x)$ for all $x \in D$ and

$$\frac{\partial u}{\partial t} = \tfrac{1}{2}\Delta u + v(x).\nabla u \quad \text{on } (0, \infty) \times D.$$

We shall study the behaviour of u over large scales in the limit $\rho \to 0$. Our analysis will rest on the following probabilistic representation of u. Let X be a diffusion process in $\mathbb{R}^d$, independent of μ with generator $\mathcal{L}$ starting from x. Such a process can be realised by solving the stochastic differential equation

$$dX_t = dW_t + v(X_t)\,dt, \quad X_0 = x \tag{1.1}$$

driven by a Brownian motion W in $\mathbb{R}^d$. Set

$$T = \inf\{t \geqslant 0 : X_t \in K + P\}.$$

Then

$$u(t, x) = \mathbb{E}_x\left(f(X_t)1_{\{T>t\}}\big|\mu\right).$$

The key step is to express the right hand side of this identity in terms of an analogue for X of the Wiener sausage. Associate to each path $\gamma \in C([0, \infty), \mathbb{R}^d)$ and to each interval $I \subseteq [0, \infty)$ a set $S_I^K(\gamma) \subseteq \mathbb{R}^d$ formed of the translates of K by γ_t as t ranges over I. Thus

$$S_I^K(\gamma) = \cup_{t \in I}(K + \gamma_t) = \{x \in \mathbb{R}^d : x - \gamma_t \in K \text{ for some } t \in I\}.$$

Write S_t^K for the random set $S_{(0,t]}^K(X)$ and write $|S_t^K|$ for the Lebesgue volume of S_t^K. We call S_t^K the *diffusion sausage* or (X, K)-*sausage* and refer to K as the *cross section*. Then $T > t$ if and only if $\mu(S_t^{\hat{K}}) = 0$, where $\hat{K} = \{-x : x \in K\}$. Hence

$$u(t, x) = \mathbb{E}_x\left(f(X_t)1_{\{\mu(S_t^{\hat{K}})=0\}}\big|\mu\right)$$

and so, by Fubini, we obtain the formulae

$$\mathbb{E}(u(t, x)) = \mathbb{E}_x\left(f(X_t)\exp(-\rho|S_t^{\hat{K}}|)\right) \tag{1.2}$$

and

$$\mathbb{E}(u(t, x)^2) = \mathbb{E}_x\left(f(X_t)f(Y_t)\exp(-\rho|S_t^{\hat{K}}(X) \cup S_t^{\hat{K}}(Y)|)\right) \tag{1.3}$$

where Y is an independent copy of X.

In the next section we review the homogenization theory for $\mathcal{L}$ in the whole space. Then, in Section 3 we show, as a straightforward application of Kingman's subadditive ergodic theorem, that the sausage volume $|S_t^K|$ has almost surely an asymptotic growth rate $\gamma(v, K)$, which is non-random. In Section 4 we make some further preparatory estimates on diffusion sausages. Then in Section 5 we identify the limiting values of $r^{2-d}\gamma(v, rK)$ as $r \to 0$ and as $r \to \infty$. In Section 6, we use the formulae (1.2), (1.3) to deduce the existence of a homogenized scaling limit for the function u, and we prove a corresponding weak limit for the diffusion process X and the hitting time T. We shall see in particular that for large obstacles it is the effective diffusivity $\bar{a}$ which accounts for the loss of heat in the obstacles. On the other hand, when the obstacles are small, the loss of heat is controlled instead by the molecular diffusivity, even over scales where the diffusive motion itself is close to its homogenized limit. Some results for non-unit molecular diffusivity are recorded in Section 7. Finally, in Section 8, we describe a new Monte Carlo algorithm to compute the volume growth rate for the (X, K)-sausage, and hence the effective long-time rate of loss of heat. We present some numerical results obtained using the algorithm which interpolate between our theoretical predictions for large and small obstacles.

2 Review of homogenization for diffusion with periodic drift

There is a well known homogenization theory for $\mathcal{L}$-diffusion in the whole space $\mathbb{R}^d$. See [1], [2], [6], [11]. We review here a few basic facts which provide the background for our treatment of the case of a perforated domain. Our hypotheses on v ensure the existence of a periodic, Lipschitz, antisymmetric 2-tensor field β on $\mathbb{R}^d$ such that $\frac{1}{2} \operatorname{div} \beta = v$. So we can write $\mathcal{L}$ in the form

$$\mathcal{L} = \tfrac{1}{2} \operatorname{div}(I + \beta(x))\nabla.$$

Then $\mathcal{L}$ has a continuous heat kernel $p : (0, \infty) \times \mathbb{R}^d \times \mathbb{R}^d$ and there exists a constant $C < \infty$, depending only on the Lipschitz constant of v, such that, for all t, x and y,

$$C^{-1} \exp\{-C|x - y|^2/t\} \leqslant p(t, x, y) \leqslant C \exp\{-|x - y|^2/Ct\}. \quad (2.1)$$

Moreover, C may be chosen so that there also holds the following Gaussian tail estimate for the diffusion process X with generator $\mathcal{L}$ starting from x: for all $t > 0$ and $\delta > 0$,

$$\mathbb{P}_x \left(\sup_{s \leqslant t} |X_s - x| > \delta \right) \leqslant C e^{-\delta^2/Ct}. \tag{2.2}$$

The preceding two estimates show a qualitative equivalence between X and Brownian motion, valid on all scales. On large scales this can be refined in quantitative terms. Consider the quadratic form q on $\mathbb{R}^d$ given by

$$q(\xi) = \inf_{\theta, \chi} \int_{\mathbb{T}^d} |\xi - \operatorname{div} \chi + \beta \nabla \theta|^2 \, dx$$

where the infimum is taken over all Lipschitz functions θ and all Lipschitz antisymmetric 2-tensor fields χ on the torus $\mathbb{T}^d = \mathbb{R}^d/\mathbb{Z}^d$. The infimum is achieved, so there is a positive-definite symmetric matrix $\bar{a}$ such that

$$q(\xi) = \langle \xi, \bar{a}^{-1} \xi \rangle.$$

The choice $\theta = 0$ and $\chi = 0$ shows that $\bar{a} \geqslant I$. As the velocity field v is scaled up, typically it is found that $\bar{a}$ also becomes large. See for example [4] for further discussion of this phenomenon.

We state first a deterministic homogenization result. Let $f \in L^2(\mathbb{R}^d)$ and $\sigma \in (0, \infty)$ be given. Denote by u the solution to the Cauchy problem for $\mathcal{L}$ in $\mathbb{R}^d$ with initial data $f(./\sigma)$ and set $u^{(\sigma)}(t, x) = u(\sigma^2 t, \sigma x)$. Then

$$\int_{\mathbb{R}^d} |u^{(\sigma)}(t, x) - \bar{u}(t, x)|^2 \, dx \to 0 \tag{2.3}$$

as $\sigma \to \infty$, for all $t \geqslant 0$, where $\bar{u}$ is the solution to the Cauchy problem for $\frac{1}{2} \operatorname{div} \bar{a} \nabla$ in $\mathbb{R}^d$ with initial data f.

In probabilistic terms, we may fix $x \in \mathbb{R}^d$ and $\sigma \in (0, \infty)$ and consider the $\mathcal{L}$-diffusion process X starting from σx. Set $X_t^{(\sigma)} = \sigma^{-1} X_{\sigma^2 t}$. Then it is known [13] that

$$X^{(\sigma)} \to \bar{X}, \quad \text{weakly on } C([0, \infty), \mathbb{R}^d) \tag{2.4}$$

where $\bar{X}$ is a Brownian motion in $\mathbb{R}^d$ with diffusivity $\bar{a}$ starting from x. The two homogenization statements are essentially equivalent given the regularity implicit in the above qualitative estimates, the Markov property, and the identity

$$u^{(\sigma)}(t, x) = \mathbb{E}(f(X_t^{(\sigma)})).$$

3 Existence of a volume growth rate for a diffusion sausage with periodic drift

Recall that the drift v is $\mathbb{Z}^d$-periodic and divergence-free.

Theorem 3.1 *There exists a constant $\gamma = \gamma(v, K) \in (0, \infty)$ such that, for all x,*

$$\lim_{t \to \infty} \frac{|S_t^K|}{t} = \gamma, \quad \mathbb{P}_x\text{-almost surely.}$$

Proof Write π for the projection $\mathbb{R}^d \to \mathbb{T}^d$. Since v is periodic, the projected process $\pi(X)$ is a diffusion on $\mathbb{T}^d$. As v is divergence-free, the unique invariant distribution for $\pi(X)$ on $\mathbb{T}^d$ is the uniform distribution. The lower bound in (2.1) shows that the transition density of $\pi(X_1)$ on $\mathbb{T}$ is uniformly positive. By a standard argument $\pi(X)$ is therefore uniformly and geometrically ergodic. Consider the case where X_0 is chosen randomly, and independently of W, such that $\pi(X_0)$ is uniformly distributed on $\mathbb{T}^d$. Then $\pi(X)$ is stationary. For integers $0 \leqslant m < n$, define $V_{m,n} = |S_{(m,n]}^K|$. Then $V_{l,n} \leqslant V_{l,m} + V_{m,n}$ whenever $0 \leqslant l < m < n$. Since Lebesgue measure is translation invariant and $\pi(X)$ is stationary, the distribution of the array $(V_{m+k,n+k} : 0 \leqslant m < n)$ is the same for all $k \geqslant 0$. Moreover $V_{m,n}$ is integrable for all m, n by standard diffusion estimates. Hence by the subadditive ergodic theorem [8] we can conclude that, for some constant $\gamma \geqslant 0$,

$$\lim_{n \to \infty} \frac{|S_n^K|}{n} = \gamma, \quad \text{almost surely.}$$

The positivity of γ follows from the positivity of $\operatorname{cap}(K)$ using Theorem 5.1 below.

Let $\mathbb{P}_x$ be the probability measure on $C([0, \infty), \mathbb{R}^d)$ which is the law of the process X starting from x. Set

$$g(x) = \mathbb{P}_x\left(\lim_{n \to \infty} \frac{|S_n^K|}{n} = \gamma\right), \quad \tilde{g}(x) = \mathbb{P}_x\left(\lim_{n \to \infty} \frac{|S_{(1,n]}^K|}{n} = \gamma\right).$$

Then g is periodic and $\tilde{g} = g$. We have shown that

$$\int_{x \in [0,1]^d} g(x)\, dx = 1.$$

Hence $g(x) = 1$ for Lebesgue-almost-all x. But then by the Markov property, for every x,

$$g(x) = \tilde{g}(x) = \int_{\mathbb{R}^d} p(1, x, y) g(y)\, dy = 1$$

which is the desired almost-sure convergence for discrete parameter n. An obvious monotonicity argument extends this to the continuous parameter t. $\qquad\square$

4 Estimates for the diffusion sausage

We prepare some estimates on the diffusion sausage which will be needed later. These are of a type well known for Brownian motion [9] and extend in a straightforward way using the qualitative Gaussian bounds (2.1) and (2.2).

Lemma 4.1 *For all $p \in [1, \infty)$ there is a constant $C(p, v, K) < \infty$ such that, for all $t \geqslant 0$ and all $x \in \mathbb{R}^d$,*

$$\mathbb{E}_x \left(|S_t^K|^p \right)^{1/p} \leqslant C(t+1).$$

Proof Reduce to the case $t = 1$ by subadditivity of volume and L^p-norms and by the Markov property. The estimate then follows from (2.2) since

$$|S_1^K| \leqslant \omega_d \left(\operatorname{rad}(K) + \sup_{t \leqslant 1} |X_t - x| \right)^d.$$

$\qquad\square$

Lemma 4.2 *There is a constant $C(v, K) < \infty$ with the following property. Let X and Y be independent $\mathcal{L}$-diffusions starting from x. For all $t \geqslant 1$ and all $x \in \mathbb{R}^d$, for all $a, b \geqslant 0$,*

$$\mathbb{P}_x \left(S_{(at,(a+1)t]}^K(X) \cap S_{(bt,(b+1)t]}^K(Y) \neq \emptyset \right) \leqslant C(a+b)^{-d/2} \qquad (4.1)$$

and, when $b \geqslant a + 1$,

$$\mathbb{P}_x \left(S_{(at,(a+1)t]}^K(X) \cap S_{(bt,(b+1)t]}^K(X) \neq \emptyset \right) \leqslant C(b-a-1)^{-d/2}. \qquad (4.2)$$

Proof We write the proof for the case $t = 1$. The same argument applies generally. There is alternatively a reduction to the case $t = 1$ by scaling. Assume that $b \geqslant a + 1$. Write $\mathcal{F}_t$ for the σ-algebra generated by $(X_s : 0 \leqslant s \leqslant t)$ and set

$$R_a = \sup_{a \leqslant t \leqslant a+1} |X_t - X_{a+1}|, \quad R_b = \sup_{b \leqslant t \leqslant b+1} |X_t - X_b|, \quad Z = X_b - X_{a+1}.$$

Then, by (2.2),

$$\mathbb{P}_x(R_b \geqslant |Z|/3 | \mathcal{F}_b) \leqslant C e^{-|Z|^2/9C}$$

so, using (2.1),

$$\mathbb{P}_x(R_b \geqslant |Z|/3) \leqslant C\mathbb{E}_x(e^{-|Z|^2/9C}) \leqslant C(b-a-1)^{d/2}.$$

On the other hand, by (2.1) again,

$$\mathbb{P}_x(R_a \geqslant |Z|/3|\mathcal{F}_{a+1}) \leqslant C(b-a-1)^{d/2}R_a^d$$

so, using (2.2),

$$\mathbb{P}_x(R_a \geqslant |Z|/3) \leqslant C(b-a-1)^{d/2}.$$

Moreover (2.1) gives also

$$\mathbb{P}_x(2\,\mathrm{rad}(K) \geqslant |Z|/3) \leqslant C(b-a-1)^{d/2}.$$

Now if $S^K_{(a,a+1]}(X) \cap S^K_{(b,b+1]}(X) \neq \emptyset$ then either $R_a \geqslant |Z|/3$ or $R_b \geqslant |Z|/3$ or $2\,\mathrm{rad}(K) \geqslant |Z|/3$. Hence the preceding estimates imply (4.2). The proof of (4.1) is similar, resting on the fact that $X_a - Y_b$ has density bounded by $C(a+b)^{-d/2}$, and is left to the reader. □

Lemma 4.3 *As $t \to \infty$, we have*

$$\sup_x \mathbb{E}_x \left(\left| \frac{|S^K_t|}{t} - \gamma \right| \right) \to 0$$

and

$$\sup_x \mathbb{E}_x \left(\frac{|S^K_t(X) \cap S^K_t(Y)|}{t} \right) \to 0$$

where Y is an independent copy of X.

Proof Note that

$$|S^K_{(1,t+1]}| \leqslant |S^K_{t+1}| \leqslant |S^K_1| + |S^K_{(1,t+1]}|.$$

Given Lemma 4.1, the first assertion will follow if we can show that, as $t \to \infty$,

$$\sup_x \mathbb{E}_x \left(\left| \frac{|S^K_{(1,t+1]}|}{t} - \gamma \right| \right) \to 0.$$

But by the Markov property and using (2.1),

$$\mathbb{E}_x \left(\left| \frac{|S^K_{(1,t+1]}|}{t} - \gamma \right| \right) = \int_{\mathbb{R}^d} p(1,x,y)\mathbb{E}_y \left(\left| \frac{|S^K_t|}{t} - \gamma \right| \right) dy$$

$$\leqslant C \int_{[0,1]^d} \mathbb{E}_y \left(\left| \frac{|S^K_t|}{t} - \gamma \right| \right) dy \to 0$$

as $t \to \infty$, where we used the almost-sure convergence $|S_t^K|/t \to \gamma$ when $\pi(X_0)$ is uniform, together with uniform integrability from Lemma 4.1 to get the final limit.

For the second assertion, choose $q \in (1, 3/2)$ and $p \in (3, \infty)$ with $1/p + 1/q = 1$. Then, for $j, k \geqslant 0$, by Lemmas 4.1 and 4.2, there is a constant $C(p, v, K) < \infty$ such that

$$\mathbb{E}_x(|S_{(j,j+1]}^K(X) \cap S_{(k,k+1]}^K(Y)|)$$

$$\leqslant \mathbb{E}_x\left(|S_{(j,j+1]}^K(X)|1_{\{S_{(j,j+1]}^K(X) \cap S_{(k,k+1]}^K(Y) \neq \emptyset\}}\right)$$

$$\leqslant \mathbb{E}_x\left(|S_1^K(X)|^p\right)^{1/p} \mathbb{P}_x\left(S_{(j,j+1]}^K(X) \cap S_{(k,k+1]}^K(Y) \neq \emptyset\right)^{1/q}$$

$$\leqslant C(j+k)^{-d/2q}.$$

So, as $n \to \infty$, we have

$$\mathbb{E}_x\left(\frac{|S_n^K(X) \cap S_n^K(Y)|}{n}\right)$$

$$\leqslant \frac{\mathbb{E}_x(|S_1^K(X)|)}{n} + \sum_{j=1}^{n-1}\sum_{k=0}^{n-1}\frac{\mathbb{E}_x(|S_{(j,j+1]}^K(X) \cap S_{(k,k+1]}^K(Y)|)}{n}$$

$$\leqslant Cn^{-1} + Cn^{-d/(2q)+1} \to 0.$$

$\square$

5 Asymptotics of the growth rate for small and large cross-sections

We investigate the behaviour of the asymptotic growth rate $\gamma(v, rK)$ of the volume of the (X, rK)-sausage S_t^{rK} in the limits $r \to 0$ and $r \to \infty$. Recall the stochastic differential equation (1.1) for X and recall the rescaled process $X^{(\sigma)}$ from Section 2. Set $W_t^{(\sigma)} = \sigma^{-1}W_{\sigma^2 t}$. Then $W^{(\sigma)}$ is also a Brownian motion and $X^{(\sigma)}$ satisfies the stochastic differential equation

$$dX_t^{(\sigma)} = dW_t^{(\sigma)} + v^{(\sigma)}(X_t^{(\sigma)})\, dt \qquad (5.1)$$

where $v^{(\sigma)}(x) = \sigma v(\sigma x)$. This makes it clear that $X^{(\sigma)} \to W$ as $\sigma \to 0$ weakly on $C([0, \infty), \mathbb{R}^d)$. Recall from Section 2 the fact that $X^{(\sigma)} \to \bar{X}$ as $\sigma \to \infty$, in the same sense, where $\bar{X}$ is a Brownian motion with diffusivity $\bar{a}$.

Take $\sigma = r$. Then

$$S_t^{rK}(X) = r S_{r^{-2}t}^K(X^{(r)})$$

so

$$|S_t^{rK}(X)| = r^d |S_{r^{-2}t}^K(X^{(r)})|.$$

Hence the limit

$$\gamma(v^{(r)}, K) := \lim_{t \to \infty} |S_t^K(X^{(r)})|/t$$

exists and equals $r^{2-d}\gamma(v, rK)$. The weak limits for $X^{(r)}$ as $r \to 0$ or $r \to \infty$ suggest the following result, which however requires further argument because the asymptotic growth rate of the sausage is not a continuous function on $C([0, \infty), \mathbb{R}^d)$. Write $\mathrm{cap}(K)$ for the Newtonian capacity of K and $\mathrm{cap}_{\bar{a}}(K)$ for the capacity of K with respect to the diffusivity matrix $\bar{a}$. Thus

$$\mathrm{cap}_{\bar{a}}(K) = \sqrt{\det \bar{a}} \; \mathrm{cap}(\bar{a}^{-1/2}K).$$

Theorem 5.1 *We have*

$$\lim_{r \to 0} r^{2-d}\gamma(v, rK) = \lim_{r \to 0} \gamma(v^{(r)}, K) = \gamma(0, K) = \mathrm{cap}(K)$$

and

$$\lim_{r \to \infty} r^{2-d}\gamma(v, rK) = \lim_{r \to \infty} \gamma(v^{(r)}, K) = \mathrm{cap}_{\bar{a}}(K).$$

Proof Fix $T \in (0, \infty)$ and write $I(j)$ for the interval $((j-1)T, jT]$. Consider for $1 \leqslant j \leqslant k$ the function $F_{j,k}$ on $C([0, \infty), \mathbb{R}^d)$ defined by

$$F_{j,k}(\gamma) = |S_{I(j)}^K(\gamma) \cap S_{I(k)}^K(\gamma)|/T.$$

Then $F_{j,k}$ is continuous, so

$$\lim_{r \to 0} \mathbb{E}(F_{j,k}(X^{(r)})) = \mathbb{E}(F_{j,k}(W)), \quad \lim_{r \to \infty} \mathbb{E}(F_{j,k}(X^{(r)})) = \mathbb{E}(F_{j,k}(\bar{X})).$$

Choose X_0 so that $\pi(X_0)$ is uniformly distributed. Then by stationarity $\mathbb{E}(F_{j,k}(X^{(r)})) = \mathbb{E}(F_{k-j}(X^{(r)}))$ where $F_j = F_{1,j+1}$. Fix r and write $S_I^K(X^{(r)}) = S_I$. Note that

$$S_{(0,nT]} = S_{I(1)} \cup \cdots \cup S_{I(n)}.$$

So, by inclusion-exclusion, we obtain

$$n\mathbb{E}(F_0(X^{(r)})) - \sum_{j=1}^{n-1}(n-j)\mathbb{E}(F_j(X^{(r)})) \leqslant \mathbb{E}(|S_{(0,nT]}|/T) \leqslant n\mathbb{E}(F_0(X^{(r)})).$$

Divide by n and let $n \to \infty$ to obtain

$$\mathbb{E}(F_0(X^{(r)})) - \sum_{j=1}^{\infty} \mathbb{E}(F_j(X^{(r)})) \leqslant \gamma(v^{(r)}, K) \leqslant \mathbb{E}(F_0(X^{(r)})).$$

Fix $q \in (1, 3/2)$ and $p \in (3, \infty)$ with $p^{-1} + q^{-1} = 1$. By Lemmas 4.1 and 4.2, there is a constant $C(p, v, K) < \infty$ such that, for all r and j,

$$\mathbb{E}(F_j(X^{(r)})) = \mathbb{E}\left(|S_{I(1)} \cap S_{I(j+1)}|/T\right)$$
$$\leqslant \mathbb{E}\left(||S_{I(1)}|/T|^p\right)^{1/p} \mathbb{P}\left(S_{I(1)} \cap S_{I(j+1)} \neq \emptyset\right)^{1/q} \leqslant 2C(j-1)^{-d/2q}.$$

Given $\varepsilon > 0$, we can choose $J(p, v, K) < \infty$ so that

$$\sum_{j=J+1}^{\infty} \mathbb{E}(F_j(X^{(r)})) \leqslant 2C \sum_{j=J}^{\infty} j^{-d/2q} \leqslant \varepsilon. \qquad (5.2)$$

We follow from this point the case $r \to \infty$. The argument for the other limit is the same. Let $r \to \infty$ to obtain

$$\mathbb{E}(F_0(\bar{X})) - \sum_{j=1}^{J} \mathbb{E}(F_j(\bar{X})) - \varepsilon \leqslant \liminf_{r \to \infty} \gamma(v^{(r)}, K)$$

$$\leqslant \limsup_{r \to \infty} \gamma(v^{(r)}, K) \leqslant \mathbb{E}(F_0(\bar{X})). \quad (5.3)$$

It is known that

$$\lim_{T \to \infty} \mathbb{E}(F_0(\bar{X})) = \lim_{T \to \infty} \mathbb{E}(|S_T^K(\bar{X})|/T) = \mathrm{cap}_{\bar{a}}(K).$$

See [9] for the case $\bar{a} = I$. The general case follows by a scaling transformation. Note that, for $j \geqslant 1$,

$$|S_{(0,T]}^K(\bar{X}) \cap S_{(jT,(j+1)T]}^K(\bar{X})| + |S_{(0,(j+1)T]}^K(\bar{X})| \leqslant \sum_{i=1}^{j+1} |S_{((i-1)T,iT]}^K(\bar{X})|.$$

Take expectation, divide by T and let $T \to \infty$ to obtain

$$(j+1)\,\mathrm{cap}_{\bar{a}}(K) + \limsup_{T \to \infty} \mathbb{E}|S_{(0,T]}^K(\bar{X}) \cap S_{(jT,(j+1)T]}^K(\bar{X})|/T$$

$$\leqslant (j+1)\,\mathrm{cap}_{\bar{a}}(K)$$

which says exactly that

$$\lim_{T \to \infty} \mathbb{E}(F_j(\bar{X})) = 0.$$

Hence the desired limit follows on letting $T \to \infty$ in (5.3). $\qquad\square$

6 Homogenization of the advection-diffusion equation in a perforated domain

Our main results are analogues to the homogenization statements (2.3), (2.4) for advection-diffusion in a perforated domain. Recall that v is a $\mathbb{Z}^d$-periodic, Lipschitz, mean-zero, divergence-free vector field on $\mathbb{R}^d$, and K is a compact subset of $\mathbb{R}^d$. The domain $D \subseteq \mathbb{R}^d$ is constructed by removing all the sets $K + p$, where p runs over the set P of atoms of a Poisson random measure μ on $\mathbb{R}^d$ of intensity $\rho = \sigma^{-2}$. Write

$$\bar{a} = \bar{a}(v, K), \quad \bar{\lambda} = \bar{\lambda}(v, K) = \gamma(v, \hat{K}).$$

Theorem 6.1 *Let $f \in L^2(\mathbb{R}^d)$ and $\sigma \in (0, \infty)$ be given. Denote by u the solution[2] to the Cauchy problem for*

$$\mathcal{L} = \tfrac{1}{2}\Delta + v(x).\nabla$$

in $[0, \infty) \times D$ with initial data $f(\cdot/\sigma)$, and with Dirichlet conditions on the boundary of D. Set $u^{(\sigma)}(t, x) = u(\sigma^2 t, \sigma x)$. Then

$$\mathbb{E} \int_{\mathbb{R}^d} |u^{(\sigma)}(t, x) - \bar{u}(t, x)|^2 \, dx \to 0$$

as $\sigma \to \infty$, for all $t \geqslant 0$, where $\bar{u}$ is the solution to the Cauchy problem for $\frac{1}{2}\operatorname{div}\bar{a}\nabla - \bar{\lambda}$ in $[0, \infty) \times \mathbb{R}^d$ with initial data f.

Proof Replace t by $\sigma^2 t$, x by σx and f by $f(\cdot/\sigma)$ in (1.2) and (1.3) to obtain

$$\mathbb{E}\left(u^{(\sigma)}(t, x)\right) = \mathbb{E}_{\sigma x}\left(f(X_t^{(\sigma)})\exp\{-\rho|S_{\sigma^2 t}^{\hat{K}}(X)|\}\right)$$

and

$$\mathbb{E}\left(u^{(\sigma)}(t, x)^2\right) = \mathbb{E}_{\sigma x}\left(f(X_t^{(\sigma)})f(Y_t^{(\sigma)})\exp\{-\rho|S_{\sigma^2 t}^{\hat{K}}(X) \cup S_{\sigma^2 t}^{\hat{K}}(Y)|\}\right)$$

where the subscript σx specifies the starting point of X and where Y is an independent copy of X. We omit from now on the superscript $\hat{K}$.

---

[2] We extend u to a function on $[0, \infty) \times \mathbb{R}^d$ by setting $u(t, x) = 0$ for any $x \notin D$.

Then[3]

$$\mathbb{E}\left(|u^{(\sigma)}(t,x) - \bar{u}(t,x)|^2\right)$$

$$= \mathbb{E}_{\sigma x}\left(f(X_t^{(\sigma)})f(Y_t^{(\sigma)})e^{-\rho|S_{\sigma^2 t}(X) \cup S_{\sigma^2 t}(Y)|}(1 - e^{-\rho|S_{\sigma^2 t}(X) \cap S_{\sigma^2 t}(Y)|})\right)$$

$$+ \left(\mathbb{E}_{\sigma x}(f(X_t^{(\sigma)}))\{e^{-\rho|S_t(X)|} - e^{-\bar{\lambda}t}\}\right.$$

$$+ \left.\left\{\mathbb{E}_{\sigma x}(f(X_t^{(\sigma)})) - \mathbb{E}_x(f(\bar{X}_t))\right\}e^{-\bar{\lambda}t}\right)^2$$

$$\leqslant \mathbb{E}_{\sigma x}\left(f(X_t^{(\sigma)})^2\right)\mathbb{E}_{\sigma x}\left(\rho|S_{\sigma^2 t}(X) \cap S_{\sigma^2 t}(Y)|\right)^{1/2}$$

$$+ 2\mathbb{E}_{\sigma x}\left(f(X_t^{(\sigma)})^2\right)\mathbb{E}_{\sigma x}\left(\rho|S_{\sigma^2 t}(X)| - \bar{\lambda}t|\right)$$

$$+ 2|u_0^{(\sigma)}(t,x) - \bar{u}_0(t,x)|^2 \qquad (6.1)$$

where $u_0^{(\sigma)}$ and $\bar{u}_0$ denote the corresponding solutions to the Cauchy problem with initial data f in the whole space, and we used Cauchy–Schwarz and $(a+b)^2 \leqslant 2a^2 + 2b^2$ and $|e^{-a} - e^{-b}|^2 \leqslant |b-a|$ to obtain the inequality. Now

$$\int_{\mathbb{R}^d} \mathbb{E}_{\sigma x}\left(f(X_t^{(\sigma)})^2\right) dx = \int_{\mathbb{R}^d} |f(x)|^2 dx < \infty$$

because dx is stationary for $X^{(\sigma)}$ and, by (2.3), as $\sigma \to \infty$

$$\int_{\mathbb{R}^d} |u_0^{(\sigma)}(t,x) - \bar{u}_0(t,x)|^2 dx \to 0.$$

So, using Lemma 4.3, on integrating (6.1) over $\mathbb{R}^d$ and letting $\sigma \to \infty$, we conclude that the right-hand side tends to 0, proving the theorem. $\square$

Theorem 6.2 *Let $x \in \mathbb{R}^d$ and $\sigma \in (0,\infty)$ be given. Let X be an $\mathcal{L}$-diffusion in $\mathbb{R}^d$ starting from σx and set*

$$T = \inf\{t \geqslant 0 : X_t \in K + P\}.$$

Set $X_t^{(\sigma)} = \sigma^{-1}X_{\sigma^2 t}$ and $T^{(\sigma)} = \sigma^{-2}T$. Write $\bar{X}$ for a Brownian motion in $\mathbb{R}^d$ with diffusivity $\bar{a}$ starting from x, and write $\bar{T}$ for an exponential random variable of parameter $\bar{\lambda}$, independent of $\bar{X}$. Then, as $\sigma \to \infty$,

$$(X^{(\sigma)}, T^{(\sigma)}) \to (\bar{X}, \bar{T}), \quad \text{weakly on } C([0,\infty), \mathbb{R}^d) \times [0,\infty).$$

Proof Write S_t for the $(X, \hat{K})$-sausage. Fix a bounded continuous function F on $C([0,\infty), \mathbb{R}^d)$ and fix $t > 0$. Then

$$\mathbb{E}\left(F(X^{(\sigma)})1_{\{T^{(\sigma)}>t\}}\right) = \mathbb{E}\left(F(X^{(\sigma)})\exp\{-\rho|S_{\sigma^2 t}|\}\right)$$

---

[3] This is an instance of the formula $\mathbb{E}(|X-a|^2) = \text{var}(X) + (\mathbb{E}(X) - a)^2$.

and

$$\mathbb{E}\left(F(\bar{X})1_{\{\bar{T}>t\}}\right) = \mathbb{E}\left(F(\bar{X})e^{-\bar{\lambda}t}\right)$$

so

$$\left|\mathbb{E}\left(F(X^{(\sigma)})1_{\{T^{(\sigma)}>t\}}\right) - \mathbb{E}\left(F(\bar{X})1_{\{\bar{T}>t\}}\right)\right|$$
$$\leqslant \|F\|_\infty \mathbb{E}_{\sigma x}|\rho|S_{\sigma^2 t}| - \bar{\lambda}t| + |\mathbb{E}(F(X^{(\sigma)})) - \mathbb{E}(F(\bar{X}))|e^{-\bar{\lambda}t}.$$

On letting $\sigma \to \infty$, the first term on the right tends to 0 by Lemma 4.3 and the second term tends to 0 by (2.4), so the left hand side also tends to 0, proving the theorem. $\qquad\square$

7 The case of diffusivity $\varepsilon^2 I$

In this section and the next we fix $\varepsilon \in (0, \infty)$ and consider the more general case of the operator

$$\mathcal{L} = \tfrac{1}{2}\varepsilon^2 \Delta + v(x).\nabla.$$

The following statements follow from the corresponding statements above for the case $\varepsilon = 1$ by scaling. Fix $x \in \mathbb{R}^d$ and let X be an $\mathcal{L}$-diffusion in $\mathbb{R}^d$ starting from x. Then

$$|S_t^K(X)|/t \to \gamma(\varepsilon, v, K), \quad \mathbb{P}_x\text{-almost surely}$$

as $t \to \infty$, where

$$\gamma(\varepsilon, v, K) = \varepsilon^2 \gamma(\varepsilon^{-2}v, K).$$

Moreover, setting $v^{(r)}(x) = rv(rx)$, as above, we have

$$\gamma(\varepsilon, v^{(r)}, K) \to \mathrm{cap}_{\varepsilon^2 I}(K) = \varepsilon^2 \mathrm{cap}(K), \quad \text{as } r \to 0$$

and

$$\gamma(\varepsilon, v^{(r)}, K) \to \mathrm{cap}_{\bar{a}(\varepsilon, v)}(K), \quad \text{as } r \to \infty$$

where

$$\bar{a}(\varepsilon, v) = \varepsilon^2 \bar{a}(\varepsilon^{-2}v).$$

Fix $\sigma \in (0, \infty)$ and suppose now that X starts at σx. Define as above $T = \inf\{t \geqslant 0 : X_t \in K + P\}$ and write $X_t^{(\sigma)} = \sigma^{-1}X_{\sigma^2 t}$ and $T^{(\sigma)} = \sigma^{-2}T$. Then, as $\sigma \to \infty$,

$$(X^{(\sigma)}, T^{(\sigma)}) \to (\bar{X}, \bar{T}), \quad \text{weakly on } C([0, \infty), \mathbb{R}^d) \times [0, \infty)$$

where $\bar{X}$ is a Brownian motion in $\mathbb{R}^d$ of diffusivity $\bar{a}(\varepsilon, v)$ starting from x, and where $\bar{T}$ is an exponential random variable independent of $\bar{X}$, of parameter $\bar{\lambda}(\varepsilon, v, K) = \gamma(\varepsilon, v, \hat{K})$.

8 Monte Carlo computation of the asymptotic growth rate

Let X be as in the preceding section. Fix $T \in (0, \infty)$. The following algorithm may be used to estimate numerically the volume of the (X, K)-sausage $S_T = S_T^K(X)$. The algorithm is determined by the choice of three parameters N, m, $J \in \mathbb{N}$.

- **Step 1:** *Compute an Euler–Maruyama solution $(X_{n\Delta t} : n = 0, 1, \ldots, N)$ to the stochastic differential equation*

$$dX_t = \varepsilon dW_t + v(X_t) \, dt, \quad X_0 = x \qquad (8.1)$$

  up to the final time $T = N\Delta t$ (Figure 8.1a).
- **Step 2:** *Calculate*

$$R_K = \max_{y \in K, \, 1 \leqslant k \leqslant d} |y^k|, \quad R_{X,T} = \max_{1 \leqslant n \leqslant N, \, 1 \leqslant k \leqslant d} |X_{n\Delta t}^k - x^k|.$$

We approximate S_T by $S_T^{(N)} = \cup_{0 \leqslant n \leqslant N}(K + X_{n\Delta t})$. Note that $S_T^{(N)}$ is contained in the cube with side-length $L = 2(R_K + R_{X,T})$ centred at x (Figure 8.1b).

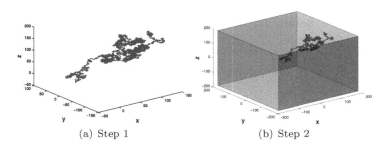

(a) Step 1   (b) Step 2

Figure 8.1 First two steps of the algorithm

- **Step 3:** *Subdivide the cube of side-length L centred at x into 2^d sub-cubes of side-length $L/2$ and check which of them have non-empty intersection with $S_T^{(N)}$. Discard any sub-cubes with empty intersection.*

Repeat the division and discarding procedure in each of the remaining sub-cubes (Figure 8.2) iteratively to obtain I sub-cubes of side-length $L/2^m$, centred at $y_1, \ldots, y_I$ say, whose union contains $S_T^{(N)}$.

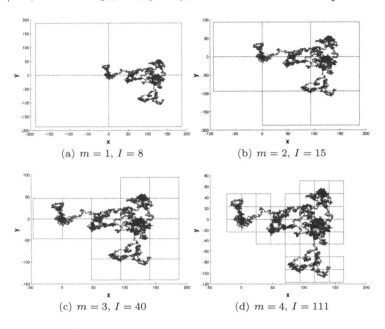

(a) $m = 1, I = 8$ (b) $m = 2, I = 15$

(c) $m = 3, I = 40$ (d) $m = 4, I = 111$

Figure 8.2 $x-y$ projection of the path and the sub-cubes for different values of m

- **Step 4:** *Generate uniform random variables $U_1, \ldots, U_J$ in $[-1/2, 1/2]^d$ and estimate $V = |S_T|$ by*

$$\hat{V} = J^{-1} 2^{-md} L^d \sum_{i=1}^{I} \sum_{j=1}^{J} \left(1 - \prod_{n=0}^{N} 1_{A_n(i,j)} \right)$$

where $A_n(i,j) = \{y_i + 2^{-m} L U_j \notin K + X_{n\Delta t}\}$.

The algorithm was tested in the case $d = 3$. We took $\varepsilon = 0.25$ and took v to be the Taylor–Green vector field in the first two co-ordinate directions; thus

$$v(x) = (-\sin x_1 \cos x_2, \ \cos x_1 \sin x_2, \ 0)^T.$$

We applied the algorithm to $X^{(r)}$, which has drift vector field $v^{(r)}(x) = rv(rx)$, for a range of choices of $r \in (0, \infty)$. We took K to be the Euclidean unit ball B and computed $|S_T^B(X^{(r)})|$ for $T = 10^4$, using parameter values $N = 10^6$, $m = 4$ and $J = 10^4$. The numerical method used to solve (8.1) was taken from [14]. The values $|S_T^B(X^{(r)})|/T$ were taken as estimates of the asymptotic volume growth rate $\gamma(0.25, v^{(r)}, B)$. These are displayed in Figure 8.3.

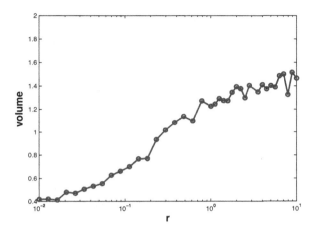

Figure 8.3 Growth rate of the sausage for different values of r

In Section 7 we stated the following theoretical limit, deduced from Theorem 5.1:

$$\lim_{r \to 0} \gamma(\varepsilon, v^{(r)}, B) = \varepsilon^2 \operatorname{cap}(B) = 2\pi\varepsilon^2 = 0.3927. \tag{8.2}$$

This is consistent with the computed values of $\gamma(0.25, v^{(r)}, B)$ when r is small.

It is known [10] that $\bar{a}(\varepsilon, v)$ has the form

$$\bar{a}(\varepsilon, v) = \begin{pmatrix} \alpha & 0 & 0 \\ 0 & \alpha & 0 \\ 0 & 0 & \varepsilon^2/2 \end{pmatrix}$$

for some $\alpha = \alpha(\varepsilon, v)$, which can be computed using Monte Carlo simulations. In [16], this was carried out for $\varepsilon = 0.25$ up to a final time $T = 10^4$, using a time step $\Delta t = 10^{-2}$, again using a numerical method from [14] to solve (8.1). The value $\alpha(0.25, v) = 0.0942$ was obtained as the sample average over 10^4 realizations of $|X_T^1|^2/T$. Using this value, we simulated $\bar{X}$ and used the volume algorithm to compute $|S_T^B(\bar{X})|/T$

as an approximation to $\mathrm{cap}_{\bar{a}(0.25,v)}(B)$, obtaining the value 1.4587. We showed theoretically that

$$\lim_{r\to\infty} \gamma(\varepsilon, v^{(r)}, B) = \mathrm{cap}_{\bar{a}(\varepsilon,v)}(B).$$

The computed value for $\mathrm{cap}_{\bar{a}(0.25,v)}(B)$ is consistent with the computed values of $\gamma(0.25, v^{(r)}, B)$ for large r.

References

[1] Aronson, D. G. 1967. Bounds for the fundamental solution of a parabolic equation. *Bull. Amer. Math. Soc.*, **73**, 890–896.

[2] Bensoussan, A., Lions, J.-L., and Papanicolaou, G. 1978. *Asymptotic Analysis for Periodic Structures*. Stud. Math. Appl., vol. 5. Amsterdam: North-Holland.

[3] Eisele, T., and Lang, R. 1987. Asymptotics for the Wiener sausage with drift. *Probab. Theory Related Fields*, **74**(1), 125–140.

[4] Fannjiang, A., and Papanicolaou, G. 1994. Convection enhanced diffusion for periodic flows. *SIAM J. Appl. Math.*, **54**(2), 333–408.

[5] Hoàng, V. H. 2000. Singularly perturbed Dirichlet problems in randomly perforated domains. *Comm. Partial Differential Equations*, **25**(1-2), 355–375.

[6] Jikov, V. V., Kozlov, S. M., and Oleĭnik, O. A. 1994. *Homogenization of Differential Operators and Integral Functionals*. Berlin: Springer-Verlag. Translated from the Russian by G. A. Yosifian [G. A. Iosif'yan].

[7] Kac, M. 1974. Probabilistic methods in some problems of scattering theory. *Rocky Mountain J. Math.*, **4**, 511–537. Notes by Macki, J., and Hersh, R. Papers arising from a Conference on Stochastic Differential Equations (Univ. Alberta, Edmonton, Alta., 1972).

[8] Kingman, J. F. C. 1968. The ergodic theory of subadditive stochastic processes. *J. Roy. Statist. Soc. Ser. B*, **30**, 499–510.

[9] Le Gall, J.-F. 1986. Sur la saucisse de Wiener et les points multiples du mouvement brownien. *Ann. Probab.*, **14**(4), 1219–1244.

[10] Majda, A. J., and Kramer, P. R. 1999. Simplified models for turbulent diffusion: theory, numerical modelling and physical phenomena. *Physics Reports*, **314**, 237–574.

[11] Norris, J. R. 1997. Long-time behaviour of heat flow: global estimates and exact asymptotics. *Arch. Rational Mech. Anal.*, **140**(2), 161–195.

[12] Papanicolaou, G. C., and Varadhan, S. R. S. 1980. Diffusion in regions with many small holes. Pages 190–206 of: *Stochastic Differential Systems (Proc. IFIP-WG 7/1 Working Conf., Vilnius, 1978)*. Lecture Notes in Control and Inform. Sci., vol. 25. Berlin: Springer-Verlag.

[13] Papanicolaou, G. C., and Varadhan, S. R. S. 1981. Boundary value problems with rapidly oscillating random coefficients. Pages 835–873 of:

Random Fields, Vol. I, II (Esztergom, 1979). Colloq. Math. Soc. János Bolyai, vol. 27. Amsterdam: North-Holland.

[14] Pavliotis, G. A., Stuart, A. M., and Zygalakis, K. C. 2009. Calculating effective diffusivities in the limit of vanishing molecular diffusion. *J. Comp. Phys*, **4**(228), 1030–1055.

[15] Rauch, J., and Taylor, M. 1975. Potential and scattering theory on wildly perturbed domains. *J. Funct. Anal.*, **18**, 27–59.

[16] Zygalakis, K. C. 2008. *Effective Diffusive Behavior For Passive Tracers and Inertial Particles: Homogenization and Numerical Algorithms*. Ph.D. thesis, University of Warwick.

18

Heavy traffic on a controlled motorway

F. P. Kelly[a] and R. J. Williams[b]

Abstract

Unlimited access to a motorway network can, in overloaded conditions, cause a loss of capacity. Ramp metering (signals on slip roads to control access to the motorway) can help avoid this loss of capacity. The design of ramp metering strategies has several features in common with the design of access control mechanisms in communication networks.

Inspired by models and rate control mechanisms developed for Internet congestion control, we propose a Brownian network model as an approximate model for a controlled motorway and consider it operating under a proportionally fair ramp metering policy. We present an analysis of the performance of this model.

AMS subject classification (MSC2010) 90B15, 90B20, 60K30

1 Introduction

The study of heavy traffic in queueing systems began in the 1960s, with three pioneering papers by Kingman [26, 27, 28]. These papers, and the early work of Prohorov [35], Borovkov [5, 6] and Iglehart [20], concerned a single resource. Since then there has been significant interest in networks of resources, with major advances by Harrison and Reiman [19], Reiman [37], Williams [43] and Bramson [7]. For discussions, further ref-

[a] Statistical Laboratory, University of Cambridge, Centre for Mathematical Sciences, Wilberforce Road, Cambridge CB3 0WB; F.P.Kelly@statslab.cam.ac.uk
[b] Department of Mathematics, University of California at San Diego, 9500 Gilman Drive, La Jolla, CA 92093–0112, USA; williams@stochastic.ucsd.edu. Research supported in part by NSF Grant DMS 0906535.

erences and overviews of the very extensive literature on heavy traffic for networks, Williams [42], Bramson and Dai [8], Harrison [17, 18] and Whitt [41] are recommended.

Research in this area is motivated in part by the need to understand and control the behaviour of communications, manufacturing and service networks, and thus to improve their design and performance. But researchers are also attracted by the elegance of some of the mathematical constructs: in particular, the multi-dimensional reflecting Brownian motions that often arise as limits.

A question that arises in a wide variety of application areas concerns how flows through a network should be controlled, so that the network responds sensibly to varying conditions. Road traffic was an area of interest to early researchers [33], and more recently the question has been studied in work on modelling the Internet. In each of these cases the network studied is part of a larger system: for example, drivers generate demand and select their routes in ways that are responsive to the delays incurred or expected, which depend on the controls implemented in the road network. It is important to address such interactions between the network and the larger system, and in particular to understand the signals, such as delay, provided to the larger system.

Work on Internet congestion control generally addresses the issue of fairness, since there exist situations where a given scheme might maximise network throughput, for example, while denying access to some users. In this area it has been possible to integrate ideas of fairness of a control scheme with overall system optimization: indeed fairness of the control scheme is often the means by which the right information and incentives are provided to the larger system [24, 38].

Might some of these ideas transfer to help our understanding of the control of road traffic? In this paper we present a preliminary exploration of a particular topic: ramp metering. Unlimited access to a motorway network can, in overloaded conditions, cause a loss of capacity. Ramp metering (signals on slip roads to control access to the motorway) can help avoid this loss of capacity. The problem is one of access control, a common issue for communication networks, and in this paper we describe a ramp metering policy, *proportionally fair metering*, inspired by rate control mechanisms developed for the Internet.

The organisation of this paper is as follows. In Section 2 we review early heavy traffic results for a single queue. In Section 3 we describe a model of Internet congestion control, which we use to illustrate the simplifications and insights heavy traffic allows. In Section 4 we describe a Brownian network model, which both generalizes a model of Section 2 and arises as a heavy traffic limit of the networks considered in Section 3. Sections 3 and 4 are based on the recent results of [21, 25]. These heavy traffic models help us to understand the behaviour of networks operating under policies for sharing capacity fairly.

In Section 5 we develop an approach to the design of ramp metering flow rates informed by the earlier Sections. For each of three examples, we present a Brownian network model operating under a proportionally fair metering policy. Our first example is a linear network representing a road into a city centre with several entry points; we then discuss a tree network, and, in Section 6, a simple network where drivers have routing choices. Within the Brownian network models we show that in each case the delay suffered by a driver at an entry point to the network can be expressed as a sum of dual variables, one for each of the resources to be used, and that under their stationary distribution these dual variables are independent exponential random variables. For the final example we show that the interaction of proportionally fair metering with choices available to arriving traffic has beneficial consequences for the performance of the system.

John Kingman's initial insight, that heavy traffic reveals the essential properties of queues, generalises to networks, where heavy traffic allows sufficient simplification to make clear the most important consequences of resource allocation policies.

2 A single queue

In this Section we review heavy traffic results for the $M/G/1$ queue, to introduce ideas that will be important later when we look at networks.

Consider a queue with a single server of unit capacity at which customers arrive as a Poisson process of rate ν. Customers bring amounts of work for the server which are independent and identically distributed with distribution G, and are independent of the arrival process. Assume the distribution G has mean $1/\mu$ and finite second moment, and that the load on the queue, $\rho = \nu/\mu$, satisfies $\rho < 1$.

Let $W(t)$ be the *workload* in the queue at time t; for a server of unit capacity this is the time it would take for the server to empty the queue if no more arrivals were to occur after time t.

Kingman [26] showed that the stationary distribution of $(1 - \rho)W$ is asymptotically exponentially distributed as $\rho \to 1$. Current approaches to heavy traffic generally proceed via a weaker assumption that the cumulative arrival process of work satisfies a functional central limit theorem, and use this to show that as $\rho \to 1$, the appropriately normalized workload process

$$\hat{W}(t) = (1 - \rho) \, W \left(\frac{t}{(1 - \rho)^2} \right), \qquad t \geq 0 \qquad (2.1)$$

can be approximated by a reflecting Brownian motion $\tilde{W}$ on $\mathbb{R}_+$. In the interior $(0, \infty)$ of $\mathbb{R}_+$, $\tilde{W}$ behaves as a Brownian motion with drift -1 and variance determined by the variance of the cumulative arrival process of work. When $\tilde{W}$ hits zero, then the server may become idle; this is where delicacy is needed. The stationary distribution of the reflecting Brownian motion $\tilde{W}$ is exponential, corresponding to Kingman's early result.

We note an important consequence of the scalings appearing in the definition (2.1), the *snapshot principle*. Because of the different scalings applied to space and time, the workload is of order $(1 - \rho)^{-1}$ while the workload can change significantly only over time intervals of order $(1 - \rho)^{-2}$. Hence the time taken to serve the amount of work in the queue is asymptotically negligible compared to the time taken for the workload to change significantly [36, 41].

Note that the workload $(W(t), t \geq 0)$ does not depend on the queue discipline (provided the discipline does not allow idling when there is work to be done), although the waiting time for an arriving customer certainly does. Kingman [29] makes elegant use of the snapshot principle to compare stationary waiting time distributions under a range of queue disciplines.

It will be helpful to develop in detail a simple example. Consider a Markov process in continuous time $(N(t), t \geq 0)$ with state space $\mathbb{Z}_+$ and non-diagonal infinitesimal transition rates

$$q(n, n') = \begin{cases} \nu & \text{if } n' = n + 1, \\ \mu & \text{if } n' = n - 1 \text{ and } n > 0, \\ 0 & \text{otherwise.} \end{cases} \qquad (2.2)$$

Let $\rho = \nu / \mu$. If $\rho < 1$ then the Markov process $(N(t), t \geq 0)$ has stationary distribution

$$\mathbb{P}\{N^s = n\} = (1 - \rho)\rho^n, \qquad n = 0, 1, 2, \ldots \qquad (2.3)$$

(here, the superscript s signals that the random variable is associated with the stationary distribution). The Markov process corresponds to an M/M/1 queue, at which customers arrive as a Poisson process of rate ν, and where customers bring an amount of work for the server which is exponentially distributed with parameter μ.

Next consider an M/G/1 queue with the processor-sharing discipline (under the processor-sharing discipline, while there are n customers in the queue each receives a proportion $1/n$ of the capacity of the server). The process $(N(t), t \geq 0)$ is no longer Markov, but it nonetheless has the same stationary distribution as in (2.3). Moreover in the stationary regime, given $N^s = n$, the amounts of work left to be completed on each of the n customers in the queue form a collection of n independent random variables, each with distribution function

$$G^*(x) = \mu \int_0^x (1 - G(z))\, dz, \quad x \geq 0,$$

a distribution recognisable as that of the forward recurrence time in a stationary renewal process whose inter-event time distribution is G. Thus the stationary distribution of W is just that of the sum of N^s independent random variables each with distribution G^*, where N^s has the distribution (2.3) [2, 23]. Let S be a random variable with distribution G. Then we can deduce that the stationary distribution of (W, N) has the property that in probability $W^s/N^s \to \mathbb{E}(S^2)/2\mathbb{E}(S)$, the mean of the distribution G^*, as $\rho \to 1$. For fixed x, under the stationary distribution for the queue, let N_x^s be the number of customers in the queue with a remaining work requirement of not more than x. Then, $N_x^s/N^s \to G^*(x)$ in probability as $\rho \to 1$. At the level of stationary distributions, this is an example of a property called state-space collapse: in heavy traffic the stochastic behaviour of the system is essentially given by W, with more detailed information about the system (in this case, the numbers of customers with various remaining work requirements) not being necessary.

The amount of work arriving at the queue over a period of time, τ, has a compound Poisson distribution, with a straightforwardly calculated mean and variance of $\rho\tau$ and $\rho\sigma^2\tau$ respectively, where $\sigma^2 = \mathbb{E}(S^2)/\mathbb{E}(S)$. An alternative approach [15] is to directly model the cumulative arrival process of work as a Brownian motion $\breve{E} = (\breve{E}(t), t \geq 0)$ with matching mean and variance parameters: thus

$$\breve{E}(t) = \rho t + \rho^{1/2}\sigma \breve{Z}(t), \qquad t \geq 0,$$

where $(\breve{Z}(t), t \geq 0)$ is a standard Brownian motion. Let

$$\breve{X}(t) = \breve{E}(t) - t, \qquad t \geq 0,$$

a Brownian motion starting from the origin with drift $-(1 - \rho)$ and variance $\rho\sigma^2$. In this approach we define the queue's workload $\breve{W}(t)$ at time t by the system of equations

$$\breve{W}(t) = \breve{W}(0) + \breve{X}(t) + \breve{U}(t), \qquad t \geq 0, \tag{2.4}$$

$$\breve{U}(t) = -\inf_{0 \leq s \leq t} \breve{X}(s), \qquad t \geq 0. \tag{2.5}$$

The interpretation of the model is as follows. While $\breve{W}$ is positive, it is driven by the Brownian fluctuations caused by arrival of work less the work served. But when $\breve{W}$ hits zero, the resource may not be fully utilized. The process $\breve{U}$ defined by equation (2.5) is continuous and non-decreasing, and is the minimal such process that permits $\breve{W}$, given by equation (2.4), to remain non-negative. We interpret $\breve{U}(t)$ as the cumulative unused capacity up to time t. Note that $\breve{U}$ can increase only at times when $\breve{W}$ is at zero.

The stationary distribution of $\breve{W}$ is exponential with mean $\rho\sigma^2/2(1-\rho)$ [15]. This is the same as the distribution of $(1 - \rho)^{-1}\tilde{W}^s$ where $\tilde{W}^s$ has the stationary distribution of the reflecting Brownian motion $\tilde{W}$ that approximates the scaled process $\hat{W}$ given by (2.1). Furthermore, the mean of the stationary distribution of $\breve{W}$ is the same as the mean of the exact stationary distribution of the workload W, calculated from its representation as the geometric sum (2.3) of independent random variables each with distribution G^* and hence mean $\mathbb{E}(S^2)/2\mathbb{E}(S)$.

In other words, for the M/G/1 queue, we obtain the same exponential stationary distribution either by (a) approximating the workload arrival process directly by a Brownian motion without any space or time scaling, or by (b) approximating the scaled workload process in (2.1) by a reflecting Brownian motion, finding the stationary distribution of the latter, and then formally unwinding the spatial scaling to obtain a distribution in the original spatial units. Furthermore, this exponential distribution has the same mean as the exact stationary distribution for the workload in the M/G/1 queue and provides a rather good approximation, being of the same order of accuracy as the exponential approximation of the geometric distribution with the same mean.

The main point of the above discussion is that, in the context of this example, we observe that for the purposes of computing approximations to the stationary workload, using a direct Brownian model for the workload arrival process (by matching mean and variance parameters) provides the same results as use of the heavy traffic diffusion approximation coupled with formal unwinding of the spatial scaling, and the approximate stationary distribution that this yields compares remarkably well with exact results. We shall give another example of this kind of fortuitously good approximation in Section 4. Chen and Yao [9] have also noted remarkably good results from using such 'strong approximations' without any scaling.

3 A model of Internet congestion

In this Section we describe a network generalization of processor sharing that has been useful in modelling flows through the Internet, and outline a recent heavy traffic approach [21, 25] to its analysis.

3.1 Fair sharing in a network

Consider a network with a finite set $\mathcal{J}$ of *resources*. Let a *route* i be a non-empty subset of $\mathcal{J}$, and write $j \in i$ to indicate that resource j is used by route i. Let $\mathcal{I}$ be the set of possible routes. Assume that both $\mathcal{J}$ and $\mathcal{I}$ are non-empty and finite, and let J and I denote the cardinality of the respective sets. Set $A_{ji} = 1$ if $j \in i$, and $A_{ji} = 0$ otherwise. This defines a $J \times I$ matrix $A = (A_{ji}, j \in \mathcal{J}, i \in \mathcal{I})$ of zeroes and ones, the *resource-route incidence matrix*. Assume that A has rank J, so that it has full row rank.

Suppose that resource j has capacity $C_j > 0$, and that there are n_i connections using route i. How might the capacities $C = (C_j, j \in \mathcal{J})$ be shared over the routes $\mathcal{I}$, given the numbers of connections $n = (n_i, i \in \mathcal{I})$? This is a question which has attracted attention in a variety of fields, ranging from game theory, through economics to political philosophy. Here we describe a concept of fairness which is a natural extension of Nash's bargaining solution and, as such, satisfies certain natural axioms of fairness [32]; the concept has been used extensively in the modelling of rate control algorithms in the Internet [24, 38].

Let $\mathcal{I}_+(n) = \{i \in \mathcal{I} : n_i > 0\}$. A *capacity allocation policy* $\Lambda = (\Lambda(n), n \in \mathbb{R}_+^I)$, where $\Lambda(n) = (\Lambda_i(n), i \in \mathcal{I})$, is called *proportionally fair* if for each $n \in \mathbb{R}_+^I$, $\Lambda(n)$ solves

$$\text{maximise} \quad \sum_{i \in \mathcal{I}_+(n)} n_i \log \Lambda_i \tag{3.1}$$

$$\text{subject to} \quad \sum_{i \in \mathcal{I}} A_{ji} \Lambda_i \leq C_j, \qquad j \in \mathcal{J}, \tag{3.2}$$

over
$$\Lambda_i \geq 0, \qquad\qquad i \in \mathcal{I}_+(n), \qquad (3.3)$$
$$\Lambda_i = 0, \qquad\qquad i \in \mathcal{I} \setminus \mathcal{I}_+(n). \qquad (3.4)$$

Note that the constraint (3.2) captures the limited capacity of resource j, while constraint (3.4) requires that no capacity be allocated to a route which has no connections.

The problem (3.1)–(3.4) is a straightforward convex optimization problem, with optimal solution

$$\Lambda_i(n) = \frac{n_i}{\sum_{j \in \mathcal{J}} q_j A_{ji}}, \qquad i \in \mathcal{I}_+(n), \qquad (3.5)$$

where the variables $q = (q_j, j \in \mathcal{J}) \geq 0$ are Lagrange multipliers (or *dual variables*) for the constraints (3.2). The solution to the optimization problem is unique and satisfies $\Lambda_i(n) > 0$ for $i \in \mathcal{I}_+(n)$ by the strict concavity on ($\Lambda_i > 0, i \in \mathcal{I}_+(n)$) and boundary behaviour of the objective function in (1.6) [25].

The dual variables q are unique if $n_i > 0$ for all $i \in \mathcal{I}$, but may not be unique otherwise. In any event they satisfy the *complementary slackness* conditions

$$q_j \left(C_j - \sum_{i \in \mathcal{I}} A_{ji} \Lambda_i(n) \right) = 0, \qquad j \in \mathcal{J}. \qquad (3.6)$$

3.2 Connection level model

The allocation $\Lambda(n)$ describes how capacities are shared for a given number of connections n_i on each route $i \in \mathcal{I}$. Next we describe a stochastic model [31] for how the number of connections within the network varies.

A connection on route i corresponds to continuous transmission of a document through the resources used by route i. Transmission is assumed to occur simultaneously through all the resources used by route i. Let the number of connections on route i at time t be denoted by $N_i(t)$, and let $N(t) = (N_i(t), i \in \mathcal{I})$. We consider a Markov process in continuous time $(N(t), t \geq 0)$ with state space $\mathbb{Z}_+^I$ and non-diagonal infinitesimal transition rates

$$q(n, n') = \begin{cases} \nu_i & \text{if } n' = n + e_i, \\ \mu_i \Lambda_i(n) & \text{if } n' = n - e_i \text{ and } n_i > 0, \\ 0 & \text{otherwise,} \end{cases} \qquad (3.7)$$

where e_i is the i-th unit vector in $\mathbb{Z}_+^I$, and $\nu_i, \mu_i > 0, i \in \mathcal{I}$.

The Markov process $(N(t), t \geq 0)$ corresponds to a model where new connections arrive on route i as a Poisson process of rate ν_i, and a connection on route i transfers a document whose size is exponentially distributed with parameter μ_i. In the case where $I = J = 1$ and $C_1 = 1$, the transition rates (3.7) reduce to the rates (2.2) of the M/M/1 queue.

Define the *load* on route i to be $\rho_i = \nu_i/\mu_i$ for $i \in \mathcal{I}$. It is known [4, 11] that the Markov process is positive recurrent provided

$$\sum_{i\in\mathcal{I}} A_{ji}\rho_i < C_j, \qquad j \in \mathcal{J}. \tag{3.8}$$

These are natural constraints: the load arriving at the network for resource j must be less than the capacity of resource j, for each $j \in \mathcal{J}$. Let $[\rho]$ be the $I \times I$ diagonal matrix with the entries of $\rho = (\rho_i, i \in \mathcal{I})$ on its diagonal, and define ν, $[\nu]$, μ, $[\mu]$ similarly.

Each connection on route i brings with it an amount of work for resource j which is exponentially distributed with mean $1/\mu_i$, for $j \in i$. The Markov process N allows us to estimate the workload for each resource: define the *workload process* by

$$W(t) = A\,[\mu]^{-1}N(t), \qquad t \geq 0. \tag{3.9}$$

3.3 Heavy traffic

To approximate the workload in a heavily loaded connection-level model by that in a Brownian network model, we view a given connection-level model as a member of a sequence of such models approaching the heavy traffic limit. More precisely, we consider a sequence of connection-level models indexed by r where the network structure, defined by A and C, does not vary with r. Each member of the sequence is a stochastic system as described in the previous section. We append a superscript of r to any process or parameter associated with the rth system that depends on r. Thus, we have processes N^r, W^r, and parameters ν^r. We suppose $\mu^r = \mu$ for all r, so that $\rho_i^r = \nu_i^r/\mu_i$, for each $i \in \mathcal{I}$. We shall assume henceforth that the following *heavy traffic* condition holds: as $r \to \infty$,

$$\nu^r \to \nu \quad \text{and} \quad r\,(A\rho^r - C) \to -\theta \tag{3.10}$$

where $\nu_j > 0$ and $\theta_j > 0$ for all $j \in \mathcal{J}$. Note that (3.10) implies that $\rho^r \to \rho$ as $r \to \infty$ and that $A\rho = C$.

We define fluid scaled processes $\bar{N}^r$, $\bar{W}^r$ as follows. For each r and $t \geq 0$, let

$$\bar{N}^r(t) = N^r(rt)/r, \qquad \bar{W}^r(t) = W^r(rt)/r. \tag{3.11}$$

What might be the limit of the sequence $\{\bar{N}^r\}$ as $r \to \infty$? From the transition rates (3.7) and the observation that $\Lambda(rn) = \Lambda(n)$ for $r > 0$, we would certainly expect that the limit satisfies

$$\frac{d}{dt} n_i(t) = \nu_i - \mu_i \Lambda_i(n(t)), \qquad i \in \mathcal{I}, \tag{3.12}$$

whenever n is differentiable at t and $n_i(t) > 0$ for all $i \in \mathcal{I}$. Indeed, this forms part of the fluid model developed in [25] as a functional-law-of-large-numbers approximation. Extra care is needed in defining the fluid model at any time t when $n_i(t) = 0$, for any $i \in \mathcal{I}$: the function $\Lambda(n)$ may not be continuous on the boundary of the region $\mathbb{R}_+^I$, and so when any component $\bar{N}_i^r(t)$ is hitting zero, $\Lambda(\bar{N}^r(t))$ may jitter.

It is shown in [25] that the set of invariant states for the fluid model is

$$\mathcal{N} = \left\{ n \in \mathbb{R}_+^I : n_i = \rho_i \sum_{j \in \mathcal{J}} q_j A_{ji}, i \in \mathcal{I} \text{ for some } q \in \mathbb{R}_+^J \right\}$$

as we would expect from formally setting the derivatives in (3.12) to zero and using relation (3.5). Call $\mathcal{N}$ the *invariant manifold*. If $n \in \mathcal{N}$, then since A has full row rank the representation of n in terms of q is unique; furthermore, $\Lambda_i(n) = \rho_i$ for $i \in \mathcal{I}_+(n)$ and then since $A\rho = C$, the vector q satisfies equation (3.5) and the complementary slackness conditions (3.6), and hence gives dual variables for the optimization problem (3.1)–(3.4).

For each $n \in \mathbb{R}_+^I$, define $w(n) = (w_j(n), j \in \mathcal{J})$, the workload associated with n, by $w(n) = A[\mu]^{-1}n$. For each $w \in \mathbb{R}_+^J$, define $\Delta(w)$ to be the unique value of $n \in \mathbb{R}_+^I$ that solves the following optimization problem:

$$\begin{aligned} \text{minimize} \quad & F(n) \\ \text{subject to} \quad & \sum_{i \in \mathcal{I}} A_{ji} \frac{n_i}{\mu_i} \geq w_j, \qquad j \in \mathcal{J}, \\ \text{over} \quad & n_i \geq 0, \qquad i \in \mathcal{I}, \end{aligned}$$

where

$$F(n) = \sum_{i \in \mathcal{I}} \frac{n_i^2}{\nu_i}, \qquad n \in \mathbb{R}_+^I.$$

The function $F(n)$ was introduced in [4] and can be used to show positive recurrence of N under conditions (3.8). In [25] the difference $F(n) - F(\Delta(w(n)))$ is used as a Lyapunov function to show that any fluid model solution $(n(t), t \geq 0)$ converges towards the invariant manifold $\mathcal{N}$. It is straightforward to check that $n \in \mathcal{N}$ if and only if $n = \Delta(w(n))$ and it turns out that

$$\Delta(w) = [\rho] A' (A[\mu]^{-1}[\nu][\mu]^{-1} A')^{-1} w.$$

Note that if $\bar{N}^r$ lives in the space $\mathcal{N}$ then $\bar{W}^r$, given by equations (3.9) and (3.11) as $A[\mu]^{-1}\bar{N}^r$, lives in the space $\mathcal{W} = A[\mu]^{-1}\mathcal{N}$, which we can write as

$$\mathcal{W} = \left\{ w \in \mathbb{R}_+^J : w = A[\mu]^{-1}[\nu][\mu]^{-1} A' q \text{ for some } q \in \mathbb{R}_+^J \right\}, \quad (3.13)$$

generally a space of lower dimension. Call $\mathcal{W}$ the *workload cone*. Let

$$\mathcal{W}^j = \Big\{ w \in \mathbb{R}_+^J : w = A[\mu]^{-1}[\nu][\mu]^{-1} A' q \\ \text{for some } q \in \mathbb{R}_+^J \text{ satisfying } q_j = 0 \Big\}, \quad (3.14)$$

which we refer to as the jth face of the workload cone $\mathcal{W}$.

We define diffusion scaled processes $\hat{N}^r$, $\hat{W}^r$ as follows. For each r and $t \geq 0$, let

$$\hat{N}^r(t) = \frac{N^r(r^2 t)}{r}, \qquad \hat{W}^r(t) = \frac{W^r(r^2 t)}{r}.$$

In the next sub-section we outline the convergence in distribution of the sequence $\{(\hat{N}^r, \hat{W}^r)\}$ as $r \to \infty$. As preparation, note that if $N(t) \in \mathcal{N}$ and $N_i(t) > 0$ for all $i \in \mathcal{I}$, then $\Lambda_i(N(t)) = \rho_i$ for all $i \in \mathcal{I}$. Suppose, as a thought experiment, that for each $i \in \mathcal{I}$ the component $\hat{N}_i^r$ behaves as the queue-length process in an independent M/M/1 queue, with a server of capacity ρ_i. Then a Brownian approximation to $\hat{N}_i^r$ would have variance $\nu_i + \mu_i \rho_i = 2\nu_i$. Next observe that if the covariance matrix of $\hat{N}^r$ is $2[\nu]$ then the covariance matrix of $\hat{W}^r = A[\mu]^{-1}\hat{N}^r$ is

$$\Gamma = 2A[\mu]^{-1}[\nu][\mu]^{-1} A'. \quad (3.15)$$

4 A Brownian network model

Let (A, ν, μ) be as in Section 3: thus A is a matrix of zeroes and ones of dimension $J \times I$ and of full row rank, and ν, μ are vectors of positive entries of dimension I. Let $\rho_i = \nu_i/\mu_i$, $i \in \mathcal{I}$, and $\rho = (\rho_i, i \in \mathcal{I})$. Let $\mathcal{W}$ and $\mathcal{W}^j$ be defined by expressions (3.13) and (3.14) respectively. Let $\theta \in \mathbb{R}^J$ and Γ be given by (3.15).

In the following, all processes are assumed to be defined on a fixed filtered probability space $(\Omega, \mathcal{F}, \{\mathcal{F}_t\}, \mathbb{P})$ and to be adapted to the filtration $\{\mathcal{F}_t\}$. Let η be a probability distribution on $\mathcal{W}$. Define a Brownian network model by the following relationships:

(i) $\tilde{W}(t) = \tilde{W}(0) + \tilde{X}(t) + \tilde{U}(t)$ for all $t \geq 0$,

(ii) $\tilde{W}$ has continuous paths, $\tilde{W}(t) \in \mathcal{W}$ for all $t \geq 0$, and $\tilde{W}(0)$ has distribution η,

(iii) $\tilde{X}$ is a J-dimensional Brownian motion starting from the origin with drift $-\theta$ and covariance matrix Γ such that $\{\tilde{X}(t) + \theta t, \mathcal{F}_t, t \geq 0\}$ is a martingale under $\mathbb{P}$,

(iv) for each $j \in \mathcal{J}$, $\tilde{U}_j$ is a one-dimensional process such that

    (a) $\tilde{U}_j$ is continuous and non-decreasing, with $\tilde{U}_j(0) = 0$,

    (b) $\tilde{U}_j(t) = \int_0^t 1_{\{\tilde{W}(s) \in \mathcal{W}^j\}} d\tilde{U}_j(s)$ for all $t \geq 0$.

The interpretation of the above Brownian network model is as follows. In the interior of the workload cone $\mathcal{W}$ each of the J resources are fully utilized, route i is receiving a capacity allocation ρ_i for each $i \in \mathcal{I}$, and the workloads $\tilde{W}$ are driven by the Brownian fluctuations caused by arrivals and departures of connections. But when $\tilde{W}$ hits the jth face of the workload cone $\mathcal{W}$, resource j may not be fully utilized. The cumulative unused capacity $\tilde{U}_j$ at resource j is non-decreasing, and can increase only on the jth face of the workload cone $\mathcal{W}$.

The work of Dai and Williams [10] establishes the existence and uniqueness in law of the above diffusion $\tilde{W} = (\tilde{W}(t), t \geq 0)$. In [21] it is shown that, if $\theta_j > 0$ for all $j \in \mathcal{J}$, then $\tilde{W}$ has a unique stationary distribution; furthermore, if $\tilde{W}^s$ denotes a random variable with this stationary distribution, then the components of $\tilde{Q}^s = 2\Gamma^{-1}\tilde{W}^s$ are independent and $\tilde{Q}_j^s$ is exponentially distributed with parameter θ_j for each $j \in \mathcal{J}$.

Now let C be a vector of positive entries of dimension J, define a sequence of networks as in Section 3.3, and suppose θ and C are related by the heavy traffic condition (3.10). In [21] it is shown that, subject to a certain local traffic condition on the matrix A and suitable convergence of initial variables $(\hat{W}^r(0), \hat{N}^r(0))$, the pair $(\hat{W}^r, \hat{N}^r)$ converges in distribution as $r \to \infty$ to a continuous process $(\tilde{W}, \tilde{N})$ where $\tilde{W}$ is the above diffusion and $\tilde{N} = \Delta(\tilde{W})$. The proof in [21] relies on both the existence and uniqueness results of [10] and an associated invariance principle developed by Kang and Williams [22]. (The local traffic condition under which convergence is established requires that the matrix A contains amongst its columns the columns of the $J \times J$ identity matrix:

this corresponds to each resource serving at least one route which uses only that resource. The local traffic condition is not needed to show that $\tilde{W}$ has the aforementioned stationary distribution; that requires only the weaker condition that A have full row rank.)

It is convenient to define $\tilde{Q} = 2\Gamma^{-1}\tilde{W}$, a process of *dual variables*. From this, the form of Δ, and the relation $\tilde{N} = \Delta(\tilde{W})$, it follows that $\tilde{N} = [\rho]A'\tilde{Q}$. The dimension of the space in which $\tilde{Q}$ lives is J, and so this is an example of state-space collapse, with the I-dimensional process $\tilde{N}$ living on a J-dimensional manifold where $J \leq I$ is often considerably less than I.

Using the stationary distribution for $\tilde{W}$, we see that $\tilde{N}^s = [\rho]A'\tilde{Q}^s$ has the stationary distribution of $\tilde{N}$. Then, after formally unwinding the spatial scaling used to obtain our Brownian approximation, we obtain the following simple approximation for the stationary distribution of the number-of-connections process in the original model described in Section 3.2:

$$N_i^s \approx \rho_i \sum_{j \in \mathcal{J}} Q_j^s A_{ji}, \quad i \in \mathcal{I}, \tag{4.1}$$

where Q_j^s, $j \in \mathcal{J}$, are independent and Q_j^s is exponentially distributed with parameter $C_j - \sum_{i \in \mathcal{I}} A_{ji}\rho_i$.

As mentioned in Section 2, an alternative approach is to directly model the cumulative arrival process of work for each route i as a Brownian motion:

$$\breve{E}_i(t) = \rho_i t + \left(\frac{2\rho_i}{\mu_i}\right)^{1/2} \breve{Z}_i(t), \quad t \geq 0,$$

where $(\breve{Z}_i(t), t \geq 0)$, $i \in \mathcal{I}$, are independent standard Brownian motions; here the form of the variance parameter takes account of the fact that the document sizes are exponentially distributed. Under this model, the potential netflow (inflow minus potential outflow, ignoring underutilization of resources) process of work for resource j is

$$\breve{X}_j(t) = \sum_{i \in \mathcal{I}} A_{ji}\breve{E}_i(t) - C_j t, \quad t \geq 0,$$

a J-dimensional Brownian motion starting from the origin with drift $A\rho - C$ and covariance matrix $2A[\rho][\mu]^{-1}A' = \Gamma$. Then the workload is modelled by a J-dimensional process $\breve{W}$ that satisfies properties (i)–(iv) above, but with $\breve{W}$ in place of $\tilde{W}$ and $A\rho - C$ in place of the drift $-\theta$; the covariance matrix remains the same. By the results of [21], if $A\rho < C$, there is a unique stationary distribution for the process $\breve{W}$

such that if $\check{W}^s$ has this stationary distribution then the components of $\check{Q}^s = 2\Gamma^{-1}\check{W}^s$ are independent and $\check{Q}^s_j$ is exponentially distributed with parameter $C_j - \sum_{i \in \mathcal{I}} A_{ji}\rho_i$ for each $j \in \mathcal{J}$. The random variable

$$\check{N}^s = [\rho]A'\check{Q}^s, \tag{4.2}$$

has the stationary distribution of $\check{N} = [\rho]A'\check{Q}$, which is the same as the distribution of the right member of (4.1). Thus, just as in the simple case considered in Section 2, in this connection-level model, using the direct Brownian model yields the same approximation for the stationary distribution of the number-of-connections process as that obtained using the heavy traffic diffusion approximation and formally unwinding the spatial scaling in its stationary distribution.

If we specialize the direct Brownian network model to the case where $I = J = 1$ and $C = 1$, then we obtain the Brownian model of Section 2, with $\Gamma = 2\nu/\mu^2 = \rho\sigma^2$ and where the stationary distribution for $\check{W}$ is exponentially distributed with mean $\rho\sigma^2/2(1 - \rho)$, yielding the same approximation as in Section 2.

A more interesting example is obtained when $I = J + 1$ and A is the $J \times (J + 1)$ matrix:

$$A = \begin{pmatrix} 1 & 0 & \cdots & 0 & 1 \\ 0 & 1 & \cdots & 0 & 1 \\ \vdots & \vdots & \ddots & \vdots & \vdots \\ 0 & 0 & \cdots & 1 & 1 \end{pmatrix},$$

so that J routes each use a single resource in such a way that there is exactly one such route for each resource, and one route uses all J resources. In this case, the stationary distribution given by (4.2) accords remarkably well with the exact stationary distribution described by Massoulié and Roberts [31]; it is again of the order of accuracy of the exponential approximation of the geometric distribution with the same mean. (We refer the interested reader to [21] for the details of this good approximation.)

In this Section and in Section 2 we have seen intriguing examples of remarkably good approximations that the direct Brownian modelling approach can yield. Inspired by this, in the next two Sections we explore the use of the direct Brownian network model as a representation of workload for a controlled motorway. Rigorous justification for use of this modelling framework in the motorway context has yet to be investigated. See the last section of the paper for further comments on this issue.

5 A model of a controlled motorway

Once motorway traffic exceeds a certain threshold level (measured in terms of density—the number of vehicles per mile) both vehicle speed and vehicle throughput drop precipitously [13, 12, 39]. The smooth pattern of flow that existed at lower densities breaks down, and the driver experiences stop-go traffic. Maximum vehicle throughput (measured in terms of the number of vehicles per minute) occurs at quite high speeds—about 60 miles per hour on Californian freeways and on London's orbital motorway, the M25 [13, 12, 39]—while after flow breakdown the average speed may drop to 20–30 miles per hour. Particularly problematic is that flow breakdown may persist long after the conditions that provoked its onset have disappeared.

Variable speed limits lessen the number and severity of accidents on congested roads and are in use, for example, on the south-west quadrant of the M25. But variable speed limits do not avoid the loss of throughput caused by too high a density of vehicles [1, 14]. Ramp metering (signals on slip roads to control access to the motorway) can limit the density of vehicles, and thus can avoid the loss of throughput [30, 34, 40, 44]. But a cost of this is queueing delay on the approaches to the motorway. How should ramp metering flow rates be chosen to control these queues, and to distribute queueing delay fairly over the various users of the motorway? In this Section we introduce a modelling approach to address this question, based on several of the simplifications that we have seen arise in heavy traffic.

5.1 A linear network

Consider the linear[1] road network illustrated in Figure 5.1. Traffic can enter the main carriageway from lines at entry points, and then travels from left to right, with all traffic destined for the exit at the right hand end (think of this as a model of a road collecting traffic all bound for a city). Let $M_1(t)$, $M_2(t)$, ..., $M_J(t)$ taking values in $\mathbb{R}_+$ be the line sizes[2] at the entry points at time t, and let C_1, C_2, ..., C_J be the respective capacities of sections of the road. We assume the road starts at the left hand end, with line J feeding an initial section of capacity C_J, and that

---

[1] We caution the reader that here we use the descriptive term 'linear network' in a manner that differs from its use in [21].

[2] The term line size is used here to mean a quantity measuring the amount of work in the queue, rather than the more restrictive number of jobs that is often associated with the term queue size.

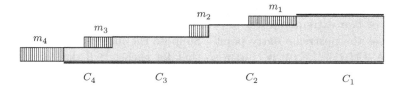

Figure 5.1 Lines of size m_1, m_2, m_3, m_4 are held on the slip roads leading to the main carriageway. Traffic on the main carriageway is free-flowing. Access to the main carriageway from the slip roads is metered, so that the capacities C_1, C_2, C_3, C_4 of successive sections are not overloaded.

$C_1 > C_2 > \ldots > C_J > 0$. The corresponding resource-route incidence matrix is the square matrix

$$A = \begin{pmatrix} 1 & 1 & \cdots & 1 \\ 0 & 1 & \cdots & 1 \\ \vdots & \vdots & \ddots & \vdots \\ 0 & 0 & \cdots & 1 \end{pmatrix}. \tag{5.1}$$

We model the traffic, or work, arriving at line i, $i \in \mathcal{I}$, as follows: let $E_i(t)$ be the cumulative inflow to line i over the time interval $(0, t]$, and assume $(E_i(t), t \geq 0)$ is an ergodic process with non-negative, stationary increments, with $\mathbb{E}[E_i(t)] = \rho_i t$, where $\rho_i > 0$, and suppose these processes are independent over $i \in \mathcal{I}$. Suppose the metering rates for lines $1, 2, \ldots, J$ at time t can be chosen to be any measurable vector-valued function $\Lambda = \Lambda(M(t))$ satisfying constraints (3.2)–(3.4) with $n = M(t)$, and such that

$$M_i(t) = M_i(0) + E_i(t) - \int_0^t \Lambda_i(M(s)) \, ds \geq 0, \qquad t \geq 0 \tag{5.2}$$

for $i \in \mathcal{I}$. Observe that we do not take into account travel time along the road: motivated by the snapshot principle, we suppose that $M(\cdot)$ varies relatively slowly compared with the time taken to travel through the system.[3]

---

[3] The time taken for a vehicle to travel through the system comprises both the queueing time at the entry point and the travel time along the motorway. If the motorway is free-flowing, the aim of ramp metering, then the travel time along the motorway may be reasonably modelled by a constant not dependent on $M(t)$, say τ_i from entry point i. A more refined treatment might insist that the rates $\Lambda(M_i(t - \tau_i), i \in \mathcal{I})$ satisfy the capacity constraints (3.2). We adopt the simpler approach, since we expect that in heavy traffic travel times along the motorway will be small compared with the time taken for $M(t)$ to change significantly.

How might the rate function $\Lambda(\cdot)$ be chosen? We begin by a discussion of two extreme strategies. First we consider a strategy that prioritises the upstream entry points. Suppose the metered rate from line J, $(\Lambda_J(M(t)), t \geq 0)$, is chosen so that for each $t \geq 0$ the cumulative outflow from line J, $\int_0^t \Lambda_J(M(s))\, ds$, is maximal, subject to the constraint (5.2) and $\Lambda_J(M(t)) \leq C_J$ for all $t \geq 0$: thus there is equality in the latter constraint whenever $M_J(t)$ is positive. For each of $j = J - 1$, $J - 2$, ..., 1 in turn define $\int_0^t \Lambda_j(M(s))\, ds$ to be maximal, subject to the constraint (5.2) and

$$\Lambda_j(M(t)) \leq C_j - \sum_{i=j+1}^{J} \Lambda_i(M(t)), \qquad t \geq 0. \tag{5.3}$$

In consequence there is equality in constraint (5.3) at time t if $M_j(t) > 0$, and by induction for each $t \geq 0$ the cumulative flow along link j, $\int_0^t \sum_{i=j}^{J} \Lambda_i(M(s))\, ds$, is maximal, for $j = J, J - 1, \ldots, 1$. Thus this strategy minimizes, for all times t, the sum of the line sizes at time t, $\sum_j M_j(t)$.

The above optimality property is compelling if the arrival patterns of traffic are exogenously determined. The strategy will, however, concentrate delay upon the flows entering the system at the more downstream entry points. This seems intuitively unfair, since these flows use fewer of the system's resources, and it may well have perverse and suboptimal consequences if it encourages growth in the load ρ_i arriving at the upstream entry points. For example, growth in ρ_J may cause the natural constraint (3.8) to be violated, even while traffic arriving at line J suffers only a small amount of additional delay.

Next we consider a strategy that prioritises the downstream entry points. To present the argument most straightforwardly, let us suppose that the cumulative inflow to line i is discrete, i.e., $(E_i(t), t \geq 0)$ is constant except at an increasing, countable sequence of times $t \in (0, \infty)$, for each $i \in \mathcal{I}$. Suppose the inflow from line 1 is chosen to be $\Lambda_1(M(t)) = C_1$ whenever $M_1(t)$ is positive, and zero otherwise. Then link 1 will be fully utilized by the inflow from line 1 a proportion ρ_1/C_1 of the time. Let $\Lambda_j(M(t)) = C_j$ whenever both $M_j(t)$ is positive and $\Lambda_i(M(t)) = 0$ for $i < j$, and let $\Lambda_j(M(t)) = 0$ otherwise. This strategy minimizes lexicographically the vector $(M_1(t), M_2(t), \ldots, M_J(t))$ at all times t. Provided the system is stable, link 1 will be utilized solely by the inflow from line

j a proportion ρ_j/C_j of the time. Hence the system will be unstable if

$$\sum_{j=1}^{J} \frac{\rho_j}{C_j} > 1,$$

and thus may well be unstable even when the condition (3.8) is satisfied. Essentially the strategy starves the downstream links, preventing them from working at their full capacity. Our assumption that the cumulative inflow to line i is discrete is not essential for this argument: the stability region will be reduced from (3.8) under fairly general conditions.

The two extreme strategies we have described each have their own interest: the first has a certain optimality property but distributes delay unfairly, while the second can destabilise a network even when all the natural capacity constraints (3.8) are satisfied.

5.2 Fair sharing of the linear network

In this sub-section we describe our preferred ramp metering policy for the linear network, and our Brownian network model for its performance.

Given the line sizes $M(t) = m$, we suppose the metered rates $\Lambda(m)$ are chosen to be proportionally fair: that is, the capacity allocation policy $\Lambda(\cdot)$ solves the optimization problem (3.1)–(3.4). Hence for the linear network we have from relations (3.5)–(3.6) that

$$\Lambda_i(m) = \frac{m_i}{\sum_{j=1}^{i} q_j}, \qquad i \in \mathcal{I}_+(m),$$

where the q_j are Lagrange multipliers satisfying

$$q_j \geq 0, \quad q_j \left(C_j - \sum_{i=j}^{J} \Lambda_i(m) \right) = 0, \qquad j \in \mathcal{J}. \tag{5.4}$$

Under this policy the total flow along section j will be its capacity C_j whenever $q_j > 0$.

Given line sizes $M(t) = m$, the ratio $m_i/\Lambda_i(m)$ is the time it would take to process the work currently in line i at the current metered rate for line i. Thus

$$d_i = \sum_{j=1}^{i} q_j, \qquad i \in \mathcal{I}, \tag{5.5}$$

give estimates, based on current line sizes, of queueing delay in each of the I lines. Note that these estimates do not take into account any change in the line sizes over the time taken for work to move through the line.

Next we describe our direct Brownian network model for the linear network operating under the above policy. We make the assumption that the inflow to line i is a Brownian motion $\breve{E}_i = (\breve{E}_i(t), t \geq 0)$ starting from the origin with drift ρ_i and variance parameter $\rho_i \sigma^2$, and so can be written in the form

$$\breve{E}_i(t) = \rho_i t + \rho_i^{1/2} \sigma \breve{Z}_i(t), \qquad t \geq 0 \tag{5.6}$$

for $i \in \mathcal{I}$, where $(\breve{Z}_i(t), t \geq 0)$, $i \in \mathcal{I}$, are independent standard Brownian motions. For example, if the inflow to each line were a Poisson process, then this would be the central limit approximation, with $\sigma = 1$. More general choices of σ could arise from either a compound Poisson process, or the central limit approximation to a large class of inflow processes.

Our Brownian network model will be a generalization of the model (2.4)–(2.5) of a single queue, and a specialization of the model of Section 4 to the case where $\mu_i = 2/\sigma^2$, $i \in \mathcal{I}$, and the matrix A is of the form (5.1).

Let

$$\breve{X}_j(t) = \sum_{i \in \mathcal{I}} A_{ji} \breve{E}_i(t) - C_j t, \quad t \geq 0;$$

note that the first term is the cumulative workload entering the system for resource j over the interval $(0, t]$. Write $\breve{X}(t) = (\breve{X}_j(t), j \in \mathcal{J})$ and $\breve{X} = (\breve{X}(t), t \geq 0)$. Then $\breve{X}$ is a J-dimensional Brownian motion starting from the origin with drift $A\rho - C$ and covariance matrix $\Gamma = \sigma^2 A[\rho]A'$. We assume the stability condition (3.8) is satisfied, so that $A\rho < C$.

Write

$$\mathcal{W} = A[\rho]A' \mathbb{R}_+^J \tag{5.7}$$

for the workload cone, and

$$\mathcal{W}^j = \{A[\rho]A'q : q \in \mathbb{R}_+^J, q_j = 0\}, \tag{5.8}$$

for the jth face of $\mathcal{W}$. Our Brownian network model for the resource level workload AM is then the process $\breve{W}$ defined by properties (i)–(iv) of Section 4 with $\breve{W}$ in place of $\tilde{W}$, $C - A\rho$ in place of θ and $\Gamma = \sigma^2 A[\rho]A'$.

The form (5.1) of the matrix A allows us to rewrite the workload cone (5.7) as

$$\mathcal{W} = \left\{ w \in \mathbb{R}^J : \frac{w_{j-1} - w_j}{\rho_{j-1}} \leq \frac{w_j - w_{j+1}}{\rho_j}, j = 1, 2, \ldots, J \right\}$$

where $w_{J+1} = 0$ and we interpret the left hand side of the inequality as 0 when $j = 1$. Under this model, at any time t when the workloads $\breve{W}(t)$ are in the interior of the workload cone $\mathcal{W}$, each resource is fully utilized. But when $\breve{W}$ hits the jth face of the workload cone $\mathcal{W}$, resource j may not be fully utilized. Our model corresponds to the assumption that there is no more loss of utilization than is necessary to prevent $\breve{W}$ from leaving $\mathcal{W}$. This assumption is made for our Brownian network model by analogy with the results reviewed in Sections 3 and 4, where it emerged as a property of the heavy traffic diffusion approximation.

In a similar manner to that in Section 4, we define a process of dual variables: $\breve{Q} = (A[\rho]A')^{-1}\breve{W}$. Since $\breve{W}$ is our model for AM, our Brownian model for the line sizes is given by

$$\breve{M} = A^{-1}\breve{W} = [\rho]A'\breve{Q}. \tag{5.9}$$

Within our Brownian model we represent (nominal) delays $\breve{D} = (\breve{D}_i, i \in \mathcal{I})$ at each line as given by

$$\breve{D} = A'\breve{Q}, \tag{5.10}$$

since these would be the delays if line sizes remained constant over the time taken for a unit of traffic to move through the line, with ρ_i both the arrival rate and metered rate at line i.[4] Relation (5.10) becomes, for the linear network,

$$\breve{D}_i = \sum_{j=1}^{i} \breve{Q}_j, \qquad i = 1, 2, \ldots, J,$$

parallelling relation (5.5). Note that when $\breve{W}$ hits the jth face of the workload cone $\mathcal{W}$, then $\breve{Q}_j = 0$ and $\breve{D}_{j-1} = \breve{D}_j$; thus the loss of utilization at resource j when $\breve{W}$ hits the jth face of the workload cone $\mathcal{W}$ is just sufficient to prevent the delay at line j becoming smaller than the delay at the downstream line $j - 1$.

If $\breve{W}^s$ has the stationary distribution of $\breve{W}$, then the components of $\breve{Q}^s = (A[\rho]A')^{-1}\breve{W}^s$ are independent and $\breve{Q}_j^s$ is exponentially distributed

---

[4] The nominal delay $\breve{D}_i(t)$ for line i at time t will not in general be the realized delay (the time taken for the amount of work $\breve{M}_i(t)$ found in line i at time t to be metered from line i). Since $A\rho < C$ the metered rate $\Lambda_i(m)$ will in general differ from ρ_i even when $m = A^{-1}\breve{W}$ and $\breve{W} \in \mathcal{W}$. Our definition of nominal delay is informed by our earlier heavy traffic results: as $A\rho$ approaches C we expect scaled realized delay to converge to scaled nominal delay. Metered rates do fluctuate as a unit of traffic moves through the line, but we expect less and less so as the system moves into heavy traffic.

with parameter

$$\frac{2}{\sigma^2} \left(C_j - \sum_{i=j}^{J} \rho_i \right), \qquad j = 1, 2, \dots, J.$$

The stationary distributions of $\breve{M}$ and $\breve{D}$ are then given by the distributions of $\breve{M}^s$ and $\breve{D}^s$, respectively, where

$$\breve{M}_i^s = \rho_i \sum_{j=1}^{i} \breve{Q}_j^s, \quad \breve{D}_i^s = \sum_{j=1}^{i} \breve{Q}_j^s \qquad i = 1, 2, \dots, J.$$

In the above example the matrix A is invertible. As an example of a network with a non-invertible A matrix, suppose that in the linear network illustrated in Figure 5.1 one section of road is unconstrained, say $C_3 = \infty$. Then, removing the corresponding row from the resource-route incidence matrix we have

$$A = \begin{pmatrix} 1 & 1 & 1 & 1 \\ 0 & 1 & 1 & 1 \\ 0 & 0 & 0 & 1 \end{pmatrix}.$$

The workload cone is the collapse of $\mathcal{W}$ obtained by setting $w_2 = w_3$, and in consequence the construction of $\breve{W}$ and $\breve{M} = [\rho]A'(A[\rho]A')^{-1}\breve{W}$ enforces the relationship $\breve{M}_2/\breve{M}_3 = \rho_2/\rho_3$. Since the matrix A is not invertible, this is no longer a necessary consequence of the network topology, but is a natural modelling assumption, motivated by the forms of state-space collapse we have seen earlier. Essentially lines 2 and 3 use the same network resources and face the same queueing delays.

A Brownian network model of the first strategy from Section 5.1 could also be constructed, but the workload cone and its faces would not be of the required form (5.7) and (5.8), but instead would be defined by

$$\mathcal{W} = \{ w \in \mathbb{R}^J : 0 \le w_J \le \dots \le w_2 \le w_1 \}, \tag{5.11}$$

and the requirement that if $w \in \mathcal{W}^j$ then $w_j = w_{j+1}$, with the interpretation $w_{J+1} = 0$. Thus face j represents the requirement that the workload for resource j comprises at least the workload for resource $j+1$, for $j = 1, 2, \dots, J$. Under this model, resource j is fully utilized except when $\breve{W}$ hits the jth face of the workload cone (5.11): it is not possible for $\breve{W}$ to leave $\mathcal{W}$, since the constraints expressed in the form (5.11) follow necessarily from the topology of the network embodied in A. The model corresponds to the assumption that there is no more loss of utilization than is a necessary consequence of the network topology. Note

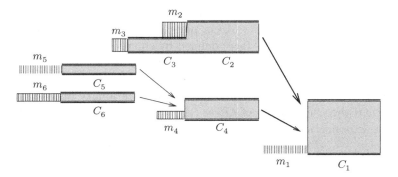

Figure 5.2 There are six sources of traffic, starting in various lines and all destined to eventually traverse section 1 of the road. Once traffic has passed through the queue at its entry point, it does not queue again.

that the proportionally fair policy may fail to fully utilize a resource not only when this is a necessary consequence of the network topology, but also when this would cause an upstream entry point to obtain more than what the policy considers a fair share of a scarce downstream resource.

5.3 A tree network

Next consider the tree network illustrated in Figure 5.2. Access is metered at the six entry points so that the capacities C_1, C_2, ..., C_6 are not overloaded. There is no queueing after the entry point, and the capacities satisfy the conditions $C_3 < C_2$, $C_5 + C_6 < C_4$, $C_2 + C_4 < C_1$.

Given the line sizes $M(t) = m$, we suppose the metered rates $\Lambda(m)$ are chosen to be proportionally fair: that is, the capacity allocation policy $\Lambda(\cdot)$ solves the optimization problem (3.1)–(3.4) where for this network

$$
A = \begin{pmatrix}
1 & 1 & 1 & 1 & 1 & 1 \\
0 & 1 & 1 & 0 & 0 & 0 \\
0 & 0 & 1 & 0 & 0 & 0 \\
0 & 0 & 0 & 1 & 1 & 1 \\
0 & 0 & 0 & 0 & 1 & 0 \\
0 & 0 & 0 & 0 & 0 & 1
\end{pmatrix}.
$$

We assume, as in the last Section, that the cumulative inflow of work to line i is given by equation (5.6) for $i \in \mathcal{I}$, where $(\breve{Z}_i(t), t \geq 0)$, $i \in \mathcal{I}$, are independent standard Brownian motions. Our Brownian network model is again the process $\breve{W}$ defined by properties (i)–(iv) of Section 4 with $\breve{W}$ in place of $\tilde{W}$, $C - A\rho$ in place of θ, $\Gamma = \sigma^2 A[\rho]A'$, and with the workload cone and its faces defined by equations (5.7)–(5.8) for the above choice of A. We assume the stability condition (3.8) is satisfied, so that all components of $C - A\rho$ are positive.

If $\breve{W}^s$ denotes a random variable with the stationary distribution of $\breve{W}$, then the components of $\breve{Q}^s = (A[\rho]A')^{-1}\breve{W}^s$ are independent and $\breve{Q}^s_j$ is exponentially distributed with parameter $2\sigma^{-2}(C_j - \sum_i A_{ji}\rho_i)$ for each $j \in \mathcal{J}$. The Brownian model line sizes and delays are again given by equations (5.9) and (5.10) respectively, each with stationary distributions given by a linear combination of independent exponential random variables, one for each section of road.

A key feature of the linear network, and its generalization to tree networks, is that all traffic is bound for the same destination. In our application to a road network this ensures that all traffic in a line at a given entry point is on the same route. If traffic on different routes shared a single line it would not be possible to align the delay incurred by traffic so precisely with the sum of dual variables for the resources to be used.[5]

6 Route choices

Next consider the road network illustrated in Figure 6.1. Three parallel roads lead into a fourth road and hence to a common destination. Access to each of these roads is metered, so that their respective capacities C_1, C_2, C_3, C_4 are not overloaded, and $C_1 + C_2 + C_3 < C_4$. There are four sources of traffic with respective loads ρ_1, ρ_2, ρ_3, ρ_4: the first source has access to road 1 alone, on its way to road 4; the second source has access to both roads 1 and 2; and the third source can access all three of the parallel roads. We assume that traffic arriving with access to more than one road distributes itself in an attempt to minimize its queueing delay, an assumption whose implications we shall explore.

We could view sources of traffic as arising in different geographical regions, with different possibilities for easy access to the motorway network and with real time information on delays. Or we could imagine a

---

[5] The tree topology of Figure 5.2 ensures that the queueing delays in the proportionally fair Brownian network model are partially ordered. A technical consequence is that a wide class of fair capacity allocations, the α-fair allocations, share the same workload cone: in the notation of [21], the cone $\mathcal{W}_\alpha$ does not depend upon α.

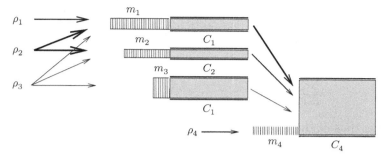

Figure 6.1 Three parallel roads lead to a fourth road and hence to a common destination. Lines of size m_1, m_2, m_3, m_4 are held on the slip roads leading to these roads. There are four sources of traffic: sources 2 and 3 may choose their first road, with choices as shown.

priority access discipline where some traffic, for example high occupancy vehicles, has a larger set of lines to choose from.

Given the line sizes $M(t) = m$, we suppose the metered rates $\Lambda(m)$ are chosen to be proportionally fair: that is, the capacity allocation policy $\Lambda(\cdot)$ solves the optimization problem (3.1)–(3.4). For this network

$$
A = \begin{pmatrix} 1 & 0 & 0 & 0 \\ 0 & 1 & 0 & 0 \\ 0 & 0 & 1 & 0 \\ 1 & 1 & 1 & 1 \end{pmatrix}, \qquad C = \begin{pmatrix} C_1 \\ C_2 \\ C_3 \\ C_4 \end{pmatrix}
$$

and so, from relations (3.5)–(3.6),

$$
\Lambda_i(m) = \frac{m_i}{q_i + q_4}, \quad i = 1, 2, 3, \qquad \Lambda_4(m) = \frac{m_4}{q_4}.
$$

We assume the ramp metering policy has no knowledge of the routing choices available to arriving traffic, but is simply a function of the observed line sizes m, the topology matrix A and the capacity vector C.

How might arriving traffic choose between lines? Well, traffic that arrives when the line sizes are m and the metered rates are $\Lambda(m)$ might reasonably consider the ratios $m_i/\Lambda_i(m)$ in order to choose which line to join, since these ratios give the time it would take to process the work currently in line i at the current metered rate for line i, for $i = 1, 2, 3$. But these ratios are just $q_i + q_4$ for $i = 1, 2, 3$. Given the choices available to the three sources, we would expect exercise of these choices to ensure that $q_1 \geq q_2 \geq q_3$, or equivalently that the delays through lines 1, 2, 3 are weakly decreasing.

Because traffic from sources 2 and 3 has the ability to make route choices, condition (3.8) is sufficient, but no longer necessary, for stability. The stability condition for the network of Figure 6.1 is

$$\sum_{i=1}^{j} \rho_i < \sum_{i=1}^{j} C_i, \ j = 1, 2, 3, \qquad \sum_{i=1}^{4} \rho_i < C_4, \qquad (6.1)$$

and is thus of the form (3.8), but with A and C replaced by $\bar{A}$ and $\bar{C}$ respectively, where

$$\bar{A} = \begin{pmatrix} 1 & 0 & 0 & 0 \\ 1 & 1 & 0 & 0 \\ 1 & 1 & 1 & 0 \\ 1 & 1 & 1 & 1 \end{pmatrix}, \qquad \bar{C} = \begin{pmatrix} \bar{C}_1 \\ \bar{C}_2 \\ \bar{C}_3 \\ \bar{C}_4 \end{pmatrix} = \begin{pmatrix} C_1 \\ C_1 + C_2 \\ C_1 + C_2 + C_3 \\ C_4 \end{pmatrix}.$$

The forms $\bar{A}$, $\bar{C}$ capture the concept of four *virtual resources* of capacities $\bar{C}_j$, $j = 1, 2, 3, 4$. Given the line sizes $m = (m_i, i = 1, 2, 3, 4)$, the workloads $w = (w_i, i = 1, 2, 3, 4)$ for the four virtual resources are $w = \bar{A}m$.

For $i = 1, 2, 3, 4$, we model the cumulative inflow of work from source i over the interval $(0, t]$ as a Brownian motion $\breve{E}_i = (\breve{E}_i(t), t \geq 0)$ starting from the origin with drift ρ_i and variance parameter $\rho_i \sigma^2$ that can be written in the form (5.6), where $(\breve{Z}_i(t), t \geq 0)$, $i = 1, 2, 3, 4$, are independent standard Brownian motions. Let $\breve{E} = (\breve{E}_i, i = 1, 2, 3, 4)$ and let $\breve{X} = (\breve{X}_j, j \in \mathcal{J})$ be defined by

$$\breve{X}(t) = \bar{A}\breve{E}(t) - \bar{C}t, \quad t \geq 0.$$

Then $\breve{X}$ is a four-dimensional Brownian motion starting from the origin with drift $\bar{A}\rho - \bar{C}$ and covariance matrix $\Gamma = \sigma^2 \bar{A} [\rho] \bar{A}'$. We assume the stability condition (6.1) is satisfied, so that all components of the drift are strictly negative. Let $\mathcal{W}, \mathcal{W}^j$ be defined by (5.7), (5.8) respectively, with A replaced by $\bar{A}$.

Our Brownian network model for $\bar{A}M$ is then the process $\breve{W}$ defined by properties (i)–(iv) of Section 4 with $\breve{W}$ in place of $\tilde{W}$ and $\bar{C} - \bar{A}\rho$ in place of θ.

Define a process of dual variables for the virtual resources: $\breve{Q} = (A[\rho]A')^{-1}\breve{W}$. Since $\breve{W}$ is our model for $\bar{A}M$, our Brownian model for the line sizes is given by

$$\breve{M} = \bar{A}^{-1}\breve{W} = [\rho]\bar{A}'\breve{Q}.$$

Our Brownian model for the delays at each line is given by

$$\breve{D} = \bar{A}'\breve{Q},$$

which from the form of $\bar{A}$ becomes

$$\breve{D}_i = \sum_{j=i}^{4} \breve{Q}_j, \qquad i = 1, 2, 3, 4.$$

Thus at any time t when $\breve{Q}_1(t) > 0$, $\breve{Q}_2(t) > 0$ then $\breve{D}_1(t) > \breve{D}_2(t) > \breve{D}_3(t)$, and the incentives for arriving traffic are such that traffic from source i joins line i. However if $\breve{Q}_1(t) = 0$, and so $\breve{D}_1(t) = \breve{D}_2(t)$, then arriving traffic from stream 2 may choose to enter line 1, and thus contribute to increments of the workload for virtual resource 1, whilst still contributing to the workload for virtual resources 2, 3 and 4. Our model corresponds to the assumption that no more traffic does this than is necessary to keep $\breve{Q}_1$ non-negative, or equivalently to keep $\breve{D}_1 \geq \breve{D}_2$. Similarly if $\breve{Q}_2(t) = 0$ then $\breve{D}_2(t) = \breve{D}_3(t)$, and arriving traffic from stream 3 may choose to enter line 2, and thus contribute to increments of the workload for virtual resource 2, whilst still contributing to the workload for virtual resources 3 and 4; we suppose just sufficient traffic does this to keep $\breve{Q}_2$ non-negative, or equivalently to keep $\breve{D}_2 \geq \breve{D}_3$. Finally if $\breve{Q}_3(t) = 0$ or $\breve{Q}_4(t) = 0$ then (real) resource 3 or 4 respectively may not be fully utilized, as in earlier examples, and our model corresponds to the assumption that there is no more loss of utilization at (real) resources 3 and 4 than is necessary to prevent $\breve{W}$ from leaving $\mathcal{W}$.

If $\breve{W}^s$ is a random variable with the stationary distribution of $\breve{W}$, then the components of $\breve{Q}^s = (A[\rho]A')^{-1}\breve{W}^s$ are independent and for $j = 1$, 2, 3, 4, $\breve{Q}_j^s$ is exponentially distributed with parameter ζ_j where

$$\zeta_1 = \frac{2}{\sigma^2}(C_1 - \rho_1),$$

$$\zeta_2 = \frac{2}{\sigma^2}(C_1 + C_2 - \rho_1 - \rho_2),$$

$$\zeta_3 = \frac{2}{\sigma^2}(C_1 + C_2 + C_3 - \rho_1 - \rho_2 - \rho_3),$$

$$\zeta_4 = \frac{2}{\sigma^2}(C_4 - \rho_1 - \rho_2 - \rho_3 - \rho_4).$$

Under the Brownian network model, the stationary distribution for line sizes and for delays at each line are given by the distributions of $\breve{M}^s$ and

$\check{D}^s$, respectively, where

$$\check{M}_i^s = \rho_i \left(\sum_{j=i}^{4} \check{Q}_j^s \right), \quad \check{D}_i^s = \sum_{j=i}^{4} \check{Q}_j^s \quad i = 1, 2, 3, 4.$$

The Brownian network model thus corresponds to natural assumptions about how arriving traffic from different sources would choose their routes. The results on the stationary distribution for the network are intriguing. The ramp metering policy has no knowledge of the routing choices available to arriving traffic, and hence of the enlarged stability region (6.1). Nevertheless, under the Brownian model, the interaction of the ramp metering policy with the routing choices available to arriving traffic has a performance described in terms of dual random variables, one for each of the virtual resources of the enlarged stability region; when a driver makes a route choice, the delay facing a driver on a route is a sum of dual random variables, one for each of the virtual resources used by that route; and under their stationary distribution, the dual random variables are independent and exponentially distributed.

7 Concluding remarks

The design of ramp metering strategies cannot assume that arriving traffic flows are exogenous, since in general drivers' behaviour will be responsive to the delays incurred or expected. In this paper we have presented a preliminary exploration of an approach to the design of ramp metering flow rates informed by earlier work on Internet congestion control. A feature of this approach is that it may prove possible to integrate ideas of fairness of a control policy with overall system optimization.

There remain many areas for further investigation. In particular, we have seen intriguing examples, in the context of a single queue and of Internet congestion control, of remarkably good approximations produced for the stationary distributions of queue length and workload by use of the direct Brownian modelling approach. Furthermore, in the context of a controlled motorway, where a detailed model for arriving traffic is not easily available, use of a direct Brownian model has enabled us to develop an approach to the design and performance of ramp metering and in the context of that model to obtain insights into the interaction of ramp metering with route choices. Nevertheless, we expect that the use of direct Brownian network models will not always produce good results. Indeed, it is possible that such models may be suitable only when the scaled workload process can be approximated in heavy traffic by a reflecting Brownian motion that has a product-form stationary distribution. We believe that understanding when the direct method is a good modelling approach and when it is not, and obtaining a rigorous understanding of the reasons for this, is an interesting topic worthy of further research.

References

[1] Abou-Rahme, N., Beale, S., Harbord, B., and Hardman, E. 2000. Monitoring and modelling of controlled motorways. Pages 84–90 of: *Tenth International Conference on Road Transport Information and Control*.

[2] Beneš, V. E. 1957. On queues with Poisson arrivals. *Ann. Math. Stat.*, **28**, 670–677.

[3] Billingsley, P. 1999. *Convergence of Probability Measures*, Second edn. New York: John Wiley & Sons.

[4] Bonald, T., and Massoulié, L. 2001. Impact of fairness on Internet performance. *Performance Evaluation Review*, **29**(1), 82–91.

[5] Borovkov, A. 1964. Some limit theorems in the theory of mass service, I. *Theory Probab. Appl.*, **9**, 550–565.

[6] Borovkov, A. 1965. Some limit theorems in the theory of mass service, II. *Theory Probab. Appl.*, **10**, 375–400.

[7] Bramson, M. 1998. State space collapse with application to heavy traffic limits for multiclass queueing networks, *Queueing Syst.*, **30**, 89–148.

[8] Bramson, M., and Dai, J. G. 2001. Heavy traffic limits for some queueing networks, *Ann. Appl. Probab.*, **11**, 49–90.

[9] Chen, H., and Yao, D. D. 2001. *Fundamentals of Queueing Networks*. New York: Springer-Verlag.

[10] Dai, J. G., and Williams, R. J. 1995. Existence and uniqueness of semimartingale reflecting Brownian motions in convex polyhedrons. *Theory Probab. Appl.*, **40**, 1–40. Correction: **50** (2006), 346–347.

[11] De Veciana, G., Lee, T. J., and Konstantopoulos, T. 2001. Stability and performance analysis of networks supporting elastic services. *IEEE/ACM Trans. on Networking*, **9**, 2–14.

[12] Gibbens, R. J., and Saatci, Y. 2008. Data, modelling and inference in road traffic networks, *Philos. Trans. R. Soc. Lond. Ser. A*, **366**, 1907–1919.

[13] Gibbens, R. J., and Werft, W. 2005. Data gold mining: MIDAS and journey time predictors. *Significance*, **2**(3), 102–105.

[14] Harbord, B., White, J., McCabe, K., Riley, A., and Tarry, S. 2006. A flexible approach to motorway control. In: *Proceedings of the ITS World Congress*.

[15] Harrison, J. M. 1985. *Brownian Motion and Stochastic Flow Systems*. New York: John Wiley & Sons.

[16] Harrison, J. M. 1988. Brownian models of queueing networks with heterogeneous customer populations. Pages 147–186 of: Fleming, W., and Lions, P. L. (eds), *Stochastic Differential Systems, Stochastic Control Theory and Their Applications*, IMA Vol. Math. Appl. 10. New York: Springer-Verlag.

[17] Harrison, J. M. 2000. Brownian models of open processing networks: canonical representation of workload. *Ann. Appl. Probab.*, **10**, 75–103. Correction: **16** (2006), 1703–1732.

[18] Harrison, J .M. 2003. A broader view of Brownian networks. *Ann. Appl. Probab.*, **13**, 1119–1150.

[19] Harrison, J. M., and Reiman, M. I. 1981. Reflected Brownian motion on an orthant. *Ann. Probab.*, **9**, 302–308.

[20] Iglehart, D. L. 1965. Limit theorems for queues with traffic intensity one. *Ann. Math. Statist.*, **36**, 1437–1449.

[21] Kang, W. N., Kelly, F. P., Lee, N. H., and Williams, R. J.. 2009. State space collapse and diffusion approximation for a network operating under a fair bandwidth sharing policy. *Ann. Appl. Probab.*, **19**, 1719–1780.

[22] Kang, W. N., and Williams, R. J. 2007. An invariance principle for semimartingale reflecting Brownian motions in domains with piecewise smooth boundaries. *Ann. Appl. Probab.*, **17**, 741–779.

[23] Kelly, F. P. 1979. *Reversibility and Stochastic Networks*. Chichester: John Wiley & Sons.

[24] Kelly, F. P., Maulloo, A., and Tan, D. 1998. Rate control in communication networks: shadow prices, proportional fairness and stability, *J. Operational Research Soc.*, **49**, 237–252.

[25] Kelly, F. P., and Williams, R. J. 2004. Fluid model for a network operating under a fair bandwidth-sharing policy. *Ann. Appl. Probab.*, **14**, 1055–1083.

[26] Kingman, J. F. C. 1961. The single server queue in heavy traffic. *Proc. Cambridge Philos. Soc.*, **57**, 902–904.

[27] Kingman, J. F. C. 1962. On queues in heavy traffic. *J. R. Stat. Soc. Ser. B*, **24**, 383–392.

[28] Kingman, J. F. C. 1963. The heavy traffic approximation in the theory of queues. Pages 137–169 of Smith, W. L., and Wilkinson, R. I. (eds), *Proceedings of the Symposium on Congestion Theory*. Chapel Hill, NC: Univ. of North Carolina.

[29] Kingman, J. F. C. 1982. Queue disciplines in heavy traffic. *Math. Oper. Res.*, **7**, 262–271.

[30] Levinson, D., and Zhang, L. 2006. Ramp meters on trial: evidence from the Twin Cities metering holiday. *Transportation Research*, **A40**, 810–828.

[31] Massoulié, L. and Roberts, J. 1998. Bandwidth sharing and admission control for elastic traffic. *Telecomm. Systems*, **15**, 185–201.

[32] Nash, J. F. 1950. The bargaining problem. *Econometrica*, **28**, 155–162.

[33] Newell, G. F. 2002. Memoirs on highway traffic flow theory in the 1950s. *Oper. Res.*, **50**, 173–178.

[34] Papageorgiou, M., and Kotsialis, A. 2002. Freeway ramp metering: an overview. *IEEE Trans. Intelligent Transportation Systems*, **3**, 271–281.

[35] Prohorov, Y. V. 1963. Transient phenomena in processes of mass service (in Russian). *Litovskii Matematiceskii Sbornik*, **3**, 199–205.

[36] Reiman, M. I. 1982. The heavy traffic diffusion approximation for sojourn times in Jackson networks. Pages 409–422 of: Disney, R. I., and Ott, T. (eds), *Applied Probability–Computer Science: The Interface*, vol. 2. Boston: Birkhauser.

[37] Reiman, M. I. 1984. Open queueing networks in heavy traffic. *Math. Oper. Res.*, **9**, 441–458.

[38] Srikant, R. 2004. *The Mathematics of Internet Congestion Control*. Boston: Birkhauser.

[39] Varaiya, P. 2005. What we've learned about highway congestion. *Access*, **27**, 2–9.

[40] Varaiya, P. 2008. Congestion, ramp metering and tolls. *Philos. Trans. R. Soc. Lond. Ser. A*, **366**, 1921–1930.

[41] Whitt, W. 2002. *Stochastic-Process Limits: an Introduction to Stochastic-Process Limits and their Application to Queues*. New York: Springer-Verlag.

[42] Williams, R. J. 1996. On the approximation of queueing networks in heavy traffic. Pages 35–56 of: Kelly, F. P., Zachary, A., and Ziedins, I. (eds), *Stochastic Networks: Theory and Applications*. Oxford: Oxford Univ. Press.

[43] Williams, R. J. 1998. Diffusion approximations for open multiclass queueing networks: sufficient conditions involving state space collapse. *Queueing Syst.*, **30**, 27–88.

[44] Zhang, L., and Levinson, D. 2004. Ramp metering and the capacity of active freeway bottlenecks. In: *Proceedings of the 83nd Annual Meeting of the Transportation Research Board*.

19

Coupling time distribution asymptotics for some couplings of the Lévy stochastic area

Wilfrid S. Kendall[a]

Abstract

We exhibit some explicit co-adapted couplings for n-dimensional Brownian motion and all its Lévy stochastic areas. In the two-dimensional case we show how to derive exact asymptotics for the coupling time under various mixed coupling strategies, using Dufresne's formula for the distribution of exponential functionals of Brownian motion. This yields quantitative asymptotics for the distributions of random times required for certain simultaneous couplings of stochastic area and Brownian motion. The approach also applies to higher dimensions, but will then lead to upper and lower bounds rather than exact asymptotics.

Keywords Brownian motion, co-adapted coupling, coupling time distribution, Dufresne formula, exponential functional of Brownian motion, Kolmogorov diffusion, Lévy stochastic area, maximal coupling, Morse–Thue sequence, non-co-adapted coupling, reflection coupling, rotation coupling, stochastic differential, synchronous coupling

AMS subject classification (MSC2010) 60J65, 60H10

Introduction

It is a pleasure to present this paper as a homage to my DPhil supervisor John Kingman, in grateful acknowledgement of the formative period which I spent as his research student at Oxford, which launched me into

[a] Department of Statistics, University of Warwick, Coventry CV4 7AL; w.s.kendall@warwick.ac.uk, http://www.warwick.ac.uk/go/wsk

a deeply satisfying exploration of the world of mathematical research. It seems fitting in this paper to present an overview of a particular aspect of probabilistic coupling theory which has fascinated me for a considerable time; given that one can couple two copies of a random process, when can one in addition couple other associated functionals of the processes? How far can one go?

Motivations for this question include: the sheer intellectual curiosity of discovering just how far one can push the notion of probabilistic coupling; the consideration that coupling has established itself as an extremely powerful tool in probability theory and therefore that any increase in its scope is of potential significance; and the thought that the challenge of coupling in extreme circumstances may produce new paradigms in coupling to complement that of the classic reflection coupling.

It has been known since the 1970s that in principle one can couple two random processes *maximally*; at first encounter this fact continues to delight and surprise researchers. I summarize this point in Section 1 and also describe the important class of *co-adapted* couplings. These satisfy more restrictive requirements than maximal couplings, are typically less efficient, but are also typically much easier to construct. Since Lindvall (1982)'s seminal preprint we have known how to couple Euclidean Brownian motion using a simple reflection argument in a way which (most unusually) is both maximal *and* co-adapted, and this has led to many significant developments and generalizations, some of which are briefly sketched in Section 2. This leads to Section 3, which develops the main content of the paper; what can we now say about the question, how to couple not just Brownian motion, say, but also associated path integrals? Of course we then need to vary our strategy, using not just reflection coupling but also so-called synchronous coupling (in which the two processes move in parallel), and even rotation coupling, which correlates different coordinates of the two processes. In Kendall (2007) I showed how to couple (co-adaptively) n-dimensional Brownian motion and all its stochastic areas; this work is reviewed in Section 3.3 using a rather more explicit coupling strategy and then new computations are introduced (in Section 3.4) which establish explicit asymptotics for the coupling time for suitable coupling strategies in the two-dimensional case, and which can be used to derive naïve bounds in higher dimensions. Section 4 concludes the paper with some indications of future research directions.

1 Different kinds of couplings

Probabilistic couplings are used in many different ways: couplings (real-izations of two random processes on the same probability space) can be constructed so as to arrange any of the following:

- for the two processes to agree after some random time (which is to say, for the coupling to be *successful*). This follows the pioneering work of Doeblin (1938), which uses this idea to provide a coupling proof of convergence to equilibrium for finite-state ergodic Markov chains;
- for the two processes to be interrelated by some *monotonicity* property —a common use of coupling in the study of interacting particle systems (Liggett, 2005);
- for one process to be linked to the other so as to provide some in-formative and fruitful *representation*, as in the case of the coalescent (Kingman, 1982);
- for one of the processes to be an illuminating *approximation* to the other; this appears in an unexpected way in Barbour et al. (1992)'s approach to Stein–Chen approximation.

These considerations often overlap. Aiming for successful coupling has historical precedence and is in some sense thematic for coupling theory, and we will focus on this task here.

1.1 Maximal couplings

The first natural question is, how fast can coupling occur? There is a remarkable and satisfying answer, namely that one can in principle construct a coupling which is *maximal* in the sense that it maximizes the probability of coupling before time t for all possible t: see Griffeath (1974, 1978), Pitman (1976), Goldstein (1978). Briefly, maximal coup-lings convert the famous Aldous inequality (the probability of coupling is bounded above by a simple multiple of the total variation between distributions) into an equality. Constructions of maximal couplings are typically rather involved, and in general may be expected to involve de-manding potential-theoretic questions quite as challenging as any prob-lem which the coupling might be supposed to solve. Pitman's approach is perhaps the most direct, involving explicit construction of suitable time-reversed Markov chains.

1.2 Co-adapted coupling

Maximal couplings being generally hard to construct, it is useful to consider couplings which are stricter in the sense of requiring the coupled processes both to be adapted to the same filtration. Terminology in the literature varies: *Markovian*, when the coupled processes are jointly Markov, with prescribed marginal kernels (Burdzy and Kendall, 2000); *co-immersed* (Émery, 2005) or *co-adapted* to emphasize the rôle of the filtration. The idea of a co-adapted coupling is simple enough, though its exact mathematical description is somewhat tedious: here we indicate the definition for Markov processes.

Definition 1.1 Suppose one is given two continuous-time Markov processes $X^{(1)}$ and $X^{(2)}$, with corresponding semigroup kernels defined for bounded measurable functions f by

$$P_t^{(i)} f(z) \quad = \quad \mathbb{E}\big[f(X_{s+t}^{(i)}) \mid X_s^{(i)} = z,\ X_u^{(i)} \text{ for } u < s\big].$$

A *co-adapted coupling* of $X^{(1)}$ and $X^{(2)}$ is a pair of random processes $\widetilde{X}^{(1)}$ and $\widetilde{X}^{(2)}$ defined on the same filtered probability space $(\Omega, \mathfrak{F}, \{\mathfrak{F}_t : t \geq 0\}, \mathbb{P})$, both adapted to the common filtration $\{\mathfrak{F}_t : t \geq 0\}$ (hence 'co-adapted') and satisfying

$$P_t^{(i)} f(z) \quad = \quad \mathbb{E}\big[f(\widetilde{X}_{s+t}^{(i)}) \mid \mathfrak{F}_s,\ \widetilde{X}_s^{(i)} = z\big]$$

for $i = 1$, 2, for each bounded measurable function f, each z, each s, $t > 0$.

Thus the individual stochastic dynamics of each $\widetilde{X}^{(i)}$ agree with those of the corresponding $X^{(i)}$ *even when the past behaviour of the opposite process is also taken into account*. (This is typically not the case for maximal couplings.) In particular, if the $X^{(i)}$ are Brownian motions then their forward increments are independent of the past given by the filtration. Moreover if the processes are specified using stochastic differential equations driven by Brownian motion then general co-adapted couplings can be represented using stochastic calculus (possibly at the price of enriching the filtration), as observed in passing by Émery (2005), and as described more formally in Kendall (2009a, Lemma 6). Briefly, any co-adapted coupling of vector-valued Brownian motions $\underline{A}$ and $\underline{B}$ can be represented by expressing $\underline{A}$ as a stochastic integral with respect to $\underline{B}$ and perhaps another independent Brownian motion $\underline{C}$: we use this later at Equation (3.1).

1.3 Coupling at different times

There are many other useful couplings falling outside this framework: for example Thorisson (1994) discusses the idea of a *shift-coupling*, which weakens the coupling requirement by permitting processes to couple at *different* times; Kendall (1994) uses co-adapted coupling of *time-changed* processes as part of an exploration of regularity for harmonic maps. However in this paper we will focus on co-adapted couplings.

2 Reflection coupling

The dominant example of coupling is *reflection coupling* for Euclidean Brownian motions $\underline{A}$ and $\underline{B}$, dating back to Lindvall (1982)'s preprint: construct $\underline{A}$ from $\underline{B}$ by reflecting $\underline{B}$ in the line segment running from $\underline{B}$ to $\underline{A}$. That this is a maximal coupling follows from an easy computation involving the reflection principle. It can be expressed as a co-adapted coupling; the Brownian increment for $\underline{A}$ is derived from that of $\underline{B}$ by a reflection in the line segment running from $\underline{B}$ to $\underline{A}$. Many modifications of the reflection coupling have been derived to cover various situations; we provide a quick survey in the remainder of this section.

2.1 Maximality and non-maximality

The reflection coupling is unusual in being both co-adapted and maximal. Hsu and Sturm (2003) point out that reflection coupling fails to be maximal even for Euclidean Brownian motion if the Brownian motion is stopped on exit from a prescribed domain. (Kuwada and Sturm 2007 discuss the manifold case; see also Kuwada 2007, 2009.) Perhaps the simplest example of an instance where no co-adapted coupling can be maximal arises in the case of the Ornstein–Uhlenbeck process (Connor, 2007, PhD thesis, Theorem 3.15). Consider the problem of constructing successful co-adapted couplings between (i) an Ornstein–Uhlenbeck process begun at 0, and (ii) an Ornstein–Uhlenbeck process run in statistical equilibrium. A direct argument shows that no such co-adapted coupling can be maximal; however in this case reflection coupling is maximal amongst all *co-adapted* couplings. (The study of couplings which are maximal in the class of co-adapted couplings promises to be an interesting field: the case of random walk on the hypercube is treated by Connor and Jacka 2008.)

2.2 Coupling for Diffusions

A variant of reflection coupling for elliptic diffusions with smooth coefficients is discussed in Lindvall and Rogers (1986) and further in Chen and Li (1989). 'Reflection' here depends on interaction between the two diffusion matrices, and in general the two coupled diffusions do not play symmetrical rôles. In the case of Brownian motion on a manifold one can use the mechanisms of stochastic development and stochastic parallel transport to define co-adapted couplings in a more symmetrical manner. The behaviour of general co-adapted Brownian couplings on Riemannian manifolds is related to curvature. Kendall (1986b) shows that successful co-adapted coupling can never be almost-surely successful in the case of a simply-connected manifold with negative curvatures bounded away from zero. On the other hand a geometric variant of reflection coupling known as *mirror coupling* will always be almost-surely successful if the manifold has non-negative Ricci curvatures (Kendall 1986a, Cranston 1992, Kendall 1988). Indeed Renesse (2004) shows how to generalize mirror coupling even to non-manifold contexts.

3 Coupling more than one feature of the process

The particular focus of this paper concerns ongoing work on the following question: is it possible co-adaptively to couple more than one feature of a random process at once? To be explicit, is it possible to couple not just the location but also some functional of the path?

On the face of it, this presents an intimidating challenge: control of difference of path functionals by coupling is necessarily indirect and it is possible that all attempts to control the discrepancy between path functionals will inevitably jeopardize coupling of the process itself.

Further thought shows that it is sensible to confine attention to cases where the process together with its functional form a *hypoelliptic diffusion*, since in such cases the Hörmander regularity theorem guarantees existence of a density, and this in turn shows that general coupling (not necessarily non-co-adapted) is possible in principle. (Hairer 2002 uses this approach to produce non-co-adapted couplings, using careful analytic estimates and a regeneration argument which corresponds to Lindvall 2002's 'γ coupling'.)

Furthermore it is then natural to restrict attention to diffusions with nilpotent group symmetries, where one may hope most easily to discover

co-adapted couplings which will be susceptible to extensive generalization, parallelling the generalizations of the Euclidean reflection coupling
which have been described briefly in Section 2.

3.1 Kolmogorov diffusion

Consider the so-called Kolmogorov diffusion: scalar Brownian motion B
plus its time integral $\int B \, dt$. This determines a simple nilpotent diffusion, and in fact it can be coupled co-adaptively by varying sequentially
between reflection coupling and *synchronous* coupling (allowing the two
Brownian motions to move in parallel) as shown in Ben Arous et al.
(1995). Jansons and Metcalfe (2007) describe some numerical investiga-
tions concerned with optimizing an exponential moment of the coupling
time.

The idea underlying this coupling is rather simple. Suppose that we
wish to couple $(B, \int B \, dt)$ with $(\widetilde{B}, \int \widetilde{B} \, dt)$. Set $U = \widetilde{B} - B$ and $V = \int \widetilde{B} \, dt - \int B \, dt$. Co-adapted couplings include stochastic integral representations such as $d\widetilde{B} = J \, dB$, for co-adapted $J \in \{-1, 1\}$; $J = 1$ yields
synchronous coupling and $J = -1$ yields reflected coupling. Suppose
$U_0 \neq 0$ and $V = 0$ (which can always be achieved by starting with a
little reflected or synchronous coupling unless $U = V = 0$ from the start,
in which case nothing needs to be done). We can cause (U, V) to wind re-
peatedly around $(0, 0)$ in ever smaller loops as follows: first use reflection
coupling till U hits $-U_0/2$, then synchronous coupling till V hits 0, then
repeat the procedure but substituting $-U_0/2$ for U_0. A Borel–Cantelli
argument combined with Brownian scaling shows that (U, V) then hits
$(0, 0)$ in finite time.

Kendall and Price (2004) present a cleaned-up version of this argument
(together with an extension to deal in addition with $\iint B \, ds \, dt$).

Curiously, this apparently artificial example can actually be applied to
the study of the tail σ-algebra of a certain relativistic diffusion discussed
by Bailleul (2008).

Remarkably it is possible to do very much better by using a completely
different and implicit method: one can couple not just the time integral,
but also any finite number of additional iterated time integrals (Kendall
and Price, 2004), by concatenating reflection and synchronous couplings
using the celebrated *Morse–Thue* binary sequence 0110100110010110 . . . .
Scaled iterations of state-dependent perturbations of the resulting concatenation of couplings can be used to deliver coupling in a finite time;

the perturbed Morse–Thue sequences encode indirect controls of higher-order iterated integrals.

3.2 Lévy stochastic areas

Moving from scalar to planar Brownian motion, the natural question is now whether one can co-adaptively couple the nilpotent diffusion formed by Brownian motion (B_1, B_2) and the Lévy stochastic area $\int (B_1 \, dB_2 - B_2 \, dB_1)$. This corresponds to coupling a hypoelliptic Brownian motion on the Heisenberg group, and Ben Arous et al. determine an explicit successful coupling based on extensive explorations using computer algebra.

Again one can do better (Kendall, 2007). Not only can one construct a simplified coupling for the 2-dimensional case based only on reflection and synchronous couplings (switching from reflection to synchronous coupling whenever a geometric difference of the stochastic areas exceeds a fixed multiple of the squared distance between the two coupled Brownian motions), but also one can successfully couple n-dimensional Brownian motion plus a $\binom{n}{2}$ basis of the various stochastic areas. In the remainder of this paper we will indicate the method used, which moves beyond the use of reflection and synchronous couplings to involve *rotation couplings* as well (in which coordinates of one of the Brownian motions can be correlated to quite different coordinates of the other).

3.3 Explicit strategies for coupling Lévy stochastic area

Here we describe a variant coupling strategy for the n-dimensional case which is more explicit than the strategy proposed in Kendall (2007). As described in Kendall (2007, Lemma 6), suppose that $\underline{A}$ and $\underline{B}$ are co-adaptively coupled n-dimensional Brownian motions. Arguing as in Kendall (2009a), and enriching the filtration if necessary, we may represent any such coupling in terms of a further n-dimensional Brownian motion $\underline{C}$, independent of $\underline{B}$;

$$d\underline{A} \;=\; \underline{J}^\top \, d\underline{B} + \widetilde{\underline{J}}^\top \, d\underline{C}, \tag{3.1}$$

where $\underline{J}$, $\widetilde{\underline{J}}$ are predictable $(n \times n)$-matrix-valued processes satisfying the constraint

$$\underline{J}^\top \underline{J} + \widetilde{\underline{J}}^\top \widetilde{\underline{J}} \;=\; \mathbb{I} \tag{3.2}$$

and where $\mathbb{I}$ represents the $(n \times n)$ identity matrix.

Note that the condition (3.2) is equivalent to the following set of symmetric matrix inequalities for the co-adapted process $\underline{J}$ (interpreted in a spectral sense):

$$\underline{J}^\top \underline{J} \leq \mathbb{I}. \tag{3.3}$$

Thus a legitimate *coupling control* $\underline{J}$ must take values in a compact convex set of $n \times n$ matrices defined by (3.3).

The matrix process $\underline{J}$ measures the correlation $(\mathrm{d}\underline{B}\,\mathrm{d}\underline{A}^\top)/\mathrm{d}t$ between the Brownian differentials $\mathrm{d}\underline{A}$ and $\mathrm{d}\underline{B}$; for convenience let $\underline{S} = \frac{1}{2}(\underline{J} + \underline{J}^\top)$ and $\underline{A} = \frac{1}{2}(\underline{J} - \underline{J}^\top)$ refer to the symmetric and skew-symmetric parts of $\underline{J}$. The coupling problem solved in Kendall (2007) is to choose an adapted $\underline{J} = \underline{S} + \underline{A}$ which brings $\underline{A}$ and $\underline{B}$ into agreement at a coupling time T_{coupling} which is simultaneously a coupling time for all the $\binom{n}{2}$ corresponding pairs of stochastic area integrals $\int (A_i\,\mathrm{d}A_j - A_j\,\mathrm{d}A_i)$ and $\int (B_i\,\mathrm{d}B_j - B_j\,\mathrm{d}B_i)$.

To measure progress towards this simultaneous coupling, set $\underline{X} = \underline{A} - \underline{B}$ and define $\underline{\mathfrak{A}}$ to be a skew-symmetric matrix of geometric differences between stochastic areas with $(i, j)^{\text{th}}$ entry

$$\mathfrak{A}_{ij} = \int (A_i\,\mathrm{d}A_j - A_j\,\mathrm{d}A_i) - \int (B_i\,\mathrm{d}B_j - B_j\,\mathrm{d}B_i) + A_iB_j - A_jB_i. \tag{3.4}$$

The nonlinear correction term $A_iB_j - A_jB_i$ is important because it converts $\mathfrak{A}_{ij}$ into a geometrically natural quantity, invariant under shifts of coordinate system, and also because it supplies a very useful contribution to the drift in the subsequent Itô analysis. Of course $\underline{\mathfrak{A}}$ and $\underline{X}$ both vanish at a given time t if and only if at that time both $\underline{A} = \underline{B}$ (so in particular the correction term vanishes) and also all the corresponding stochastic areas agree.

Some detailed Itô calculus (originally carried out in an implementation of the Itô calculus procedures *Itovsn3* in *Axiom*, Kendall, 2001, but now checked comprehensively by hand) can now be used to derive the following system of stochastic differential equations for the squared distance $V^2 = \|\underline{X}\|^2$ and the 'squared areal difference' $U^2 = \text{trace}(\underline{\mathfrak{A}}^\top \underline{\mathfrak{A}}) = \sum\sum_{ij} \mathfrak{A}_{ij}^2$:

$$
\begin{aligned}
(\mathrm{d}(V^2))^2 &= 8\underline{\nu}^\top \left(\mathbb{I} - \underline{S}\right) \underline{\nu}\, V^2\,\mathrm{d}t, \\
\text{Drift } \mathrm{d}(V^2) &= 2\,\text{trace} \left(\mathbb{I} - \underline{S}\right)\,\mathrm{d}t, \\
\mathrm{d}(V^2) \times \mathrm{d}(U^2) &= -16\underline{\nu}^\top \underline{Z}^\top \underline{A}\,\underline{\nu}\, UV^2\,\mathrm{d}t, \\
(\mathrm{d}(U^2))^2 &= 32\underline{\nu}^\top \underline{Z}^\top \left(\mathbb{I} + \underline{S}\right) \underline{Z}\,\underline{\nu}\, U^2V^2\,\mathrm{d}t,
\end{aligned}
$$

$$\text{Drift } d(U^2) \quad = \quad 4\operatorname{trace}\left(\underline{\underline{Z}}^{\top}\underline{\underline{A}}\right) U \, dt +$$
$$+ \; 4\left(\operatorname{trace}\left(\underline{\underline{\mathbb{I}}}+\underline{\underline{S}}\right) - \underline{\nu}^{\top}\left(\underline{\underline{\mathbb{I}}}+\underline{\underline{S}}\right)\underline{\nu}\right) V^2 \, dt \, . \tag{3.5}$$

Here the vector $\underline{\nu}$ and the matrix $\underline{\underline{Z}}$ encode relevant underlying geometry: respectively $V\underline{\nu} = \underline{X}$ and $U\underline{\underline{Z}} = \underline{\underline{\mathfrak{A}}}$. Note that $\underline{\nu}$ is a unit vector and $\underline{\underline{Z}}$ has unit Hilbert–Schmidt norm: $\operatorname{trace}\underline{\underline{Z}}^{\top}\underline{\underline{Z}} = 1$.

The strategy is to consider V^2 and U^2 on a log-scale: further stochastic calculus together with suitable choice of coupling control $\underline{\underline{J}}$ then permits comparison to two Brownian motions with constant negative drift in a new time-scale, and a stochastic-calculus argument shows that $K = \frac{1}{2}\log(V^2)$ and $H = \frac{1}{2}\log(U^2)$ reach $-\infty$ at finite time in the original time-scale measured by t. In fact further stochastic calculus, based on the martingale differential identity

$$d\log Z \quad = \quad \frac{dZ}{Z} - \tfrac{1}{2}\left(\frac{dZ}{Z}\right)^2 ,$$

shows that

$$(dK)^2 \quad = \quad \tfrac{1}{2}\,\underline{\nu}^{\top}\left(\underline{\underline{\mathbb{I}}}-\underline{\underline{S}}\right)\underline{\nu}\,d\widetilde{\tau} ,$$
$$\text{Drift}(dK) \quad = \quad \tfrac{1}{4}\left(\operatorname{trace}\left(\underline{\underline{\mathbb{I}}}-\underline{\underline{S}}\right) - 2\,\underline{\nu}^{\top}\left(\underline{\underline{\mathbb{I}}}-\underline{\underline{S}}\right)\underline{\nu}\right)\,d\widetilde{\tau} ,$$
$$dK \times dH \quad = \quad -\underline{\nu}^{\top}\underline{\underline{Z}}^{\top}\underline{\underline{A}}\,\underline{\nu}\,\tfrac{d\widetilde{\tau}}{W} ,$$
$$(dH)^2 \quad = \quad 2\,\underline{\nu}^{\top}\underline{\underline{Z}}^{\top}\left(\underline{\underline{\mathbb{I}}}+\underline{\underline{S}}\right)\underline{\underline{Z}}\,\underline{\nu}\,\tfrac{d\widetilde{\tau}}{W^2} ,$$
$$\text{Drift}(dH) \quad = \quad \tfrac{1}{2}\operatorname{trace}\left(\underline{\underline{Z}}^{\top}\underline{\underline{A}}\right)\tfrac{d\widetilde{\tau}}{W} +$$
$$+ \tfrac{1}{2}\left(\operatorname{trace}\left(\underline{\underline{\mathbb{I}}}+\underline{\underline{S}}\right) - \underline{\nu}^{\top}\left(\underline{\underline{\mathbb{I}}}+\underline{\underline{S}}\right)\underline{\nu} - 4\,\underline{\nu}^{\top}\underline{\underline{Z}}^{\top}\left(\underline{\underline{\mathbb{I}}}+\underline{\underline{S}}\right)\underline{\underline{Z}}\,\underline{\nu}\right)\tfrac{d\widetilde{\tau}}{W^2} . \tag{3.6}$$

Here $d\widetilde{\tau} = 4\frac{dt}{V^2}$ determines the new time-scale $\widetilde{\tau}$, and $W = U/V^2 = \exp\left(H - 2K\right)$. It is clear from the system (3.6) that the contribution of $\frac{1}{2}\operatorname{trace}(\underline{\underline{Z}}^{\top}\underline{\underline{A}})\,\frac{d\widetilde{\tau}}{W}$ (deriving ultimately from the areal difference correction term in (3.4)) is potentially a flexible and effective component of the control if $1/W^2$ is small. On the other hand if this is to be useful then $\underline{\underline{\mathbb{I}}} - \underline{\underline{S}}$ must be correspondingly reduced so as to ensure that H and K are subject to dynamics on comparable time-scales.

Kendall (2007) considers the effect of a control $\underline{\underline{J}}$ which is an affine mixture of *reflection* $\underline{\underline{\mathbb{I}}} - 2\,\underline{\nu}\,\underline{\nu}^{\top}$ and *rotation* $\exp\left(-\theta\,\underline{\underline{Z}}\right)$. Second-order Taylor series expansion of the matrix exponential is used to overcome analytical complexities at the price of some mild asymptotic analysis.

Here we indicate an alternative route, replacing the second-order truncated expansion $\underline{\underline{J}}' = \underline{\underline{\mathbb{I}}} - \theta\,\underline{\underline{Z}} - \frac{1}{2}\theta^2\,\underline{\underline{Z}}^\top\underline{\underline{Z}}$ (which itself fails to satisfy (3.3) and is not therefore a valid coupling control) by $\underline{\underline{J}}'' = \underline{\underline{\mathbb{I}}} - \theta\,\underline{\underline{Z}} - \theta^2\,\underline{\underline{Z}}^\top\underline{\underline{Z}}$, which does satisfy (3.3) when $\theta^2 \leq 2$ (using the fact that non-zero eigenvalues of $\underline{\underline{Z}}^\top\underline{\underline{Z}}$ have multiplicity 2, so that $\mathrm{Trace}(\underline{\underline{Z}}^\top\underline{\underline{Z}}) = 1$ implies that $\underline{\underline{0}} \leq \underline{\underline{Z}}^\top\underline{\underline{Z}} \leq \frac{1}{2}\underline{\underline{\mathbb{I}}}$). Thus we can consider

$$\underline{\underline{J}} \;=\; \underline{\underline{\mathbb{I}}} - 2p\underline{\nu}\,\underline{\nu}^\top - (1-p)\theta\,\underline{\underline{Z}} - (1-p)\theta^2\,\underline{\underline{Z}}^\top\underline{\underline{Z}}, \qquad (3.7)$$

which is a valid coupling control satisfying (3.3) when $0 \leq p \leq 1$ and $\theta^2 \leq 2$.

This leads to the following stochastic differential system, where terms which will eventually be negligible have been separated out:

$$
\begin{aligned}
(\mathrm{d}K)^2 &= \left(p + \tfrac{1}{2}\left\|\underline{\underline{Z}}\,\underline{\nu}\right\|^2\theta^2\right)\mathrm{d}\widetilde{\tau} - \tfrac{p}{2}\left\|\underline{\underline{Z}}\,\underline{\nu}\right\|^2\theta^2\,\mathrm{d}\widetilde{\tau},\\
\mathrm{Drift}(\mathrm{d}K) &= -\tfrac{1}{2}\left(p - \tfrac{1}{2}(1 - 2\|\underline{\underline{Z}}\,\underline{\nu}\|^2)\theta^2\right)\mathrm{d}\widetilde{\tau}\\
&\quad - \tfrac{p}{4}\left(1 - 2\|\underline{\underline{Z}}\,\underline{\nu}\|^2\right)\theta^2\,\mathrm{d}\widetilde{\tau},\\
\mathrm{d}K \times \mathrm{d}H &= \|\underline{\underline{Z}}\,\underline{\nu}\|^2\theta\,\tfrac{\mathrm{d}\widetilde{\tau}}{W} - p\|\underline{\underline{Z}}\,\underline{\nu}\|^2\theta\,\tfrac{\mathrm{d}\widetilde{\tau}}{W},\\
(\mathrm{d}H)^2 &= 4\,\|\underline{\underline{Z}}\,\underline{\nu}\|^2\,\tfrac{\mathrm{d}\widetilde{\tau}}{W^2} - 2(1-p)\|\underline{\underline{Z}}^\top\underline{\underline{Z}}\,\underline{\nu}\|^2\theta^2\,\tfrac{\mathrm{d}\widetilde{\tau}}{W^2},\\
\mathrm{Drift}(\mathrm{d}H) &= -\tfrac{1}{2}\theta\,\tfrac{\mathrm{d}\widetilde{\tau}}{W} + \left(n - 1 - 4\|\underline{\underline{Z}}\,\underline{\nu}\|^2\right)\tfrac{\mathrm{d}\widetilde{\tau}}{W^2}\\
&\quad + \tfrac{1}{2}p\theta\,\tfrac{\mathrm{d}\widetilde{\tau}}{W} - \tfrac{1-p}{2}\left(1 - \left\|\underline{\underline{Z}}\,\underline{\nu}\right\|^2 - 4\,\|\underline{\underline{Z}}^\top\underline{\underline{Z}}\,\underline{\nu}\|^2\right)\theta^2\,\tfrac{\mathrm{d}\widetilde{\tau}}{W^2}.
\end{aligned}
$$
$$(3.8)$$

In order to ensure comparable dynamics for K and H, set $p = \alpha^2/W^2$ and $\theta = \beta/W$ (valid when $W \geq \max\{\tfrac{1}{\sqrt{2}}\beta, \alpha\}$): writing $\mathrm{d}\tau = \mathrm{d}\widetilde{\tau}/W^2 = 4(V/U)^2\,\mathrm{d}t$ leads to

$$
\begin{aligned}
(\mathrm{d}K)^2 &= \left(\alpha^2 + \tfrac{1}{2}\left\|\underline{\underline{Z}}\,\underline{\nu}\right\|^2\beta^2 - \tfrac{1}{2}\left\|\underline{\underline{Z}}\,\underline{\nu}\right\|^2\tfrac{\alpha^2\beta^2}{W^2}\right)\mathrm{d}\tau,\\
\mathrm{Drift}(\mathrm{d}K) &= -\mu_1\,\mathrm{d}\tau,\\
\mathrm{d}K \times \mathrm{d}H &= \|\underline{\underline{Z}}\,\underline{\nu}\|^2\left(1 - \tfrac{\alpha^2}{W^2}\right)\beta\,\mathrm{d}\tau,\\
(\mathrm{d}H)^2 &= 4\left(\|\underline{\underline{Z}}\,\underline{\nu}\|^2 - \tfrac{1}{2}(1 - \tfrac{\alpha^2}{W^2})\|\underline{\underline{Z}}^\top\underline{\underline{Z}}\,\underline{\nu}\|^2\tfrac{\beta^2}{W^2}\right)\mathrm{d}\tau,\\
\mathrm{Drift}(\mathrm{d}H) &= -\mu_2\,\mathrm{d}\tau,
\end{aligned}
$$
$$(3.9)$$

where μ_1 and μ_2 are given by

$$
\mu_1 = \tfrac{1}{2}\alpha^2 - \tfrac{1}{4}\left(1 - 2\|\underline{Z}\,\underline{\nu}\|^2\right)\beta^2 + \tfrac{1}{4W^2}\left(1 - 2\|\underline{Z}\,\underline{\nu}\|^2\right)\alpha^2\beta^2,
$$

$$
\mu_2 = \tfrac{1}{2}\beta - n + 1 + 4\|\underline{Z}\,\underline{\nu}\|^2
$$
$$
\quad - \tfrac{1}{2W^2}\left(\alpha^2\beta - \left(1 - \tfrac{\alpha^2}{W^2}\right)\left(1 - \|\underline{Z}\,\underline{\nu}\|^2 - 4\,\|\underline{Z}^{\mathsf{T}}\,\underline{Z}\,\underline{\nu}\|^2\right)\beta^2\right).
$$

$$(3.10)$$

In order to fulfil the underlying strategy, μ_1 and μ_2 should be chosen to accomplish the following:

1. $W = \exp(H - 2K)$ must remain large; this follows by a strong-law-of-large-numbers argument if μ_1, μ_2 are chosen so that $2\mu_1 - \mu_2 =$ Drift $\mathrm{d}(H - 2K)/\mathrm{d}\tau$ is positive and bounded away from zero and yet $(\mathrm{d}H)^2/\mathrm{d}\tau$ and $(\mathrm{d}K)^2/\mathrm{d}\tau$ are bounded.
2. Both of $-\mu_1 =$ Drift $\mathrm{d}(K)/\mathrm{d}\tau$ and $-\mu_2 =$ Drift $\mathrm{d}(H)/\mathrm{d}\tau$ must remain negative and bounded away from zero, so that K and H both tend to $-\infty$ as $\tau \to \infty$.
3. If coupling is to happen in finite time on the t-time-scale then

$$
T_{\text{coupling}} = \frac{1}{4}\int_0^\infty \left(\frac{U}{V}\right)^2 \mathrm{d}\tau
$$
$$
= \frac{1}{4}\int_0^\infty \exp\left(2H - 2K\right)\mathrm{d}\tau < \infty.
$$

This follows almost surely if μ_1, μ_2 are chosen so that $2\mu_1 - 2\mu_2 =$ Drift $\mathrm{d}(2H - 2K)/\mathrm{d}\tau$ is negative and bounded away from zero.

Now the inverse function theorem can be applied to show that for any constant $L > 0$ there is a constant $L' > 0$ such that if $W^2 \geq L'$ then (3.10) can be solved for any prescribed $0 \leq \mu_1, \mu_2 \leq L$ using $\alpha^2, \beta^2 \leq L'$ (incidentally thus bounding $(\mathrm{d}H)^2/\mathrm{d}\tau$ and $(\mathrm{d}K)^2/\mathrm{d}\tau$). Moreover by choosing L' large enough it then follows that $p = \alpha^2/W^2 \leq 1$ and $\beta^2/W^2 \leq 2$, so that the desired μ_1, μ_2 can be attained using a valid coupling control.

A comparison with Brownian motion of constant drift now shows that if initially $W \geq 2L'$ then there is a positive chance that $W \geq L'$ for all τ, and thus that coupling happens at τ-time infinity, and actual time $T_{\text{coupling}} < \infty$. Should W drop below L', then one can switch to the pure reflection control ($p = 1$) and run this control until $W = U/V^2$ rises again to level $2L'$. This is almost sure to happen eventually, since otherwise $V \to 0$ and thus $U = WV^2 \to 0$, which can be shown to have probability 0 under this control.

3.4 Estimates for coupling time distribution

In the planar case $n = 2$ the stochastic differential system (3.6) under mixed rotation-reflection controls can be simplified substantially. In this section we go beyond the work of Ben Arous et al.(1995) and Kendall (2007) by using this simplification to identify limiting distributions for suitable scalings of the coupling time T_{coupling}. The simplification arises because the non-zero eigenvalues of the skew-symmetric matrix $\underline{\underline{Z}}$ have even multiplicity; consequently in the two-dimensional case we may deduce from $\text{trace}(\underline{\underline{Z}}^\top \underline{\underline{Z}}) = 1$ and $\|\underline{\nu}\| = 1$ that $\left\|\underline{\underline{Z}}\,\underline{\nu}\right\|^2 = \frac{1}{2}$ and $\|\underline{\underline{Z}}^\top \underline{\underline{Z}}\,\underline{\nu}\|^2 = \frac{1}{4}$. Moreover an Euler formula follows from $\underline{\underline{Z}}\,\underline{\underline{Z}} = -\frac{1}{2}\underline{\underline{\mathbb{I}}}$:

$$\exp\left(-\sqrt{2}\,\theta\,\underline{\underline{Z}}\right) \quad = \quad \cos\theta\,\underline{\underline{\mathbb{I}}} - \sqrt{2}\,\sin\theta\,\underline{\underline{Z}}\,.$$

Accordingly in the planar case the mixed coupling control

$$\begin{aligned}
\underline{\underline{J}} \quad &= \quad p\left(\underline{\underline{\mathbb{I}}} - 2\,\underline{\nu}\,\underline{\nu}^\top\right) + q\exp\left(-\sqrt{2}\,\theta\,\underline{\underline{Z}}\right) \\
&= \quad \underline{\underline{\mathbb{I}}} - 2p\,\underline{\nu}\,\underline{\nu}^\top - q\,(1 - \cos\theta)\,\underline{\underline{\mathbb{I}}} - \sqrt{2}\,q\,\sin\theta\,\underline{\underline{Z}} \quad (3.11)
\end{aligned}$$

(for $0 \leq p = 1 - q \leq 1$, unrestricted θ) renders the system (3.6) as

$$\begin{aligned}
(\,\mathrm{d}K)^2 \quad &= \quad (p + \tfrac{1}{2}q(1 - \cos\theta)\ \mathrm{d}\widetilde{\tau}\,, \\
\text{Drift}(\,\mathrm{d}K) \quad &= \quad -\tfrac{1}{2}p\ \mathrm{d}\widetilde{\tau}\,, \\
\mathrm{d}K \times \mathrm{d}H \quad &= \quad \tfrac{1}{\sqrt{2}}q\sin\theta\ \tfrac{\mathrm{d}\widetilde{\tau}}{W}\,, \\
(\,\mathrm{d}H)^2 \quad &= \quad (2 - q(1 - \cos\theta))\ \tfrac{\mathrm{d}\widetilde{\tau}}{W^2}\,, \\
\text{Drift}(\,\mathrm{d}H) \quad &= \quad -\tfrac{1}{\sqrt{2}}q\sin\theta\ \tfrac{\mathrm{d}\widetilde{\tau}}{W} - (1 - \tfrac{1}{2}q(1 - \cos\theta))\ \tfrac{\mathrm{d}\widetilde{\tau}}{W^2}\,. \quad (3.12)
\end{aligned}$$

If we set $\theta = 0$ (so that the coupling control is a mixture of reflection coupling and *synchronous* couplings) then the result can be made to yield a successful coupling: choosing $p = \min\{1, \alpha^2/W^2\}$ the stochastic differential system becomes

$$\begin{aligned}
(\,\mathrm{d}K)^2 \quad &= \quad \min\{W^2, \alpha^2\}\ \mathrm{d}\tau\,, \\
\text{Drift}(\,\mathrm{d}K) \quad &= \quad -\tfrac{1}{2}\min\{W^2, \alpha^2\}\ \mathrm{d}\tau\,, \\
\mathrm{d}K \times \mathrm{d}H \quad &= \quad 0\,, \\
(\,\mathrm{d}H)^2 \quad &= \quad 2\ \mathrm{d}\tau\,, \\
\text{Drift}(\,\mathrm{d}H) \quad &= \quad -\,\mathrm{d}\tau\,, \quad\quad\quad\quad\quad (3.13)
\end{aligned}$$

where $\mathrm{d}\tau = 4\,\mathrm{d}t/V^2$ as before. Accordingly, for fixed α^2, once $W^2 \geq \alpha^2$ then H and K behave as uncorrelated Brownian motions with constant negative drifts in the τ time-scale; moreover if $\alpha^2 > 1$ then

$$\text{Drift } d(\log W)/d\tau \quad = \quad \alpha^2 - 1$$

is strictly positive and so $W \to \infty$ almost surely. In order to argue as before we must finally show almost-sure finiteness of

$$T_{\text{coupling}} \quad = \quad \frac{1}{4} \int_0^\infty \exp\left(2H - 2K\right) d\tau.$$

Now if we scale using the ratio $(U_0/V_0)^2$ at time zero then we can deduce the following convergence-in-distribution result as $W_0 = U_0/V_0^2 \to \infty$:

$$\left(\frac{V_0}{U_0}\right)^2 T_{\text{coupling}} \quad = \quad \frac{1}{4} \int_0^\infty \exp\left(2H - 2K - (2H_0 - 2K_0)\right) d\tau$$

$$\to \quad \frac{1}{4} \int_0^\infty \exp\left(2\sqrt{2 + \alpha^2}\, \widetilde{B} - (2 - \alpha^2)\tau\right) d\tau \quad (3.14)$$

where $\widetilde{B}$ is a standard real Brownian motion begun at 0. This integral is finite when $\alpha^2 < 2$, so finally we deduce that coupling occurs in finite time for this simple mixture of reflection and synchronous coupling if we choose $1 < \alpha^2 < 2$.

However we can now say much more, since the stochastic integral in (3.14) is one of the celebrated and much-studied exponential functionals of Brownian motion (Yor, 1992, or Yor, 2001, p. 15). In particular Dufresne (1990) has shown that such a functional

$$\int_0^\infty \exp\left(a\widetilde{B}\right)_s - bs\right) ds$$

has the distribution of $2/(a^2\Gamma_{2b/a^2})$, where Γ_κ is a Gamma-distributed random variable of index κ. In summary,

Theorem 3.1 *Let T_{coupling} be the coupling time for two-dimensional Brownian motion plus Lévy stochastic area under a mixture of reflection coupling and synchronous coupling. Using the notation above, let $\min\{1, \alpha^2/W^2\}$ be the proportion of reflection coupling. Scale T_{coupling} by the square of the ratio between initial areal difference U_0 and initial spatial distance V_0; if $1 < \alpha^2 < 2$ then as $W_0 = U_0/V_0^2 \to \infty$ so a re-scaling of the coupling time has limiting Inverse Gamma distribution*

$$\left(\frac{V_0}{U_0}\right)^2 T_{\text{coupling}} \quad \to \quad \frac{2}{2 + \alpha^2} \frac{1}{\Gamma_{(2-\alpha^2)/(2(2+\alpha^2))}}.$$

Thus the tail behaviour of T_{coupling} is governed by the index of the Gamma random variable, which here cannot exceed $\frac{1}{6}$ (the limiting case when $\alpha^2 \to 1$). This index is unattainable by this means since $H - 2K$ behaves like a Brownian motion when $\alpha^2 = 1$, so we cannot have $W \to \infty$. Kendall (2007, Section 3) exhibits a similar coupling for the planar case in which there is a state-dependent switching between reflection and synchronous coupling, depending on whether the ratio W exceeds a specified threshold; it would be interesting to calculate the tail-behaviour of the inverse of the coupling time in this case.

Note that scaling by $(U_0/V_0)^2$ rather than $W_0 = U_0/V_0^2$ quantifies something which can be observed from detailed inspection of the stochastic differential system (3.5); the rate of evolution of the areal distance U is reduced if the spatial distance V is small. The requirement $W_0 \to \infty$ is present mainly to remove the effect of higher-order terms $(\,d\tau/W^2)$ in systems such as (3.9), and in particular to ensure in (3.13) that $\min\{W^2, \alpha^2\} = \alpha^2$ for all time with high probability.

In fact one can do markedly better than the reflection-synchronous mixture coupling of Theorem 3.1 by replacing synchronous coupling by a rotation coupling for which $\sin\theta = \sqrt{2}\,\beta/W$: similar calculations then show the index of the inverse of the limiting scaled coupling time can be increased up to the limit of $\frac{1}{2}$, which remarkably is the index of the inverse of the coupling time for reflection coupling of Brownian motion alone! However this limit is not attainable by these mixture couplings, as it corresponds to a limiting case of $\beta = 2$, $\alpha^2 = 3$. At this choice of parameter values $H - 2K$ again behaves like a Brownian motion so we do not have $W \to \infty$.

In higher dimensions similar calculations can be carried out, but the geometry is more complicated; in the planar case the form of $\underline{Z}$ is essentially constant, whereas it will evolve stochastically in higher dimensions and relate non-trivially to $\underline{\nu}$. This leads to correspondingly weaker results: the inverse limiting scaled coupling time can be bounded above and below using Gamma distributions of different indices.

We should not expect these mixture couplings to be maximal, even within the class of co-adapted couplings. Indeed Kendall (2007) gives a heuristic argument to show that maximality amongst co-adapted couplings should be expected only when one Brownian differential is a (state-dependent) rotation or rotated reflection of the other. The interest of these mixture couplings lies in the ease with which one may derive limiting distributions for them, hence gaining a good perspective on how rapidly one may couple the stochastic area.

4 Conclusion

After reviewing aspects of coupling theory, we have indicated an approach to co-adapted coupling of Brownian motion and its stochastic areas, and shown how in the planar case one can use Dufresne's formula to derive asymptotics of coupling time distributions for suitable mixed couplings. Aspects of these asymptotic distributions indicate the price that is to be paid for coupling stochastic areas as well as the Brownian motions themselves; however it is clear that these mixed couplings should not be expected to be maximal amongst all co-adapted couplings. Accordingly a very interesting direction for future research is to develop these methods to derive estimates for coupling time distributions for more efficient couplings using state-dependent coupling strategies as exemplified in Kendall (2007, Section 3). Progress in this direction would deliver probabilistic gradient estimates in the manner of Cranston (1991, 1992) (contrast the analytic work of Bakry et al. 2008).

A further challenge is to develop these techniques for higher-dimensional cases. Here the two-dimensional approach extends naïvely to deliver upper and lower bounding distributions; a more satisfactory answer with tighter bounds will require careful analysis of the evolution under the coupling of the geometry as expressed by the pair $(\nu, \underline{Z})$.

A major piece of unfinished business in this area is to determine the extent to which these co-adapted coupling results extend to higher-order iterated path integrals (simple Itô calculus demonstrates that it suffices to couple Lévy stochastic areas in order to couple all possible non-iterated path integrals of the form $\int B_i \, dB_j$). Some tentative insight is offered by the rôle played by the Morse–Thue sequence for iterated time-integrals (Kendall and Price, 2004). Moreover it is possible to generalize the invariance considerations underlying (3.4) for the areal difference, so as to produce similarly invariant differences of higher-order iterated path integrals. But at present the closing question of Kendall (2007) still remains open, whether one can co-adaptively couple Brownian motions together with all possible iterated path and time-integrals up to a fixed order of iteration.

References

Bailleul, I. 2008. Poisson boundary of a relativistic diffusion. *Probab. Theory Related Fields*, **141**(1-2), 283–329.

Bakry, D., Baudoin, F., Bonnefont, M., and Chafaï, D. 2008. On gradient bounds for the heat kernel on the Heisenberg group. *J. Funct. Anal.*, **255**(8), 1905–1938.

Barbour, A. D., Holst, L., and Janson, S. 1992. *Poisson Approximation*. Oxford Studies in Probability, vol. 2. New York: Oxford Univ. Press. Oxford Science Publications.

Ben Arous, G., Cranston, M., and Kendall, W. S. 1995. Coupling constructions for hypoelliptic diffusions: two examples. Pages 193–212 of: Cranston, M., and Pinsky, M. (eds), *Stochastic Analysis: Summer Research Institute July 11-30, 1993*. Proc. Sympos. Pure Math., vol. 57. Providence, RI: Amer. Math. Soc.

Burdzy, K., and Kendall, W. S. 2000. Efficient Markovian couplings: examples and counterexamples. *Ann. Appl. Probab.*, **10**(2), 362–409.

Chen, M. F., and Li, S. F. 1989. Coupling methods for multidimensional diffusion processes. *Ann. Probab.*, **17**(1), 151–177.

Connor, S. B. 2007. *Coupling: Cutoffs, CFTP and Tameness*. Ph.D. thesis, Department of Statistics, University of Warwick.

Connor, S. B., and Jacka, S. D. 2008. Optimal co-adapted coupling for the symmetric random walk on the hypercube. *J. Appl. Probab.*, **45**(1), 703–713.

Cranston, M. 1991. Gradient estimates on manifolds using coupling. *J. Funct. Anal.*, **99**(1), 110–124.

Cranston, M. 1992. A probabilistic approach to gradient estimates. *Canad. Math. Bull.*, **35**(1), 46–55.

Doeblin, W. 1938. Exposé de la théorie des chaînes simples constants de Markoff á un nombre fini d'états. *Revue Math. de l'Union Interbalkanique*, **2**, 77–105.

Dufresne, D. 1990. The distribution of a perpetuity, with applications to risk theory and pension funding. *Scand. Actuar. J.*, (1-2), 39–79.

Émery, M. 2005. On certain almost Brownian filtrations. *Ann. Inst. H. Poincaré Probab. Statist.*, **41**(3), 285–305.

Goldstein, S. 1978. Maximal coupling. *Z Wahrscheinlichkeitstheorie verw. Gebiete*, **46**(2), 193–204.

Griffeath, D. 1974. A maximal coupling for Markov chains. *Z Wahrscheinlichkeitstheorie verw. Gebiete*, **31**, 95–106.

Griffeath, D. 1978. Coupling methods for Markov processes. Pages 1–43 of: *Studies in Probability and Ergodic Theory*. Adv. in Math. Suppl. Stud., vol. 2. New York: Academic Press.

Hairer, M. 2002. Exponential mixing properties of stochastic PDEs through asymptotic coupling. *Probab. Theory Related Fields*, **124**(3), 345–380.

Hsu, E. P., and Sturm, K.-T. 2003. *Maximal Coupling of Euclidean Brownian Motions*. SFB-Preprint 85. Universität Bonn.

Jansons, K. M., and Metcalfe, P. D. 2007. Optimally coupling the Kolmogorov diffusion, and related optimal control problems. *J. Comput. Math.*, **10**, 1–20.

Kendall, W. S. 1986a. Nonnegative Ricci curvature and the Brownian coupling property. *Stochastics and Stochastic Reports*, **19**, 111–129.

Kendall, W. S. 1986b. Stochastic differential geometry, a coupling property, and harmonic maps. *J. Lond. Math. Soc. (2)*, **33**, 554–566.

Kendall, W. S. 1988. Martingales on manifolds and harmonic maps. Pages 121–157 of: Durrett, R., and Pinsky, M. (eds), *The Geometry of Random Motion*. Contemp. Math., vol. 73. Providence, RI: Amer. Math. Soc.

Kendall, W. S. 1994. Probability, convexity, and harmonic maps II: Smoothness via probabilistic gradient inequalities. *J. Funct. Anal.*, **126**, 228–257.

Kendall, W. S. 2001. Symbolic Itô calculus: an ongoing story. *Stat. Comput.*, **11**, 25–35.

Kendall, W. S. 2007. Coupling all the Lévy stochastic areas of multidimensional Brownian motion. *Ann. Probab.*, **35**(3), 935–953.

Kendall, W. S. 2009a. Brownian couplings, convexity, and shy-ness. *Electron. Commun. Probab.*, **14**, 66–80.

Kendall, W. S., and Price, C. J. 2004. Coupling iterated Kolmogorov diffusions. *Electron. J. Probab.*, **9**, 382–410.

Kingman, J. F. C. 1982. The coalescent. *Stochastic Process. Appl.*, **13**(3), 235–248.

Kuwada, K. 2007. On uniqueness of maximal coupling for diffusion processes with a reflection. *J. Theoret. Probab.*, **20**(4), 935–957.

Kuwada, K. 2009. Characterization of maximal Markovian couplings for diffusion processes. *Electron. J. Probab.*, **14**, 633–662.

Kuwada, K., and Sturm, K.-T. 2007. A counterexample for the optimality of Kendall-Cranston coupling. *Electron. Commun. Probab.*, **12**, 66–72.

Liggett, T. M. 2005. *Interacting Particle Systems*. Classics Math. Berlin: Springer-Verlag. Reprint of the 1985 original.

Lindvall, T. 1982. *On Coupling of Brownian Motions*. Tech. rept. 1982:23. Department of Mathematics, Chalmers University of Technology and University of Göteborg.

Lindvall, T. 2002. *Lectures on the Coupling Method*. Mineola, NY: Dover Publications. Corrected reprint of the 1992 original.

Lindvall, T., and Rogers, L. C. G. 1986. Coupling of multidimensional diffusions by reflection. *Ann. Probab.*, **14**(3), 860–872.

Pitman, J. 1976. On coupling of Markov chains. *Z Wahrscheinlichkeitstheorie verw. Gebiete*, **35**(4), 315–322.

Renesse, M. von. 2004. Intrinsic coupling on Riemannian manifolds and polyhedra. *Electron. J. Probab.*, **9**, 411–435.

Thorisson, H. 1994. Shift-coupling in continuous time. *Probab. Theory Related Fields*, **99**(4), 477–483.

Yor, M. 1992. Sur certaines fonctionnelles exponentielles du mouvement brownien réel. *J. Appl. Probab.*, **29**(1), 202–208.

Yor, M. 2001. *Exponential Functionals of Brownian Motion and Related Processes*. Springer Finance. Berlin: Springer-Verlag.

20
Queueing with neighbours

Vadim Shcherbakov[a] and Stanislav Volkov[b]

Abstract

In this paper we study asymptotic behaviour of a growth process generated by a semi-deterministic variant of the cooperative sequential adsorption model (CSA). This model can also be viewed as a particular example from queueing theory, to which John Kingman has contributed so much. We show that the quite limited randomness of our model still generates a rich collection of possible limiting behaviours.

Keywords cooperative sequential adsorption, interacting particle systems, max-plus algebra, queueing, Tetris

AMS subject classification (MSC2010) Primary 60G17, 62M30; Secondary 60J20

1 Introduction

Let $\mathcal{M} = \{1, 2, \ldots, M\}$ be a lattice segment with periodic boundary conditions (that is, $M + 1$ will be understood as 1 and $1 - 1$ will be understood as M), where $M \geq 1$. The growth process studied in this paper is defined as a discrete-time Markov chain $(\xi_i(t), i \in \mathcal{M}, t \in \mathbb{Z}_+)$,

[a] Laboratory of Large Random Systems, Faculty of Mechanics and Mathematics, Moscow State University, 119991 Moscow, Russia;
v.shcherbakov@mech.math.msu.su

[b] Department of Mathematics, University of Bristol, University Walk, Bristol BS8 1TW; S.Volkov@bristol.ac.uk

with values in $\mathbb{Z}_+^M$ and specified by the following transition probabilities:

$$\mathbb{P}\left(\xi_i(t+1) = \xi_i(t) + 1, \ \xi_j(t+1) = \xi_j(t) \ \forall j \neq i \mid \xi(t)\right)$$
$$= \begin{cases} 0, & \text{if} \quad u_i(t) > m(t), \\ 1/N_{\min}(t), & \text{if} \quad u_i(t) = m(t), \end{cases} \tag{1.1}$$

for $i \in \mathcal{M}$, where

$$u_i(t) = \sum_{j \in U_i} \xi_j(t), \ i \in \mathcal{M},$$

U_i is a certain neighbourhood of site i,

$$m(t) = \min_{k \in \mathcal{M}} u_k(t) \tag{1.2}$$

and $N_{\min}(t) \in \{1, 2, \ldots, M\}$ is the number of $u_i(t)$ equal to $m(t)$. The quantity $u_i(t)$ is called the *potential* of site i at time t.

The growth process describes the following random sequential allocation procedure. Arriving particles are sequentially allocated at sites of $\mathcal{M}$ such that a particle is allocated uniformly over sites with minimal potential. Then the process component $\xi_k(t)$ is the number of particles at site k at time t. The growth process can be viewed as a certain limit case of a growth process studied in Shcherbakov and Volkov (2010). The growth process in Shcherbakov and Volkov (2010) is defined as a discrete-time Markov chain $(\xi_i(t), i \in \mathcal{M}, t \in \mathbb{Z}_+)$, with values in $\mathbb{Z}_+^M$ and specified by the following transition probabilities

$$\mathbb{P}\{\xi_i(t+1) = \xi_i(t) + 1, \ \xi_j(t+1) = \xi_j(t) \ \forall j \neq i \mid \xi(t)\} = \frac{\beta^{u_i(t)}}{Z(\xi(t))} \tag{1.3}$$

where

$$Z(\xi(t)) = \sum_{j=1}^M \beta^{u_j(t)},$$

β is a positive number and the other notations are the same as before. It is easy to see that the process defined by transition probabilities (1.1) is the corresponding limit process as $\beta \to 0$. In turn, the growth process specified by the transition probabilities (1.3) is a particular version of the cooperative sequential adsorption model (CSA). CSA is a probabilistic model which is widely used in physics for modelling various adsorption processes (see Evans (1993), Privman (2000) and references therein). Some asymptotic and statistical studies of similar CSA in a continuous setting were undertaken in Shcherbakov (2006), Penrose and Shcherbakov (2009) and Penrose and Shcherbakov (2010).

In Shcherbakov and Volkov (2010) we consider the following variants of neighbourhood:

(**A1**) $U_i = \{i\}$ (empty),

(**A2**) $U_i = \{i, i+1\}$ (asymmetric),

(**A3**) $U_i = \{i-1, i, i+1\}$ (symmetric),

where, due to the periodic boundary conditions, $U_M = \{M, 1\}$ in case (**A2**) and $U_1 = \{M, 1, 2\}$, $U_M = \{M-1, M, 1\}$ in case (**A3**) respectively. It is easy to see that for the growth process studied in this paper the case (**A1**) is trivial. Therefore, we will consider cases (**A2**) and (**A3**) only.

A stochastic process $u(t) = (u_1(t), \ldots, u_M(t))$ formed by the sites' potentials plays an important role in our asymptotic study of the growth process. It is easy to see that $u(t)$ is also a Markov chain, with transition probabilities given by

$$\mathbb{P}\left(u_i(t+1) = u_i(t) + 1_{\{i \in U_k\}}\right) = \begin{cases} 0, & \text{if } u_k(t) > m(t), \\ 1/N_{\min}(t), & \text{if } u_k(t) = m(t), \end{cases} \quad (1.4)$$

for $k \in \mathcal{M}$. This process has the following rather obvious queueing interpretation (S. Foss, personal communications) explaining the title of the paper (originally titled 'Random sequential adsorption at extremes'). Namely, consider a system with M servers, with the clients arriving in bunches of 2 in case (**A2**) and of 3 in case (**A3**). The quantity $u_i(t)$ is interpreted as the number of clients at server i at time t. In case (**A2**), of the two clients in the arriving pair, one is equally likely to join any of the shortest queues, while the other joins its left neighbouring queue. In case (**A3**), of the three clients in an arriving triple, one joins the shortest queue, the others its left and right neighbouring queues.

Our goal is to describe the long time behaviour of the growth process, or, equivalently, to describe the limiting queueing profile of the network. It should be noted that the method of proof in this paper is purely combinatorial. This is in contrast with Shcherbakov and Volkov (2010), where the results are proved by combining the martingale techniques from Fayolle et al. (1995) with some probabilistic techniques used in the theory of reinforced processes from Tarrès (2004) and Volkov (2001).

Observe that the model considered here can be viewed as a randomized *Tetris* game, and hence it can possibly be analyzed using the techniques of max-plus algebra as well; see Bousch and Mairesse (2002) and Section 1.3 of Heidergott et al. (2006) for details.

For the sake of completeness, let us mention another limit case of the growth process specified by transition probabilities (1.3): namely, the limit process arising as $\beta \to \infty$. It is easy to see that the limit process in this case describes the allocation process in which a particle is allocated with equal probabilities to one of the sites with maximal potential. The asymptotic behaviour (as $t \to \infty$) of this limit process is briefly discussed in Section 5.

2 Results

Theorem 2.1 *Suppose $U_i = \{i, i+1\}$, $i \in \mathcal{M}$. Then, with probability 1, there is a $t_0 = t_0(\omega)$ (depending also on the initial configuration) such that for all $t \geq t_0$*

$$|\xi_i(t) - \xi_{i+2}(t)| \leq 2 \tag{2.1}$$

for $i \in \mathcal{M}$. Moreover,

$$\xi_i(t) = \frac{t}{M} + \eta_i(t) + \begin{cases} 0, & \text{when } M \text{ is odd,} \\[2mm] (-1)^i \, Z(t), & \text{when } M \text{ is even,} \end{cases} \tag{2.2}$$

where $|\eta_i(t)| \leq 2M$ and for some $\sigma > 0$

$$\lim_{n \to \infty} \frac{Z(\lfloor sn \rfloor)}{\sigma\sqrt{n}} = B(s),$$

where $\lfloor x \rfloor$ denotes the integer part of x and $B(s)$ is a standard Brownian motion.

Theorem 2.2 *Suppose $U_i = \{i-1, i, i+1\}$, $i \in \mathcal{M}$. Then with probability 1 there exists the limit $\mathbf{x} = \lim_{t \to \infty} \xi(t)/t$, which takes a finite number of possible values with positive probabilities. The set of limiting configurations consists of those $\mathbf{x} = (x_1, \ldots, x_M) \in \mathbb{R}^M$ which simultaneously satisfy the following properties:*

- *there exists an $\alpha > 0$ such that $x_i \in \{0, \alpha/2, \alpha\}$ for all $i \in \mathcal{M}$; also $\sum_{i=1}^{M} x_i = 1$;*
- *if $x_i = 0$, then $x_{i-1} > 0$ or $x_{i+1} > 0$, or both;*
- *if $x_i = \alpha/2$, then*

$$(x_{j-3}, x_{j-2}, x_{j-1}, x_j, x_{j+1}, x_{j+2}) = (\alpha, 0, \alpha/2, \alpha/2, 0, \alpha),$$

where $j \in \{i, i+1\}$;

- *if $x_i = \alpha$, then $x_{i-1} = x_{i+1} = 0$;*
- *if $M = 3K$ is divisible by 3, then*

$$\min\{x_j, x_{j+3}, x_{j+6}, \ldots, x_{j+3(K-1)}\} = 0,$$

for $j = 1, 2, 3$.

Moreover, the adsorption eventually stops at all $i \in \mathcal{M}$ where $x_i = 0$, that is

$$\sup_{t \geq 0} \xi_i(t) = \infty \text{ if and only if } x_i > 0.$$

*Additionally, if the initial configuration is empty, then for each $x_i = 0$ we must have that **both** $x_{i-1} > 0$ **and** $x_{i+1} > 0$.*

Table 2.1 *Limiting configurations for symmetric interaction*

M	Limiting configurations (up to rotation)	No. of limits
4	$\left(\frac{1}{2}, 0, \frac{1}{2}, 0\right)$	2
5	$\left(\frac{1}{4}, \frac{1}{4}, 0, \frac{1}{2}, 0\right), \left(\frac{1}{2}, 0, 0, \frac{1}{2}, 0\right)^*$	5 (10*)
6	$\left(\frac{1}{3}, 0, \frac{1}{3}, 0, \frac{1}{3}, 0\right)$	2
7	$\left(\frac{1}{6}, \frac{1}{6}, 0, \frac{1}{3}, 0, \frac{1}{3}, 0\right), \left(\frac{1}{3}, 0, \frac{1}{3}, 0, \frac{1}{3}, 0, 0\right)^*$	7 (14*)
8	$\left(\frac{1}{4}, 0, \frac{1}{4}, 0, \frac{1}{4}, 0, \frac{1}{4}, 0\right),$ $\left(\frac{1}{6}, \frac{1}{6}, 0, \frac{1}{3}, 0, 0, \frac{1}{3}, 0\right)^*, \left(0, 0, \frac{1}{3}, 0, 0, \frac{1}{3}, 0, \frac{1}{3}\right)^*$	2 (18*)
9	$\left(\frac{1}{8}, \frac{1}{8}, 0, \frac{1}{4}, 0, \frac{1}{4}, 0, \frac{1}{4}, 0\right), \left(0, 0, \frac{1}{4}, 0, \frac{1}{4}, 0, \frac{1}{4}, 0, \frac{1}{4}\right)^*$	9 (18*)
10	$\left(\frac{1}{8}, \frac{1}{8}, 0, \frac{1}{4}, 0, \frac{1}{8}, \frac{1}{8}, 0, \frac{1}{4}, 0\right), \left(\frac{1}{5}, 0, \frac{1}{5}, 0, \frac{1}{5}, 0, \frac{1}{5}, 0, \frac{1}{5}, 0\right),$ $\left(0, \frac{1}{4}, 0, 0, \frac{1}{4}, 0, \frac{1}{4}, 0, \frac{1}{8}, \frac{1}{8}\right)^*, \left(0, \frac{1}{4}, 0, \frac{1}{4}, 0, 0, \frac{1}{4}, 0, \frac{1}{8}, \frac{1}{8}\right)^*,$ $\left(0, 0, \frac{1}{4}, 0, 0, \frac{1}{4}, 0, \frac{1}{4}, 0, \frac{1}{4}\right)^*, \left(0, 0, \frac{1}{4}, 0, \frac{1}{4}, 0, 0, \frac{1}{4}, 0, \frac{1}{4}\right)^*$	7 (42*)

Table 2.2 *Numbers of limiting configurations for symmetric interaction for larger M*

M	11	12	13	14	15	16
Distinct conf.	1(4*)	2(7*)	1(8*)	3(12*)	2(16*)	3(20*)
All conf.	11(44*)	14(74*)	13(104*)	23(142*)	20(220*)	34(290*)

We will derive the asymptotic behaviour of the process $\xi(t)$ from the asymptotic behaviour of the process of potentials. In turn the study of the process of potentials is greatly facilitated by analysis of the following auxiliary process

$$v_k(t) = u_k(t) - m(t), \quad k = 1, \ldots, M. \tag{2.3}$$

Observe that $v(t)$ also forms a Markov chain on $\{0, 1, 2, \ldots\}$ and for each t there is a k such that $v_k(t) = 0$. Loosely speaking, the quantities $v_k(t)$, $k = 1, \ldots, M$, represent what happens 'on the top of the growth profile'.

It turns out that in the asymmetric case there is a single class of recurrent states to which the chain eventually falls and then stays in forever. As we show later, this class is a certain subset of the set $\{0, 1, 2\}^M$ containing the origin $(0, \ldots, 0)$. Thus a certain 'stability' of the process of potentials is observed as time goes to infinity.

In particular, it yields, as we show, that there will not be long queues in the system if M is odd; however, this does not prevent occurrence of relatively long queues if M is even. For instance, if M is even, then one can easily imagine the profile with peaks at even sites, and zeros at odd sites. Besides, observe that the sum of the potentials of the even sites equals the sum of the potentials of the odd sites (see Proposition 3.1); therefore the difference between the total queue to the even sites, and the total queue to the odd ones, behaves similarly to the zero-mean random walk. It means that there are rather long periods of time during which much longer queues are observed at the even sites in comparison with the queues at the odd sites, and vice versa. Thus, in the case of the asymmetric interaction we observe in the limit $t \to \infty$ a 'comb pattern' when M is even, and a 'flat pattern' when M is odd.

The picture is completely different in the symmetric case. The Markov chain $v(t)$ is transient for any M; moreover, there can be only finitely many types of paths along which the chain escapes to infinity. By this we mean that if the chain reaches a state belonging to a particular *escape path*, then it will never leave it and will go to infinity along this path, and we will show that there can be only a finite number of limit patterns. An escape path can be viewed as *an attractor*, since similar effects are observed in neural network models studied in Karpelevich et al. (1995) and Malyshev and Turova (1997). In fact, the Markov chain $v(t)$ describes the same dynamics as the neural network models in Karpelevich et al. (1995) and Malyshev and Turova (1997) though in a slightly different set-up. The difference seems to be technical but it results in quite different model behaviour. We do not investigate this issue in depth here.

Table 2.1 contains the list of all possible limiting configurations (for proportions of customers/particles) for small M, while in Table 2 we provide only the numbers of possible limiting configurations for some larger M. Note that in the tables the symbol $*$ stands for the configurations which cannot be achieved from the empty initial configuration. Unfortunately, we cannot compute exact numbers of possible limiting configurations for any M; nor can we predict which of them will be more likely (though it is obvious that if we start with the empty initial configuration, all possible limits which can be obtained by a rotation of the same $\mathbf{x}$ will have the same probability.)

3 Asymmetric interaction

In the asymmetric case the potential of site k at time t is

$$u_k(t) = \xi_k(t) + \xi_{k+1}(t), \ k \in \mathcal{M}.$$

The transition probabilities of the Markov chain $u(t) = (u_1(t), \ldots, u_M(t))$ are given by

$$\mathbb{P}\left(u_i(t+1) = u_i(t) + 1_{i \in \{k-1,k\}}, i = 1, \ldots, M | u(t)\right)$$
$$= \left\{ \begin{array}{ll} 0, & \text{if} \quad u_k(t) > m(t), \\ N_{\min}^{-1}(t), & \text{if} \quad u_k(t) = m(t), \end{array} \right.$$

for $k \in \mathcal{M}$, where $N_{\min}(t) \in \{1, 2, \ldots, M\}$ is the number of $u_i(t)$ equal to $m(t)$.

Proposition 3.1 *If M is odd, then for any $u = (u_1, u_2, \ldots, u_M)$ the system*

$$
\begin{aligned}
u_1 &= \xi_1 + \xi_2 \\
u_2 &= \xi_2 + \xi_3 \\
&\vdots \\
u_M &= \xi_M + \xi_1
\end{aligned}
\tag{3.1}
$$

has a unique solution. On the other hand, if M is even, system (3.1) has a solution if and only if

$$u_1 + u_3 + \cdots + u_{M-1} = u_2 + u_4 + \cdots + u_M. \tag{3.2}$$

Proof For a fixed ξ_1 we can express the remaining ξ_k as

$$\xi_k = u_{k-1} - u_{k-2} + u_{k-3} - \cdots + (-1)^{k-1}\xi_1, \qquad (3.3)$$

for any $k = 2, \ldots, M$. Now, when M is odd, there is a unique choice of

$$\xi_1 = \frac{1}{2}\left(u_M - u_{M-1} + u_{M-2} + \cdots - u_2 + u_1\right).$$

When M is even, by summing separately odd and even lines of (3.1) we obtain condition (3.2). Then it turns out that we can set ξ_1 to be any real number, with ξ_k, $k \geq 2$, given by (3.3). $\qquad \square$

The following statement immediately follows from Proposition 3.1.

Corollary 3.2 *If M is even, then*

$$v_1(t) + v_3(t) + \cdots + v_{M-1}(t) = v_2(t) + v_4(t) + \cdots + v_M(t) \quad \forall t. \quad (3.4)$$

In the following two Propositions we will show that when either M is odd *or* M is even and condition (3.4) holds, the state $\mathbf{0} = (0, \ldots, 0)$ is recurrent for the Markov chain $v(t)$. First, define the following stopping times

$$t_0 = 0,$$
$$t_j = \min\{t > t_{j-1} : m(t_j) > m(t_{j-1})\}, \ j = 1, 2, \ldots. \quad (3.5)$$

Let

$$S(j) = \sum_{k=1}^{M} v_k^*(t_j)$$

where

$$v_k^*(t_j) = \begin{cases} v_k(t_j), & \text{if } v_k(t_j) \geq 2, \\ 0, & \text{otherwise,} \end{cases}$$

where the stopping times are defined by (3.5).

Proposition 3.3 $S(j+1) \leq S(j)$. *Moreover, there is an integer $K = K(M)$ and an $\varepsilon > 0$ such that*

$$\mathbb{P}(S(j + K) - S(j) \leq -1 \mid v(t_j)) \geq \varepsilon$$

on the event $S(j) > 0$.

Proof For simplicity, let us write v_k for $v_k(t_j)$. Take some non-zero element $a \geq 1$ in the sequence of v_k at time t_j. Whenever it is followed by a consecutive chunk of 0s, namely

$$\dots a \underbrace{0\ 0\ \dots\ 0}_{m} \dots$$

at time t_{j+1} this becomes either

$$\dots a \underbrace{z_1\ z_2\ \dots\ z_m}_{m} \dots$$

or

$$\dots a - 1 \underbrace{z_1\ z_2\ \dots\ z_m}_{m} \dots,$$

where $z_i \in \{0,1\}$, and the latter occurs if the second 0 from the left is chosen before the first one. On the other hand, if a is succeeded by a non-zero element, say '$\dots a\ b\ \dots$' at time t_{j+1} this becomes either '$\dots\ a - 1\ b\ \dots$' or '$\dots\ a - 1\ b - 1\ \dots$'. In all cases, this leads to $S(j + 1) \leq S(j)$.

Secondly, from the previous arguments we see that if there is at least one $a \geq 2$ in the sequence of $(v_1 \dots v_M)$ followed by a non-zero element, then this element becomes $a - 1$ at t_{j+1} and hence $S(j + 1) \leq S(j) - 1$.

Now let us investigate what happens if the opposite occurs. Then each element $a \geq 2$ in $(v_1 \dots v_M)$ is followed by a sequence of 0s and 1s such that we observe either

$$\dots a\ 0\ b \dots$$

or

$$\dots a\ 0\ 1 \underbrace{z_3\ z_4\ \dots\ z_m}_{m-2}\ b \dots,$$

where $m \geq 2$, $b \geq 2$ and $z_i \in \{0,1\}$. Because of Corollary 3.2, we cannot have an alternating sequence of 0s and non-zero elements; therefore, we must be able to find somewhere in the sequence of vs a chunk which looks either like

$$\dots a\ \underbrace{\underbrace{0\ 1}\ \underbrace{0\ 1}\ \dots\ \underbrace{0\ 1}}_{2l\ \text{elements}}\ 0\ 1\ c \dots \text{ where } c \geq 1 \qquad \text{(A)}$$

or

$$\dots a\ \underbrace{\underbrace{0\ 1}\ \underbrace{0\ 1}\ \dots\ \underbrace{0\ 1}}_{2l\ \text{elements}}\ \underbrace{0\ 0\ ?}_{i\ \ i+1\ i+2}\ \dots \qquad \text{(B)}$$

where $l \geq 0$. Note that a configuration of type (A) at time t_{j+1} with probability 1 becomes a configuration of type (B). At the same time, in configuration (B), with probability of at least $\frac{1}{3}$ the 0 located at position $i + 1$ is chosen before either the 0 at position i or (possibly) the 0 at position $i + 2$. On this event, the configuration in (B) at time t_{j+1} becomes

$$\ldots a \; \underbrace{\underbrace{0\;1}\;\underbrace{0\;1}\;\ldots\;\underbrace{0\;1}}_{2l-2 \text{ elements}} \; 0\;0\;0\;? \ldots \qquad (\mathrm{B'})$$

By iterating this argument until $l = 0$, we conclude that eventually there will be a chunk '$\ldots a\;0\;0\ldots$' on some step $t_{j'}$ which in turn at time $t_{j'+1}$ will become '$\ldots a - 1\;0\;? \ldots$'with probability at least $\frac{1}{3}$, resulting in $S_{j'+1} \leq S_{j'} - 1$. This yields the statement of Proposition 3.3 with $K = M$ and $\varepsilon = 3^{-M}$. $\qquad\square$

Proposition 3.4 *With probability 1, there is a $j_0 = j_0(\omega)$ such that*

$$S(j) = 0 \text{ for all } j \geq j_0.$$

Additionally, the state $\mathbf{0} = (0, 0, \ldots, 0)$ is recurrent for the Markov chain $v(t)$.

Proof The first statement trivially follows from Proposition 3.3. Next observe that at times $t_j \geq t_{j_0}$ the sequence $(v_1(t_j), \ldots, v_M(t_j))$ consists only of 0s and 1s locally looking either like

$$\ldots 1 \; \underbrace{\underbrace{0\;0}\;\underbrace{0\;0}\;\ldots\;\underbrace{0\;0}}_{2l \text{ elements}} \; 1 \ldots \qquad (\mathrm{C})$$

or

$$\ldots 1 \; \underbrace{\underbrace{0\;0}\;\underbrace{0\;0}\;\ldots\;\underbrace{0\;0}}_{2l \text{ elements}} \; 0\;1 \ldots \qquad (\mathrm{D})$$

With positive probability even-located 0s are picked before odd-located 0s, hence at time t_{j+1} configuration (C) becomes

$$\ldots 0 \; \underbrace{\underbrace{0\;0}\;\underbrace{0\;0}\;\ldots\;\underbrace{0\;0}}_{2l \text{ elements}} \; ? \ldots \qquad (\mathrm{C'})$$

while configuration (D) becomes

$$\ldots 0 \; \underbrace{\underbrace{0\;0}\;\underbrace{0\;0}\;\ldots\;\underbrace{0\;0}}_{2l-2 \text{ elements}} \; 0\;1\;0\;? \ldots \qquad (\mathrm{D'})$$

In both cases (C) $\rightarrow$ (C′) and (D) $\rightarrow$ (D′) the number of 1s among the v_i does not increase, and in the first case it goes down by 1. However, it is easy to see that whether M is odd or even (in the latter case due to Corollary 3.2) there will be at least one chunk of type (C), and hence with positive probability $v(t)$ reaches state $\mathbf{0}$ in at most M^2 steps (since $t_{j+1} - t_j \leq M$). The observation that after t_{j_0} the Markov chain $v(t)$ lives on a finite state space $\{0, 1, 2\}^M$ finishes the proof. $\qquad\square$

Proof of Theorem 2.1 The first part easily follows from Proposition 3.4 and the definition of potentials v. Indeed, for $j \geq j_0$ and all i we have $v_i(t_j) \in \{0, 1\}$, while for $t \in (t_j, t_{j+1})$ we have $v_i(t) \in \{0, 1, 2\}$. On the other hand, omitting (t), we can write $v_{i+1} - v_i = u_{i+1} - u_i = \xi_{i+2} - \xi_i$, $i \in \mathcal{M}$, yielding (2.1).

Next, iterating this argument, we obtain $|\xi_{i+2l} - \xi_i| \leq 2l$. Because of the periodic boundary condition, in the case when M is odd, this results in $|\xi_i - \xi_j| \leq 2M$ for all i and j, while in the case when M is even this is true only whenever $i - j$ is even. The observation that $\sum_j \xi_j(t) = t$ thus proves (2.2) for odd M, since

$$|t - M\xi_i(t)| = \left| \sum_{j=1}^{M} [\xi_j(t) - \xi_i(t)] \right| \leq (M - 1) \times 2M.$$

Now, when $M = 2L$ is even, denote

$$H(t) = \frac{\sum_{j=1}^{L} \xi_{2j}(t) - \sum_{j=1}^{L} \xi_{2j-1}(t)}{M}.$$

Suppose $i \in \mathcal{M}$ is even. Then

$$|t - M\xi_i(t) + MH(t)| = \left| 2 \sum_{j=1}^{L} \xi_{2j}(t) - 2L\xi_i(t) \right|$$

$$= \left| 2 \sum_{j=1}^{L} [\xi_{2j}(t) - \xi_i(t)] \right| \leq 4M(L - 1) < 2M^2.$$

A similar argument holds for odd i. Hence we have established (2.2) for even M as well as for odd M.

To finish the proof, denote by τ_m, $m \geq 0$, the consecutive renewal times of the Markov chain $v(t)$ after t_{j_0}, that is

$$\tau_0 = \inf\{t \geq t_{j_0} : v_1(t) = v_2(t) = \cdots = v_M(t) = 0\},$$
$$\tau_m = \inf\{t \geq \tau_{m-1} : v_1(t) = v_2(t) = \cdots = v_M(t) = 0\}, \quad m \geq 1.$$

By Proposition 3.4, these stopping times are well-defined; moreover, $\tau_{m+1} - \tau_m$ are i.i.d. and have exponential tails. Let $\zeta_{m+1} = H(\tau_{m+1}) - H(\tau_m)$. Then the ζ_m are also i.i.d.; moreover, their distribution is symmetric around 0, and $|\zeta_{m+1}| \leq \tau_{m+1} - \tau_m$ hence the ζ_m also have exponential tails. The rest follows from the standard Donsker–Varadhan invariance principle; see e.g. Durrett et al. (2002), pp. 590–592, for a proof in a very similar set-up. $\qquad\square$

4 Symmetric interaction

In the symmetric case, the potential of site k at time t is

$$u_k(t) = \xi_{k-1}(t) + \xi_k(t) + \xi_{k+1}(t), \ k \in \mathcal{M}, \qquad (4.1)$$

and the transition probabilities of the Markov chain $u(t)$ are now given by

$$\mathbb{P}\left(u_i(t+1) = u_i(t) + 1_{i \in \{k-1, k, k+1\}}, i = 1, \ldots, M | u(t)\right)$$

$$= \begin{cases} 0, & \text{if} \quad u_k(t) > m(t), \\ \\ N_{\min}^{-1}(t), & \text{if} \quad u_k(t) = m(t), \end{cases}$$

for $k \in \mathcal{M}$, where, as before, $N_{\min}(t) \in \{1, 2, \ldots, M\}$ is the number of $u_i(t)$ equal to $m(t)$.

Proposition 4.1 *If $(M \bmod 3) \neq 0$, then for any $u = (u_1, u_2, \ldots, u_M)$ the system*

$$u_1 = \xi_M + \xi_1 + \xi_2$$
$$u_2 = \xi_1 + \xi_2 + \xi_3$$
$$\vdots \qquad (4.2)$$
$$u_M = \xi_{M-1} + \xi_M + \xi_1$$

has a unique solution. On the other hand, if M is divisible by 3, system (4.2) has a solution if and only if

$$u_1 + u_4 + \cdots + u_{M-2} = u_2 + u_5 + \cdots + u_{M-1}$$
$$= u_3 + u_6 + \cdots + u_M. \qquad (4.3)$$

Proof If M is not divisible by 3, then the determinant of the matrix

$$\begin{pmatrix} 1 & 1 & 0 & 0 & \cdots & 0 & 0 & 1 \\ 1 & 1 & 1 & 0 & \cdots & 0 & 0 & 0 \\ 0 & 1 & 1 & 1 & \cdots & 0 & 0 & 0 \\ \vdots & \vdots & \vdots & \vdots & \ddots & \vdots & \vdots & \vdots \\ 0 & 0 & 0 & 0 & \cdots & 1 & 1 & 1 \\ 1 & 0 & 0 & 0 & \cdots & 0 & 1 & 1 \end{pmatrix}$$

corresponding to the equation (4.2) is ± 3 (which can be easily proved by induction). Hence the system must have a unique solution.

When M is divisible by 3, by summing separately the 1st, 4th, 5th, ... lines of (4.2), and then repeating this for the 2nd, 5th, ... or 3th, 6th, ... lines, we obtain condition (4.3).

Then it turns out that we can set both ξ_1 and ξ_2 to be any real numbers, so $\xi_3 = u_2 - \xi_1 - \xi_2$, and ξ_k, $k \geq 4$, are given:

$$\xi_{k+1} = [u_k - u_{k-1}] + [u_{k-3} - u_{k-4}] + \cdots + \xi_{(k \bmod 3)+1}.$$

□

Similarly to the asymmetric case, consider the Markov chain $v(t)$ on $\{0, 1, 2, \dots\}$ and recall the definition of t_j from (2.3). The following statement is straightforward.

Proposition 4.2 *For any $k \in \mathcal{M}$*

$$v_k(t_{j+1}) \leq v_k(t_j)$$

unless both $v_{k-1}(t_j) = 0$ and $v_{k+1}(t_j) = 0$.

Proposition 4.3 *For j large enough, in the sequence of $v_k(t_j)$, $k \in \mathcal{M}$, there are no more than two non-zero elements in a row, that is*

if $v_k(t_j) > 0$ then either $v_{k-1}(t_j) = 0$ or $v_{k+1}(t_j) = 0$, or both.

Proof Fix some $k \in \mathcal{M}$. Then $v_k(t_j)$ is either 0 or positive. In the first case, unless both of the neighbours of point k are zeros at time t_j, by Proposition 4.2 we have $v_k(t_{j+1}) = 0$. On the other hand, if $(v_{k-1}(t_j), v_k(t_j), v_{k+1}(t_j)) = (0, 0, 0)$, then at time t_{j+1} either this triple becomes $(0, 1, 0)$ if both $k - 1$ and $k + 1$ are chosen, or $v_k(t_{j+1}) = 0$.

Now suppose that $v_k(t_j) > 0$. If both $v_{k-1}(t_j) = 0$ and $v_{k+1}(t_j) = 0$, then from Proposition 4.2 applied to $k - 1$ and $k + 1$, we conclude $v_{k-1}(t_{j+1}) = v_{k+1}(t_{j+1}) = 0$, hence point k remains surrounded by 0s.

Similarly, if $v_k(t_j) > 0$ and $v_{k+1}(t_j) > 0$ but $v_{k-1}(t_j) = v_{k+2}(t_j) = 0$, then points $\{k, k+1\}$ remain surrounded by 0s at time t_{j+1}.

Finally, if point k is surrounded by non-zeros on both sides, that is $v_{k-1}(t_j)$, $v_k(t_j)$ and $v_{k+1}(t_j)$ are all positive, we have $v_k(t_{j+1}) = v_k(t_j) - 1$.

Consequently, all sequences of non-zero elements of length ≥ 3 are bound to disappear, and no such new sequence can arise as j increases. □

Proposition 4.4 *For any $k \in \mathcal{M}$, if for some s*

$$v_{k-1}(s) > 0, \ \ v_k(s) = 0, \ \ v_{k+1}(s) > 0$$

then for all j such that $t_j \geq s$

$$v_{k-1}(t_j) > 0, \ \ v_k(t_j) = 0, \ \ v_{k+1}(t_j) > 0.$$

Proof This immediately follows from the fact that there must be a particle adsorbed at point k during the time interval $(t_{j_0}, t_{j_0+1}]$ where $j_0 = \max\{j : \ t_j \leq s\}$, and that would imply that $v_{k\pm1}(t_{j_0+1}) \geq v_{k\pm1}(t_{j_0})$ while $v_k(t_{j_0+1}) = 0$. Now an induction on j finishes the proof. □

Proposition 4.5 *For j large enough, in the sequence of $v_k(t_j)$, $k \in \mathcal{M}$, there are no more than two 0s in a row, that is*

if $v_k(t_j) = 0$ then either $v_{k-1}(t_j) > 0$ or $v_{k+1}(t_j) > 0$, or both.

Proof Suppose j is so large that already there are no consecutive subsequences of positive elements of length ≥ 2 in $(v_1, \ldots, v_M)$ (see Proposition 4.3). Let

$$Q(j) = |\{k : \ v_{k-1}(t_j) > 0, v_k(t_j) = 0, v_{k-1}(t_j) > 0, \}|.$$

Proposition 4.4 implies that $Q(j)$ is non-decreasing; since $Q(j) < M$ it means that $Q(j)$ must converge to a finite limit.

Let A_j be the event that at time t_j there are 3 or more zeroes in a row in $v(t_j)$. On A_j there is a $k \in \mathcal{M}$ such that $v_k(t_j) = v_{k+1}(t_j) = v_{k+2}(t_j) = 0$ but $v_{k-1}(t_j) > 0$, (unless all $v_k = 0$ but then the argument is similar). Then, with a probability exceeding $1/M$, at time $t_j + 1$ new particle gets adsorbed at $k + 2$, yielding by Proposition 4.4 that for all $j' > j$ we have $v_{k-1}(t_{j'}) > 0$, $v_k(t_{j'}) = 0$, $v_k(t_{j'}) > 0$, hence the event $B_j := \{Q(j+1) \geq Q(j) + 1\}$ occurs as well. Therefore,

$$\mathbb{P}(B_j \mid \mathcal{F}_{t_j}) \geq \frac{1}{M} \mathbb{P}(A_j \mid \mathcal{F}_{t_j}),$$

where $\mathcal{F}_{t_j}$ denotes the sigma-algebra generated by $v(t)$ by time t_j. Combining this with the second Borel–Cantelli lemma, we obtain

$$\{A_j \text{ i.o.}\} = \left\{ \sum_j \mathbb{P}(A_j \mid \mathcal{F}_{t_j}) = \infty \right\} \subseteq \left\{ \sum_j \mathbb{P}(B_j \mid \mathcal{F}_{t_j}) = \infty \right\}$$
$$= \{B_j \text{ i.o.}\} = \{Q(j) \to \infty\}$$

leading to a contradiction. □

Proposition 4.6 *Let*

$$W(j) = |\{k: \ v_{k-1}(t_j) = 0, v_k(t_j) > 0, v_{k+1}(t_j) > 0, v_{k-1}(t_j) = 0, \}|$$

be the number of 'doubles'. Then $W(j)$ is non-increasing.

Proof Let us investigate how we can obtain a subsequence $(0, *, *, 0)$ starting at position $k - 1$ at time t_{j+1}, where $*$ stands for a positive element. One possibility is that at time t_j we already have such a subsequence there; this does not increase $W(j)$. The other possibilities at time t_j are

$$(0,0,0,0), \ (0,0,0,*), \ (0,0,*,0), \ (0,0,*,*),$$
$$(0,*,0,0), \ (0,*,0,*), \ (*,*,0,0), \ (*,*,0,*).$$

By careful examination of all of the configurations above, we conclude that the subsequence $(0, *, *, 0)$ cannot arise at time t_{j+1}. Consequently, $W(j)$ cannot increase. □

Proposition 4.7 *For j large enough, in the sequence of $v_k(t_j)$, $k \in \mathcal{M}$, there are no consecutive subsequences of the form $(*, *, 0, 0)$ or $(0, 0, *, *)$ where each $*$ stands for any positive number; that is there is no k such that*

$$v_k(t_j) = v_{k+1}(t_j) = 0 \text{ and either}$$
$$v_{k+2}(t_j) > 0 \text{ and } v_{k+3}(t_j) > 0$$
$$\text{or } v_{k-1}(t_j) > 0 \text{ and } v_{k-2}(t_j) > 0.$$

Proof Omitting (t_j), without loss of generality suppose $v_k > 0$, $v_{k+1} > 0$, $v_{k+2} = v_{k+3} = 0$. Then either at some time $j_1 > j$ we will have $v_{k+3}(t_{j_1}) > 0$ (hence the configuration $(*, *, 0, 0)$ gets destroyed), or with probability at least $\frac{1}{3}$ for each $j' \geq j$ we have adsorption at position $k + 3$ at some time during the time interval $(t_{j'}, t_{j'+1}]$. This would imply that $v_{k+1}(t_{j'+1}) = v_{k+1}(t_{j'}) - 1$. Hence, in a geometrically distributed number of times, we obtain 0 at position $k+1$, and thus the configuration

$(*, *, 0, 0)$ gets destroyed. On the other hand, by Proposition 4.6, the number of doubles is non-increasing, so no new configurations of this type can arise. Consequently, eventually all configurations $(*, *, 0, 0)$ and $(0, 0, *, *)$ will disappear. □

Proposition 4.8 *For j large enough, in the sequence of $v_k(t_j)$, $k \in \mathcal{M}$, there are no consecutive subsequences of the form $(0, 0, *, 0, 0)$ where $*$ stands for any positive number; that is there is no k such that*

$$v_{k-2}(t_j) = v_{k-1}(t_j) = 0 = v_{k+1}(t_j) = v_{k+2}(t_j) \text{ and } v_k(t_j) > 0.$$

Proof Propositions 4.3 and 4.5 imply that for some (random) J large enough for all $j \geq J$ consecutive subsequences of zero (non-zero resp.) elements have length ≤ 2, and Proposition 4.7 says that two consecutive 0s must be followed (preceded resp.) by a single non-zero element. Therefore, $(0, 0, *, 0, 0)$ must be a part of a longer subsequence of form $(0, *, 0, 0, *, 0, 0, *, 0)$. This, in turn, implies for the middle non-zero element located at k that

$$v_k(t_{j+1}) = \begin{cases} v_k(t_j) + 1, & \text{with probability } 1/4, \\ v_k(t_j), & \text{with probability } 1/2, \\ v_k(t_j) - 1, & \text{with probability } 1/4. \end{cases}$$

Hence, by the properties of simple random walk, for some $j' > J$ we will have $v_k(t_{j'}) = 0$ (suppose that j' is the first such time). On the other hand, by Proposition 4.2,

$$v_{k-2}(t_{j'}) = v_{k-1}(t_{j'}) = v_{k+1}(t_{j'}) = v_{k+2}(t_{j'}) = 0$$

as well. This yields a contradiction with the choice of J (see Proposition 4.5). □

Proof of Theorem 2.2 Let a *configuration of the potential* be a sequence $\bar{v} = (\bar{v}_1, \ldots, \bar{v}_M)$ where each $\bar{v}_i \in \{0, *\}$. Then we say that $v = (v_1, v_2, \ldots, v_M)$ with the following property has type $\bar{v}$:

$$v_i = 0 \text{ if } \bar{v}_i = 0,$$
$$v_i > 0 \text{ if } \bar{v}_i = *.$$

Propositions 4.3, 4.5, 4.7, and 4.8 rule out various types of configurations for all j large enough. On the other hand, it is easy to check that all remaining configurations for $v(t_j)$ are possible and stable, that is, once you reach them, you stay in them forever.

Call a configuration $\bar{v}$ *admissible*, if there is a collection $\xi_1, \xi_2, \ldots,$ ξ_M such that the system (4.2) has a solution for some $u = (u_1, \ldots, u_M)$

having type $\bar{v}$. If M is not divisible by 3, according to Proposition 4.2 all configurations $\bar{v}$ are admissible. On the other hand, it is easy to see that if $M = 3K$ then a necessary and sufficient condition for a non-zero configuration $\bar{v}$ to be admissible is

$$\bar{v}_i = * \text{ for some } i \text{ such that } i \bmod 3 = 0, \text{ and}$$
$$\bar{v}_j = * \text{ for some } j \text{ such that } j \bmod 3 = 1, \text{ and}$$
$$\bar{v}_k = * \text{ for some } k \text{ such that } k \bmod 3 = 2.$$

This establishes all possible stable configurations for v and hence the potential u, thus determining the subset of $\mathcal{M}$ where points are adsorbed for sufficiently large times, namely, $\xi_i(t) \to \infty$ if and only if $v_i(t_j) = 0$ for all large j.

Moreover, whenever we see a subsequence of type $(v_{k-1}, v_k, v_{k+1}) = (*, 0, *)$, we have

$$0 \leq \lim_{j \to \infty} [t_j - u_k(t_j)] < \infty,$$

and for a subsequence of type $(v_{k-1}, v_k, v_{k+1}, v_{k+2}) = (*, 0, 0, *)$ we have

$$\lim_{j \to \infty} \frac{u_k(t_j)}{t_j} = \lim_{j \to \infty} \frac{u_{k+1}(t_j)}{t_j} = \frac{1}{2}$$

by the strong law. Setting

$$\alpha = \frac{1}{\lim_{j \to \infty} |\{i \in \mathcal{M} : v_i(t_j) > 0, v_{i+1}(t_j) = 0\}|}$$

finishes the proof of the first part of the Theorem.

Finally, note that if the initial configuration is empty, the conditions of Proposition 4.6 are fulfilled with no 'doubles' at all, i.e. $W(0) = 0$. Consequently, for all $j \geq 0$ we have that there are no consecutive non-zero elements in $v_k(t_j)$, yielding the final statement of the Theorem. $\square$

5 Appendix

In this section we briefly describe the long-time behaviour of the growth process generated by the dynamics, where a particle is allocated at random to a site with maximum potential. The process is trivial in both the symmetric and asymmetric cases. Consider the symmetric case, i.e. $U_i = \{i - 1, i, i + 1\}$, $i \in \mathcal{M}$. It is easy to see that with probability 1, there exists k such that either

$$\lim_{t\to\infty} \frac{\xi_k(t)}{t} = 1 \quad \text{and} \quad \sup_{i\neq k} \xi_i(t) < \infty \tag{5.1}$$

or

$$\lim_{t\to\infty} \frac{\xi_k(t)}{t} = \lim_{t\to\infty} \frac{\xi_{k+1}(t)}{t} = \frac{1}{2} \quad \text{and} \quad \sup_{i\notin\{k,k+1\}} \xi_i(t) < \infty. \tag{5.2}$$

Indeed, recall the formula for the potential given by (4.1). Then $u(t)$ is a Markov chain with transition probabilities given by

$$\mathbb{P}\left(u_i(t+1) = u_i(t) + 1_{i\in\{k-1,k,k+1\}}\right) = \frac{1_{\{k\in S_{\max}(t)\}}}{|S_{\max}(t)|}$$

for $k \in \mathcal{M}$, where

$$S_{\max}(t) = \left\{i : \; u_i(t) = \max_{i\in\mathcal{M}} u_i(t)\right\} \subseteq \mathcal{M}$$

is the set of those i for which $u_i(t)$ equals the maximum value.

Observe that if at time s the adsorption/allocation occurs at point i, then $S_{\max}(s+1) \subseteq \{i-1, i, i+1\}$. In particular, if the maximum is unique, that is, $S_{\max}(s+1) = \{i\}$, then for all times $t \geq s$ this property will hold, and hence all the particles from now on will be adsorbed at i only.

If, on the other hand, $|S_{\max}(s+1)| = 2$, without loss of generality say $S_{\max}(s+1) = \{i, i+1\}$, then this property will be also preserved for all $t > s$ and each new particle will be adsorbed with probability $\frac{1}{2}$ at either i or $i+1$.

Finally, if $|S_{\max}(s+1)| = 3$, say $S_{\max}(s+1) = \{i, i+1, i+2\}$, then at time $s+2$ either $S_{\max}(s+2) = \{i, i+1, i+2\}$ if the adsorption occurred at $i+1$, or $S_{\max}(s+2) = \{i+1, i+2\}$ or $\{i, i+1\}$ otherwise. By iterating this argument we obtain that after a geometric number of times we will arrive at the situation where $|S_{\max}(t)| = 2$, and then the process will follow the pattern described in the previous paragraph.

A similar simple argument shows that in the case of the asymmetric interaction only the outcome (5.1) is possible.

References

Bousch, T., and Mairesse, J. 2002. Asymptotic height optimization for topical IFS, Tetris heaps, and the finiteness conjecture. *J. Amer. Math. Soc.*, **15**, 77–111.

Durrett, R., Kesten, H., and Limic, V. 2002. Once edge-reinforced random walk on a tree. *Probab. Theory Related Fields*, **122**, 567–592.

Evans, J. W. 1993. Random and cooperative sequential adsorption, *Rev. Modern Phys.*, **65**, 1281–1329.

Fayolle, G., Malyshev, V. A., and Menshikov, M. V. 1995. *Topics in the Constructive Theory of Countable Markov Chains.* Cambridge: Cambridge Univ. Press.

Heidergott, B., Oldser, G. J., and Woude, J. van der. 2006. *Max Plus at Work. Modeling and Analysis of Synchronized Systems: a Course on Max-Plus Algebra and its Applications.* Princeton Ser. Appl. Math. Princeton, NJ: Princeton Univ. Press.

Karpelevich, F. I., Malyshev, V. A., and Rybko, A. N. 1995. Stochastic evolution of neural networks, *Markov Process. Related Fields*, **1**(1), 141–161.

Malyshev, V. A., and Turova, T. S. 1997. Gibbs measures on attractors in biological neural networks, *Markov Process. Related Fields*, **3**(4), 443–464.

Penrose, M. D., and Shcherbakov, V. 2009. Maximum likelihood estimation for cooperative sequential adsorption. *Adv. in Appl. Probab. (SGSA)*, **41**(4), 978–1001.

Penrose, M. D., and Shcherbakov, V. 2010. *Asymptotic Normality of Maximum Likelihood Estimator for Cooperative Sequential Adsorption.* Preprint.

Privman, V., ed. 2000. Special issue of *Colloids and Surfaces A*, **165**.

Shcherbakov, V. 2006. Limit theorems for random point measures generated by cooperative sequential adsorption, *J. Stat. Phys.*, **124**, 1425–1441.

Shcherbakov, V., and Volkov, S. 2010. Stability of a growth process generated by monomer filling with nearest-neighbour cooperative effects. *Stochastic Process. Appl.*, **doi:10.1016/j.spa.2010.01.020**, in press.

Tarrès, P. 2004. Vertex-reinforced random walk on $\mathbb{Z}$ eventually gets stuck on five points. *Ann. Probab.*, **32**, 2650–2701.

Volkov, S. 2001. Vertex-reinforced random walk on arbitrary graphs, *Ann. Probab.*, **29**, 66–91.

21

Optimal information feed

Peter Whittle[a]

Abstract

The paper considers the situation of transmission over a memoryless noisy channel with feedback, which can be given a number of interpretations. The criteria for achieving the maximal rate of information transfer are well known, but examples of a simple and meaningful coding meeting these are few. Such a one is found for the Gaussian channel.

Keywords feedback channel, Gaussian channel

AMS subject classification (MSC2010) 94A24

1 Interrogation, transmission and coding

In this section we set out material which is classic in high degree, harking back to Shannon's seminal paper (1948), and presented in some form in texts such as those of Blahut (1987), Cover and Thomas (1991) and MacKay (2003). However, some exposition is necessary if we are to distinguish clearly between three versions of the model.

Suppose that an experimenter wishes to determine the value of a random variable U of which he knows only the probability distribution $P(U)$. The formal argument is conveyed well enough for the moment if we suppose all distributions discrete and use the notation $P(\cdot)$ generically for such distributions. We may also abuse this convention by

[a] Statistical Laboratory, Centre for Mathematical Sciences, Cambridge University, Wilberforce Road, Cambridge CB3 0WB; P.Whittle@statslab.cam.ac.uk

occasionally using $P(U)$ to denote the function of U defined by the distribution.

The outcome of the experiment, if errorless, might be written $x(U)$, where the form of the function $x(U)$ reflects the design of the experiment. The experimenter will choose this, subject to practical constraints, so as to make the experiment as informative as possible. The experiment may also be subject to error, in which case the experimenter observes a random variable y whose value is conditioned by the value of the random variable $x(U)$. The presence of this error will affect design and optimisation of the experiment.

The experimenter may in fact consider a sequence of experiments with outcomes $x_t(U)$ for $t = 1, 2, \ldots, n$. The design problem is then sequential, in that x_t can now be chosen in the light of past observations as well as of dependence upon U. It will then have the functional form $x_t(U, Y_{t-1})$, where Y_{t-1} is the sequence of observations $\{y_1, y_2, ..., y_{t-1}\}$ received up time $t - 1$.

This situation could equally well be regarded as one of *interrogation*, in which the tth question is chosen to elicit the answer $x_t(U, Y_{t-1})$, if U represents the true state of affairs and questions are answered truthfully.

A variant of the problem which leaves it changed in interpretation, although not in design, is that in which an agent who knows the true value of U (the 'sender') tries to convey this to the experimenter (now the 'receiver') by sending the sequence of signals $X_n(U) = \{x_t(U, Y_{t-1}); 1 \leq t \leq n\}$, received in the error-corrupted form $Y_n = \{y_t : 1 \leq t \leq n\}$. This choice of signal sequence is now a 'coding', more specifically a 'channel coding', which adapts the transmitted signal to the distorting effects of the noisy channel which converts X_n to Y_n.

We shall refer to this inverted view of the interrogation problem as 'information feed'. If the coding is known to the receiver, if the criterion remains one of efficient information transfer (a term yet to be defined) and if the sender knows the values of past received signals Y_{t-1} when he forms x_t, then the problem is essentially unchanged—the sender now chooses the coding on the same information as the receiver before chose it. This is the situation of information transmission 'with feedback', in that the sender is supposed to know what signals the receiver has received at any given time. In most practical cases there is no such feedback. Transmission is then 'blind', in that sender and receiver must agree a coding $X(U)$ in advance. Otherwise expressed, the procedure is non-sequential.

The   optimisation   criterion   effectively   taken   for   the   signals

$\{x_t(U, Y_{t-1}); t = 1, 2, ..., n\}$ is that these should be chosen, subject to practical constraints, to minimise the conditional Shannon information measure $h(U|Y_n)$, where

$$h(U|Y) = -\sum_U \sum_Y P(U, Y) \log_2 P(U|Y).$$

This criterion has implications for the size of the error in the estimation of U for a given value of n, although one must go into rate distortion theory if the nature of the error is to be clarified.

Let us now specialise to the case of a *memoryless* channel, for which

$$P(Y_n|X_n) = \prod_{t=1}^{n} P(y_t|x_t) \tag{1.1}$$

for any n, with the factors in the product all having the same functional form. This is a specialisation which is unnecessary on formal grounds, but makes for a considerable simplification and is accepted as realistic for many actual channels. Under this condition one finds easily for the interrogative version of the model that the information gain at step t has the upper bound

$$h(U|Y_{t-1}) - h(U|Y_t) \leq C, \tag{1.2}$$

where

$$C = \max_{P(x)} \sum_x \sum_y P(x)P(y|x) \log[P(y|x)/P(y)]. \tag{1.3}$$

Here $P(y) = \sum_x P(x)P(y|x)$ and the maximisation in (1.3) is over all distributions on the set of input letters x permitted by the channel. Equality holds in (1.2) if and only if the responses $x_t(U, Y_{t-1})$ can be so chosen that these random variables are independent and follow the extremising distribution, $\overline{P}(x)$ say, determined in (1.3).

Inequality (1.2) must hold *a fortiori* in the case of blind transmission, when the sender does not know the receiver state and sequential operation is not permitted. A key conclusion of Shannon's classical 1948 analysis was that asymptotic equality in (1.2) can in fact still be attained for large n, the condition for equality being just that stated above. The expression C is then justifiably termed the *capacity* of the channel. The formal passage from discrete to continuous variables is now immediate because, as is well recognised, expression (1.3) is just a difference of entropies.

In the interrogative case the coding problem is one of achieving an appropriate 'information split'. That is, at stage t one will split the values of U into classes, those in class $\mathcal{U}_x$ to be coded on to the value x, in such a way that, on the distribution $P(U|Y_{t-1})$, class $\mathcal{U}_x$ has total probability as near as possible to $\overline{P}(x)$. The principle is thus clear, but the procedure is messy and the solution highly non-unique. In the next section we shall consider a case for which a tidy and natural determination is possible.

The determination of near-optimal codings in the case of blind transmission faces difficulties of another order. If one reckons from the time of Shannon's radical insights, sixty years of determined and ingenious attack have only recently closed the gap between the practicable and the ideal, by the introduction of the techniques of turbo-coding and low-density parity-check coding (see MacKay, 2003). It is likely, however, that there are further insights to be gleaned.

2 A tractable infinite-horizon case

In this section we shall consider information transmission with perfect feedback, a variant formulation of the interrogation problem which we shall refer to for brevity as 'information feed'. As indicated in the last section, the problem is then one of achieving appropriate information splits, a task whose solution becomes ever more indeterminate as the size of the set $\mathcal{U}$ of possible U-values increases.

Other criteria must be sought to provide guidance. Beyond seeking to maximise the information gain at each stage, one must do so in such a way that the information gained is relevant. Physical considerations may well be helpful here. Beyond relevance, one may have urgency. To take a common (and indeed canonical) case, suppose that the information source generates a message U in the form of a sequence $\{u_\tau; \tau = 1, 2, 3, ...\}$, binary in that the 'letters' u_τ are independent random variables, taking the values 0 or 1 equiprobably. If we say that the sequence extends into the indefinite future then we imply an identification of τ with 'time', but t and τ are not to be identified—information may be transmitted at a rate different from that at which it is supplied. However, the obvious prioritisation is to seek information on the earlier letters first, while maintaining the maximal rate of information gain.

In this case the information in U is infinite, so one is mining an infinite and uniform information source. There is clearly a need to have a well-defined mining face, which a time-prioritisation would give. There is also an expectation that mining could and should follow a fixed practice.

That is, that one follows a fixed information-gaining procedure, which moves forward at a constant rate in time.

For the binary source described above, define the variable

$$\mathbf{u} = \sum_{\tau=1}^{\infty} u_\tau 2^{-\tau}. \tag{2.1}$$

This takes a set of values dense in $(0,1)$, and is in fact uniformly distributed over this interval. The value of $\mathbf{u}$ determines the whole future sequence of letters u_t in principle. It can do so in fact only if one has infinite resolution, in that from the value of $\mathbf{u}$ one can generate this sequence.

The reformulation (2.1) gives priority to the immediate future: two values of $\mathbf{u}$ are close only if they agree in the early letters of the sequence $\{u_\tau\}$. It also gives greater determinacy in treatment: the natural U-classes to consider are $\mathbf{u}$-intervals.

The initially uniform distribution of $\mathbf{u}$ changes as one acquires information by observation, so that its posterior density $f(\mathbf{u})$ changes. One can query whether a density can represent the infinitely fine detail which the distribution of $\mathbf{u}$ might develop—the hope must be that smooth operations on $\mathbf{u}$ will yield smooth distributions. This hope turns out to be justified in the particular case considered below.

Suppose that the 'true' value of $\mathbf{u}$ in a given case is v. Suppose also that one is transmitting through a Gaussian channel, so that if one requests information on the value of a function $x(\mathbf{u})$, say, one receives a signal

$$y = x(v) + \epsilon, \tag{2.2}$$

where ϵ is distributed normally with mean zero and variance c. We suppose the channel subject to the power constraint

$$E[x(v)^2] = A^2 \tag{2.3}$$

on the transmitted signal, and suppose it also memoryless, in that the errors ϵ for distinct queries are independent.

Suppose also that the receiver's posterior distribution of v after t observations is normal with mean v_t and variance σ_t^2. The criterion of maximal information gain demands in this case that the coding should minimise $\sigma_t^2 = E(v - v_t)^2$. We are then led into familiar linear/quadratic calculations, which conserve the property of normality at each stage, with familiar updatings for the parameters v_t and σ_t.

In fact, the assumption of normality is contradicted by the fact that the prior distribution of v is rectangular. However, a conservative procedure would be to neglect this prior information and simply transmit the value of λv, yielding an observation y_0 at the receiver. Here λ is a scalar multiplier assuring observance of the power constraint (2.3). The receiver then begins with a normal prior of mean $v_0 = \lambda^{-1} y_0$ and variance $\sigma_0^2 = \lambda^{-2} c$. The subsequent shrinkage of the posterior distribution on to an arbitrarily small neighbourhood of the true value v essentially confines the distribution to the interval $[0, 1)$. Otherwise expressed, the procedure determined below establishes an optimal course which soon makes up the initial sacrifice of prior information.

Of course, one will not continue by simply repeating the transmission of $\lambda^{-1} v$. To do so would yield a σ_t^2 declining as t^{-1} whereas one expects an exponential decline with increasing t, at a maximal rate determined by the capacity of the channel.

Suppose that at stage $t + 1$ one transmits the signal

$$x_{t+1}(v) = \lambda_t(v - v_t). \tag{2.4}$$

That is, one transmits the *error* in the receiver's current estimate. The power constraint (2.3) implies the evaluation (2.5) of the scaling factor. This evaluation reveals the reason for the effectiveness of rule (2.4): that sensitivity λ_t to error increases as mean square error σ_t^2 decreases.

Theorem	(i) *With the evaluation*

$$\lambda_t = A/\sigma_t \tag{2.5}$$

of the scaling factor implied by the power constraint (2.3) *the coding* (2.4) *conserves normality of the posterior distribution of v and is optimal. The mean square error of the estimate v_t then declines exponentially as*

$$\sigma_t^2 = (1 + \rho)^{-t} \sigma_0^2 = 2^{-2tC} \sigma_0^2, \tag{2.6}$$

where

$$\rho = A^2/c \tag{2.7}$$

is the signal-to-noise power ratio and

$$C = (1/2) \log_2(1 + \rho) \tag{2.8}$$

is the capacity of the Gaussian channel in bits.

(ii) *More explicitly, the scaling factor has the determination*

$$\lambda_t = (A/\sigma_0)(1 + \rho)^{t/2} = (A/\sigma_0) 2^{tC}, \tag{2.9}$$

and the sequence of estimates is generated by the recursion

$$v_{t+1} = v_t + (A\sigma_0/c)(1 + \rho)^{-(t/2)-1}y_{t+1}. \tag{2.10}$$

Proof The updated estimate v_{t+1} is the value of v minimising the expression

$$\sigma_t^{-2}(v - v_t)^2 + c^{-1}[y_{t+1} - \lambda_t(v - v_t)]^2,$$

whence we find that

$$\sigma_{t+1}^{-2} = \sigma_t^{-2} + \lambda_t^2/c = \sigma_t^{-2}(1 + \rho) \tag{2.11}$$

and

$$v_{t+1} = v_t + \sigma_{t+1}^2\lambda_t y_{t+1}/c. \tag{2.12}$$

With these determinations, relations (2.6), (2.9) and (2.10) follow.

It follows from (2.11) and (2.8) that the information gain between stages t and $t+1$ is

$$\log_2(\sigma_t/\sigma_{t+1}) = C. \tag{2.13}$$

The procedure is thus optimal (after the initial normalising sacrifice of information). □

We see from (2.4) that λ_t is the factor by which **u**-space is expanded about the true value v at the nth stage, and so from (2.9) that this expansion factor is 2^C per stage. The exponential growth of **u**-space is equivalent to the passage of time, and its relation (2.9) to the capacity of the channel indicates that this is the fastest rate of expansion that could be achieved, consistent with convergence of the estimates v_t to v.

So, the familiar linear estimation rules which emerge from a Gaussian model imply optimal information splits in this case, in a form so elegant as to be invisible. Relation (2.1) further implies that priority has been given to the determination of the earlier source letters u_τ. The procedure settles to a time-invariant one, in which the passage of transmission time at unit rate induces expansion of **u**-space at a constant rate 2^C about the true value v.

One now asks whether this simple treatment transfers to the case of blind transmission. It does not. As Witsenhausen (1968) demonstrated in another context: multi-stage games with a linear/quadratic solution can lose this property if the players must base their decisions on differing information.

I am grateful to the referee for drawing my attention to the paper by Bressler, Parekh and Tse (2008), who in their equation (3) propose exactly the transformation of our equation (2.1). They do this in the context of a quasi-deterministic evaluation of the capacity of a network of communicating sources; the so-called multiple-input multiple-output model.

References

Blahut, R. E. 1987. *Principles and Practice of Information Theory.* Reading, MA: Addison–Wesley.

Bressler G., Parekh A., and Tse, D. 2008. *The Approximate Capacity of the Many-to-One and One-to-Many Gaussian Interference Channels.* http://arXiv.org/abs/0809.3554v1

Cover, T. M., and Thomas, J. A. 1991. *Elements of Information Theory.* New York: John Wiley & Sons.

MacKay, D. J. C. 2003. *Information Theory, Inference, and Learning Algorithms.* Cambridge: Cambridge Univ. Press.

Shannon, C. E. 1948. A mathematical theory of communication. *Bell Sys. Tech. J.*, **27**, 379–423, 623–656.

Witsenhausen, H. 1968. A counterexample in stochastic optimal control. *SIAM J. Control*, **6**, 131–146.

22

A dynamical-system picture of a simple branching-process phase transition

David Williams[a]

Abstract

This paper proves certain results from the 'appetizer for non-linear Wiener–Hopf theory' [5]. Like that paper, it considers only the simplest possible case in which the underlying Markov process is a two-state Markov chain. Key generating functions provide solutions of a simple two-dimensional dynamical system, and the main interest is in the way in which Probability Theory and ODE theory complement each other. No knowledge of either ODE theory or Wiener–Hopf theory is assumed. Theorem 1.1 describes one aspect of a phase transition which is more strikingly conveyed by Figures 4.1 and 4.2.

AMS subject classification (MSC2010) 60J80, 34A34

1 Introduction

This paper is a development of something I mentioned briefly in talks I gave at Bristol, when John Kingman was in the audience, and at the Waves conference in honour of John Toland at Bath. I thanked both John K and John T for splendid mathematics and for their wisdom and kindness.

The main point of the paper is to prove Theorem 1.1 and related results in a way which emphasizes connections with a simple dynamical system. The phase transition between Figures 4.1 and 4.2 looks more dramatic than the famous 1-dimensional result we teach to all students.

[a] Mathematics Department, Swansea University, Singleton Park, Swansea SA2 8PP; dw@reynoldston.com

The model studied here is a special case of the model introduced in Williams [5]. I called that paper, which contained no proofs, an 'appetizer'; but before writing a fuller version, I became caught up in Jonathan Warren's enthusiasm for the relevance of *complex* dynamical systems (in $\mathbb{C}^2$). See Warren and Williams [4]. This present paper, *completely independent of the earlier appetizer* and of [4], can, I hope, provide a more tempting appetizer for what I called 'non-linear Wiener–Hopf theory'. No knowledge of any kind of Wiener–Hopf theory is assumed here.

I hope that Simon Harris and I can throw further light on the models considered here, on the other models in [5], and on still other, quite different, models.

Our model. A particle moving on the real line can either be of type $+$ in which case it moves right at speed 1 or of type $-$ in which case it moves left at speed 1.

Let q_- and q_+ be fixed numbers with $q_- > q_+ > 0$, and let β be a positive parameter. We write $K_\pm = q_\pm + \beta$. So, to display things, we have

$$q_- > q_+ > 0, \quad \beta > 0, \quad K_+ = q_+ + \beta, \quad K_- = q_- + \beta. \tag{1.1}$$

We define

$$\beta_c := \tfrac{1}{2} \left(\sqrt{q_-} - \sqrt{q_+} \right)^2.$$

A particle of type $\pm$ can flip to the 'opposite' type at rate $q_\pm$ and can, at rate β, die and at its death give birth to two daughter particles (of the same type and position as their 'parent'). This is why β is a 'birth rate'. The usual independence conditions hold.

Theorem 1.1 *Suppose that our process starts at time 0 with just 1 particle of type $+$ at position 0.*

(a) *Suppose that $\beta > \beta_c$. Then, with probability 1, each of infinitely many particles will spend time to the left of 0.*

(b) *Suppose instead that $\beta \leq \beta_c$. Then, with probability not less than $1 - \sqrt{q_+/q_-}$, there will never be any particles to the left of 0.*

Large-deviation theory (of which the only bit we need is proved here) allows one to prove easily that if $\beta < \beta_c$, then, almost surely, only a finite number of particles are ever to the left of 0.

The interplay between the Probability and the ODE theory is what is most interesting. We shall see that β_c plays the rôle of a critical parameter in several ways, some probabilistic, some geometric. The 'balance' which occurs when $\beta = \beta_c$ is rather remarkable.

The paper poses a tantalizing problem which I cannot yet solve.

2 Wiener–Hopferization

2.1 The processes $\{N^{\pm}(\varphi) : \varphi \geq 0\}$

For any particle i alive at time t, we define $\Phi_i(t)$ to be its position on the real line at time t, and we extend the definition of Φ_i by saying that at any time s before that particle's birth, $\Phi_i(s)$ is the position of its unique ancestor alive at time s.

So far, so sane! But we are now going to Wiener–Hopferize everything with a rather clumsy definition which defines for each $\varphi \geq 0$ two subsets, $S^+(\varphi, \beta)$ and $S^-(\varphi, \beta)$, of particles.

We put particle i in set $S^+(\varphi, \beta)$ if there is some t in $[B(i), D(i))$ where $B(i)$ and $D(i)$ are, respectively, the times of birth and death of particle i, such that

- $\Phi_i(t) = \varphi$,
- $\Phi_i(t) \geq \max\{\Phi_i(s) : s \leq t\}$, and
- Φ_i grows to the right of t in that, for $\varepsilon > 0$, there exists a δ with $0 < \delta < \varepsilon$ such that $\Phi_i(\cdot) > \varphi$ throughout $(t, t + \delta)$.

At the risk of labouring things, let me describe $S^-(\varphi, \beta)$ for $\varphi \geq 0$. We put particle i in set $S^-(\varphi, \beta)$ if there is some t in $[B(i), D(i))$ such that $\Phi_i(t) = -\varphi$, $\Phi_i(t) \leq \min\{\Phi_i(s) : s \leq t\}$, and Φ_i decreases to the right of t in that, for $\varepsilon > 0$, there exists a δ with $0 < \delta < \varepsilon$ such that $\Phi_i(\cdot) < -\varphi$ throughout $(t, t + \delta)$.

Of course, there may be particles not in $\bigcup_{\varphi \geq 0} \{S^+(\varphi, \beta) \cup S^-(\varphi, \beta)\}$.

We define $N^+(\varphi, \beta)$ [respectively, $N^-(\varphi, \beta)$] to be the number of particles in $S^+(\varphi, \beta)$ [resp., $S^-(\varphi, \beta)$].

We let $\mathbb{P}_\beta^+$ [respectively, $\mathbb{P}_\beta^-$] be the probability law of our model when it starts with 1 particle of type $+$ [resp., $-$] at position 0 at time 0; and we let $\mathbb{E}_\beta^+$ [resp., $\mathbb{E}_\beta^-$] be the associated expectation. *We often suppress the 'β' in the notation for $\mathbb{P}_\beta^{\pm}$, $\mathbb{E}_\beta^{\pm}$, $S_\beta^{\pm}$, $N_\beta^{\pm}$.*

Then, under $\mathbb{P}^+$, $N^+ = \{N^+(\varphi) : \varphi \geq 0\}$ is a standard branching process, in which a particle dies at rate K_+ and is replaced at the 'Φ-time' of its death by a random non-negative number, possibly 1 and possibly

∞, of children, the numbers of children being independent, identically distributed random variables. I take this as intuitively obvious, and I am not going to ruin the paper by spelling out a proof.

Note that in the $\mathbb{P}^-$ branching process $N^- = \{N^-(\varphi) : \varphi \geq 0\}$, a particle may die without giving birth.

For $0 \leq \theta < 1$, define

$$g^{++}(\varphi, \theta) := \mathbb{E}^+ \theta^{N^+(\varphi)}, \quad h^{-+}(\varphi, \theta) := \mathbb{E}^- \theta^{N^+(\varphi)}.$$

Clearly, for $0 \leq \theta < 1$,

$$
\begin{aligned}
h^{-+}(\varphi, \theta) &= \mathbb{E}^- \mathbb{E}^- \left[\theta^{N^+(\varphi)} \mid N^+(0) \right] = \mathbb{E}^- g^{++}(\varphi, \theta)^{N^+(0)} \\
&= H^{-+} \left(g^{++}(\varphi, \theta) \right),
\end{aligned}
$$

where

$$H^{-+}(\theta) = \mathbb{E}^- \theta^{N^+(0)} = \sum h_n^{-+} \theta^n,$$

where

$$h_n^{-+} := \mathbb{P}^- [N^+(0) = n].$$

It may well be that $h_\infty^{-+} := \mathbb{P}^- [N^+(0) = \infty] > 0$. Note that

$$H^{-+}(1-) := \lim_{\theta \uparrow 1} H^{-+}(\theta) = \mathbb{P}^- [N^+(0) < \infty].$$

2.2 The dynamical system

We now take θ in $(0, 1)$ and derive the backward differential equations for

$$x(\varphi) := g^{++}(\varphi, \theta), \quad y(\varphi) := h^{-+}(\varphi, \theta),$$

in the good old way in which we teach Applied Probability, and then study the equations. Paper [5] looks a bit more 'rigorous' here.

Consideration of what happens between times 0 and dt tells us that

$$x(\varphi + d\varphi) = \{1 - K_+ \, d\varphi\} x(\varphi) + \{q_+ \, d\varphi\} y(\varphi) + \{\beta \, d\varphi\} x(\varphi)^2 + o(d\varphi).$$

The point here is of course that if we started with 2 particles in the $+$ state, then $\mathbb{E}\theta^{N^+(\varphi)} = x(\varphi)^2$. We see that, with x' meaning $x'(\varphi)$,

$$x' = q_+(y - x) + \beta(x^2 - x). \tag{2.1a}$$

Similarly, remembering that Φ starts to run backwards when the particle starts in state $-$, we find that

$$y(\varphi - d\varphi) = \{1 - K_- \, d\varphi\} y(\varphi) + \{q_- \, d\varphi\} x(\varphi) + \{\beta \, d\varphi\} y(\varphi)^2 + o(d\varphi),$$

whence

$$-y' = q_-(x - y) + \beta(y^2 - y). \tag{2.1b}$$

Of course, $y = H^{-+}(x)$ must represent the track of an integral curve of the *dynamical system* (2.1), and since $y' = H^{-+\prime}(x)x'$, we have an *autonomous equation for* H^{-+} which we shall utilize below.

Note that the symmetry of the situation shows that $x = H^{+-}(y)$ must also represent the track of an integral curve of our dynamical system, though one traversed in the 'φ-reversed' direction.

Probability Theory guarantees the existence of the 'probabilistic solutions' of the dynamical system tracking curves $y = H^{-+}(x)$ and $x = H^{+-}(y)$.

Lemma 2.1 *There can be no equilibria of our dynamical system in the interior of the unit square.*

Proof For if (x, y) is in the interior and

$$q_+(y - x) + \beta(x^2 - x) = 0, \quad q_-(x - y) + \beta(y^2 - y) = 0,$$

then $y \geq x$ from the first equation and $x \geq y$ from the second. Hence $x = y$ and $x^2 - x = y^2 - y = 0$. ☐

2.3 Change of θ

We need to think about how a change of θ would affect things. Suppose that $\alpha = \mathbb{E}^+ \theta^{N^+(\psi)}$ where $0 < \alpha < 1$. Then

$$\mathbb{E}^+ \mathbb{E}^+ \left[\theta^{N^+(\varphi+\psi)} \mid N^+(\varphi) \right] = \mathbb{E}^+ \alpha^{N^+(\varphi)} = g^{++}(\varphi, \alpha),$$

where $\alpha = g^{++}(\psi, \theta)$. So, we have the *probabilistic-flow relation*

$$g^{++}(\varphi + \psi, \theta) = g^{++}\left(\varphi, g^{++}(\psi, \theta)\right). \tag{2.2}$$

Likewise, $h^{-+}(\varphi + \psi, \theta) = h^{-+}\left(\varphi, g^{++}(\psi, \theta)\right)$. Thus, changing from θ to $\alpha = \mathbb{E}^+ \theta^{N^+(\psi)}$ just changes the starting point of the motion along the probabilistic curve from $(\theta, H^{-+}(\theta))$ to the point $(\alpha, H^{-+}(\alpha))$ still on the probabilistic curve. This is why we may sometimes seem not to care about θ, and why it is not in our notation for $x(\varphi)$, $y(\varphi)$. But we shall discuss θ when necessary, and the extreme values 0 and 1 of θ in Subsection 4.6.

If for *any* starting point $\mathbf{v}_0 = (x_0, y_0)$ within the unit square, we write $\mathbf{V}(\varphi, \mathbf{v}_0)$ for the value of $(x(\varphi), y(\varphi))$, then, for values of φ and ψ

in which we are interested, we have (granted existence and uniqueness theorems) the *ODE-flow relation*

$$\mathbf{V}(\varphi + \psi, \mathbf{v}_0) = \mathbf{V}(\varphi, \mathbf{V}(\psi, \mathbf{v}_0))$$

which generalizes (2.2). (The possibility of explosions need not concern us: we are interested only in what happens within the unit square.) For background on ODE flows, see Arnol'd [1].

3 How does ODE theory see the phase transition?

3.1 An existence theorem

Even if you skip the (actually quite interesting!) proof of the following theorem, do not skip the discussion of the result which makes up the next subsection.

Theorem 3.1 *There exist constants $\{a_n : n \geq 0\}$ with $a_0 = 0$, all other a_n strictly positive, and*

$$\sum a_n \leq q_+/(q_- + \beta), \tag{3.1}$$

and a solution $(x(\varphi), y(\varphi))$ of the 'φ-reversed' dynamical system

$$-x' = q_+(y - x) + \beta(x^2 - x), \qquad y' = q_-(x - y) + \beta(y^2 - y),$$

such that $x(\varphi) = A(y(\varphi))$, where we now write $A(y) = \sum a_n y^n$.

Proof Assume that constants a_n as described exist. Since $x'(\varphi) = A'(y(\varphi))y'(\varphi)$, we have

$$- \{q_+ y + \beta A(y)^2 - K_+ A(y)\} = A'(y) \{q_- A(y) + \beta y^2 - K_- y\}.$$

Comparing coefficients of y^0,

$$\beta a_0^2 - K_+ a_0 = -a_1 q_- a_0,$$

and we are guaranteeing this by taking $a_0 = 0$. Comparing coefficients of y^1, we obtain

$$q_- a_1^2 - (K_- + K_+)a_1 + q_+ = 0.$$

We take

$$a_1 = \frac{K_- + K_+ - \sqrt{(K_- + K_+)^2 - 4q_- q_+}}{2q_-}$$

$$= \frac{2q_+}{K_- + K_+ + \sqrt{(K_- + K_+)^2 - 4q_- q_+}}$$

from which it is obvious that $0 < a_1 < 1$.

On comparing coefficients of y^n we find that, for $n \geq 2$,

$$\{K_+ + nK_- - (n+1)q_-a_1\}a_n$$

$$= \beta \sum_{k=1}^{n-1} a_k a_{n-k} + q_- \sum_{k=1}^{n-2} (k+1)a_{k+1}a_{n-k} + \beta(n-1)a_{n-1}.$$

We now consider the a_n as being *defined* by these recurrence relations (and the values of a_0 and a_1). It is clear that the a_n are all positive.

Temporarily fix $N > 2$, and define

$$A_N(y) := \sum_{n=0}^{N} a_n y^n,$$

$$L(y) := \sum_{n=0}^{N} \ell_n y^n := -q_+y - \beta A_N(y)^2 + K_+ A_N(y),$$

$$R(y) := \sum r_n y^n := A_N'(y)\{q_- A_N(y) + \beta y^2 - K_- y\}.$$

For $n \leq N$ we have $\ell_n = r_n$ by the recurrence relations. It is clear that for $n > N$ we have $\ell_n \leq 0$ and $r_n \geq 0$. Hence, for all y in $(0,1)$,

$$-q_+y - \beta A_N(y)^2 + K_+ A_N(y) \leq A_N'(y)\{q_- A_N(y) + \beta y^2 - K_- y\}. \quad (3.2)$$

Suppose for the purpose of contradiction that there exists y_0 in $(0,1)$ with $A_N(y_0) = y_0$. Then

$$-q_+ - \beta y_0 + K_+ \leq A_N'(y_0)\{q_- + \beta y_0 - K_-\}.$$

However, the left-hand side is positive while the right-hand side is negative.

Because $A_N(0) = 0$ and $A_N'(0) = a_1 < 1$, the contradiction establishes that $A_N(y) < y$ for $y \in (0,1)$, so that $A_N(1) \leq 1$. Since this is true for every N, and each a_n $(n > 1)$ is strictly positive, we have $A_N(1) < 1$ for every N.

By inequality (3.2), we have

$$D_N q_-(A_N - 1) + q_+ + \beta A_N^2 - K_+ A_N \geq 0,$$

where $A_N := A_N(1) < 1$ and $D_N := A_N'(1)$. Because each a_n $(n > 0)$ is positive it is clear that $A_N < D_N$. We therefore have

$$q_+ + \beta A_N^2 - K_+ A_N \geq D_N q_-(1 - A_N) \geq q_- A_N(1 - A_N),$$

which simplifies to

$$(1 - A_N)\{q_+ - (\beta + q_-)A_N\} \geq 0.$$

Since $(1 - A_N) > 0$, we have $(\beta + q_-)A_N \leq q_+$, and result (3.1) follows.

It is clear that we now need to consider the autonomous equation for $y = y(\varphi)$:

$$y' = q_-[A(y) - y] + \beta[y^2 - y], \qquad y(0) = \theta.$$

But we can describe $y(\varphi)$ as $\mathbb{E}\theta^{Z(\varphi)}$ where $\{Z(\varphi) : \varphi \geq 0\}$ is a classical branching process in which (with the usual independence properties) a particle dies at rate K_- and at the moment of its death gives birth to C children where

$$\mathbb{P}(C = n) = \begin{cases} q_- a_n/K_- & \text{if } 1 \leq n \leq \infty \text{ and } n \neq 2, \\ (\beta + q_- a_2)/K_- & \text{if } n = 2. \end{cases}$$

Of course, $a_\infty = 1 - \sum a_n$.

Then $(x(\varphi), y(\varphi)) = (A(y(\varphi)), y(\varphi))$ describes the desired solution starting from $(A(\theta), \theta)$.   □

3.2 Important discussion

Of course, ODE theory cannot see what we shall see later: namely that $A(\cdot) = H^{+-}(\cdot)$ when $\beta > \beta_c$ but $A(\cdot) \neq H^{+-}(\cdot)$ when $\beta \leq \beta_c$. When $\beta > \beta_c$, the curve $x = H^{+-}(y)$ is the steep bold curve $x = A(y)$ at the *left*-hand side of the picture as in Figure 4.1. But when $\beta \leq \beta_c$, the curve $x = H^{+-}(y)$ is the steep bold curve at the *right*-hand side of the picture as in Figure 4.2. Ignore the shaded triangle for now.

What ODE theory must see is that whereas there is only one integral curve linking the top and bottom of the unit square when $\beta > \beta_c$, there are infinitely many such curves when $\beta \leq \beta_c$ of which two, the curves $x = A(y)$ and $x = H^{+-}(y)$, derive from probability generating functions (pgfs).

It does not seem at all easy to prove by Analysis that, when $\beta \leq \beta_c$, there is an integral curve linking the bottom of the unit square to the point $(1, 1)$, of the form $x = F(y)$ where F is the pgf of a random variable which can perhaps take the value ∞. Methods such as that used to prove Theorem 3.1 will not work.

Moreover, it is not easy to compute $H^{+-}(0)$ when β is equal to, or close to, β_c. If for example, $q_+ = 1$, $q_- = 4$ and $\beta = 0.4$, then one can be certain that $H^{+-}(0) = 0.6182$ to 4 places, and indeed one can easily calculate it to arbitrary accuracy. But the critical nature of β_c shows itself in unstable behaviour of some naïve computer programs when β is equal to, or close to, β_c. I believe that in the critical case when $q_+ = 1$, $q_- = 4$ and $\beta = 0.5$, $H^{+-}(0)$ is just above 0.6290.

Mathematica is understandably extremely cautious in regard to the non-linear dynamical system (2.1), and drives one crazy with warnings. If forced to produce pictures, it can produce some rather crazy ones, though usually, but not absolutely always, under protest. Its pictures can be coaxed to agree with those in the earlier appetizer [5] which were produced from my own 'C' Runge–Kutta program which yielded Postscript output. Sadly, that program and lots of others were lost in a computer burn-out before I backed them up.

4 Proof of Theorem 1.1 and more

4.1 When $\beta > \beta_c$

Lemma 4.1   *When* $\beta > \beta_c$,

(a) $H^{-+}(0) = \mathbb{P}^-[N^+(0) = 0] = 0$,
(b) $H^{-+}(1-) = \mathbb{P}^-[N^+(0) < \infty] < 1$,
(c) $H^{+-}(1-) = \mathbb{P}^+[N^-(0) < \infty] < 1$,
(d) $H^{+-}(0) = \mathbb{P}^+[N^-(0) = 0] = 0$.

It is clearly enough to prove the lemma under the assumption

$$\tfrac{1}{2}\left(\sqrt{q_-} - \sqrt{q_+}\right)^2 < \beta < \tfrac{1}{2}\left(\sqrt{q_-} + \sqrt{q_+}\right)^2,$$

and this is made throughout the proof.

Proof   Result (a) is obvious.

The point $(1, 1)$ is an equilibrium point of our dynamical system, and we consider the linearization of the system near this equilibrium. We put $x = 1 + \xi$, $y = 1 + \eta$ and linearize by ignoring terms in ξ^2 and η^2:

$$\begin{pmatrix} \xi' \\ \eta' \end{pmatrix} = \begin{pmatrix} -q_+ + \beta & q_+ \\ -q_- & q_- - \beta \end{pmatrix} \begin{pmatrix} \xi \\ \eta \end{pmatrix},$$

the matrix being the linearization matrix of our system at $(1, 1)$. The characteristic equation for the eigenvalues of this matrix is

$$\lambda^2 + (q_+ - q_-)\lambda + (q_- + q_+)\beta - \beta^2 = 0.$$

The discriminant '$B^2 - 4AC$' is

$$\{2\beta - (q_+ + q_-)\}^2 - 4q_-q_+.$$

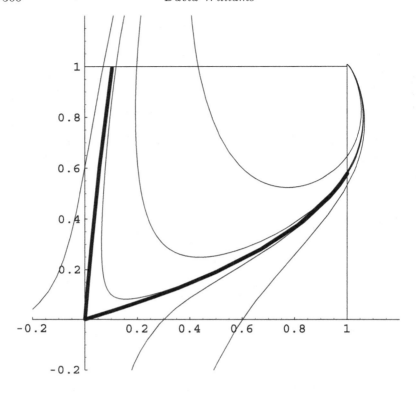

Figure 4.1 *Mathematica* picture of the supercritical case when $q_+ = 1$, $q_- = 4$, $\beta = 4$

This expression is zero if $\beta = \frac{1}{2}(\sqrt{q_-} \pm \sqrt{q_+})^2$. So, in our case, the eigenvalues λ have non-zero imaginary parts. Any solution of our system converging to $(1,1)$ as $\varphi \to \pm\infty$ must spiral, and cannot remain inside the unit square. Hence results (b) and (c) hold.

It is now topologically obvious (since there are no equilibria inside the unit square) that we must have $\mathbb{P}^+[N^-(0) = 0] = 0$; otherwise how could the curve $x = H^{+-}(y)$ link the top and bottom edges of the square? Thus result (d) holds. $\square$

Of course, we can now deduce from result (3.1) that (when $\beta > \beta_c$)

$$H^{+-}(1-) = \mathbb{P}^+[N^-(0) < \infty] \leq q_+/(q_- + \beta).$$

Figure 4.1, which required 'cooking' beyond choosing different φ-ranges for different curves, represents the case when $q_+ = 1$, $q_- = 4$

and $\beta = 4$. The lower bold curve represents $y = H^{-+}(x)$ and the upper $x = H^{+-}(y)$. As mentioned previously, $H^{+-}(y) = A(y)$ where A is the function of Theorem 3.1.

The motion along the lower probabilistic curve $y = H^{-+}(x)$ will start at $(\theta, H^{-+}(\theta))$ and move towards $(0,0)$ converging to $(0,0)$ as $\varphi \to \infty$ since $N^{\pm}(\varphi) \to \infty$. If we fix φ and let $\theta \to 1$, we move along the curve towards the point $(1, H(1-))$. Of course, we could alternatively leave θ fixed and run φ backwards. It is clear because of the spiralling around $(1,1)$ that the power series $H^{-+}(x)$ must have a singularity at some point x not far to the right of 1.

Motion of the dynamical system along the steep probabilistic curve $x = H^{+-}(y)$ on the left of the picture will be upwards because it is the φ-reversal of the natural probabilistic motion. Now you understand the sweep of the curves in the top right of the picture.

4.2 A simple large-deviation result

Let $\{X(t) : t \geq 0\}$ be a Markov chain on $\{+, -\}$ with Q-matrix

$$Q = \begin{pmatrix} -q_+ & q_+ \\ q_- & -q_- \end{pmatrix}.$$

Let V be the function on $\{+, -\}$ with $V(+) = 1$ and $V(-) = -1$ and define $\Phi_X(t) = \int_0^t V(X_s)\,ds$. Almost surely, $\Phi_X(t) \to \infty$. We stay in 'dynamical-system mode' to obtain the appropriate Feynman–Kac formula.

Let $\mu > 0$, and define (with the obvious meanings of $\mathbb{E}^{\pm}$)

$$u(t) = \mathbb{E}^+ \exp\{-\mu\Phi_X(t)\}, \quad v(t) = \mathbb{E}^- \exp\{-\mu\Phi_X(t)\}.$$

Then

$$u(t + dt) = \{1 - q_+\,dt\}e^{-\mu\,dt}u(t) + q_+\,dt\,v(t) + o(dt),$$

with a similar equation for v. We find that

$$\begin{pmatrix} u' \\ v' \end{pmatrix} = \begin{pmatrix} -q_+ - \mu & q_+ \\ q_- & -q_- + \mu \end{pmatrix} \begin{pmatrix} u \\ v \end{pmatrix}, \quad \begin{pmatrix} u(t) \\ v(t) \end{pmatrix} = \exp\{t(Q - \mu V)\} \begin{pmatrix} 1 \\ 1 \end{pmatrix},$$

where V also denotes the operator $\begin{pmatrix} 1 & 0 \\ 0 & -1 \end{pmatrix}$ of multiplication by the function V.

Lemma 4.2 *If $\beta < \beta_c := \frac{1}{2}\left(\sqrt{q_-} - \sqrt{q_+}\right)^2$, then there exist positive* *constants ε, κ, A such that*

$$e^{\beta t}\mathbb{P}^{\pm}[\Phi_X(t) \leq \varepsilon t] \leq Ae^{-\kappa t}.$$

Proof We have just shown that

$$\mathbb{E}^{\pm}[e^{-\mu\Phi_X(t)}f(X_t)] = \exp\{t(Q - \mu V)\}f.$$

Now $Q - \mu V$ has larger eigenvalue

$$\gamma = -\tfrac{1}{2}(q_- + q_+) + \tfrac{1}{2}\sqrt{(q_- + q_+)^2 - 4(q_- - q_+)\mu + 4\mu^2}.$$

We fix μ at $\tfrac{1}{2}(q_- - q_+)$ to obtain the minimum value $\tfrac{1}{2}\left(\sqrt{q_-} - \sqrt{q_+}\right)^2$ of γ. Hence, for $\varepsilon > 0$ and some constant A_ε,

$$\mathbb{P}^{\pm}[\Phi_X(t) \leq \varepsilon t] = \mathbb{P}^{\pm}[\mu(\varepsilon t - \Phi_X(t)) \geq 0] \leq \mathbb{E}^{\pm}\exp\{\mu\varepsilon t - \mu\Phi_X(t)\}$$

$$\leq A_\varepsilon \exp\left\{\tfrac{1}{2}\varepsilon(q_- - q_+)t - \tfrac{1}{2}\left(\sqrt{q_-} - \sqrt{q_+}\right)^2 t\right\}.$$

The lemma follows.   □

For a fine paper proving very precise large-deviation results for Markov chains via explicit calculation, see Brydges, van der Hofstad, and König [2].

4.3 When $\beta < \beta_{\mathrm{c}}$

Lemma 4.3   *When $\beta \leq \beta_{\mathrm{c}}$,*

(a) $H^{-+}(0) = \mathbb{P}^-[N^+(0) = 0] = 0$,
(b) $H^{-+}(1-) = \mathbb{P}^-[N^+(0) < \infty] = 1$,
(c) $H^{+-}(0) = \mathbb{P}^+[N^-(0) = 0] > 0$,
(d) $H^{+-}(1-) = \mathbb{P}^+[N^-(0) < \infty] = 1$.

Note Though Figure 4.2 relates to the case when $\beta = \beta_{\mathrm{c}}$, pictures for the subcritical case when $\beta < \beta_{\mathrm{c}}$ look very much the same.

Proof Result (a) remains trivial.

By Lemma 4.2, there exist $\varepsilon > 0$, $\kappa > 0$ and $A > 0$ such that, for a single particle moving according to Q-matrix Q, we have

$$e^{\beta t}\mathbb{P}[\Phi_X(t) \leq \varepsilon t] \leq Ae^{-\kappa t}.$$

For the branching process, the expression on the left-hand side is the expected number of particles with Φ-value less than or equal to εt at real time t. So the probability that some particle has Φ-value less than or equal to t is at most $Ae^{-\kappa t}$.

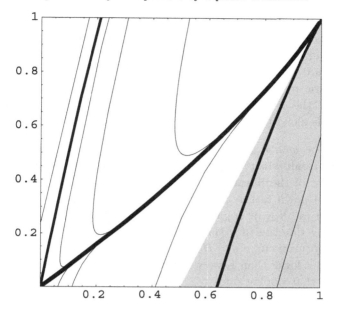

Figure 4.2 *Mathematica* picture of the critical case when $q_+ = 1$, $q_- = 4$, $\beta = 0.5$

By the Borel–Cantelli Lemma, there will almost surely be a random positive integer n_0 such that for all $n \geq n_0$, every particle alive at real time n will have Φ-value greater than εn. Since Φ can only move left at speed 1, there must almost surely come a time after which no particle has a positive Φ-value. Hence $\mathbb{P}^-[N^+(0) = \infty] = 0$, and result (b) is proved.

Now suppose for the purpose of contradiction that $\mathbb{P}^+[N^-(0) > 0] = 1$. Since a particle started at state $+$ can remain there for an arbitrary long time without giving birth, it follows that any particle in the $+$ state and with any positive Φ-value will have a descendant for which Φ will become negative. This contradicts what we proved in the previous paragraph, so result (c) is established.

Since the $y = H^{-+}(x)$ curve connects $(1,1)$ to $(0,0)$ and the other probabilistic curve $x = H^{+-}(y)$ starts at $(H^{+-}(0), 0)$ where $H^{+-}(0) > 0$, and since these curves cannot cross at an interior point of the unit square, it must be the case that $H^{+-}(1-) = 1$, so that property (d) holds. $\qquad\square$

In the analogue of Figure 4.2 for a subcritical case (which, as I have said, looks very much like Figure 4.2), motion along the higher probabilistic curve $y = H^{-+}(x)$ will again start at $(\theta, H^{-+}(\theta))$ and move towards $(0,0)$, this because $N^+(\varphi) \to \infty$. Since $N^-(\varphi) \to 0$, the natural probabilistic motion of the lower curve $x = H^{+-}(y)$ will converge to $(1,1)$; but the φ-reversal means that the dynamical system will move downwards along this curve.

Sketch of geometric proof that $H^{-+}(1-) = 1$ if $\beta \le \beta_c$ It is enough to prove the result when $\beta = \beta_c$. Let $m = \sqrt{q_-/q_+}$, the slope of the unique eigenvector of the linearity matrix at $(1,1)$. Draw the line of slope m from $(1,1)$ down to the y-axis, the sloping side of the shaded triangle in the picture. Now it is particularly easy to check that at any point of the sloping side the $(\mathrm{d}y/\mathrm{d}x)$-slope of an integral curve is greater than m. If the convex curve $y = H^{-+}(x)$ hit the vertical side of the triangle at any point lower than 1, we would have 'contradiction of slopes' where it crossed that sloping side.   □

4.4 Nested models and continuity at phase transition

Take $\beta_0 > \beta_c$ and let M_{β_0} be our model with initial law $\mathbb{P}^+_{\beta_0}$ (in the obvious sense). Label birth-times T_1, T_2, T_3, ... in the order in which they occur, and for each n call one of the two children born at T_n 'first', the other 'second'. Let U_1, U_2, U_3, ... be independent random variables each with the uniform distribution on $[0,1)$. We construct a nested family of models $\{M_\beta : \beta \le \beta_0\}$ as follows.

Fix $\beta < \beta_0$ for the moment. If $U_n > \beta/\beta_0$, erase the whole family tree in M_{β_0} descended from the second child born at time T_n. Of course, this family tree may already have been erased earlier. In this way, we have a model M_β with desired law $\mathbb{P}^+_\beta$. The set $S^-(\varphi, \beta)$ will now denote the $S^-(\varphi)$ set for the 'nested' model M_β, and $N^-(\varphi, \beta)$ will denote its cardinality.

A particle i contributing to $S^-(0, \beta)$ determines a path in $\{+, -\} \times [0, \infty)$:

$$\{(\text{Ancestor}_i(t), \Phi_i(t)) : t < \rho_i\}$$

where ρ_i is the first time after which Φ_i becomes negative. Along that M_β-path, there will be finitely many births. Now, for fixed β it is almost surely true that $U_n \ne \beta$ for all n. It is therefore clear that, almost surely, for $\beta' < \beta$ and β' sufficiently close to β, the M_β-path will also be a path of $M_{\beta'}$. In other words, we have the left-continuity property

$$S^-(0, \beta) = \bigcup_{\beta' < \beta} S^-(0, \beta'), \text{ almost surely.}$$

It therefore follows from the Monotone-Convergence Theorem that

$$\mathbb{E}^+ N^-(0, \beta_c) = \uparrow \lim_{\beta \uparrow \beta_c} \mathbb{E}^+ N^-(0, \beta). \tag{4.1}$$

Clearly, something goes seriously wrong in regard to right-continuity at β_c. Suppose we have a path which contributes to $S^-(0, \beta)$ for all $\beta > \beta_c$. Then, for all birth-times T_n along that path we have $U_n \leq \beta / \beta_0$ for all $\beta > \beta_c$ and hence $U_n \leq \beta_c / \beta_0$. Hence

$$S^-(0, \beta_c) = \bigcap_{\beta > \beta_c} S^-(0, \beta).$$

But it is clearly possible to have a decreasing sequence of infinite sets with finite intersection. And recall that (more generally) the Monotone-Convergence Theorem is guaranteed to work 'downwards' (via the Dominated-Convergence Theorem) only when one of the random variables has finite expectation.

4.5 Expectations and an embedded discrete-parameter branching process

If either of the curves $y = H^{-+}(x)$ or $x = H^{+-}(y)$ approaches $(1, 1)$, it must do so in a definite direction and it is well known (and an immediate consequence of l'Hôpital's rule) that that direction must be an eigenvector of the linearity matrix at $(1, 1)$. When $\beta = \beta_c$, there is (as we have seen before) only one eigenvector $\binom{m}{1}$ with $m = \sqrt{q_-/q_+}$. Thus

$$\mathbb{E}^- N^+(0, \beta_c) = (H^{-+})'(1, \beta_c) = m = \sqrt{q_-/q_+}. \tag{4.2}$$

We know that if $\beta < \beta_c$, then $H^{+-}(1-) = 1$ and we can easily check that (as geometry would lead us to guess)

$$\mathbb{E}^+ N^-(0, \beta) = (H^{+-})'(1, \beta) \leq 1/m = \sqrt{q_+/q_-},$$

and now, by equation (4.1) we see that

$$\mathbb{E}^+ N^-(0, \beta_c) \leq 1/m = \sqrt{q_+/q_-}. \tag{4.3}$$

In particular, $\mathbb{P}^+[N^-(0, \beta_c) = \infty] = 0$, and so, in fact, $H^{+-}(1) = 1$ and we have equality at (4.3), whence

$$\mathbb{P}^+[N^-(0, \beta_c) \geq 1] \leq \mathbb{E}^+ N^-(0, \beta_c) = \sqrt{q_-/q_+}, \tag{4.4}$$

part of Theorem 1.1.

Now, for $\varphi > 0$, let

$$b(\varphi) = \mathbb{E}^+ N^+(\varphi), \quad c(\varphi) = \mathbb{E}^- N^+(\varphi).$$

Then

$$b'(\varphi) = q_+\{c(\varphi) - b(\varphi)\} + \beta b(\varphi), \quad \text{etc.},$$

so that the linearization matrix at $(1,1)$ controls expectations. We easily deduce the following theorem.

Theorem 4.4 *When $\beta = \beta_{\mathrm{c}}$,*

$$\mathbb{E}^+ N^+(\varphi) = e^{\frac{1}{2}(q_- - q_+)\varphi}, \qquad\qquad \mathbb{E}^- N^+(\varphi) = \sqrt{\tfrac{q_-}{q_+}}\, e^{\frac{1}{2}(q_- - q_+)\varphi},$$

$$\mathbb{E}^+ N^-(\varphi) = \sqrt{\tfrac{q_+}{q_-}}\, e^{-\frac{1}{2}(q_- - q_+)\varphi}, \qquad \mathbb{E}^- N^-(\varphi) = e^{-\frac{1}{2}(q_- - q_+)\varphi}.$$

For any β, we can define the discrete-parameter branching processes $\{W^\pm(n) : n \geq 0\}$ as follows. Let B_i be the birth time and D_i the death time of particle i. Recall that Φ_i is defined on $[0, D_i)$. Let $\sigma_i(0) = 0$ and, for $n \geq 1$, define

$$\sigma_i(n) := \inf\{t : B_i \leq t < D_i : t > \sigma_i(n-1); (-1)^n \Phi_i(t) > 0\},$$

with the usual convention that the infimum of the empty set is ∞. Let

$$W^+(n) := \sharp\{i : \sigma_i(2n) < \infty\}, \quad W^-(n) := \sharp\{i : \sigma_i(2n+1) < \infty\}.$$

The 'W' notation is suggested by 'winding operators' in linear Wiener–Hopf theory.

Theorem 4.5 *$W^\pm$ is a classical branching process under $\mathbb{P}_\beta^+$, and is critical when $\beta = \beta_{\mathrm{c}}$.*

The proof (left to the reader) obviously hinges on the case when $\varphi = 0$ of Theorem 4.4. And do have a think about the consequences of the 'balance'

$$\mathbb{E}^- N^+(\varphi, \beta_{\mathrm{c}})\, \mathbb{E}^+ N^-(\varphi, \beta_{\mathrm{c}}) = 1$$

in that theorem.

4.6 When $\theta = 0$ or 1

If we take $\theta = 0$ and set

$$b(\varphi) := \mathbb{P}^+[N^-(\varphi) = 0], \quad c(\varphi) := \mathbb{P}^-[N^-(\varphi) = 0],$$

then $\{(b(\varphi), c(\varphi)) : \varphi \geq 0\}$ is a solution of the φ-reversed dynamical system such that

$$b(\varphi) = H^{+-}(c(\varphi)).$$

When $\beta > \beta_c$, this solution stays at equilibrium point $(0,0)$. When $\beta \leq \beta_c$, $(b(\varphi), c(\varphi))$ moves (as $0 \leq \varphi \uparrow \infty$) from $(H^{+-}(0), 0)$ to $(1,1)$ tracing out the right-hand bold curve in the appropriate version of Figure 4.2.

When $\theta = 1$, $\{(B(\varphi), C(\varphi)) : \varphi \geq 0\}$, where

$$B(\varphi) := \mathbb{P}^+[N^-(\varphi) < \infty], \quad C(\varphi) := \mathbb{P}^-[N^-(\varphi) < \infty],$$

gives a solution of the φ-reversed dynamical system. When $\beta \leq \beta_c$, this solution stays at the equilibrium point $(1,1)$. When $\beta > \beta_c$, $(B(\varphi), C(\varphi))$ moves (as $0 < \varphi \uparrow \infty$) from $(H^{+-}(1-), 1)$ to $(0,0)$ tracing out the bold upper curve in the appropriate version of Figure 4.1. Of course, there is an appropriate version ('with $+$ and $-$ interchanged') for the lower curve.

4.7 A tantalizing question

When $\beta > \beta_c$, we have for the function $A(\cdot) = A(\cdot, \beta)$ of Theorem 3.1:

$$A(\theta, \beta) = \mathbb{E}^+ \theta^{N^-(0,\beta)}.$$

When $\beta \leq \beta_c$, we have, for $0 < \theta < 1$, $A(\theta, \beta) = \mathbb{E}^+ \theta^{Y_\beta}$ for some random variable Y_β. Can we find such a Y_β which is naturally related to our model? In particular, can we do this when $\beta = \beta_c$? It would be very illuminating if we could.

What is true for all $\beta > 0$ is that if X and Φ_X are as at the start of Subsection 4.2 and $\tau_X^-(0) := \inf\{t : \Phi_X(t) < 0\}$, then

$$\mathbb{E}^+ \exp\{-\beta \tau_X^-(0)\} = a_1,$$

with a_1 as in Theorem 3.1; and this tallies with $a_1 = \mathbb{P}^+[N^-(0) = 1]$ when $\beta > \beta_c$. Proof of the statements in this paragraph is left as an exercise.

Acknowledgements I certainly must thank a referee for pointing out some typos and, more importantly, a piece of craziness in my original version of a key definition. The referee also wished to draw our attention (mine and yours) to an important survey article on tree-indexed processes [3] by Robin Pemantle. The process we have studied is, of course, built on the tree associated with the underlying branching process. But our Wiener–Hopferization makes for a rather unorthodox problem.

I repeat my thanks to Chris Rogers and John Toland for help with the previous appetizer. My thanks to Ian Davies and Ben Farrington for helpful discussions.

And I must thank the Henschel String Quartet—and not only for wonderful performances of Beethoven, Haydn, Mendelssohn, Schulhoff and Ravel, and fascinating discussions on music and mathematics generally and on the connections between the Mendelssohns and Dirichlet, Kummer, Hensel, Hayman. If I hadn't witnessed something of the quartet's astonishing dedication to music, it is very likely that I would have been content to leave things with that early appetizer and not made even the small advance which this paper represents. So, Happy music making, Henschels!

References

[1] Arnol'd, V. I. 2006. *Ordinary Differential Equations*, transl. from Russian by R. Cooke. Universitext. Berlin: Springer-Verlag.

[2] Brydges, D., Hofstad, R. van der, and König, W. 2007. Joint density for the local times of continuous-time Markov chains. *Ann. Probab.*, **35**, 1307–1332.

[3] Pemantle, R. 1995. Tree-indexed processes. *Statist. Sci.*, **5**, 200–213.

[4] Warren, J., and Williams, D. 2000. Probabilistic study of a dynamical system. *Proc. Lond. Math. Soc. (3)*, **81**(3), 618–650.

[5] Williams, D. 1995. Non-linear Wiener–Hopf theory, 1: an appetizer. Pages 155–161 of: Azema, J., Émery, M., Meyer, P.-A., Yor, M. (eds), *Séminaire de probabilités* **29**. Lecture Notes in Math. **1613**. New York: Springer-Verlag.

Index